Eri Culture –
Rearing of Ahimsa Silk

Eri Culture –
Rearing of Ahimsa Silk

V. Lakshmi Narayanamma

BS Publications

An Imprint of **BSP Books Pvt., Ltd.**
4-4-309/316, Giriraj Lane,
Sultan Bazar, Hyderabad - 500 095

Eri Culture - Rearing of Ahimsa Silk
by **V. Lakshmi Narayanamma**

© 2024, *by Publisher*

Disclaimer: The authors and the publishers have taken due care to provide the authentic, reliable and up to date information related to the subject. However, neither the authors nor the publisher shall be responsible for any liability for any damage caused as a result of use of this book. The respective user must check the accuracy from other sources too.

Published by:

 **BS Publications**

An Imprint of **BSP Books Pvt. Ltd.**
4-4-309/316, Giriraj Lane, Sultan Bazar,
Hyderabad - 500 095
Phone: 040 - 23445688
e-mail : info@bspbooks.net
www.bspbooks.net

ISBN: 978-93-95038-12-6 (Paperback)

Dedication

This book is humbly dedicated to

My Mother V. Gangamma
Father V. Venkateswarlu
Husband N. C. Nagesh

and

My Guru Dr. N. V. Krishnaiah

Who are solely responsible for what all I am today and stood solidly behind me in all tough times that I faced in life.

With reverence
Lakshmi Narayanamma

Preface

It fills my heart with admiration and joy to title this book as "Eri culture-Rearing of Ahimsa Silk" for more reasons than one. India (Bharath) the land of the most ancient civilization on mother earth is also the land that stood for Divine Principles of *"Satya"* (Truth), *"Dharma"* (Righteousness), *"Shanthi"* (Peace), *"Prema"* (Love) and *"Ahimsa"* (Non-Violence in Thought, Word and Deed). *Ahimsa* is the principle of identifying oneself with all other living beings and respecting their existence as the same Divinity dwells in humans as well as all other creatures.

If we turn the pages of history of Bharath, this country never invaded any other country during thousands of years in spite of its high scientific wealth and economic dominance. But itself suffered from invasions of Greeks, Muslims, Mongols and more recently the British. Every attempt was made to wreck into pieces its rich cultural heritage, economic dominance and scientific wealth. Still the nation has survived and preserved its treasure of culture and presently moving on the path of recovery in economic and scientific fields. Even in most recent history, it is the fight based on *Ahimsa* principle that brought us political freedom from the clutches of mighty British Empire. All these points inspired me to include the word *"Ahimsa"* and call "Eri-culture as Ahimsa silk rearing".

Eri silkworm, *Samia cynthia ricini* Boisduval is a polyphagous insect species that can eat and survive on fresh leaves of about fifty kinds of plants including the castor oil plant. Eri silkworm spins silken cocoon as a cover to its pupal case. But the silken thread is not continuous unlike Mulberry silk worm *Bombyx mori*. Hence the silken thread can be spun without killing the pupa. Thus it is appropriate to call it Ahimsa silk rearing. The eri silk is mainly useful for making winter wear, chaddars, shawls etc. unlike Mulberry silk which is mainly used for saris. If we consider the quality and durability of eri silk it is no way inferior to Mulberry silk. However, eri silk did not receive the attention it deserves due to its low cost and less sophistry compared to Mulberry silk.

As the father of the nation Mahatma Gandhi says "India lives in villages". In 1951, just 3 years after independence, rural population in

India was 29.9 crore constituting 82.7% of the total population of 36.1 crores. According to 2011 census, i.e even 60 years later, the rural population in India is 83.3 crores which is 2.8 times that of 1951, in spite of heavy urbanization and growth of towns and cities. Due to present highly fragile state of agriculture, there is no dependable and sustainable source of income for rural households. Even dairying is currently facing serious problems of exploitation of the farmers. Under these circumstances eri silk production can be undertaken in all possible areas to augment the income of rural households in general, poorer sections and women in particular. Therefore, detailed knowledge on the subject is the necessity of the day.

The first book on Eri-culture in India by Sarkar D. C. was published in 1980, by Central Silk Board, Bangalore. Later in 2002, Reddy D N R, Manjunath Gowda and Narayanaswamy K C. wrote briefly (82 Pages) on Ericulture-An Insight, published by Zen Publishers, Bangalore. In the same year, Advances in Ericulture was authored by Sannappa B, Jayaramaiah M, Govindan R and Chinnaswamy K P. and published by Seri Scientific Publishers, Bangalore. In 2005, Singh and Suryanarayana wrote the book on "Principles of Ericulture". Recent book on "Eri Culture" has come in 2013 by Ravindra Nath Singh and Beera Saratchandra, published by APH Publishing Society. During the last 5 years some major advancements have taken place on scientific and social aspects of ericulture. Hence Indian Council of Agricultural Research has instructed me to bring out a comprehensive up to date publication on the subject. I earnestly hope that this book will be of great help for researchers, trainers, students, educated farmers, administrators and all others who are concerned and interested on eri culture.

This book contains 13 chapters on different aspects like Nomenclature, Host plants, Life cycle, Artificial diet for rearing of eri silkworm, Exploitation of Genetic Diversity in Eri Silkworm for Rearing Performance and Amenability, chapters related to Role of environment on biology and rearing, Participatory profile of women in eri culture and Role of Extension strategies in addition to the regular chapters on Introduction, Byproducts, Rearing performance of eri silkworm on castor etc. My experience in ericulture is limited and this is my first attempt to write a book of this comprehension. The readers will sympathize with me and I request them to point out the corrections and lapses in the book. That will be of great help for me and to the scientific community dealing with the subject and will enable me to bring out a revised edition in future date.

"ఆరంభింపరునీచమానవుల్ఉరువిఘ్నాయతాసంత్రస్తులై,

ఆరంభించిపరిత్యజింతురువిఘ్నాయాసులైమధ్యముల్,

ధీరుల్విఘ్ననిహన్యమానులగుచున్, ధృత్యున్నతోత్సాహులై

ప్రరబ్ధార్థములుజ్జగింపరుగదా,ప్రజ్ఞానిధుల్లావునన్"

meaning

"Persons with low level of will power and determination do not start an activity with the fear of obstacles;

Those persons with moderate will power start an activity but stop it the movement they encounter difficulties,

The persons with very high will power and determination do continue the started activity until it reaches its successful end by fighting against all unbearable odds."

- Lakshmi Narayanamma

Acknowledgements

At the outset, I owe my obeisance to the Almighty for showing his grace and blessings on me for successful completion of the Book titled "Eri culture – Rearing of Ahimsa Silk".

I am grateful to the Authorities of Indian Council of Agricultural Research, New Delhi, India for giving me this opportunity of writing a comprehensive publication on "Eri Culture-Rearing of Ahimsa Silk", which is not just a subject of my research but it is close to my heart as a woman researcher and hailing from the background of the farmers who can utilize the information contained in the book.

I am highly thankful to all the authorities of Professor Jayashankar Telangana Sate Agricultural University (PJTSAU) for permitting me to avail the opportunity. My special thanks are due to Dr. V. Praveen Rao Garu, Hon'ble Vice Chancellor of the University, Dr. D. Raji Reddy Garu, Director of Extension and Sri R. Sudhakar Reddy Garu, Co-ordinator, Electronic Wing of the University in which I am presently serving.

I am very much thankful to National Bank for Agriculture & Rural Development (NABARD) and Dr. P. Mohanaiah Garu, The Chief General Manager of that time for providing the financial support to conduct the project on "Eri culture – An additional income generation source for the resource poor farmers of Mahabubnagar district of Andhra Pradesh". Without this project, the present book would have not come into existence.

With bound less affection and love I would heartily offer my warm thanks to Dr. K. Dharma Reddy Garu, Professor (Entomology) (Retired) for his affection, encouragement and support on me and whose contribution behind the scene has enabled the successful completion of my project work at RARS, Palem.

My special reverence to my superiors and staff of entomology *viz.,* Dr. A. Vishnuvardhan Reddy, Director, IIOR, Dr. S. J. Rahman, Principal Scientist and University Head of Department of Entomology, Dr. Srinivas, Professor and Head, Dept. of Entomology, Dr. T. Uma Maheswari, Colleagues, Dr. A. V. Ramanjaneyulu, Dr. P. Bindu Priya, Dr. V. Swarna Latha, Dr. K. Kavitha and Dr. S. Upendar.

I am thankful to the staff of electronic wing *viz.,* Dr. P. Prashanth, Sri K. R. Sanjeeth Kumar, Sri, J. Rajaram, Preethi, Jahnavi, Srinivas, Srikanth, Yadaiah, Andalamma and Rama Chander for their support during the period.

Diction is not enough to express my deep and profound sense of gratitude and indebtedness, reverence and heartfelt thanks to my Guru, teacher, guide, mentor Dr. N.V. Krishnaiah, Principal Scientist (Entomology)(Retired), who taught me the way to look into the things regarding subject matter and manner to understand, interpret and present them. His magnanimous attitude, meticulous guidance, transcendent suggestions, kind treatment and constant encouragement have instilled in me the spirit of respect and gratitude.

I am also thankful to my sweet kids Master Haratejas and Baby Gayathri, brother V. Rama Krishna for their love and affection.

I thank immensely all the authors of the research papers for using their publications in the journey of my writing.

- Lakshmi Narayanamma

Contents

Abbreviations used in the Book

- MT – Metric Tonnes
- B.C – Before Christ
- A.D – Anno Damini
- GDP – Gross Domestic Production
- CSB – Central Silk Board
- Kgs – Kilograms
- Dfls – Disease free layings
- % – Per cent
- Et al., – Others
- Viz., – Namely
- & – And
- XII – Twelth
- Km – Kilometer
- Mm – Milli meter
- SRI – *Samiar ricini*
- RERS – Regional Eri Research Station
- CMERTI – Central Muga Eri Research & Training Institute
- MEG – Meghalaya
- ASM – Assam
- NAL – Nagaland
- MAN – Manipur
- ARP – Arunachal Pradesh
- O – Old
- RCU – Race in current use
- N – New
- OR – Original race
- GCV – Genotypic co-efficient of variation

- PCV – Phenotypic co-efficient of variation
- GA – Genetic Advance
- g – Gram
- wt – Weight
- d – Days
- no – Number
- ERR (%) – Effective Rate of Rearing
- $^{\circ}$C – Degree Centigrade
- i.e. – That is
- YP – Yellow Plain
- YS – Yellow Spotted
- YZ – Yellow Zebra
- GBP – Green Blue Plain
- GBS – Green Blue Spotted
- GBZ – Green Blue Zebra
- WP – White Plain
- GP – Green Plain
- WS – White Spotted
- GS – Green Spotted
- WSZ – White Semi Zebra
- YSZ – Yellow Semi Zebra
- F_1 – First Generation
- DNA – Deoxyribo Nucleic Acid
- Acc. – Accession
- ISSR – Inter Simple Sequence Repeat
- GST – Goods and Services Tax
- Kb – Kilo bands
- RAPD – Randomly Amplified Polymorphic DNA
- SDS-PAGE – Sodium dodecyl sulphate polyacrylamide gel electrophoresis

➢ MT – Million Tonnes
➢ °N – Degree North
➢ °S – Degree South
➢ m – Meter
➢ > – Grater than
➢ < – Less than
➢ ISF – Interspersed Staminate Flowers
➢ Ha – Hectare
➢ USA – United States of America
➢ ID – Irrigated dry
➢ @ – At the rate of
➢ FYM – Farm Yard Manure
➢ DAS – Days after sowing
➢ IW/CPE – Irrigation Water/ Cumulative Pan Evaporation
➢ t/ha – Tonne/Hectare
➢ qt – Quintal
➢ PSB – Phosphate Solubilising Bacteria
➢ RDF – Recommended dose of fertilizer
➢ SSP – Single Super Phosphate
➢ MOP – Muriate of Potash
➢ N – Nitrogen
➢ P – Phosphorus
➢ K – Potassium
➢ NaCl – Sodium chloride
➢ Cm – Centimetre
➢ IWM – Integrated weed management
➢ IC – Intercultivation
➢ HW – Hand weeding
➢ Ml – Milli liter

- Ltr – Liter
- IPM – Integrated Pest Management
- NSKE – Neem seed kernel extract
- NPV – Nuclear Polyhedrosis Virus
- ETL – Economic Threshold Level
- IDM – Integrated Disease Management
- Eg – Example
- ALA – Alpha linolenic acid
- Ca – Calcium
- S – Sulphur
- mg – Milli gram
- E. Ext – Ether extracts
- N.F.E – Nitrogen free extracts
- Y – Year
- UAS – University Agricultural Sciences
- BC – Benefit Cost
- AEFP – Augmentation of Eri food plant
- CDP – Catalytic Development Programme
- CPP – Cluster Promotion Programme
- CMER&TI – Central Muga Eri Research & Training Institute
- CAD – Co-efficient of apparent digestibility

Formulae used in the Book

- **Sericin content:** Dry weight of cocoon shell – Dry weight of cocoon shell after alkali treatment

- **Fibroin content:** Dry weight of cocoon shell – Sericin content.

- **Ingesta** = Dry weight of leaf fed - Dry weight of left over leaf

- **Digesta** =Dry weight of leaf ingested - Dry weight of litter

- **Excreta** =Ingesta - Digesta

- **Mean larval weight** = (Final weight - Initial weight)/2 + Initial weight

- **Approximate Digestibility (%)** = Dry weight of Digesta × 100/Dry weight of food ingested

- **RR (%)** = Dry weight of food ingested × 100/Dry weight of excreta

- **RCI** = Ingesta/(Mean Dry weight of larva × Larval duration in days)

- **RGR** = Dry weight gain of the larva/(Larval duration in days x Mean dry weight of the larva)

- **ECI to larva** = Maximum dry weight of larva × 100/Dry weight of ingesta

- **ECD to larva** = Maximum dry weight of larva × 100/Dry weight of digesta

- **ECI to cocoon** = Dry weight of cocoon × 100/Dry weight of ingesta

- **ECD to larva** = Maximum dry weight of larva × 100/Dry weight of digesta

- **ECI to shell** = Dry weight of shell × 100/Dry weight of ingesta

- **ECD to Shell** = Dry weight of shell × 100/Dry weight of digesta

- **Food consumption** = weight of fresh food offered to larvae – weight of fresh remnants

- **Consumption Index**

$$= \frac{\text{Weight of food offered}}{\text{Duration of feeding} \times \text{Mean weight of larvaduring the feeding period}}$$

- **Approximate digestibility (%)** $= \dfrac{\text{Weight of food offered}}{\text{Weight of food ingested}} \times 100$

- **Nutritional ratio (NR)** $=$

$$\frac{\text{Per cent digestible carbohydrate} + \left(\text{Per centdigestible fat} \times 2.25\right)}{\text{Per cent digestible crude protein}}$$

- **% mortality** $= \dfrac{\text{Number of worms dead}}{\text{Totalnumber of worms taken}} \times 100$

- **Weight of food consumed** $=$ Weight of food offered $-$ Weight of food remained

- **Leaf provided to cocoon ratio:** It is the ratio of gross quantity of leaves supplied to produce one unit of cocoon.

- **Leaf consumed to cocoon ratio:** It is the ratio of quantity of leaves consumed to produce one unit of cocoon and were calculated adopting the formulae.

- **Leaf offered to cocoon ratio** $=$

$$\frac{\text{Average weight of food offered to a larva}}{\text{Average weight of a cocoon}}$$

- **Leaf consumed to cocoon ratio** $=$

$$\frac{\text{Average weight of food consumed}}{\text{Average weight of a cocoon}}$$

- **Leaf offered to Egg ratio:** It is the number of eggs produced for the unit of leaves offered

- **Leaf consumed to Egg ratio:** It is the number of eggs produced for the unit of leaves consumed

- **Leaf offered to egg ratio** $=$

$$\frac{\text{Average fecundity of a moth}}{\text{Average weight of food offered to a larva}}$$

- **Leaf consumed to egg ratio** =

$$\frac{\text{Average fecundity of a moth}}{\text{Average weight of food consumed by a larva}}$$

- **Assimilation (A)** = Consumption (C) – Egestion (F)

- **Evaluation Index (EI)** = $\dfrac{A-B}{C} \times 10 + 50$

where, A = Individual value of the genotype

B = average value of the particular trait of the genotypes

C = standard deviation of the particular trait

10 = standard unit, 50 = fixed value.

- **Metabolism** = Larval growth – Assimilation

- **Consumption rate (Cr)** = $\dfrac{\text{Consumption}}{\text{Mid body wt.} \times \text{days}}$ g/g/day

- **Assimilation rate (Ar)** = $\dfrac{\text{Assimilation}}{\text{Mid body wt.} \times \text{days}}$ g/g/day

- **Growth rate (Pr)** = $\dfrac{\text{Larval growth}}{\text{Mid body wt.} \times \text{days}}$ g/g/day

- **Metabolic rate (Mr)** = $\dfrac{\text{Metabolism}}{\text{Mid body wt.} \times \text{days}}$ g/g/day

- **Leaf Silk Conversion Rate (%)** =

$$\frac{\text{Single shell weight}}{\text{V} - \text{instars larvae ingesta} \ / \ 400} \times 100$$

- **Training Need Index** = $\dfrac{\text{Total training need score obtained}}{\text{Total training score obtainable}} \times 100$

- **Knowledge Index** =

$$\frac{\text{Total knowledge score obtained by each respondent}}{\text{Total training score obtainable}} \times 100$$

CHAPTER 1

Introduction

Silk, "The queen of textiles" is a natural protein fibre secreted by arthropods especially lepidopteran silkworms. Silk is a way of life in India. Over thousands of years, it has become an inseparable part of Indian culture and tradition. No ritual is complete without silk being used as a wear in some form or the other. Silk is the undisputed queen of textiles over the centuries. Silk provides much needed work in several developing and labour rich countries. Sericulture refers to the conscious mass scale rearing of silk producing organisms in order to obtain silk from them. Silkworms are broadly classified as mulberry and wild or non-mulberry silkworms. Among non-mulberry silks, eri silk production is in increasing trend. Ericulture is a viable agro-based industry, plays a significant role in rural livelihood security especially among the marginalized and weaker sections of the society. Ericulture though relatively a less remunerative occupation but it has many advantages. Several socio-economic studies have affirmed that the benefit-cost ratio in ericulture is higher among comparable agricultural crops. This chapter discusses in detail about the different types of silks available in India, how it has been introduced, the global scenario of silk marketing, position of ericulture in silk industry, its significance as Ahimsa silk and trends of growth in eri silk industry in different states of India.

If fashion is a fine art, then silk is its biggest canvas, and if silk is the canvas, then all its weavers, dyers, designers, embroiderers are the greatest artists. Silk is nature's gift to mankind and a commercial fibre of animal origin other than wool. Being an eco-friendly, biodegradable and self-sustaining material, silk has assumed special relevance in present age. Silk is a natural protein fibre secreted by arthropods especially lepidopteran silkworms (Chowdhary, 2006). Sericulture refers to the conscious mass scale rearing of silk producing organisms in order to obtain silk from them (Ganga and Chetty, 1997). The word sericulture has been derived from the Greek word "Sericos" means silk and the English word culture means 'rearing'. Sericulture is both an art and science of raising silk worms for silk production. Silk is the most elegant textile in the world with unparalleled grandeur, natural sheen, and inherent affinity for dyes with high absorbance, light weight, soft touch, smooth, strong and highly durable than any natural or artificial fibre. Hence silk is known as "The Queen of Textiles". The insects that produce silk of economic value are termed as sericigenous insects. The natural silk producing insects are broadly classified as mulberry and wild or non-mulberry silkworms. The mulberry silk moths are represented by domesticated *Bombyx mori* Linnaeus. Non-mulberry sericulture is universally known as forest or wild sericulture that provides an important source of employment for the native population in forest areas.

Sericulture includes raising of host plants, rearing of silkworm, production of silk yarn. Sericulture requires low investment, gives quick returns, provides employment opportunities and earns foreign exchange. Sericulture suits both for marginal and small land holders because of its high returns, short gestation period, and it creates opportunity for own family employment round the year. Sericulture serves as an important tool for rural reconstruction, benefiting the weaker sections of the society (Lakshmanan *et al.*, 1997 and Goswami *et al.*, 2015).

Types of Silks

Sericulture was introduced in India about 2000 years ago and the silkworm producing yellow silk was known since the ancient time (Reddy, 2000). India is a home to a vast variety of silk secreting fauna which also includes an amazing diversity of silk moths. This has enabled India to achieve the unique distinction of being a producer of all the five commercially traded varieties of natural silks namely, Mulberry, Tropical Tasar, Oak Tasar, Eri and Muga. Silk obtained from sources other than mulberry are generally termed as non-mulberry or vanya silks. The term 'Vanya' is of Sanskrit origin, meaning untamed, wild or forest based. The bulk of the commercial silk produced in the world is mulberry silk that comes from the domesticated silkworm, *Bombyx mori* L. which feeds solely on the leaves of the mulberry (*Morus alba*) plant. In India, mulberry silk is produced mainly in the states of Karnataka, Andhra Pradesh, Tamil Nadu, Jammu & Kashmir and West Bengal, while the non-mulberry silks are produced in Jharkhand, Chattisgarh, Orissa and North-Eastern states. Tasar silk is copperish in colour, coarse in nature and is mainly used for furnishing and interiors. It is secreted by the Tropical Tasar silkworm, *Antheraea mylitta* which thrives on plants like Asan and Arjun (*Terminalia* sp.). Tasar silkworm can be reared on naturally growing trees in the forests and is the main stay for many tribal communities in the states of Jharkhand, Chhattisgarh, Orissa, Maharashtra, West Bengal and Andhra Pradesh. Oak Tasar is a finer variety of Tasar produced by the temperate tasar silkworm, *Antheraea proylei* which feeds on natural oak plants (*Quercus* sp.) and is found in abundance in the sub-Himalayan belt. Eri silk is a silk spun from open-ended cocoons and secreted by the domesticated silkworm, *Samia cynthia ricini* Boisduval that feeds mainly on castor (*Ricinus communis* Linn) leaves. Muga silk is golden yellow in colour and is preferred mainly in the state of Assam during festivities. Muga silk is secreted by *Antheraea assama* that feeds on aromatic leaves of naturally growing Som (*Persia bombycina*) and Soalu (*Litsea polyantha*) plants. Among the four varieties of silks produced, mulberry accounted for 71.8% (20,434 MT), tasar's share was 9.9% (2,818 MT), eri accounted for 17.8% (5,054 MT) and muga silk accounted for 0.6% i.e. 166 MT of the total raw silk production of 28,472 MT during 2015-16 (Fig. 1.1).

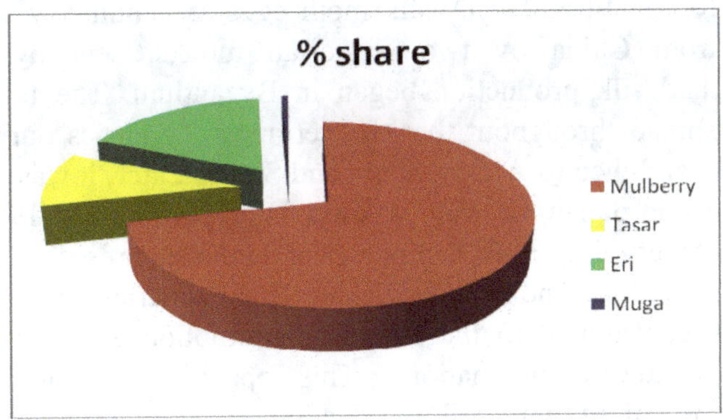

Figure 1.1 Variety wise raw silk production in India (2015-16)
Source : Central Silk Board, Bangalore

Different Versions about History of Sericulture

Silk production has a long history. Silk, the weavable fiber was first discovered during 2640 BC by the Chinese Empress Xi Ling Shi, the 14 year old bride of the China's 3rd Emperor, Huang Ti, also called "Yellow Emperor". One day she was sitting under mulberry tree, drinking a cup of tea into which a silkworm cocoon fell from above. She found that the delicate fibres started to unravel in the hot liquid and could be twisted together to make a thread that was strong enough to be woven into cloth. Thereafter, Xi Ling Shi discovered not only the means of raising silkworms, but also the manners of reeling the silk and of employing it to make garments. Silk was profitable trade commodity in China. The Chinese kept the secret of the beautiful and value-added material that they were producing, from the rest of the world for more than 30 centuries. Traders from ancient Persia (now, Iran) used to bring richly coloured and fine textured silks from Chinese merchants through hazardous routes interspersed with dangerous mountainous terrains, difficult passes, dry deserts and thick forests. Commodities like amber, glass, spices and tea were also traded along with silk which indeed rapidly became one of the principal elements of the Chinese economy and hence, the trade route got the name "Silk Route".

After 1200 B.C. Chinese immigrants who had settled in Korea helped in the emergence of silk industry in Korea. During the third century B.C. Semiramus, a general of the army of Empress Singh-Kongo, invaded and conquered Korea. In 550 A.D. silk moth eggs and mulberry seeds were smuggled from China by two Nestorian monks, sent by Emperor, Justinian-1 and silk production began in Byzantium. The technique of sericulture spread throughout the Mediterranean countries during the 7th century A.D. and then to Africa, Spain and Sicily. Later, it was introduced into Europe and Japan as well. During later part of the 19th century, modern machinery, improved techniques and intensive research helped the growth of sericulture industry in Japan. The industrial and commercial uses of silk contributed to the silkworm promotion all over the world especially in developing nations. This opens ways for integrating agricultural practices with silk production as with animal husbandry, dairying, fisheries and horticulture, which will improve the overall productivity of the societies (Sahay *et al.*, 1997; Lamelu, 1998; Hiwar, 2001, Tzenov, 2007 and Kedir Shifa *et al.,* 2014). The silkworms are very important economic insect which contributes substantially to the national economy and Gross Domestic Production (GDP) of many countries like China, India, Thailand etc. (Chen, 2003; Chen and Gu, 2006). In Ethiopia, rearing of eri and mulberry silkworm is practiced sparsely (Metaferia *et al.*, 2007). Geographically, Asia is the main producer of silk in the world and manufactures over 95 % of the total global output. At present, China is the major producer of silk followed by India. Other countries like Korea, Italy, Soviet Union, France, Brazil Japan also contribute to silk production in the world.

Table 1.1: Global Silk production (In Metric Tonnes) from 2010-11 to 2015-16

Country	2010-11	2011-12	2012-13	2013-14	2014-15	2015-16	% Share
China	115000	104000	126000	130000	146000	170000	84.12
India	21005	23060	23679	26480	28708	28523	14.11
Uzbekistan	940	940	940	980	1100	1200	0.59
Thailand	655	655	655	680	692	698	0.34
Brazil	770	558	614	550	560	600	0.29
Vietnam	550	500	450	475	420	450	0.22
North Korea	-	300	300	300	320	350	0.17
Iran	75	120	123	123	110	120	0.05
Bangladesh	40	38	42.50	43	44.5	44	0.02
Japan	54	42	30	30	30	30	0.01
Turkey	18	22	22	25	32	30	0.01
Indonesia	20	20	20	16	10	8	0.003
Bulgaria	9.4	6	8.5	8.5	8	8	0.003
Madagascar	16	16	18	18	15	5	0.002
Tunisia	0.12	3	3.95	4	4	3	0.001
Philippines	1	1	0.89	1	1.1	1.2	-
South Korea	3	3	1.5	1.6	1.2	1	-
Egypt	0.3	0.7	0.7	0.7	0.82	0.83	-
Colombia	0.6	0.6	0.6	0.6	0.5	0.5	-
Syria	0.6	0.5	0.5	0.7	0.5	0.3	-
Total	139100.02	129661.80	152845.64	159737.10	178057.62	202072.83	100.00

Introduction of Silk Industry in India

A story is that a Chinese princess married an Indian prince. She carried silkworm eggs/ mulberry cocoons in her elaborate head dress. She disclosed the secret of raising silkworms thus, silk production spread in India. According to Western historians, mulberry-tree cultivation spread to India through Tibet during 140 BC and cultivation of mulberry trees, rearing of silkworms began in the areas flanking the Brahmaputra and Ganges rivers. According to some Indian scholars silkworms (*Bombyx mori*) were first domesticated in the foothills of the Himalayas. Evidences in ancient Sanskrit literature reveal that certain kind of wild silks were cultivated in India since time immemorial. When British came to India, the flourishing silk trade was exploited and they developed silk centres in many parts of the country. The Company exported large quantities of silk produced in West Bengal to England. The Company's monopoly was abolished in 1836 and the entire trade turned over to private enterprise.

Slowly, due to improper organized system, the silk industry in West Bengal declined. By the time other silk producing states in the country *viz.,* Jammu & Kashmir, Mysore have developed the industry. According to reports available, sericulture industry flourished in India as an agro-industry till 1857, with an annual production of two million pounds of silk fibre. The industry survived the attack of the Pebrine disease during the period from 1857 to 1895. However, after 1928, the sericulture industry showed a decline in its production owing to the fierce competition from advanced sericulture countries, such as Japan, China and European countries. After Independence in India, the industry again started flourishing as an agro-industry, giving employment to rural population in the Country (Anitha, 2011).

Sericulture in India

Silk is a way of life in India. Over thousands of years, it has become an inseparable part of Indian culture and tradition. No ritual is complete without silk being used as a wear in some form or the other. Silk is the undisputed queen of textiles over the centuries. Silk provides much needed work in several developing and labour rich countries. Sericulture is one of the most labour intensive sectors of the Indian economy combining both agriculture and industry, which provides means of livelihood to a large section of the population i.e. crop cultivator, co-operative rearer, silkworm seed producer, farmer-cum-rearer, reeler, twister, weaver, hand spinners of silk waste, trade etc. It is the only one cash crop in agriculture sector that gives returns within 30 days. The silk industry has a distinctive position in India, and plays a significant role in textile industry and export. This industry provides employment nearly to 35 million people in our country. Out of which, 60% of the people are women respondents. Thus, in contrast to any other agro-based profession the role of women in sericulture industry is dominating which will be helpful for improving the status of women in family enterprises. In the light of women welfare through sericulture industry, the Central Silk Board, a statutory organization, under the Ministry of Textiles, Government of India has established a special component of assistance to women and NGO's in the National Sericulture Project.

Sericulture is cultivated in Karnataka, Bengal, Tamil Nadu, Telangana, Andhra Pradesh, Jammu & Kashmir, Gujarat, Kerala, Maharashtra, Uttar Pradesh, Rajasthan, Bihar, Orissa etc. Though India is the second largest silk producer in the World after China, it accounts for just 5% of the global silk market, since the bulk of Indian silk thread and silk cloth are consumed domestically. The Central Silk Board, Ministry of Textiles, Govt. of India has been acting as a facilitator for planning, development and monitoring of sericulture industry between the States and Central Govt. The silk production, which is 16,500 MT during 2004-05, has increased to 28,523 MT in 2015-16, contributes to 14% of the total world raw silk production (Fig. 1.2). The state wise raw silk production during the 12th plan period is shown in the table 1.2.

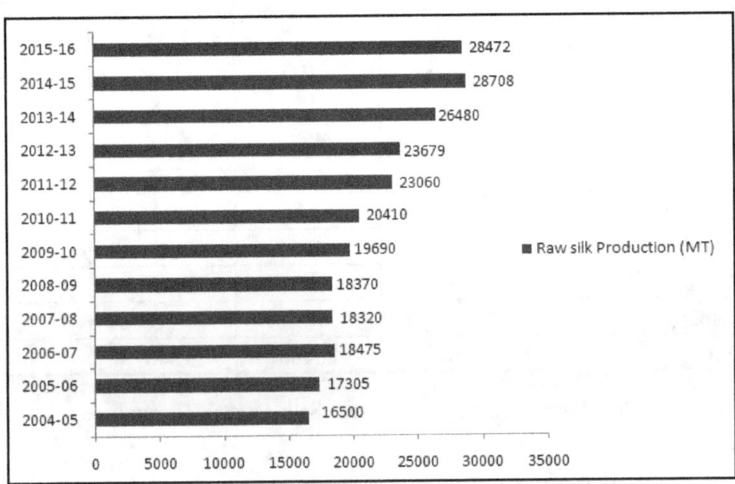

Fig. 1.2 Raw silk production in India (MT)

Table 1.2: State wise raw silk production during the XII Plan period (2012-13 to 2015-16)

S.No	State	Raw silk production (MTs)			
		2012-13	2013-14	2014-15	2015-16 (P)
1	Karnataka	8219	8574	9645	9823
2	Andhra Pradesh	6550	6912	6485	5086
3	Telangana	--	--	100	116
4	Tamil Nadu	1185	1120	1602	1898
5	Kerala	6	4	7	9
6	Maharashtra	97	122	222	274
7	Uttar Pradesh	157	188	236	249
8	Madhya Pradesh	190	195	177	214
9	Chhattisgarh	391	391	200	261
10	West Bengal	2070	2079	2500	2391
11	Bihar	22	52	53	67
12	Jharkhand	1090	2003	1946	2284

Table 1.2 *Contd...*

S.No	State	Raw silk production (MTs)			
		2012-13	2013-14	2014-15	2015-16 (P)
13	Orissa	104	53	98	117
14	Jammu & Kashmir	145	136	147	127
15	Himachal Pradesh	23	25	30	32
16	Uttarakhand	17	22	29	30
17	Haryana	0.13	0.13	0.3	0.6
18	Punjab	5	4	4	0.8
19	Assam & Bodoland	2068	2766	3222	3325
20	Arunachal Pradesh	22	15	28	37
21	Manipur	418	487	369	522
22	Meghalaya	517	644	655	857
23	Mizoram	40	44	50	64
24	Nagaland	324	606	619	631
25	Sikkim	3	0.20	8	6
26	Tripura	15	40	36	52
	Total	23679	26480	28467	28472

P=Provisional (Source: Central Silk Board, Bangalore)

Export potential of India

Indian silk industry has registered an impressive growth, both horizontally and vertically over the last six decades because of the favourable climate (Virk *et al.,* 2009). Plans and schemes implemented by central and state agencies and relentless efforts of thousands of dedicated persons in the fields of research and extension have helped in this context. The sericulture industry has witnessed a quantum jump in raw silk productivity. As a result, the export potential of India, which is 1422.85 crores in 1997-98, has reached to 4351.23 crores during 2015-16. The Indian silk goods are being exported to the traditional major markets like USA, European countries and small markets of Asia region. For instance, the age old univoltine hybrids have been replaced by bivoltine and multivoltine hybrids. The new technology besides doubling yields has also led to qualitative improvements in cocoon production with considerably reduced renditta and has also helped to break the climate barrier. The average yield of 25 kgs of cocoons/100 dfls in the recent past has increased and currently the average yields are in the range of 60-65 kgs/100 dfls.

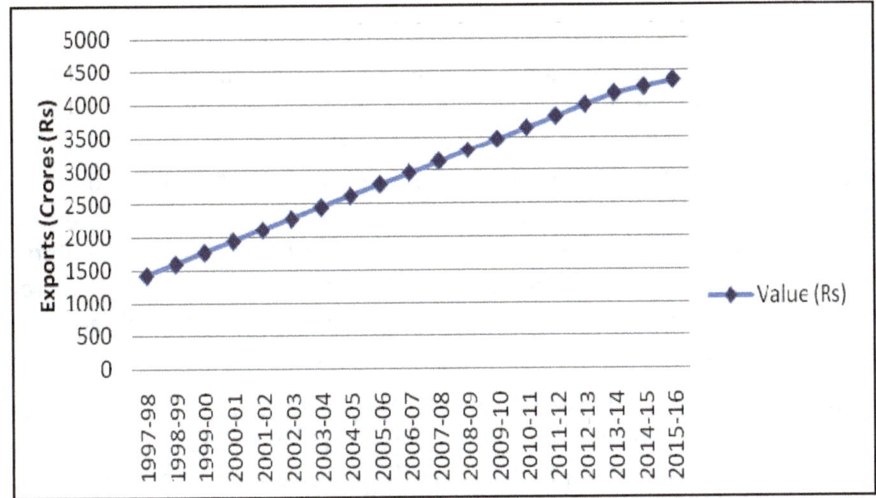

Fig. 1.3 Export of silk from India

Trends in Indian Sericulture in recent years

As we have already seen, in the global scenario, India is the world leader in tropical sericulture and stands second in raw silk production in the world next to China. It has a strong tradition and culture bound domestic market of silk. India produces 20,434 MT of mulberry raw silk along with 8,038 MT of non-mulberry silk (2015-16). Total annual consumption of silk in the country is around 29,300 MT. The additional requirement of silk is imported mainly from China. Therefore, there is scope for production of additional quantity of silk in the country to meet the domestic demand. Among non-mulberry silk, eri silk production is increasing very fast.

Position of Ericulture in silk industry

Ericulture plays significant role in rural livelihood security especially among marginalized and weaker section of the society covering more than 0.18 million families of the country (Ahmed *et al.,* 2015) with an annual production of 4236 metric tonnes. It is a viable agro-based industry, which was first introduced into India about 400 years back. Since then the industry is flourishing as an agro-based industry. Compared with the other non-mulberry silkworms, eri silkworm has a very rich harvest. Among the commercially exploited non-mulberry silkworms, the eri silkworm, *Samia cynthia ricini* (Boisduval) is the only species domesticated completely and adopted to indoor rearing all through the year (Reddy *et al.,* 2000 and

Debaraj *et al.,* 2002). Apart from the benefits of indoor rearing, silkworm itself is extremely hardy and less susceptible to diseases. Eri silk is commonly considered as poor man's silk and its production in India is limited to backyard venture. However, it offers tremendous scope and opportunities for developing the same into an industry with immense potentialities for self-employment and can play a vital role in poverty alleviation, besides generating additional income to the farmers (Siddiqui *et al.,* 1993). The silk produced by the eri silkworm is considered economically the third most important silk in the world after mulberry silk and Chinese tasar.

The word 'Eri' is derived from the Sanskrit term "Erranda", which refers to the castor plant, *Ricinus communis* Linnaeus (Euphorbiaceae), which is the primary host plant of eri silkworm (Patil and Savanurmath, 1994 and Saratchandra, 2003). Eri silk is also known as endi or errandi in India. Ericulture involves rearing of eri silkworm and production of eri silk. Eri silk is popular in some Indian states because of its qualities like texture, lustre, tensile qualities, comfort, adaptability to all climatic conditions, royal look, natural shine, soft, inherent affinities for dyes and vibrant colours with high absorbance and light weight (Ganga and Sulochana Chetty, 1995; Anonymous 1984, 1994, 2004). Further, the castor can be grown on semi-arable and degraded soil. It can also be grown on waste lands and in rain fed conditions. Castor plant does not attract either cattle or wild animals. This makes cultivation of castor more suitable for tribal areas, adjacent to the forests.

Ericulture is prevalent mainly amongst the tribal women in hill districts of North Eastern region. Apart from North-Eastern region ericulture is also practiced in the States of West Bengal, Bihar, Orissa, Jharkhand, Andhra Pradesh, and Telangana. Recently its culture has been spread to Uttarakhand, Chhattisgarh, Maharashtra, Gujarat, UP etc. Ericulture though relatively a less remunerative occupation but it has its own advantages. Just like castor plant eri silkworm is also considered as hardier, and it is highly nutritious. It does not make annoying sound or odours. It also does not need water to drink and has less mortality rate than other silkworms. Eri silkworms require comparatively minimum care as they are easy to handle. With feeding on the castor leaves, it is possible to obtain 75-85% effective yield. In India ericulture is essentially village-based industry providing employment to a sizable activity falling under the cottage and small scale sector. The cocoon of eri silkworm cannot be reeled but are being spun into thread like cotton and the weaving can be done on hand looms under cottage industry. It particularly suits rural based farmers, entrepreneurs and artisans, as it requires low investment but have potential for relative higher return. It provides income and employment to the rural poor specially farmers with small land holding and marginalized weaker section of the society. Several socio-economic studies have affirmed that the benefit-cost ratio in ericulture is higher among comparable agricultural crops.

Origin and History of Eri silkworm with Special Reference to Indian Conditions

Eri silkworm is biologically named as *Ailanthus* silk moth. This silkworm is known as a caterpillar and is prominently found in the South Asian regions especially in China and Japan. These silk moths are predominantly found over the wild trees and shrubs of *Shorea robusta* and *Terminalia*. Eri silk being the most textured silk needs a huge amount of preservation and care strategy. It has shorter fibers' than the usual cultured silks. The eri silk is so popular because texture of the fabric is coarse, fine and dense. It is very strong, durable and has elasticity, darker, heavier than other silks. It blends well with wool and cotton. Hence it is widely used in home furnishing. It is indeed one of the softest and purest forms of silk which is fancied by almost all the silk lovers and is a staple in every fashionists' wardrobe. The eri silkworms give the silk with a dull yellow or gold like sheen. Eri Silk has become the face of Indian silk. India is the largest producer of eri silk in the world as 96% of eri silk is produced in India of the total eri silk produced in the world (Rajesh Kumar and Gangwar, 2010). Eri silk production in India in the year 2007-08 was 1530 tones this made up 73% of the total vanya silk production of 2075 tonnes. Around forty per cent of eri silk is produced in Nagaland, Meghalaya, Manipur, Bihar, Orissa, Karnataka, Assam, Andhra Pradesh, Telangana, and Jharkhand. The bulk of Eri Silk production gives Assam the name of eri silk state. In North eastern region eri silk production is 1714.0 MT in Assam, 222.0 MT in Manipur, 480 MT in Meghalaya, 280 MT in Nagaland, 16 MT in Arunachal Pradesh, 6.50 MT in Mizoram. Ericulture production in other states is 5.0 MT in Andhra Pradesh, 9.0 MT in west Bengal, 5 MT in Bihar, 3.0 MT in Chhattisgarh, 4.5 MT in Madhya Pradesh, 5.0 MT in Orissa. The details are presented in the Table 1.3.

Table 1.3: State wise Eri raw silk production in India

State	Raw silk in MT			
	2007-08	2008-09	2009-10	2010-11
Andhra Pradesh	5.00	7.00	8.00	5.00
West Bengal	10.00	11.23	13.00	9.00
Assam	837.00	1141.00	1410.00	1714.00
Arunachal Pradesh	11.00	14.00	15.00	16.00
Bihar	2.00	2.00	2.50	5.00
Chhattisgarh	2.00	1.50	2.00	3.00
Jharkhand	1.30	0.10	0.50	0.00
Madhya Pradesh	6.50	4.00	4.00	4.50
Manipur	213.00	240.00	280.00	222.00
Mizoram	3.00	6.00	6.00	6.50
Meghalaya	303.00	435.00	450.00	480.00
Nagaland	124.00	160.00	250.00	280.00
Orissa	5.00	8.50	9.00	5.00
Sikkim	0.20	0.50	2.00	1.00
Punjab	0.00	0.00	0.00	0.50
Uttarakhand	1.00	1.40	2.00	0.50
Uttar Pradesh	6.00	5.78	6.00	8.00
Total	1530	2038	2460	2760

(Source: Central Silk Board, Bangalore)

Eri silk is believed to have originated in India and its history dates back to 1600 B.C. in vedic literature. Eri silk was traded from Assam to Northern India as early as 600-650 B.C. during the period of King Bhaskar Burman. The great Ahom king of Sibsagar patronized silk industry in Assam during 1492-1520 A.D. Ericulture is one of the oldest professions adopted by the people of North Eastern region of India for production of eri silk as well as use of eri pupa as food material which is highly nutritive. Ericulture is purely traditional and a leisure time occupation limited to meet sericulture family's clothing and food needs. Sarkar (1980) mentioned that the earliest available reference to ericulture in India has been documented in 1779 according to which vast quantity of eri silk was produced in the country in the environs of Ghoraghat within the then undivided Bengal and since then, a few British agents recorded various references about eri silk in their diaries from time to time. Sankardev named it as "Vrindavania Vastra". During the British rule, Bengal and Myanmar were important consumers of silk fabrics produced in Assam. Indeed, eri silk has been known more to the people of the North-Eastern region of India than others for many centuries. Many communities in Assam, especially the Bodos considered eri silk as a part of their heritage and culture. Yet, the fact remains that the use of eri silk over the centuries is confined to only a few pockets in India, Tibet, Bhutan and Myanmar. For years, it has remained a locally available substitute for wool in North-East India for making winter clothing. Primitive techniques of spinning, crude ways of processing and production of fabric on very basis looms, all lead to eri becoming almost a synonym for chaddar. The popularity of eri pupae as a tasty and nutritious food item also ensured that silk was relegated more as a by-product in ericulture.

The eri silk production of 85,000lb during the early decades of 19th century is increased tremendously to 5,054 MT during 2015-16 in India (Table 1.4). Captain Jenkins and Mr. Michael Atkonson of Jangipur, Murshidabad (1771) remarked that eri silk, though inferior to muga silk, was of incredible durability. Attempts were made to introduce ericulture, on large scale during the turn of the 19th century by the planters in Assam. However, it did not succeed as they found it uneconomical due to high labour cost involved in it. Since then ericulture hardly made any headway as an industry and remained confined to Assam. Ministry of Textiles, Government of India has established a national organization in the year 1949 to look after the overall development of ericulture and silk industry in India. Central Silk Board studied the feasibility of ericulture as an additional income and employment generating opportunity to castor growers of all states in India. The increasing demand for eri silk products outmatches the production and to match this central silk board, of late has planned development of ericulture units in several states like Andhra Pradesh, Bihar, Jharkhand, West Bengal, Orissa, Madhya Pradesh, Chhatisgarh, Uttar Pradesh, Punjab, Tamil Nadu, Pondicherry and Kerala under various schemes of central silk board and Ministry of Rural Development, Government of India. Limited efforts were taken to introduce ericulture in Rajasthan and Gujarat as a subsidiary to castor or cassava cultivation aiming at some additional income. Recently, ericulture is being introduced in Madhya Pradesh, Delhi, Punjab, Karnataka and Maharashtra (Kshama Giridhar *et al.,* 2007).

Table 1.4: Trend of growth in Eri raw silk production in India
(1951-52 to 2015-16)

Year	Production of raw silk (MT)	Year	Production of raw silk (MT)
1951-52	100	1984-85	279
1952-53	102	1985-86	352
1953-54	99	1986-87	392
1954-55	104	1987-88	522
1955-56	127	1988-89	565
1956-57	130	1989-90	589
1957-58	143	1990-91	624
1958-59	143	1991-92	704
1959-60	112	1992-93	-
1960-61	110	1993-94	-
1961-62	132	1994-95	-
1962-63	137	1995-96	745
1963-64	194	1996-97	745
1964-65	204	1997-98	814
1965-66	201	1998-99	970
1966-67	208	1999-00	974
1967-68	200	2000-01	1089
1968-69	213	2001-02	1160
1969-70	218	2002-03	1316
1970-71	161	2003-04	1352
1971-72	168	2004-05	1448
1972-73	143	2005-06	1442
1973-74	141	2006-07	1485
1974-75	115	2007-08	1530
1975-76	123	2008-09	2038
1976-77	106	2009-10	2460
1977-78	56	2010-11	2645
1978-79	120	2011-12	3072
1979-80	183	2012-13	3116
1980-81	135	2013-14	4237
1981-82	147	2014-15	4726
1982-83	213	2015-16	5054
1983-84	270		

The recent advances in ericulture technologies and their dissemination have increased the quality and productivity levels in all stages of production including host plant cultivation with high yielding varieties. Added to this, the prospects of intercropping of eri host plants with grams, pulses, oilseeds and vegetables increase the return from unit area. Castor, primary host of eri silkworm can be exploited both for seed and cocoon production. The technologies developed by research and development institutes of CSB have been popularized among the farmers to maximize yield and returns which in turn resulted in bringing out vertical growth of the industry. The constant efforts made by CSB and state sericulture department have resulted in the overall increase in the silk production and quality. Dissemination of advanced technologies supplemented by the quality seed supply, extension programmes, training, technical assistance etc. have resulted in improvement of production, productivity and quality of eri cocoons in the field. There has been considerable impact on increase in uptake of silkworm seed and enhancement of raw silk production per unit area. There has been up gradation of infrastructure of farmers, improvement of skills and knowledge of the farmers that paved way for strict discipline in sericulture practices. This has been made possible through technology demonstration programmes, adoption of new technologies in host plant cultivation, silkworm rearing and disease management. Besides this other extension activities *viz.,* kisan melas/ field days/ farmers days/ exhibitions/ training programmes/ audio visual/ vichar gosthi/ work shop/ film shows are undertaken to reach more number of beneficiaries and keep abreast of the latest methodologies among the farmers. The growth and development of ericulture depends on the rapidity with which the new technologies and practices are developed. Indian scientists have significantly contributed to the development of technologies for silkworm rearing practices. Ericulture provides livelihood to 1.83 lakh farm families. The increasing production is due to the horizontal expansion of plantation and not due to the increase in productivity. Development of improved breeds along with adoption of available improved technologies is the important tools for enhancement of productivity and employment generation among rural poor.

Eri silkworm being multivoltine, about 5-6 crops can be reared in a year and the worms can be fed on various food plants by virtue of its polyphagous nature. Although ericulture has been closely associated with the tradition and culture of North East, the overall productivity of the region is poor mainly due to lack of systematic plantation, small scale rearing and plasticising of the culture in conventional manner. The whole gamut of ericulture activities right from plantation, rearing, seed production to post cocoon are being carried out in unorganized manner there by affecting the overall production and productivity of the sector. Eri silk shares about 73% of the total vanya or non-mulberry raw silk produced in the country and offers vast scope for its development and expansion in traditional as well as non- traditional states by virtue of its thermal property and blending abilities with other natural and synthetic fibres. Though North East India contributes more than 96% of the total raw silk produced in the country, it has been steadily expanding in the states of Orissa, West Bengal, Andhra Pradesh, Madhya Pradesh, Uttar Pradesh, and Chhattisgarh also due to the existing potential for its development. Ericulture can be easily practiced with less investment and the crops are more assured as compared to other vanya silks such as muga and tasar due to their outdoor nature of rearing coupled with severe attack by pests and predators. Evaluation on rearing and grainage performance of eri silkworm in non-traditional states revealed encouraging results and adoption of improved rearing package has increased the quality and productivity of cocoons substantially. By products in the form of protein rich eri pupae, silkworm litters and excreta also form integral and important economic components of the culture. The introduction of new and improved spinning devices has also enabled to produce finer yarn and paved the way to multiplicity of design and value added products with ample marketing avenues. Exploitation of eri pupa as protein rich food enhances the income level of rural poor from Rs. 2500 to Rs. 14,500, out of 100 disease free layings (Ahmed and Rajan, 2011).

In order to cater research and development needs of eri silk industry, studies on various aspects pertaining to improvement of host plants, effect of host plants on rearing performance and economic traits of eri silkworm, rearing methods, inter-changeability of food plants, crop improvement including breeding studies of eri silk moths were initiated by the erstwhile Central Muga & Eri Research Station (CMERS), Titabar, Central Eri Research & Training Institute (CER&TI), Mendipathar, Meghalaya. Of late Central Silk Board established Central Muga & Eri Research & Training Institute (CMER&TI), Lahdoigarh, Jorhat to carry out studies on various aspects of host plants and silkworm crop improvement. The institute was mandated to serve as the apex research institute for eri development *viz.,* to collect and conserve eri germplasm host plants and eco races of silkworms, to evolve suitable package of practices for eri silkworm rearing for its geographical area and to improve various seed technological aspects connected with ericulture.

Eri silkworm rearing is being practiced traditionally in rural areas of India. It contributes to approximately 14% of the total raw silk production in India. Of late eri silk production is increasing steadily and there was remarkable growth during the X to XII plan period. Eri silk industry gained a new dimension due to systematic and scientific interventions, introduction of high yielding plantation, replacement of traditional spinning devices with improved ones and development of diversified products having potential markets in India and abroad. Farmers adopted new technologies and modern practices as they yield better returns. The main contribution to increase the production is attributed to the production of eri silk in main land states of the country. During XII plan period as against the target of 4.10 lakh dfl rearing, we have achieved 5.69 lakh dfl rearings.

While rearing mulberry silkworm *Bombyx mori*, exclusively for silk, it is always essential to kill the pupae to extract the silk from the cocoons, whereas the rearing of eri silkworm (*Samia cynthia ricini*) is just opposite because its cocoons are used for processing spun silk wherefrom moths have already emerged. Therefore, culturing eri silkworm has been favoured by the people whose religious practices forbid the taking of silkworm life such as Buddhists in Sri Lanka (Suryanarayana and Choaba Singh, 2003; Choaba Singh and Suryanarayana, 2003). Hence the eri silk can be called "Ahimsa silk".

Significance of Eri silk and why it is called as 'Ahimsa silk' or fabric of peace

Eri silk is known as poor person's silk (As it is not so enormously priced compared to other silk types and its cost of production is very less than other silks) (Sarkar, 1988). It is also known as non-violence silk (There is no need to kill the pupae inside the cocoon as in the case of other silk because eri silk is spun into thread like cotton) and cocoon of eri silkworm is open mouthed at one end. Texture of eri fabric is coarse, fine and dense. It is very strong and durable and has elasticity, darker, heavier than other silk. It blends well with wool and cotton. The thermal property (warm in winter and cool in summers) of eri silk makes it a suitable fabric for shawls, jackets and blankets. Eri silk is valued for the strong, supple fabric it produces. The fibre is woven into chaddars or wraps and used as a substitute for woollens. Baby dresses are also made from eri silk because of its soft texture and moisture absorbent qualities (Vaidya and Yadav, 2014). However, the British called it as "Palma Christi" or "Divine silk" as it is obtained without killing the insect. Eri silk is one of the purest forms of silk that is a true and genuine product of the *Samia cynthia ricini* worm. Eri silk is called the father of all forms of cultured and textured silks. It is the only domesticated silk produced in India, as the process doesn't involve any killing of the silk worm, also naming Eri silk as 'Ahimsa silk' or fabric of peace. Eri silkworms are relatively cheap and easy to maintain and their cultivation forms a small scale, cottage industry in Assam and Meghalaya that produce around 95% of the world's Eri silk. Here are some reasons why Eri silk earned its moniker 'ahimsa silk' and why it is one of the most sustainable fabrics in the world.

1. **It is a free gift from *Samia c. ricini***

   The Eri Silk moth spins open-ended cocoons as it transitions into beautiful moths. Because these cocoons are open-ended, allowing the moth to leave and enter the cocoon through the opening. It also means that the worm does not need to be killed in order to obtain the silk threads and that it can complete the process of metamorphosing into a moth, hatch and breed. Unlike other silks, the moth is allowed to leave the cocoon before the Eri silk is extracted earning the fabric many names like ahimsa, non-violence, peace or vegan silk. These are yellowish-white or golden in colour and have an almost divine sheen to them.

2. **Its cultivation is sustainable**

   The *Samia cynthia ricini* worms feed off castor leaves (*era* in Assamese). Unlike mulberry silkworm rearing which is known to be land intensive, cultivation of castor is easier, possible in drought-prone regions on small plots of land alongside other crops, thus providing small and marginal farmers an alternative economic activity and income source.

3. **It has a small water footprint and produces zero waste**

   Most plant based natural fibres like cotton and linen require a lot of water during the growing stage and need very good irrigation infrastructure. Since the eri silk fibre is technically a waste in itself (after the worm becomes a moth), no additional resources are spent by nature in its making, apart from helping a life transition into another version of itself. Hundred % of the cocoon can be used to make yarns and, in turn, fabrics. It is in fact nature's very way of teaching us up cycling.

4. **Its production empowers small and marginal tribal farmers**

   Almost all eri silk is produced in North Eastern part of our country. The communities rearing the eri silkworms, mostly indigenous tribal communities, are incentivized by Ericulture as they get good profits for this valuable material. The Central Silk Board runs many clusters in the region to ensure that the activity serves as a sustainable alternate source of income for small and marginal farmers in the region.

Summary and Conclusions

➢ Silk is a natural protein fibre secreted by arthropods especially lepidopteran silkworms.

➢ The word sericulture has been derived from the Greek word "Sericos" means silk and the English word culture means 'rearing'.

➢ Silk is the most elegant textile in the world with unparalleled grandeur, natural sheen, and inherent affinity for dyes with high absorbance, light weight, soft touch, smooth, strong and high durable than any natural or artificial fibre. Hence silk is known as "The Queen of Textiles".

➢ The insects that produce silk of economic value are termed as sericigenous insects. The natural silk producing insects are broadly classified as mulberry and wild or non-mulberry silkworms.

➢ The mulberry silk moths are represented by domesticated *Bombyx mori* Linnaeus. Non-mulberry sericulture is universally known as forest or vanya or wild sericulture. The term 'Vanya' is of Sanskrit origin, meaning untamed, wild or forest based.

➢ Sericulture includes raising of host plants, rearing of silkworm, production of silk yarn. Sericulture requires low investment, gives quick return, provides employment opportunities and earns foreign exchange.

➢ India is a home to a vast variety of silk secreting fauna which also includes an amazing diversity of silk moths.

➢ India is the producer of all the five commercially traded varieties of natural silks namely, Mulberry, Tropical Tasar, Oak Tasar, Eri and Muga.

➢ Tasar silk is copperish in colour, coarse in nature and is secreted by the Tropical Tasar silkworm, *Antheraea mylitta* which thrives on plants like Asan and Arjun (*Terminalia* sp.).

➢ Oak Tasar is a finer variety of Tasar produced by the temperate tasar silkworm, *Antheraea proylei* which feeds on natural oak plants (*Quercus* sp.) and is found in abundance in the sub-Himalayan belt.

➢ Eri silk is a silk spun from open-ended cocoons and secreted by the domesticated silkworm, *Samia cynthia ricini* Boisduval that feeds mainly on castor (*Ricinus communis* Linn) leaves.

➢ Muga silk is golden yellow in colour and is preferred mainly in the state of Assam during festivities. Muga silk is secreted by *Antheraea assama* that feeds on aromatic leaves of naturally growing Som (*Persia bombycina*) and Soalu (*Litsea polyantha*) plants.

➢ Silk was first discovered during 2640 BC by the Chinese Empress Xi Ling Shi, the 14 year old bride of the China's 3rd Emperor, Huang Ti, also called "Yellow Emperor".

➢ The Chinese kept the secret of the beautiful and value added material from the rest of the world that they were producing, for more than 30 centuries.

➢ After 1200 B.C. Chinese immigrants who had settled in Korea helped in the emergence of silk industry in Korea. From there onwards the silk industry spread to Africa, Spain and Sicily. Later, it was introduced into Europe and Japan as well by 7th century AD.

➢ The Chinese princess married to Indian prince disclosed the secret of raising silkworms thus, silk production spread in India.

➢ According to Western historians, mulberry-tree cultivation spread to India through Tibet during 140 BC.

➢ According to some Indian scholars silkworms (*Bombyx mori*) were first domesticated in the foothills of the Himalayas.

➢ The flourishing Indian silk trade was exploited by the British people and they exported large quantities of silk produced in West Bengal to England.

➢ After Independence in India, the industry again started flourishing as an agro-industry, giving employment to 35 million people in the country.

➢ The silk production, which is 16,500 MT during 2004-05, has increased to 28,523 MT in 2015-16, contributes to 14% of the total world raw silk production.

➢ Indian silk industry has registered an impressive growth, both horizontally and vertically over the last six decades because of the favourable climate.

➢ As a result, the export potential of India, which is 1422.85 crores in 1997-98, has reached to 4351.23 crores during 2015-16.

➢ The average yield of 25 kgs of cocoons/100 dfls in the recent past has increased to 60-65 kgs/100 dfls because of the improved practices in the field of sericulture.

- ➤ Among the commercially exploited non-mulberry silkworms, the eri silkworm, *Samia cynthia ricini* (Boisduval) is the only species domesticated completely and adopted to indoor rearing all through the year.

- ➤ Eri silk is commonly considered as poor man's silk and its production in India is limited to backyard venture. However, it offers tremendous scope and opportunities for developing the same into an industry with immense potentialities for self-employment and can play a vital role in poverty alleviation, besides generating additional income to the farmers.

- ➤ The word 'Eri' is derived from the Sanskrit term "Erranda", which refers to the castor plant, the primary host of eri silkworm.

- ➤ Ericulture though relatively a less remunerative occupation but it has its own advantages. It is highly nutritious, does not make annoying sound or odours. It also does not need water to drink and has less mortality rate than other silkworms.

- ➤ The cocoon of eri silkworm cannot be reeled but are being spun into thread like cotton and the weaving can be done on hand looms under cottage industry.

- ➤ The eri silk is so popular because texture of the fabric is coarse, fine and dense. It is very strong, durable and has elasticity, darker, heavier than other silks. It blends well with wool and cotton. Hence it is widely used in home furnishing.

- ➤ Eri Silk has become the face of Indian silk. India is the largest producer of eri silk in the world as 96% of eri silk is produced in India.

- ➤ Ericulture is one of the oldest professions adopted by the people of North Eastern region of India for production of eri silk as well as use of eri pupa as food material which is highly nutritive.

- ➤ Ericulture is purely traditional and a leisure time occupation limited to meet sericulture family's clothing and food need.

- ➤ The eri silk production of 85,000lb during the early decades of 19th century is increased tremendously to 5,054 MT during 2015-16 in India.

- ➤ Eri silk industry gained a new dimension due to systematic and scientific interventions, introduction of high yielding plantation, replacement of traditional spinning devices with improved ones and development of diversified products having potential markets in India and abroad.

➤ During XII plan period as against the target of 4.10 lakh dfl rearing, we have achieved 5.69 lakh dfl rearings.

➤ While rearing mulberry silkworm, exclusively for silk, it is always essential to kill the pupae to extract the silk from the cocoons, whereas the rearing of eri silkworm is just opposite because its cocoons are used for processing spun silk wherefrom moths have already emerged.

➤ Hence eri silk is also known as non-violence silk or ahimsa silk or fabric of peace or vegan silk.

References

Ahmed S A and Rajan R K. 2011. Exploration of *vanya* silk biodiversity in North Eastern Region of India: Sustainable livelihood and poverty alleviation. *Proceedings of International Conference on Management Economics and Social Sciences* (ICMESS 2011), December 23-24, 2011 at Bangkok, Thailand pp. 485-489.

Ahmed S A, Sarkar C R, Sarmah M C, Ahmed M and Singh N I. 2015. Rearing Performance and Reproductive Biology of Eri Silkworm, *Samia ricini* (Donovan) Feeding on *Ailanthus* Species and Other Promising Food Plants. *Advances in Biological Research* **9**(1): 07-14.

Anitha R. 2011. Indian silk industry in the global scenario. *International Journal of Multidisciplinary Management Studies* **1**(3): 100-110.

Anonymous. 1984. A few guidelines to Eri silkworm rearing. *Regional Sericultural Research Station, Central Silk Board*, Titabar, Assam pp. 6.

Anonymous. 1994. Package of practices for higher yields. *University of Agricultural Sciences and State Department of Agriculture*, Bangalore pp. 98.

Anonymous. 2004. Package of practices for eri host plant cultivation and silkworm rearing. *Central Muga, Eri Research and Training Institute*, Lahdoigarh, Jorhat, Assam.

Chen C H and Gu S H. 2006. Stage dependent effects of starvation on the growth, metamorphosis and ecdysteriodogenesis by the prothoracic glands during last larval instar of silkworm *B. mori. Journal of Insect Physiology* **52**: 968-974.

Chen Y. 2003. Variable tolerance of the silkworm *Bombyx mori* to atmospheric fluoride pollution. *Fluoride* **36**: 157-162.

Chaoba Singh K and Suryanarayana N. 2003. Wild silkmoth – Wealth of India. *National Conference on Tropical Sericulture for Global Competitiveness*, 5-7th Nov, 2003 at Central Sericultural Research and Training Institute, Mysore. Pp. 71-78.

Chowdhary S N. 2006. Host Plant species of Non-mulberry (Vanya) Silkworms. *Non-Mulberry Silkworm and Host Plants Germplasm* pp: 67-73.

Debaraj Y, Datta R N, Das P K and Benchamin K V. 2002. Eri silkworm crop improvement – A Review. *Indian Journal of Sericulture* **41**(2): 100-105.

Ganga G and Chetty S J. 1997. An Introduction to Sericulture, 2nd Edition, *Oxford and IBH Publishing Co. Ltd*, New Delhi.

Ganga G and Sulochana Chetty 1995. An introduction to sericulture hand book. *Central Silk Board* pp. 45.

Goswami D, Singh N I, Mustaq Ahmed, Rajesh Kumar, Mech D and Giridhar K. 2015. Impact of Integrated Chawki Rearing Technology on cocoon production of Muga Silkworm, *Antheraea assamensis* Helfer. *Biological Forum – An International Journal* **7**(1): 146-151.

Hiwar C J. 2001. Agro Cottage Industry: Sericulture. *Daya Publishing House*, New Delhi pp. 1-117.

Kedir Shifa, Waktole Sori and Emana Getu. 2014. Feed utilization efficiency of eri-silkworm (*Samia cynthia ricini* Boisduval) (Lepidoptera: Saturniidae) on eight castor (*Ricinus communis* L.) genotypes. *International Journal of Innovative and Applied Research* 2(4): 26-33.

Kshama Giridhar, Mathanta J C and Deole A L. 2007. Raw silk production 2006-07. *Indian Silk* **46**(6): 43-44.

Lakshmanan S, Ganapathy Rao R, Jayram H and Geetha Devi R G. 1997. Labour composition in Sericulture. *Indian Silk* **35**(12): 19-21.

Lamelu A G. 1998. Illustrated textbook on sericulture-translated from Japanese. *Sciences pulishing Inc. Enfield*, New Hampshiere pp.10-35.

Metaferia H Y, Amanuel T and Kedir S. 2007. Scaling up of silk production technologies for employment and income generation in Ethiopia. In: *Success with Value Chain: proceedings of scaling up and scaling out of agricultural technologies in Ethiopia, an international conference, 9-11 May 2006* (Tsedeke Abate ed). Ethiopian Institute of Agricultural Research, Addis Ababa.

Patil G M and Savanurmath C J. 1994. Eri silkworm – The poor man's friend. *Indian Silk* **33** (4): 41-45.

Rajesh Kumar and Gangwar S K. 2010. Impact of varietal feeding on *Samia ricini* Donovan in spring and autumn season of Uttar Pradesh. *ARPN Journal of Agricultural and Biological Science* **5**(3): 46-51.

Reddy D N R, Narayanaswamy K C and Devaiah M C. 2000. The eri silkworm egg. *Zen Publishers*, Bangalore P. 41.

Reddy D N R. 2000. On the nomenclature of eri silkworm. *Sericologia* **40**: 665-667.

Sahay A, Singh B K, Deori S and Mukherjee P J. 1997. Ericulture: Nature's gift. *Indian Silk* **3**:12-15.

Saratchandra. 2003. A Thought for development of Ericulture in India. *Indian Silk* **41**: 25 – 28.

Sarkar D C. 1980. Ericulture in India. *Central Silk Board*, Bangalore, India P 21-23.

Sarkar D C. 1980. Ericulture in India. *Central Silk Board*, Bangalore, India P 1-49.

Siddiqui A A, Rajaram and Sengupta A K. 1993. Eri common man's silk. *Indian Silk* **32**(4): 34-36.

Suryanarayana N and Chaoba Singh K. 2003. Muga & Ericulture and Forestry in North Eastern India: Presented in *National Workshop on wild silks Culture and Forestry* 21-22 April (2003), Forest Research Institute, Dehradun.

Tzenov P. 2007. Present status and utilization of sericulture germplasm and comparative studies of different silkworm hybrids performance for sericultural enterprise development in the Black, Caspian seas and Central Asia (BACSA) region. In: *Proceedings of International Conference on Sericulture Challenges in the 21st Century and the 3rd BACSA meeting*, September 2007, Vratza.

Vaidya S and Yadav U. 2014. Rearing performance of *Philosamia ricini* (Eri silkworm) in different seasons of Ujjain district. *Environment Conservation Journal* **15**(3): 109-113.

Virk J S, Rabinder Kaur and Parwinder Kaur. 2009. Performance of eri silkworm, *Samia cynthia ricini* Boisduval in different seasons of Punjab. *Indian Journal of Sericulture* **48**(1): 78-80.

CHAPTER 2

Nomenclature and Biodiversity of ERI Silkworm

Just like human names and names for other animals and plants each insect group or species has got a distinct name or otherwise called scientific name of that particular group of insects. Unless there is absolute clarity about the scientific name existing at present for an insect species and its comparative and evolutionary history is clear to the reader he cannot grasp about that particular insect species comprehensively. Any amount of details about the insect species without this clarity the whole of the information can be futile at times. Now let us understand how the particular insect species "Samia cynthia ricini (Boisduval)", the hero of our book which is commonly called as castor or eri silk worm has come into existence into the present and final form and shape in this particular chapter. The genus Samia Hubner is a monophyletic group of large and beautiful moths that range in tropical Asia, across a span of over 6000 kilometres. The structure of the genitalia, wing pattern and chromosome number demonstrate that Samia ricini is derived from its wild progenitor, Samia canningi. There has been lot of confusion and inconsistency regarding the authorship and taxonomic status of this insect. Peigler & Naumann (2003) made considerable efforts in clarifying and applying the correct name to the insect without making total generic revision. Many authors used the generic name 'Philosamia' for this group. As per Arora and Gupta (1979), the current valid name of the domesticated Eri silkworm is "Samia cynthia ricini (Boisduval)". The commercially exploited S. c. ricini is multivoltine and has several eco races based on the morphology and physiology. These 26 eri silkworm germplasms are maintained at Central Eri & Muga Research and Training Institute, Central Silk Board, Ladoigarh, Assam. The eri silkworm genetic resources along with primary and secondary host plants were conserved at Central Sericulture Germplasm Resources Centre i.e. Eri P2 Basic Seed Farm in South India, in Hosur, Tamil Nadu. This chapter discusses in detail about the geographical and ecological distribution of eri silkworm, its diagnosis, different strains/varieties and their molecular characterisation.

Saturniidae, the family of wild silk moths consists of many silk producing insects including *Samia cynthia ricini*. The genus *Samia* is distributed along the Himalayas from Pakistan to Vietnam, covering all tropical South Eastern Asia. In China, it is available in half of the South Eastern Region particularly in one third of its eastern parts. It is also spread along the Palaearctic territory of Korea, Japan, Philippine islands, whole of Indonesia except Western New Guinea, this latter zone is now the part of Papuan sub-region of the Indo-Australian region. Thus, the genus *Samia* can biogeographically be described as covering all of the oriental region (including India, Southern Asia, East Indies and Philippine islands) and eastern Palaearctic region (Europe, North Africa and Asian North of the tropic of cancer).

Origin of the genus "*Samia*"

It is really astonishing that Linnaeus and Fabricius did not cite a group of common and wide ranging insects such as *Samia*, though they acquired insect specimens from all around the world. On *Samia*, the first published reference was by Drury (1773). Even though the generic name *Samia* was proposed by Hubner (1819), the name was reviewed only scattered and contradictory usage for more than a century. Even today many workers insist or remain in using the incorrect name *Philosamia*. Peigler & Naumann (2003) stated that "this is remarkably amateur lepidopterists, sericultursits and those who study insect physiology and biochemistry i.e. groups who traditionally have little or no training and understanding of Zoological nomenclature. Many of the authors in India even in 21st century name the genus as *Philosamia* due to lack of professional expertise.

The genus *Samia* Hubner is a monophyletic group of large and beautiful moths that range in tropical Asia, across a span of over 6000 kilometres. Regarding the nomenclature, the name *Philosamia* has been widely used for the species of *Samia*, but the two names are objective synonyms, both being based on the same type species (*cynthia*). Therefore it is incorrect to use the name *Philosamia*, because it is an invalid name (Peigler and Naumann, 2003). The generic name *Samia* was proposed by Hubner in 1819. The type species *Phalaena Attacus cynthia* Drury (1773), first designated by Grote (1874). The name *Samia* derives from "*Samian*" and has no appropriateness to the moths. *Samian* (from the Latin Samius) refers to an object or native or inhabitant of Samos, a Greek island in the Aegean Sea, near the coast of Turkey (Eliot and Soule, 1902). Later, the name served as a root for more generic names proposed in Saturniidae, such as Callosamia, Philosamia, Pltysamia and Metosamia. The word *Samia* is also used occasionally as a woman's given name, and it is the name of a town in Western Kenya.

Nomenclature of eri silkworm

The eri silkmoth (*Samia c. ricini*) is the third most important silk producer in the world. The moth exists only in captivity, having been artificially selected from a wild progenitor, generally recognized as *Samia canningi* (Hutton) of the Himalayas. In the taxonomic and sericultural literature, there has been considerable confusion and inconsistency regarding the correct authorship of the name *Phalaena ricini* as originally described. The author of *P. ricini* has most often been cited as Boisduval, but other researchers have attributed authorship to Anderson, Jones, Donovan or Hutton. The original description was located thus revealing that *P. ricini* should be credited to Sir William Jones. In turn, the date of publication fixes the name *P. ricini* as the senior subjective synonym for both the wild and cultivated entities, thereby forcing *Saturnia canningi* into synonymy (Peigler and Calhoun, 2013).

There has been confusion regarding the taxonomic status of the genus *Samia* Hubner especially on its nomenclature. Packard (1914) treated *Samia* and *Philosamia* Grote as separate genera. Seitz (1926) considered *Samia* as a valid genus and treated *Philosamia* and *Drepanoptera* Rothchild as its counterparts representing Indian and African fauna of *Samia*, respectively. Subsequently, Draudt (1927) referred to *Drepanoptera* as a subgenus of *Epiphora* but Schiissler (1933) treated *Drepanoptera* as an independent genus. He also revalidated both *Samia* and *Philosamia* as separate genera, representing African, American and Oriental fauna. Michener (1952) set at rest the controversy and regarded the genus *Philosamia* as a junior synonym of *Samia* and stated "this (*Samia*) is an Oriental genus represented in the Eastern United States by a single introduced species". Ferguson (1972) followed Michener (1952) and was also of the view that "*Samia* is indigenous to the oriental region, but the type species was introduced into the United States, where it has not been naturalized for years". The genus *Samia* belongs to the old world group which includes Asiatic genera *Attacus*, *Archaeoattacus* and the African genus *Drepanoptera*. The genus *Samia* is more closely allied to *Archaeoattacus* than *Attacus* in respect of characters of frons, labial palpi and spurs but can be distinguished by the olive-green coloration, long lunate discal spots and by the abdominal markings (Reddy, 2000).

Chronology of Nomenclature of the genus *Samia* Hubner:

1820. *Samia* : Hubner, Verz. Bek. Schmett : 156

1874. *Philosamia* : Grote, proc. Americal phil. Soc., 14: 258.

1914. *Desgodinisia* Oberthur, Ethud. Lepid. Comp., 9(2): 56

1972. *Samia* : Ferguson, The Moths of America North of Mexico., 20(2B): 212-214.

Samia cynthia (Drury) one of the most economically important non-mulberry silk moths, popularly called "*Ailanthus* or Eri silkmoth" is represented by a single species *cynthia* Drury with several sub species, including *ricini* Boisduval which is a domesticated race (Arora and Gupta, 1979). These two (*cynthia* and *ricini*) were treated as separate species by Hampson (1892), who differentiated *ricini* from *cynthia* on the basis of abdomen having segmented band of white hairs instead of tufts. Its fore wing with an ante medial line is more angled and generally joining the post medial band, the lunule is much shorter, the post medial band of both wings with fuscous replacing the pink. Seitz (1926), however treated *ricini* as a form of *cynthia* as considered by Hampson (1892).

Peigler (1992) stated that this species is distinctive, although is a domesticated form and not really a species, the correct identity has apparently always been clear to all authors. The larvae are reared in open trays similar to those of *Bombyx mori*, and the large puffy cocoons indicate a long history of selection by humans for silk production (United Nations, 1980). The worms are commonly reared on castor (*Ricinus communis*) leaves resulting in naming the insect after the vernacular name of the plant, eri or endi. The grey brown colour and whitish abdomen make it easily recognizable. Peigler (1992) opined that it was probably derived from either the Japanese *pryeri* because of the grey colouration and the chromosome number or the Himalayan *canningi*. Having been cultivated for many centuries in China, India and Japan, he suspects that genes from *cynthia*, *walkeri*, *pryeri* and *canningi* have continually been bred into the *ricini* stock at various times and localities through centuries. Esaki *et al.,* (1973) showed a Japanese *ricini* and a *pryeri*, which agree closely with each other in the pink colour of the post median band and the shape of the hind wing crescent. Peigler (1992) observed that the specimens of *pryeri* from Sakai, Osaka prefecture, and *ricini* from Fukuoka, showed similarities, which were not shared by *ricini* from Cuba, and Assam and the *ricini* he had from Assam look more like *canningi* (Choudhury, 1982). Thus, the systematic position of the domesticated eri silk moth stands as super family: Bombycoidea Latreill, 1802; Family: Saturniidae Boisduval (1837) 1834; sub family: Saturniinae Boisduval (1837) 1834; Tribe: *Attacini* Blanchard, 1840; Genus: *Samia* Hubner (1819) 1816; Species: *ricini* Boisduval, 1798.

Chronology of nomenclature of Eri silkmoth:

1773. *Phalanea attacus cynthia* Drury Illust. Nat. Hist. Exot. Ins., 2:10 (Type locality China)

1790. *Bombyx cynthia*: Olivier, Encyclop. Meth. Hist. Nat. Ins., 5:30.

1820. *Samia cynthia*: Hubner, Verz. Bek. Schmett: 156.

1837. *Saturnia cynthia*: West wood. Illust. Exot. Ent., 2:12.

1837. *Phalanea cynthia*: Helfer, J. Asiat. Soc. Bengal., 6:45.

1854. *Saturnia ricini* Boisduval, Annls. Soc Ent. France., 8 (2): 755.

1855. *Attacus cynthia*: Walker, Cat. Lep. Het. Brit. Mus., 5:1220-1221.

1858. *Bombyx querini* Moore, Cat. Lep. Het. E.I House, 2:409.

1859. *Attacus ricini*: Moore, Proc. Zool. Soc. London: 267.

1860. *Attacus canningi*: Huntton, J. Agric. Soc. India: 10.

1874. *Philosamia cynthia*: Grote, Proc. American Phil. Soc., 14:258.

1879. *Attacus obscurus* Butler, Trans. Ent. Soc London: 5.

1887. *Attacus vesta* Cotes and Swinhoe, Cat. Moths India: 225

1887. *Philosamia ricini*: Wardle, Royal Jubilee Exhibition: 23.

1933. *Philosamia cynthia*: Schiissler, Lepid. Cat., 55: 24-41.

1972. *Samia cynthia* Ferguson, The moths of America North of Mexico, 20 (2B): 215-218.

1979. *Samia cynthia ricini*: Arora and Gupta, Taxonomic studies on some of the Indian non mulberry silk moths (Lepidoptera: Saturniidae: Saturniinae). Mem. Zool. Survey India., 16(1) 49-53.

The current valid name of the domesticated Eri silkworm is "*Samia cynthia ricini* (Boisduval)". It is distributed in India (Himachal Pradesh, Uttar Pradesh, Arunachal Pradesh, Assam, Meghalaya, West Bengal, Bihar, Sikkim, and South Andaman), Bhutan, Bangladesh, Sri Lanka (Ceylon), Myanmar (Burma), Indonesia (Java) and China (Arora and Gupta, 1979).

Nomenclature of the genus in different countries is as follows

Samia cynthia (Drury, 1773) P.R. China, Korea and Russia

Samia canningi (Hutton, 1859) SE Asian mainland

Samia pryeri (Butler, 1878) Japan, Hokkaido, Honshu, Shikoku and Kyushu

Samia watsoni (Oberthur, 1914) China and N. Vietnam

Samia vandenberghi (Watson, 1915) Central Indonesia

Samia insularis (Snellen van Vollenhoven, 1862) Indonesian islands of Java and Sumatra

Samia luzonica (Watson, 1913) Philippines

Samia peigleri (Naumann & Nassig, 1995) Central Indonesia

Samia treadawayi (Naumann, 1998) W S Philippines

Samia naumanni Paukstadt, Peigler & Paukstadt, 1998 North Central Indonesia

Samia abrerai Naumann & Peigler, 2001 Java and Bali

Samia naessigi Naumann & Peigler, 2001 Eastern Indonesia

Samia kohlli Naumann & Peigler, 2001 SE Asia

Samia wangi Naumann & Peigler, 2001 SE China and N. Vietnam

Samia tetrica (Rebel, 1924) Singapore, Brunei and Indonesia

Samia yuyukae Paukstadt, Peigler & Paukstadt, 1998 Sunda Islands of Indonesia

Samia fulva Jordan, 1911 Andaman Islands

Samia ceramensis (Bouvier, 1927) Central Moluccas of Eastern Indonesia

Samia ricini (Donovan, 1798) Esaetern Gangetic Plain, Bangladesh, Assam and adjacent sections (Singh and Suryanarayana, 2005).

The Vernacular names of *Samia cynthia ricini* are:

- En silk moth
- Bi ma can (Chinese: castor silkworm)
- Bi ma mei win wang er (Taiwanese: castor crescent emperor moth)
- Pimojoo nooe nabee (Korean: castor silkworm butterfly)

- Pigmajoo nooe nabang (Korean: castor silkworm moth)
- Le bombyx du ricin (French)
- Ver a soie du ricin (French)
- Erisan (Japanese)
- Himasan (Japanese: castor silkworm)
- Eri spinner (German)
- Eri-seiden spinner (German)
- Rizinus spinner (German)
- Ricinussein raupe (German: castor silkworm)
- Bombice del ricino (Italian)
- Bicho-da-seda da mamona (Portuguese)
- Enia, arrindi, andi
- Endi (Bhutan) (Velayudhan *et al.,* 2014)

Diagnosis of *Samia cynthia ricini* (Biosduval)

Samia cynthia ricini always has a solidly white dorsal surface on the abdomen. This is derived from the white tufts seen in related species being expanded, flattened, and fused. The ground colour is usually grey or greyish brown, rarely reddish, but occasionally like grey. The wing pattern is heavily marked by white in the ante median and post median lines. The post median area is often much darkened, which is probably the easiest and quickest way to recognize this species.

Description. Adult male: Head. Frons and occipit dark brown, with lighter edges. Antennae 12 mm long, 4 mm wide. **Thorax.** Tergum and sternum brown to dark grey. **Fore wing.** Length 55-75 mm. Ground colour dark grey or brown; coastal margin ground colour; ante median area ground colour, sometimes darker along veins; ante median line prominent with more white than black; median area brown or grey; post median line black, white, rose, and lavender, sometimes greyish purple instead of rose, drawn inward above and below cescents; post median area strongly blackened; distance between ante median area post median lines in cell Cu_{1a} 4 mm; sub marginal line prominent, often separating darker area on outside from lighter area inside; outer margin ochre brown; crescent short, thick, with more white than yellow; apex very shortened; eyespot black, outer edge flattened distally; underside paler overall, but with very dark post median area. **Hind wing.** Length 44-52 mm. Ante median area lighter brown or grey; ante median line prominent and white; median area of ground colour; post median line coloured as in forewing, with white at tornus of hind wing; post median area blackened; sub marginal line obsolete; sub marginal ornamentation contrasting; outer margin rounded; crescent shorter and thicker that in forewing; tornus weakly pointer. **Abdomen.** Tergum suffused with white scales, tufts flattened; spiracular ornamentation well developed, with white, dark brown, and golden ochre, sternum with mainly white tufts, but dark medially.

Adult female. Differing from male as follows. **Head.** Antennae 12 mm long, 3 mm wide. **Forewing**. Length 60-70 mm; distance between ante median and post median lines in cell Cu_{1a} 3-4mm; crescent short; apex very shortened and rounded. **Hind wing.** Length 42-55 mm; crescent short and thick; tornus rounded. **Abdomen.** Tergum white; spiracular ornamentation white and dark brown; sternum mainly golden ochre (Plate 1).

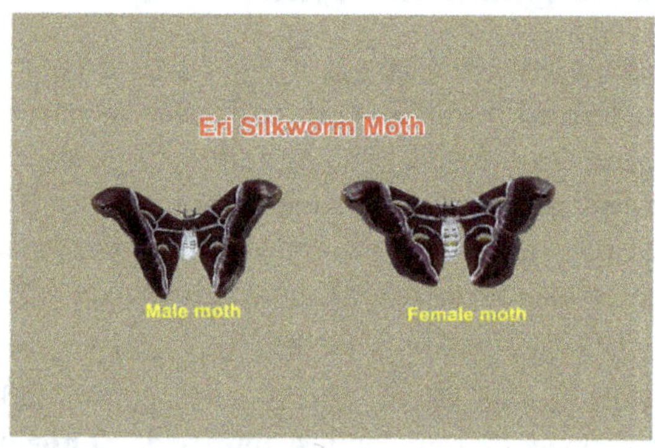

Plate 1 Adult male and female moths of *S. c. ricini.*

Eco races of Eri silkworm

Biodiversity is the complete variability in all living organisms and the ecological complexes that they inhabit and has three levels in diversity namely ecosystem, species and genetic diversity. India is blessed with rich natural beneficial insect fauna. The North East region of India is an ideal natural home for a variety of silkworms. A total of 47 species of silkworms are recorded from India, out of which 24 are reported from North East region (Singh and Suryanarayana, 2005). Out of these Eri and muga Silkworms are predominantly cultured in NE region. There are total

19 species of Eri (genus=*Samia*) all over the world of which only three species are reported from India and out of which two from NE region they are *Samia canningi* which is a wild species and *Samia ricini,* a totally domesticated species (Pigler and Naumann, 2003). The structure of the genitalia, wing pattern and chromosome number demonstrate that *Samia ricini (*Danovan) is derived from its wild progenitor, *samia canningi* (Hutton).

The commercially exploited *S. c. ricini* is multivoltine and has several eco races like Nongpoh, Borduar, Titabar, Sille, Dhanubhanga, Mendipathar and Khanapara etc. (Singh *et al.,* 2003). Siddiqui *et al.,* (2000) evaluated these eco races during 1995-97. These eco-races were collected from different locations of North East India and maintained at the Regional Research Station at Mendipathar. Later, Sarmah *et al.,* (2002) and Sharmah *et al.,* (2012) studied the diversity in eri silkworm eco-races and their utilization for sustainable development in North East India. Detailed characterization of the eri silkworm accessions and their documentation based on descriptor has been done (Chakravorty *et al.,* 2008). Since the establishment of the Central Silk Board, Research and development set up in Assam, several workers studied the different pre-breeding and breeding aspects of eri silkworm. Based on the morphology 26 eco-races of *Samia c. ricini* have been identified. Background information on the germplasm for future reference along with samples (stage of specimens) is collected at the time of silkworm germplasm collection. These include passport data, collection data and specimen data. Collection data and specimen data are required mainly for taxonomic study, while the passport data are generally required for the documentation of the germplasm. The six important parameters provided to all the 26 eri silkworm germplasm accessions are listed in the Table 2.1. A unique number is assigned by the Central Muga & Eri Research and Training Institute to particular collection. The accession number is prefixed by SRI which stands for *Samia ricini*. Total number of accessions available with accession codes is presented hereunder.

Table 2.1: List of Eri silkworm Germplasm Passport data.

S.No	Acc.No.	Race Name	Donor	Origin	Class	Parentage
1	SRI-001	Borduar	RERS, MEG	ASM	O (RCU)	OR
2	SRI-002	Titabar	RERS, MEG	ASM	O (RCU)	OR

Table 2.1: *Contd…*

S.No	Acc.No.	Race Name	Donor	Origin	Class	Parentage
3	SRI-003	Khanapara	RERS, MEG	ASM	O (RCU)	OR
4	SRI-004	Nongpoh	RERS, MEG	MEG	O (RCU)	OR
5	SRI-005	Mendipathar	RERS, MEG	MEG	O (RCU)	OR
6	SRI-006	Dhanubhanga	RERS, MEG	MEG	O (RCU)	OR
7	SRI-007	Chuchuyimlang	CMERTI, ASM	NAL	N	OR
8	SRI-008	Lahing	CMERTI, ASM	ASM	N	OR
9	SRI-009	Barpathar	CMERTI, ASM	ASM	N	OR
10	SRI-010	Diphu	CMERTI, ASM	ASM	N	OR
11	SRI-011	Adokgiri	CMERTI, ASM	MEG	N	OR
12	SRI-012	Lakhimpur	CMERTI, ASM	ASM	N	OR
13	SRI-013	Dhemaji	CMERTI, ASM	ASM	N	OR
14	SRI-014	Kokrajhar	CMERTI, ASM	ASM	N	OR
15	SRI-015	Imphal	CMERTI, ASM	MAN	N	OR
16	SRI-016	Cachar	CMERTI, ASM	ASM	N	OR
17	SRI-017	Dhakuakhana	CMERTI, ASM	ASM	N	OR
18	SRI-018	Genung	RERS, MEG	MEG	N	OR
19	SRI-019	Jonai	CMERTI, ASM	ASM	N	OR
20	SRI-020	Dhansiripar	CMERTI, ASM	NAL	N	OR
21	SRI-021	Sadiya	CMERTI, ASM	ASM	N	OR
22	SRI-022	Tura	CMERTI, ASM	MEG	N	OR
23	SRI-023	Jona Kachari	CMERTI, ASM	ARP	N	OR
24	SRI-024	Barpeta	CMERTI, ASM	ASM	N	OR
25	SRI-025	Ambagaon	CMERTI, ASM	ASM	N	OR
26	SRI-026	Rongpipi	CMERTI, ASM	ASM	N	OR

SRI – *Samia ricini*, RERS- Regional Eri Research Station, CMERTI – Central Muga Eri Research & Training Institute, MEG – Meghalaya, ASM – Assam, NAL – Nagaland, MAN – Manipur, ARP – Arunachal Pradesh, O- Old, RCU- Race in current use, N- New, OR- Original race.

These 26 eri silkworm germplasm are maintained at Central Eri & Muga Research and Training Institute, Central Silk Board, Ladoigarh, Assam. The characterization of eri silkworms is mainly based on the heritable morphological characters of *Samia c. ricini*. Ten descriptions have been utilized for investigation of the 26 eri silkworm germplasm accessions based on the larval colour with rearing performance and cocoon colour with economic traits. These include larval colour, cocoon colour, fecundity, hatching percentage, larval weight, larval period, effective rate of rearing, cocoon weight, shell weight and shell ratio (Tables 2.2 and 2.3). Number of observations was recorded at different stages *viz.,* egg, larva, pupa, cocoon and adult stage based on the descriptors. However, only the heritable characters are listed for the purpose of documentation *viz.,* large black mottles on the integument, blue vs. white integument and the white and brick red cocoons. Analysis of the growth and economic traits of cocoon of different eri silkworm races revealed that eri silkworm accessions *viz.,* SRI-001, SRI-010 and SRI-024 are the most promising eri silkworm races for commercial exploitation in agro climatic condition of North eastern region of India (Sarkar *et al.,* 2008). Breed SRI-010 showed the highest cocoon yield (20.88 kg) followed by SRI-024 (20.01 kg) and SRI-001 (18.82 kg). Cocoon yield is positively correlated with larval period (0.201), effective rate of rearing (0.302) and fecundity (0.668) and negatively correlated with cocoon weight (-0.061), cocoon shell weight (-0.002). Genotypic co-efficient of variation (GCV) and Phenotypic co-efficient of variation (PCV) showed closeness for the characters like ERR, cocoon yield and hatching indicated minimal influence on the expression of these traits. High heritability coupled with high genetic advance (GA) per cent mean and high GCV in traits shown in cocoon yield, hatching percentage and pupal weight (Sarkar and Gogoi, 2010).

Table 2.2: List of morphological traits of eri germplasm

S.No	Acc.No.	Larval body colour	Cocoon colour
1	SRI-001	Plain & Zebra on Yellow and blue	White
2	SRI-002	Plain & Zebra on Yellow and blue	White
3	SRI-003	Plain yellow and blue	White
4	SRI-004	Plain yellow and blue	White
5	SRI-005	Plain blue	White
6	SRI-006	Plain yellow and blue	White
7	SRI-007	Plain yellow	White& brick red
8	SRI-008	Plain & Zebra on Yellow and blue	White
9	SRI-009	Plain & Zebra on Yellow and blue	White& brick red
10	SRI-010	Plain & Zebra on Yellow and blue	White

Table 2.2: *Contd...*

S.No	Acc.No.	Larval body colour	Cocoon colour
11	SRI-011	Plain yellow and blue	White
12	SRI-012	Plain & spotted on yellow & blue	White
13	SRI-013	Plain & zebra on yellow and blue	White& brick red
14	SRI-014	Plain yellow and blue	Brick red
15	SRI-015	Plain yellow and blue	White
16	SRI-016	Plain yellow and blue	Brick red
17	SRI-017	Plain yellow and blue	White& brick red
18	SRI-018	Plain yellow and blue	White
19	SRI-019	Spotted and yellow	White
20	SRI-020	Plain yellow	White
21	SRI-021	Plain yellow	White
22	SRI-022	Plain yellow	White
23	SRI-023	Plain yellow	White
24	SRI-024	Plain yellow and blue	Brick red
25	SRI-025	Plain yellow	White
26	SRI-026	Plain yellow and blue	Brick red

In their natural habitats in North East India the silkworms are geographically isolated and have morphological differences (Singh *et al.,* 2003). The eco races (types) of eri silkworm are commercially exploited in these places because of their high silk yield potential. Many scientists worked on eco races of eri silkworms (Chakravorty, 2004, Singh *et al.,* 2003, Jaiswal *et al.,* 2006, Debaraj *et al.,* 2002, Sakthivel *et al.,* 2004, Mahobia *et al.,* 2005, Singh *et al.,* 2011). To develop high yielding breed of eri silkworm in terms of shell weight and fecundity, breeding work was started by Singh *et al.,* (2003) utilizing Borduar and Genang races. Very recently, Central Muga and Eri Research Training Institute, Lahdoigarh, Jorhat, Assam has developed a high productive eri silkworm breed with higher shell weight and fecundity through hybridization programme and named as C2. This breed was maintained at Regional Eri Research Station, Mendipathar (Singha, 2010). This breed showed better performance over the local eco races on the specified characters listed in the Table 2.4.

Table 2.3: Evaluation data of eri silkworm germplasm

S.No	Acc.No.	Fecundity	Hatching (%)	Larval wt (g)	Larval period (d)	ERR (%)	Cocoon wt (g)	Shell wt (g)	Shell ratio (%)
1	SRI-001	441.99	95.04	8.27	23	90.06	3.64	0.50	13.74
2	SRI-002	458.91	94.35	8.25	22	90.52	3.70	0.48	12.97
3	SRI-003	435.46	94.05	7.31	23	88.12	3.55	0.48	13.52
4	SRI-004	442.74	92.56	8.47	21	87.31	3.52	0.46	13.07
5	SRI-005	455.44	91.77	8.42	23	89.25	3.61	0.47	13.02
6	SRI-006	459.50	92.61	8.43	22	88.24	3.56	0.47	13.20
7	SRI-007	385.44	76.89	7.97	22	85.22	3.79	0.42	11.08
8	SRI-008	413.00	78.90	7.73	23	82.00	3.13	0.39	12.46
9	SRI-009	345.00	71.20	7.92	22	80.00	3.32	0.40	12.05
10	SRI-010	418.50	90.42	8.00	23	87.50	3.10	0.38	12.26
11	SRI-011	366.50	89.33	8.11	21	85.25	3.56	0.45	12.64
12	SRI-012	318.50	86.78	8.14	22	90.50	3.10	0.36	11.61
13	SRI-013	385.00	84.92	7.62	22	81.50	3.24	0.37	11.42
14	SRI-014	348.00	79.50	7.80	23	76.65	3.87	0.46	11.89
15	SRI-015	414.75	91.40	8.19	21	79.50	3.62	0.44	12.15
16	SRI-016	340.40	89.56	8.17	21	88.65	3.08	0.37	12.01
17	SRI-017	371.50	87.45	7.79	23	86.50	3.17	0.38	11.99
18	SRI-018	635.79	92.20	8.48	22	89.96	4.51	0.59	13.08
19	SRI-019	357.50	85.04	7.64	23	90.06	2.95	0.37	12.54
20	SRI-020	472.00	84.35	8.19	24	90.52	3.09	0.38	12.30
21	SRI-021	353.00	84.05	7.48	25	88.12	3.62	0.49	13.54
22	SRI-022	442.74	72.56	8.14	25	87.31	3.11	0.40	12.86
23	SRI-023	255.44	91.77	8.11	25	89.25	3.24	0.41	12.65
24	SRI-024	359.00	82.61	8.00	25	88.24	2.85	0.35	12.28
25	SRI-025	430.00	94.00	8.43	18	85.00	3.01	0.45	14.95
26	SRI-026	380.00	92.00	6.65	20	92.00	2.00	0.30	12.55

Table 2.4: Performance of Eri silkworm C2 breed

Sl. No.	Particulars	Norms fixed by Hybrid Authorization Committee of Central Silk Board	Local eco-races	C2 breed
1	Fecundity (No)	350	322	356
2	Hatching (%)	85	79.65	83.91
3	Cocoon yield by number/dfl	250	203	247
4	Cocoon yield by weight/dfl (kg)	0.750	0.590	0.900
5	ERR (%)	85	79.38	84.02
6	Single Cocoon weight (g)	3.0	2.89	3.67
7	Single Shell weight (g)	0.45	0.38	0.54
8	Cocoon shell ratio (%)	14.00	13.09	14.80

Effect of environment on the performance of eco races of eri silkworm:

Seasonal change in performance of silkworm races is of vital importance to understand the action of environment and genetic potentiality. The performance and economic characters of eri silkworm eco races are not only controlled by genes, but also known to be influenced by different climatic factors such as temperature, relative humidity, photoperiodic cycle etc. to which it is exposed during its life time (Jaiswal *et al.*, 2003 and Harada, 1961). In the study conducted by Sarmah *et al.*, (2002) stated that, rearing parameters such as single cocoon weight, shell weight, shell ratio, larval weight, larval duration, yield (by number and weight), cocoon shape variability, pupal period, pupation rate and leaf silk conversion rate showed significant differences among the eco races. These are of great importance in the field of ericulture.

Hota *et al.*, (2005) reported under performance by the eri silkworm in odisha with regards to the larval characters like silk yield, pupation rate, moth emergence and fecundity. Pure line eco races of eri silkworm show significant increase over the mixed type in terms of morphological and quantitative characters (Ray *et al.*, 2010). It is well known that the dynamic environmental conditions prevailing during various seasons bring about profound changes in the growth, development and the expression of economic characters in different silkworm races (Krishnaswami *et al.*, 1970 and Rajesh kumar and Elangovan, 2012). Singh *et al.*, (2003) and Chakravorty and Neog (2006) reported that Titabar eco race showed better performance compared to others on hatching (%), larval weight, larval duration, single cocoon weight, yield (by number and weight), shell weight, shell ratio, cocoon shape variability, pupal period, pupation rate and leaf silk conversion rate of rearing parameters. Silk ratio varied with the type of host and breed used in eri silkworm rearing (Dookia, 1980).

It was reported that the changing conditions of different seasons of the year bring profound variance in growth and development and the expression of characters in *Samia cynthia ricini* (Kumar and Elangovan, 2010). It was found that the winter season is the most favourable and summer was the least favourable season for the rearing of eco races for the production of better silk. Also, the performance of the silkworm larvae was best at temperature range of 15°C to 23°C. At very high temperature such as 43°C during summer, performance of the eco races was significantly lower which might be due to the decline in feeding rate of the larvae and their low conversion efficiencies resulting in low silk yield. Further, significant differences were also observed in the productivity parameters of the eco races with respect to different seasons of rearing.

Conservation of eri silkworm germplasm

For conservation of biodiversity we require intensive co-operation of all concerned who utilizes the products arising from biodiversity. It was decided in the global biodiversity conservation conference that the concerned parties would adopt strategic plan effectively and coherent implementation of the objectives of CSB and to prevent biodiversity loss at global, national and regional level. In order to conserve the biodiversity and to augment the Eri silkworm seed sector and its primary and secondary host plants Castor (*Ricinus communis*) and Tapioca (*Manihot utilissima*), Central Silk Board, Ministry of Textiles, Government of India, has established Central Sericulture Germplasm Resources Centre i.e. Eri P_2 Basic Seed Farm in South India i.e. in Hosur, Tamil Nadu. The prime function of the Basic Seed Farm is to conserve the eri silkworm genetic resources. Presently it is conserving 443 eri silkworm genetic resources, which includes 26 eco races along with primary host plant Castor and secondary host plant Tapioca.

Ecological Distribution

The ecological distribution of *Samia* species was reported by Rebel (1925). The identity of *Samia cynthia* and the status of its introduced populations was reported by Richard Peigler. The form *ricini* was introduced from Montevideo, Uruguay. However, it was also reported from Cuba. Since *ricini* cannot exist outside of captivity, it was assumed no *Samia* now exist on the Caribbean Island of Cuba. Riotte (1975) reported that the form *ricini* has long been and still is much cultivated in Egypt and perhaps to nearby countries. Seitz (1918) reported a population of *cynthia* lived in Germany early in the 1900s but became extinct later on. Scherdlin (1906) provided a detailed summary of introduced population of *cynthia* in France. Apparently, *cynthia* has had a long and successful establishment in Vienna, Austria as mentioned by many European authors. A population of *cynthia* was established in Barcelona, Spain just before the turn of the century (Gmez Bustillo and Fernandez Rubio, 1976). Spanish authors have declared this population to be gravely, endangered. The population was also introduced in the several parts of Eastern United States. Ferguson (1972) and Pyle (1975) gave good summaries of the history and ecology of this moth in the United States. Two more recent works by Frank, (1986) provided greater insight and detail into the factors that limit the habit and distribution of the moth, and also account for its decline in recent years. The moth may become extinct in North America because of urban renewal that destroys the unique and primary habitat of *Ailanthus* trees in railway roads and industrial parks. Many collectors and dealers report in recent years that *cynthia* is getting harder to find. It is probably extinct by now in New Haven and has disappeared in some sections of Philadelphia (Frank, 1986). Reasons for the decline may be the result of urban renewal. Another severe pressure on populations is attacks by the parasite *Spilochalcis mariae* (Hymenoptera: Chalcidae), a small yellow wasp that occurs gregariously in pupae. It attacks other large native saturniids including *Rothschildia lebeau*, *Antheraea polyphemus* and *Callosamia promethea*. The bagworm moth (*Thyridopteryx ephemeraeformis*), an extremely common pest species, serves as an alternate host for this parasite. Another parasite that kills large numbers of *cynthia* is *Enicospilus americanus* (Hymenoptera: Ichneumonidae), which is known to heavily parasitize native *American attacini*. Probably the European population do not suffer from attack by such effective native parasites since no species of *attacini* is native to Europe. Frank (1986) additionally postulated that the introduction of a bird called the European Starling (*Sturnus vulgaris*) may have become a new and effective predator to *cynthia* in the urban environment. The other limiting factor is that of winter temperature. Pupae of this moth cannot withstand temperature below -15°C (Danilevsky, 1940). The mild climate in certain regions of Europe and North coastal America has allowed *cynthia* to establish and persist for over a century. *Samia caniggia* is a wild eri silk producing insect found in Assam but is not commercially exploited to a great extent (Kalita and Dutta, 2014). Genus *Samia* has many species in oriental countries and only the species *ricini* is reared for production of silk.

Eri silkworm Strains/ Varieties

Thangavelu (1991) revealed that saturniids exhibit genetic diversity and natural variation in the wild population indicating natural adaptations to specific niches. The eri silkworm strains are quite different from each other both in morphology (colour polymorphism) and genetic traits. Phenotypic diversity and characterization of six different strains of eri silkworm may be useful for selection of breeding components for developing high silk productive breeds of silkworm. Sharma and Kalita, 2013 and Singh *et al.,* 2011 carried out study on morphological characters of eco races and 6 strains of eri silkworm and found out their rearing performance. Gogoi and Goswami (1998) revealed that genetically useful and important traits of wild silk moth may be a sound basis for future breeding programme in evolving commercially economically desirable improved strains of the species. Based on the morphological characters twelve varieties of eri silkworm have been reported by Vijayan *et al.,* (2006) and Sarkar *et al.,* (2008). Six pure line strains were isolated from Borduar and Titabar eco-races on the basis of larval markings and colour. These are yellow plain (YP), yellow spotted (YS), yellow zebra (YZ), Green blue plain (GBP), green blue spotted (GBS) and Green blue zebra (GBZ) (Plate 2). The remaining six strains *viz.,* white plain (WP), green plain (GP), white spotted (WS), green spotted (GS), white semi zebra (WSZ) and yellow semi zebra (YSZ) were isolated from wild type, *P. ricini* through selection and breeding over 10 generations and reared under standard laboratory condition. The details are presented hereunder.

Plate 2 Variability in eri silkworm larvae

Yellow Plain: it is pure multivoltine eri silkworm variety without undergoing diapause. This race undergoes four moults and has five instars with the larval duration of 23-24 days. The male and female larvae are yellow plain and the cocoons spun by this race are brick red in colour (Plate 3).

Plate 3 Yellow plain strain of eri silkworm

Yellow Zebra: It is also pure multivoltine eri silkworm variety without undergoing diapause. This variety undergoes four moults and has five instars. The male and female larvae possess yellow zebra markings. The cocoons are brick red in colour. Different agro climatic zones produced several regional silkworm races with different characteristics. The results of the extensive genetic investigations together with those of silkworm breeding experiments have contributed much to the establishment of contemporary silkworm breeds of high economic value (Plate 4).

Plate 4 Yellow zebra strain of eri silkworm

Yellow Spotted: It is also a pure multivoltine variety without undergoing diapause. This variety undergoes four moults and has five instars with larval duration of 24 days. The male and female larvae possess yellow spots. It gives brick red colour cocoons (Plate 5).

Photo 5 Yellow spotted strain of eri silkworm

Blue Green Plain: This is the multivoltine silkworm variety without undergoing diapause. This race undergoes 4 moults and has five instars with larval duration ranges between 23-24 days. The male and female larvae are bluish green plain in colour and spins white colour cocoons (Plate 6).

Plate 6 Blue green plain strain of eri silkworm

Blue Green Spotted: This is a pure multivoltine variety, which undergoes four moults. Larval duration varies between 23-24 days. The male and female larvae possess bluish green spots. It spins brick red colour cocoons (Plate 7).

Plate 7 Blue green spotted strain of eri silkworm

Blue Green Zebra: It is a pure multivoltine variety without undergoing diapause. It undergoes 4 moults with five instars having larval duration of 22-23 days. The phenotype of male and female larvae possess blue green zebra markings. It spins white cocoons (Plate 8).

Plate 8 Blue green zebra strain of eri silkworm

White plain: It is a multivoltine pure race of eri silkworm and can be reared 5 to 6 crops in a year. It is popular variety used for commercial production of eri silk in India. It undergoes four moults and has five instars with larval duration of 23 days. The larvae are resistant to diseases like flacherie and grasserie. The cocoons are white in colour and the quantitative characters are better than any other varieties.

Green plain: It is pure multivoltine eri silkworm variety without undergoing diapause. This race undergoes four moults and has five instars with larval duration of 23-24 days. The male and female larvae are green plain in colour and spins white colour cocoons

White spotted: It is a pure multivoltine variety undergoes four moults and five instars with larval duration of 22-23 days. The phenotype of male and female larvae exhibits white spots. The cocoons of these varieties are white in colour.

Green spotted: This is a pure multivoltine variety, undergoes four moults with larval duration of 23-24 days. The phenotypes of male and female larvae possess green spots. It spins white cocoons.

White semi zebra: It is a pure race of multivoltine eri silkworm and can be reared 5 to 6 times a year. The race undergoes four moults and has five instars. The larval duration lasts for 22-23 days. The male and female larvae possess white semi zebra markings on their body. It spins white colour cocoons.

Yellow semi zebra: This variety of eri silkworm does not undergo diapause. This race undergoes four moults and has five instars with larval duration of 24 days. The male and female larvae possess yellow semi zebra markings on their body. It spins brick red colour cocoons.

It was observed that except fecundity there was not much variation in other parameters. The fecundity varied from 435 to 459 in different stocks in different seasons but there was not much variation in hatching percentage, larval weight, ERR (%), cocoon weight and shell weight (Singh *et al.*, 2003). Debaraj *et al.*, (2001) performed elite crosses utilizing these lines of eri silkworm. Sannappa and Jayaramaiah (1999) recorded higher cocoon weight (1.967g), shell weight (0.277g), shell ratio (13.97%), normal moths formation (98.66%) and hatchability of eggs (97.68%) and slower growth rate in yellow plain and greenish blue plain strains. Marghitas *et al.*, 2011, Singh *et al.*, 2012, Sharma and Kalita, 2013 stated that yellow zebra strain is best in terms of rearing performance and green blue spotted strain is best in effective rate of rearing, fiber content, shell weight and sericin content. This clearly indicates that environmental conditions prevailing in different seasons and food plants fed to the silkworms bring profound changes in the growth, development and the expression of economic characters in different silkworm races (Krishnaswami *et al.,* 1970 and Rajesh kumar and Elangovan, 2012). These strains can be reared in bulk and can be utilized in breeding programme for high silk production. Nagaraja *et al.,* (1996) tested the five parental breeds of *Samia cynthia ricini* i.e. WP, WZ, GP, GS and GZ differed in their general combining ability for rate of pupation, pupal weight, pupal duration, moth emergence, sex ratio, fecundity and hatching and stated that the white plain breed was a good general combiner. Additive gene action was predominant for all the characters except moth emergence as the estimates of GCA variance were greater than SCA variance. Debaraj *et al.,* (2001) conducted diallel crossing with these strains following Griffing's method and two eri crosses viz., ES-1 (YZ x GBS) and ES-2 (GBS x GBZ) have been developed after combining ability. Field trial of two eri crosses revealed better performance of ES-1 in terms of fecundity, hatching, cocoon weight, shell weight, shell ratio and yield (Table 2.5).

Table 2.5: Performance of two elite crosses of eri silkworm

Particulars	ES-1	ES-2	Control
Fecundity (nos)	473.55	468.99	447.44
Hatching (%)	92.49	90.27	89.77
Cocoon weight (g)	3.52	3.42	3.30
Shell weight (g)	0.51	0.47	0.45
Shell ratio (%)	14.44	13.99	13.83
ERR (%)	90.17	87.33	86.04
Cocoon yield nos/100 dfl	39,540	37,213	34,658

The genetic analysis of the 42 combinations involving all the parental lines revealed that negative cross of white plain male component with white zebra (WZ) female component are the top general combiners to obtain superior F_1 hybrids for all the quantitative traits when compared to other crosses. Therefore, this parental combination can be used as superior breeding material to produce substantial improvement in quality and quantity of eri silk.

Molecular characterization

Molecular characterization work pertaining to eri silkworm accessions maintained in the germplasm bank is very limited due to lack of adequate manpower and infrastructure facilities. A preliminary study on the DNA fingerprinting by Vijayan *et al.,* (2006) using the six eco races (Acc. No. 001 to 006) suggested that the gene flow between the populations of Acc. 003 and Acc. 005 is quite high (0.9035) and lowest (0.2172) is between Acc.002 and Acc. 003, but have similar in phenotypic traits, such as cocoon colour. Due to over exploitation coupled with rapid deforestation, most of the natural populations of *S. cynthia ricini* are dwindling rapidly and its preservation has become an important goal. Assessment of the genetic structure of each population is a prerequisite for a sustainable conservation program. DNA fingerprinting has been used in different insect species to detect genetic variation not only between populations, but also between individuals within a population. Since, information on the genetic basis of phenotypic variability and genetic diversity within the *S. cynthia ricini* populations is scanty, inter simple sequence repeat (ISSR) system was used to assess genetic diversity and differentiation among six commercially exploited *S. cynthia ricini* populations. Twenty ISSR primers produced 87% of inter population variability among the six populations. Genetic distance was lowest between the populations of Khanapara (E5) and Mendipathar ecorace (E6) (0.0654) and highest between Dhanubhanga (E4) and Titabar ecorace (E3) (0.3811). Within population, heterozygosity was higher in Borduar (E2) (0.1093) and lowest in Titabar (E3) (0.0510). Regression analysis showed positive correlation between genetic distance and geographic distance among the populations. The high GST value (0.657) among the populations combined with low gene flow contributes significantly to the genetic differentiation among the *S. cynthia ricini* populations. Based on genetic diversity, these populations can be considered as different ecotypes and in situ conservation of them is recommended (Vijayan *et al.,* 2006). The high phenotypic and genetic similarity as well as gene flow between populations of Acc. 002 and Acc. 005 suggests its common origin and later progression into different populations by adapting to the varying climatic conditions.

Zuo ZhengHong *et al.,* (2001) used, Random amplified polymorphic DNA to analyse the genetic variance and diversity of five eri silkworm breeds. The results showed that 27 of 40 arbitrary primers with 243 fragments could amplify repeatable bands clearly. Each primer had 4-17 bands (length = 0.33-3.0 kb) with an average of 9. The genetic distance value (0.0683-0.1603) between different breeds of eri silkworm was used to construct a dendrogram using the unweighted pair group method with arithmetic means. Later molecular study by using Randomly Amplified Polymorphic DNA (RAPD) marker conducted by Shivshankar *et al.,*

(2013) and on protein profile by sodium dodecyl sulphate polyacrylamide gel electrophoresis (SDS-PAGE) by Singh *et al.,* (2011), reported differences in DNA and protein profiles among the eri strains.

Hou Cheng Xiang *et al.,* (2005) applied the ISSR (inter-simple sequence repeat) markers to the analysis of genetic correlation among the germplasms of eri silkworm. Nineteen primers containing simple sequence repeat motifs were tested for amplification on a panel of 8 strains. They were Minbai, Min-B, Gaoyi, Baihuang, Haihuang, Hualan, Xuyi and 754. Among the 19 primers, 9 could amplify distinctive and reproducible bands. In the 8 strains, 76 ISSR marker bands were amplified with the 9 primers. Among the 76 bands, 29 (37.66%) were polymorphic. These results showed that the dinucleotide motifs had a higher abundance than that of trinucleotide and tetranucleotide motifs in eri-silkworm. The poly (CA) motif was more abundant than that of poly (CT) in the dinucleotide motif. The genetic distance among the 8 strains was 0.088-0.2703.

Zhu *et al.,* (1983) performed repetitious crossing of the silkworms *Samia cynthia* and *S. c. ricini* in Guangdong, China, in 1968-73 and developed 9 poly hybrids, namely 110, Guanghuahuang, 656, Guangsi, 670, 684, LC-1, LAr-1 and 681, which are all of the semi-feeding type and characterized by heat resistance, omnivory and high silk yield. When the poly hybrids were compared with the old variety Shenlan in studies in 1980, survival rate, cocoon weight, cocoon shell weight, cocoon shell percentage, egg production and silk quality were all generally higher for the poly hybrids.

High genotypic coefficients of variability and high heritability estimates in different genotypes indicated that a wide range of genetic diversity existed which could be used in a breeding programme. It also indicated the importance of additive gene effects in these characters and phenotypic selection of these characters would be effective. A number of characters showed high heritability but low genetic advance due to intra- or inter-allelic interactions (Rahman and Rahman, 1990).

Summary and Conclusions

> Saturniidae, the family of wild silk moths consists of many silk producing insects including *Samia cynthia ricini.*

> The genus *Samia* Hubner is a monophyletic group of large and beautiful moths that range in tropical Asia, across a span of over 6000 kilometres.

- Regarding the nomenclature, the name *Philosamia* has been widely used for the species of *Samia*, but the two names are objective synonyms, both being based on the same type species (*cynthia*).
- The eri silkmoth (*Samia c. ricini*) is the third most important silk producer in the world. The moth exists only in captivity, having been artificially selected from a wild progenitor, *Samia canningi* (Hutton).
- The structure of the genitalia, wing pattern and chromosome number demonstrate that *Samia ricini* (Danovan) is derived from its wild progenitor, *Samia canningi* (Hutton).
- There has been lot of confusion and in-consistency regarding the authorship and taxonomic status of the insect, *Samia* Hubner.
- As per Arora and Gupta (1979), the current valid name of the domesticated Eri silkworm is "*Samia cynthia ricini* (Boisduval)".
- It is distributed in India, Bhutan, Bangladesh, Sri Lanka (Ceylon), Myanmar (Burma), Indonesia (Java) and China.
- *Samia cynthia ricini* always has a solidly white dorsal surface on the abdomen. This is derived from the white tufts seen in related species being expanded, flattened and fused.
- Complete variability in all living organisms and the ecological complexes that they inhabit is termed as biodiversity.
- 26 eco-races of *Samia c. ricini* were identified based on the heritable morphological characters of *Samia c. ricini*. A unique number is assigned by the Central Muga & Eri Research and Training Institute to these ecoraces. The accession number is prefixed by SRI which stands for *Samia ricini*.
- These 26 eri silkworm germplasms are maintained at Central Eri & Muga Research and Training Institute, Central Silk Board, Ladoigarh, Assam.
- The eri silkworm accessions *viz.,* SRI-001, SRI-010 and SRI-024 are the most promising eri silkworm races for commercial exploitation because of their highest cocoon yield.
- The performance and economic characters of eri silkworm eco races are not only controlled by genes, but also known to be influenced by different climatic factors such as temperature, relative humidity, photoperiodic cycle etc. to which it is exposed during its life time.
- In order to conserve the biodiversity and to augment the Eri silkworm seed sector and its primary and secondary host plants Castor (*Ricinus communis*) and Tapioca (*Manihot utilissima*), Central Silk Board,

Ministry of Textiles, Government of India, has established a Central Sericulture Germplasm Resources Centre i.e. Eri P_2 Basic Seed Farm in South India i.e. in Hosur, Tamil Nadu.

➢ Presently it is conserving 443 eri silkworm genetic resources, which includes 26 eco races along with primary host plant Castor and secondary host plant Tapioca.

➢ Eri silkworm has six pureline strains *viz.,* yellow plain (YP), yellow spotted (YS), yellow zebra (YZ), Green blue plain (GBP), green blue spotted (GBS) and Green blue zebra (GBZ). These were isolated from Borduar and Titabar eco-races on the basis of larval markings and colour.

➢ The eri silkworm strains are quite different from each other both in morphology (colour polymorphism) and genetic traits.

➢ The six more strains *viz.,* white plain (WP), green plain (GP), white spotted (WS), green spotted (GS), white semi zebra (WSZ) and yellow semi zebra (YSZ) were isolated from wild type, *P. ricini* through selection and breeding over 10 generations.

➢ Based on the research conducted by the authors, yellow zebra strain is best in terms of rearing performance and green blue spotted strain is best in effective rate of rearing, fiber content, shell weight and sericin content.

➢ Molecular characterization work pertaining to eri silkworm accessions maintained in the germplasm bank is very limited due to lack of adequate manpower and infrastructure facilities.

➢ A preliminary study conducted on DNA finger printing using the six eco races (Acc. No. 001 to 006) suggested that the gene flow between the populations of Acc. 003 and Acc. 005 is quite high (0.9035) and lowest (0.2172) was observed between Acc.002 and Acc. 003, but have similar in phenotypic traits, such as cocoon colour.

➢ The genetic distance value (0.0683-0.1603) between different breeds of eri silkworm was used to construct a dendrogram using the unweighted pair group method with arithmetic means.

References

Arora G S and Gupta I J. 1979. Taxonomic studies of some of the Indian non-mulberry silk moths (Lepidoptera: Saturniidae : Saturniinae). *Memoirs of Zoological Survey of India* **16**(1): 49-53.

Chakravorty R and Neog K. 2006. Food plants of eri silkworm, *Samia ricini* Donovan- Their rearing performance and prospects for exploitation. Proceeding of National workshop on Eri food plants, 11-12 October, 1-7.

Chakravorty R, Singh K C, Sarkar B N, Neog K, Mech D, Sarmah M C, Barah A and Dutta P. 2008. In: *Catalogue on Eri Silkworm (Samia ricini) Germplasm*, published by CMER&TI, Jorhat, Assam.

Chakravorty R. 2004. Diversity of er and muga sericulture and its prospects in Himalayan states. *Proceedings, National workshop on Potential and strategies for sustainable development of vanya silk in the Himalayan States*, Dehradun Pp. 9-16.

Choudhury S N. 1982. Eri silk industry. Directorate of sericulture and weaving, Assam.

Danilevsky A S. 1940. Experiments of the ecological basis of geographical distribution and acclimatisation of *Philosamia cynthia*. *Zoologicheskii Zhurnal* **19**: 26-44.

Debaraj Y, Datta R N, Das P K and Benchamin K V. 2002. Eri silkworm crop improvement –Review. *Indian Journal of Sericulture* **41**(2): 100-105.

Debraj Y, Sarmah M C, Dutta R N, Singh L S, Das P K and Benchamin K V. 2001. Field trail of elite crosses of eri silkworm, *Philosamia ricini* Hutt. *Indian Silk* **40**(2): 15-16.

Dookia B R. 1980. Varied silk ratio in cocoons of eri silkworms (*Philosamia ricini* Hutt.) reared on different castor varieties in Rajasthan. *Indian journal of Sericulture* **19**: 38-40.

Draudt M. 1927. Saturniidae (American) In: *"The Macrolepidoptera of the world"* **6**: 713-827.

Drury 1773. *Phalanea attacus cynthia*. Illustrations of Natural History (Type locality China) **2**:10.

Eliot I M and Soule C G. 1902. Caterpillars and their moths. Century company, New York **8**: 302.

Esaki T, Mutuura A, Inoue H, Ogata M, Okagaki H and Kurudo H. 1973. Icones heterocerorum Japonicorum in Coloribus Naturalibus, Hoikusha, Osaka **2**: 65-136.

Ferguson D C. 1972. The moths of America North of Mexico (Bombycoidea: Saturniidae). *Fascicle* **20**(2B): 155-275.

Frank K D. 1986. History of the *Ailanthus* silk moth (Lepidoptera: Saturniidae) in Philadelphia: a case study in urban ecology. *Entomological News* **97**: 41-51.

Gmez Bustillo M R and Fernandez Rubio F. 1976. Mariposas de la Peninsula Iberica, Madrid **18**: 17.

Gogoi B and Goswami B C. 1998. Studies on certain aspects of wild Eri silkworm (*Philosamia cynthia* Drury) with special reference to rearing performance. *Sericologia* **38**(3): 465-468.

Grote 1874. *Philosamia cynthia*. *Proceedings of American Philosophical Society 14*: 258.

Hampson G F. 1892. The fauna of British India including Ceylon and Burma. *Moths* **1**: 12-29.

Harada C. 1961. Heterosis of the quantitative characters in the silkworm. *Bulletin of the Sericultural Experiment Station* **17**(1): 50-52.

Hota M M, Patil P K, Pradhan K C and Sharma Mona. 2005. Ericulture in Orissa-Potential exists. *Indian Silk* **44**:12-13.

Hou Cheng Xiang, Li MuWang, Miao XueXia, Xu AnYing, Sun PingJiang, Zhang YueHua, Huang YongPing 2005. Application of ISSRs to researches of genetic diversity in *Philosamia cynthia ricini* strains. *Acta Sericologica Sinica* **31**(3): 358-361.

Hubner J. 1819. Verzeichnis bekannter Schmettlinge. *J. Hubner Augsburg* 17-176.

Jaiswal K, Dhami S S, Gangwar S K and Jain A. 2003. Performance of some bivoltine silkworm (*Bombyx mori* L.) hybrids for cocoon shape uniformity in summer season. *Proceedings of recent researches of Indian sericulture industry*, 1-2 February, Lucknow, 122-125.

Jaiswal K, Gangwar S K and Rajesh Kumar 2006. Comparative study on rearing performance of different eco-race of eri silkworm (*Philosamia ricini*) in monsoon season of Uttar Pradesh. In: *Proceedings of National seminar on Prospects and problems of sericulture as an economic enterprise in North West India*, Dehradun Pp. 483-485.

Kalita T and Dutta K. 2014. Biodiversity of Sericigenous insects in Assam and their role in employment generation. *Journal of Entomology and Zoology Studies* **2**(5): 119-125.

Krishnaswami S, Asan M and Sriharan T P. 1970. Studies on the quality of mulberry leaves and silkworm cocoon crop production. *Indian Journal of Sericulture* **9**(1): 1-25.

Kumar R and Elangovan V. 2010. Rearing Performance of Eri Silkworm in Monsoon Season of Uttar Pradesh. *Asian Journal of Biological Sciences* **1**(2): 303-310.

Mahobia C P, Shankarrao K V, Verma R S, Roy G C and Suryanarayan N. 2005. Potential of eri silkworm rearing in Chattis Garh. *Indian Silk* **8**: 12-14.

Marghitas L A, Dezmirean D, Pasca I, Gherman B, Alexandra Matei, Emilia, Furdui M. 2011. Comparative study of biological and technological parameters regarding silkworm breed in Transylvania, 10th International Symposium Prospects for the 3rd Millennium Agriculture, USAMV Cluj-Napoca, Animal Sciences **68**(12): 22- 26.

Michener C D. 1952. The Saturniidae (Lepidoptera) of the Western Hemisphere – Morphology, Philogeny and Classification. *Bulletin of the American Museum of Natural History*, New York **98**(5): 335-420.

Nagaraja M, Govindan R and Narayanaswamy T K. 1996. Estimation of combining ability in eri silkworm *Samia cynthia ricini* Boisduval for pupal and allied traits. *Mysore Journal of Agricultural Sciences* **30**(1): 48-51.

Packard A S. 1914. Monograph of the Bombycine moths of North America, including their transformation and origin of the larval markings and armature. *Memoirs of the National Academy of Science, Washington* **12**(1): 276.

Peigler R S and Calhoun J V. 2013.Correct authorship of the name *Phalaena ricini* and the nomenclatural status of the name *Saturnia canningi* (Lepidoptera: Saturniidae). *Tropical Lepidoptera Research* **23**(1): 39-43.

Peigler R S and Naumann S. 2003. A revision of the silkmoth genus *Samia*. University of Incarnate World, San Antonio, Texas pp. 230.

Peigler R. 1992. The identity of *Samia cynthia* and the status of its introduced populations. In: Wild Silkmoths Ed: Akai H, kato Y, Kiuchi M and Kobayashi J. *International Society for Wild Silkmoths* 164-167.

Pyle R M. 1975. Silk moth of the railroad yards. *Natural History* **84**: 45-51.

Rahman S and Rahman S M.1990. Estimates of variability and some genetic parameters in eri silkworm *Philosamia ricini* Boisd. *Bangladesh Journal of Zoology* **18**(2): 239-244.

Rajesh Kumar and Elangovan V. 2012. Rearing performance of different ecoraces of eri silkworm (*Philosamia ricini*) Donvan during Summer season of Uttar Pradesh. *Journal of Experimental Zoology, India* **15**(1): 163-168.

Ray P P, Rao T V and Dash P. 2010. Performance of promising ecoraces of Eri (*Philosamia ricini*) in agroclimatic conditions of Western Odisha. *The Bioscan* **5**(2): 201-205.

Rebel H. 1925. Revision des Formenkreises von *Philosamia cynthia* Drury (Saturniidae). *Annalen des Naturhistorischen Museums in Wien* 39: 154-176.

Reddy D N R. 2000. On the nomenclature of Eri silkworm. *Sericologia* **40**: 665-667.

Riotte J C E. 1975. Book review of the moths of American North of Mexico. *The Canadian Field Naturalist* **89**: 87-88.

Sakthivel N, Qadri S M H and Krishnamoorthy T S. 2004. A new technique for commercial eri silkworm (*Samia c. ricini* Boisd.) seed production. *Proceedings, National workshop on Potential strategies for sustainable development of vanya silk in the Himalayan States*, Dehradun Pp. 171-173.

Sannappa B and Jayaramaiah M. 1999. Influence of tannins and phenols in leaves of castor genotypes on economic parameters of eri silkworm, *Samia cynthia ricini* Boisduval (Lepidoptera: Saturniidae). *Mysore Journal of Agricultural Sciences* **33**: 297-300.

Sarkar B N and Gogoi S N. 2010. Genetic variability studies in Eri silkworm (*Samia ricini* Donovan) of North Eastern Region of India. *Indian Journal of Genetics and Plant Breeding* **70**(1): 76-79.

Sarkar B N, Sarmah M C and Chakravorty R 2008. Trimoult in eri silkworm. *Indian Silk* **46**(9): 14-15.

Sarmah M C, Ahmed S A and Sarkar B N. 2012. *Munis Entomology and Zoology* **7**(2):1006-1016.

Sarmah M C, Datta R N, Das P K and Benchamin K V. 2002. Evaluation of certain castor genotypes for improving ericulture. *Indian Journal of Sericulture* **41**(1): 62-63.

Sharma P and Kalita J C. 2013. A comparative study on six strains of eri silkworm (*Samia ricini* Donovan) based on morphological traits. *Global Journal of Bio-Science and Biotechnology* **2**(4): 506-511.

Scherdlin P. 1906. *Attacus cynthia* Drury in Strassburg and Umgebung. *Entomologist's Gazette* **20**(77): 78-83.

Schiissler H. 1933. Saturniidae-1. Sub-Family Attacinae, Saturniinae. *Lepidopteran catalogue, Berlin* **55**: 85-324.

Seitz A. 1918. Die Seidenzucht in Deutschland: Eine kritische Untersuchung. *A Kernen, Stuttgart* **7**: 320.

Seitz A. 1926. Saturniidae (Indo-Australian) The Macrolepidoptera of the world. **10**: 497-520.

Shivashankar M N, Chandan and Nagananda G S. 2013. *Journal of Chemical, Biological and Physical Sciences* **3**(1):326-335.

Siddiqui A A, Lal B, Bhattacharya A and Das P K. 2000. Genetic Variability and correlation studies of some quantitative traits in eri silkworm. *International Journal of Wild Silkmoth and Silk* **5**: 234-236.

Singh B K, Debaraj Y, Sarmah M C, Das P K and Suryanarayan N. 2003. Eco-races of eri silkworm. *Indian Silk* **5**: 7-10.

Singh C K and Suryanarayana N. 2005. Principles of Ericulture. Zen Publishers, 1-56.

Singh H R, Unni B G, Neog K and Bhattacharya M. 2011. Sodium dodecyl sulfate polyacrylamide gel electrophoresis (SDS-PAGE) and random amplified polymorphic DNA (RAPD) based genetic variation studies in eri silkworm *Samia cynthia ricini* (Lepidoptera: Saturniidae). *African Journal of Biotechnology* **10**(70): 15684-15690.

Singh L S, Debaraj Y, Singh N I, Ray B C and Singh R. 2012. Studies on the combining ability analysis of six inbred lines of eri silkworm, *Samia ricini* Donovan. *Indian Journal of Sericulture* **51**(2): 167-172.

Singha B B. 2010. Development of eri silkworm *Samia ricini* (Donovan) breeds with higher fecundity and shell weight, *Annual Report*, CMER&TI, Lahdoigarh 46-52.

Thangavelu K. 1991. Wild sericigenous insetcs of India. A need for conservation. *Wild Silkmoths* **91**: 71-77.

United Nations. 1980. China: Sericulture report on a FAO/UNDP study tour to the people's Republic of China. Food & Agric. Argo United-Nations, Servo Bull. 42. Rome.

Velayudhan K, Balachandran N, RadhaKrishnan S, Singh B K and Jayaprakash P. 2014. Biodiversity in eri silkworm *samia ricini* (donovan) genetic resources and its conservation. *Journal of Aquatic Biology and Fisheries* **2**: 817 to 824.

Vijayan K, Anuradha H J, Nair C V, Pradeep A R, Awasthi A K, Saratchandra B, Rahman S A S, Singh K C, Chakraborti R and Urs S R. 2006. Genetic diversity and differentiation among populations of the Indian eri silkworm, *Samia cynthia ricini*, revealed by ISSR markers. *Journal of Insect Science* (Tucson) **6**: 30.

Zhu S L, Chen D L and Poo M S.1983. Studies on the hybrid strains of the eri-silkworm bred by repetitious crossing with *Philosamia cynthia ricini*, B. and *P. cynthia*, W. of tropical and subtropical regions. *Science of Sericulture* **9**(4): 249-252.

Zuo ZhengHong, Gui MuYan, Wang XueMin and Chen YuanLin 2001. Application of the RAPD technique in a genetic relationship study of the silk insect. I. Genetic variance in the eri silkworm. *Hereditas* (Beijing) **23**(2): 128-130.

CHAPTER 3

Host Plants of ERI Silkworm and their Exploitation for Enhanced Silk Production

North-Eastern region of India is a homeland of about a dozen sericigenous insects. Among them, Eri silkworm, Samia cynthia ricini (Boisduval) is a polyphagous vanya species, which feeds on host leaves belonging to family Euphorbiaceae, Araliaceae, Apocynaceae and Simaroubiceae. Ericulture broadly comprises two inter-linked activities i.e. food plant cultivation and maintenance to feed eri silkworm (rearing). Eri silk production and productivity depends highly on feeds consumed by eri silkworms, which will be a function of feed sources. Preferred feed of eri silkworms is castor (Ricinus communis Linn), however, there is an important difference in feed utilization efficiency when different castor genotypes were used as feed source for this silkworm. Though castor is the main host plant of eri silkworm, but it is mainly annual in nature, requires huge investment in maintenance and has to be grown a fresh every six months. Castor leaf is not available throughout the year. There are several secondary host plants also viz., tapioca (Manihot esculenta), Barpat (Ailanthus grandis), Barkesseru (Ailanthus excelsa), Payam etc. which are used for rearing of eri silkworms during the scarcity of primary host plants. On the other hand, eri host plants are interchangeable at 3rd instar to 5th instar especially during the scarcity of castor leaves. The host plant selection behaviour or feeding preferences of the insects are largely mediated by the presence and distribution of secondary metabolites in plants. Ericulture has been proven beneficial and gainful employment to the castor farmers, as 30% of defoliation will not affect the seed yield. This chapter discusses in detail about the suitability of different castor genotypes, defoliation studies, different alternate hosts that were available to eri silkworm, phaostimulants and feeding deterrents etc.

Eri is the most popular and rapidly expanding sericulture in the vanya silk map of India and is now getting National as well as International limelight. North East India is rich in Seri biodiversity being a natural abode for a number of sericigenous insects and their host plants (FAO Manual, 1987). Ericulture once confined to the hilly, tribal districts of North Eastern Region of India has spread to several other states *viz.*, Andhra Pradesh, Telangana, Madhya Pradesh, Tamil Nadu, Karnataka, Maharashtra, Uttaranchal, Uttar Pradesh, Jharkhand, Bihar, West Bengal, Orissa, Punjab, Uttarakhand, Chhattisgarh, Maharashtra, Gujarat, Rajasthan etc. (Sahu *et al.*, 2006). The reason being, eri silkworm, *Samia cynthia ricini* Boisduval is a multivoltine and polyphagous species and it can be reared throughout the year depending on the availability of feed (Thangavelu and Phukon, 1983; Debaraj *et al.*, 2003; Chowdhury, 1982; Krishnaswami *et al.*, 1971 and Kapil, 1967). The eri food plants are abundantly found in natural forests in plains and hilly areas and leaves of these plants are available in one or the other season for eri silk production. Eri host plants are interchangeable at rearing during scarcity of one host. These plant species are distributed all over India in both natural as well as in cultivated forms and are generally perennial in nature (Patil and Savanurmath, 1994).

Ericulture primarily comprises two activities *viz.*, food plant cultivation and silkworm rearing. Silkworm nutrition is the major factor which affects the growth, development and overall production. Silkworms feed to derive and store adequate energy, nutrients and water from the food they consume (Krishnaswami, 1978). Thus, the quality of leaves fed to the larvae of eri-silkworm does have a lot of bearing effect on the quantity and quality of the cocoons produced (Sinha *et al.*, 1986). The better the quality of leaves, the greater will be cocoon production and productivity (Sannappa and Jayaramaiah, 1999). The food plant of different species influences the larval growth, larval duration, cocoon weight, pupal weight, shell weight, Shell ratio % (SR), Effective rate of rearing % (ERR) and fecundity. The effect of host plant species on the growth and development in the insects has been reported by several workers (Reddy *el al.*, 1989). Therefore, the properties of the feed including their morphological features, susceptibility to enzymatic hydrolysis, the action of inhibitory substances like glucosides, tannins, alkaloids, etc., and the total processing capacity of the insect's digestive system will determine the value of the feed. As a result, chemical composition and digestibility of the feeds have significant importance (Maynard and Loosli, 1962).

Almost all insects including the eri silkworm are host specific and select their most preferred food in order to extract the maximum benefit out of it. It is quite obvious that the larva attains its maturity and finally into adulthood by consuming its fullest feeding requirements, and utilizes it at a particular rate to achieve its proper growth. Thus, the energy stored in the body helps the insect later on in performing various metabolic activities which have to be completed in the non-feeding period. Moreover, the quality of food which, the insect feeds directly influences on the ultimate quantity of product produced by them. The highly nutritious and nutrient balanced food are the prime factors responsible for healthy growth and development of any insect, as it provides the ultimate source of energy to the insect. Thus both these aspects of nutrition (quantitative and qualitative aspects) unanimously contribute a much more prominent knowledge and understanding of the insect plant relationship that exists between them because undernourished silkworms are prone to diseases and crop losses (Lakshmanan and Geethadevi, 2005). So it is of great importance for improvement of diet of the insect through selection of food plants processing for superior nutritive value (Waldbauer, 1968).

Studies on quantitative aspect of nutrition in any insect are of fundamental importance for understanding the insect plant relationship. Studies on different aspects of eri silkworm and their host plants began way back in 1974. Fukuda (1963) stated that eri silkworm eat fresh leaves of about fifty kinds of plants. Arora and Gupta (1979) reported that eri silkworm is known to feed on more than 30 host plant species but only on a few host plants developmental biology and other related aspects have been studied. Patil and Savanurmath (1994) reported that there are a number of host plants for eri silkworm, but they have been left unexploited due to lack of technology.

Among the different host plants, Castor (*Ricinus communis* Linn) is the primary, potential and highly preferred host plant of eri silkworm and castor also plays an important role in the oilseed production of the country (Chowdhury, 1982; Preetirekha *et al.,* 2014; Dutta *et al.,* 2015; Joshi, 1987; Subramanianan *et al.,* 2013; Sarmah *et al.,* 2015; Ahmed *et al.,* 2015; Deori *et al.,* 2016; Severino *et al.,* 2010; Sarkar *et al.,* 2015; Sarmah *et al.,* 2002; Debaraj *et al.,* 2003; Sarmah *et al.,* 2011; Lakshmi Narayanamma *et al.,* 2013; 2014; Lakshmi Narayanamma and Dharma Reddy, 2016; Rajasri and Lakshmi Narayanamma, 2015; Rajesh Kumar and Gangwar, 2010; Chandrappa, 2003; Dasari Prasanna *et al.,* 2013; Govindan *et al.,* 1978; Arora and Gupta, 1979; Dayashankar, 1982; Devaiah *et al.,* 1985; Gogoi, 1998; Sannappa *et al.,* 2007; Narayanaswamy *et al.,*2006; Hazarika and Hazarika, 1996; Ghose, 1949 and Fukuda *et al.,* 1961). Castor is rich in varietal composition and also abundantly available to the growers during all the seasons to rear the eri silkworms. It has been observed that growth, development and cocoon yield of eri silkworm are influenced by the castor genotype and quality of leaves fed to the larvae

(Chandrashekhar and Govindan, 2010). Rearing on castor leaves produced cocoons of higher quality in terms of silk per cent in almost all the seasons (Rajadurai *et al.,* 2010). Hence it is essential to look into the details of castor cultivation.

Castor cultivation details

Among the nine cultivated oilseed crops, castor (*Ricinus communis* Linn) is the most important and is grown across the world in tropical, sub-tropical and warm temperate regions. Castor is known as *Amudam* in Telugu, belongs to the family *Euphorbiaceae*. It is an important crop of dry lands in semiarid zones and small farmers and tribals. It grows as shrubs or small trees in all the regions. Based on the historical evidences, it is believed to be originated both in Africa (Ethiopia) and in India. Castor is cultivated on a commercial scale in an area of 1.525 million ha in 30 countries with a production of 1.581 million tonnes of seed. India, China, Brazil, Russia, Thailand, Ethiopia and Philippines are the major castor growing countries in the world (Damodaram and Hegde, 2011). India ranks first, contributing nearly 65% of the total global production (Ramanjaneyulu *et al.,* 2013). More than 11.65 lakh hectares of land in India is covered under castor plantation in different states like Gujarat, Andhra Pradesh, Karnataka, Madhya Pradesh, Tamil Nadu, Orissa, and Maharashtra. Andhra Pradesh is leading in 2nd position in the area under castor cultivation next to Gujarat. Castor is an important non edible oilseed crop and occupies an important place in India's vegetable oil industry (Rajadurai *et al.,* 2010).

Climate

Castor can be grown in areas lying between 40°N and 40°S and 300 m to 1800 m above mean sea level. It requires 20-26°C temperature with low humidity during the growing season. High temperature above 41°C even for short period results in blasting of flowers thus resulting in poor seed set. While low temperature delays the emergence of seed and prolongs vegetative period. It can be grown in areas with 600-700 mm rainfall with 100 mm evenly distributed rainfall per month during first four months of growing period. Continuous drizzling associated with high humidity (>90%) promotes incidence of *Botrytis* grey rot. A frost free growing period of 130-190 days is required for satisfactory yields. Being a long day plant, it requires a day length of 12-18 hours. Temperature above 31°C promotes production of interspersed staminate flowers (ISF) while lower temperature results in fully female racemes in hybrid seed production.

Suitable areas for castor cultivation in India and Telangana

India accounts for nearly 66.5 and 82.9% of world's castor area and production, respectively. Our country ranks first both in area (13.17 lakh ha), production (21.77 lakh tonnes) and productivity (1653 kg ha^{-1}) (2014-15). In India, castor is grown in 13 states. However, only four states *viz.*, Gujarat (78%), Rajasthan (15.7%), Andhra Pradesh and Telangana (4.73%) together contribute 98.4% of the total castor production. India has significant place in castor oil trade as it produces more than one third of the total world's castor oil and is holding about 80–90% share of the international market. India also exports vegetable oil to European Union, USA, Japan and Thailand. Our country is the biggest exporter of castor oil trade. Castor oil is the second entrant in international commodities trading after pepper. India is earning more than Rs. 3500 crores of foreign exchange through export of castor oil and other by-products.

Among various oilseed crops grown in Telangana state, castor is one of the most important non edible and industrial crops. It is traditionally cultivated in Mahabubnagar, Ranga Reddy and Nalgonda districts. Mahabubnagar district occupies lions' share in area and production (Dasari Prasanna *et al.,* 2013). However, it can also be cultivated in other districts in the state where shallow to deep red soils and shallow to medium black soils are present, under both rain fed and irrigated conditions. Further, due to its' C_3 nature, the plant can adapt to climate change in the event of global warming. Thus, castor is an efficient and resilient crop with many mechanisms for adaptability in diversified environments. It can be grown as catch crop or contingent crop under late sown conditions due to late onset of monsoon.

Castor growing conditions/environments in Telangana are summarized below:

S.No.	Environment	Districts
1.	Traditional rainfed environment	Nalgonda, Mahabubnagar and Ranga Reddy
2.	Non-traditional rainfed environment	Warangal, Karimnagar, Khammam
3.	Irrigated environment	Mahabubnagar, Warangal, Karimnagar, Khammam

The productivity of castor in Telangana is very low hovering around 500-600 kg ha^{-1} against the national average of 1592 kg ha^{-1}. The reasons for low productivity include cultivation of the crop in marginal and sub marginal lands, rainfed conditions, vagaries of monsoon mainly late onset of monsoon and mid season or terminal dry spell during crop growth period, incidence of *Botrytis* due to incessant rainfall coupled with high relative humidity (> 90%), poor adoption of improved agro techniques and plant health management practices.

Table 3.1: Castor varieties/hybrids suitable for Commercial cultivation and Seed production

S.No.	Variety/ Hybrid	Year of release	Duration (days)	Yield (q/ha)	Special Characters	Resistant/ tolerant to
Varieties						
1.	Jyothi (DCS-9)	1995	90-150	15-20	Red stem, double bloom, early duration	Resistant to wilt
2.	Haritha (PCS-124)	2002	90-180	20-25	Green stem, double bloom, spiny	Resistant to wilt and drought
3.	Jwala (48-1)	2007	90-180	20-25	Red stem, double bloom and non-spiny	Resistant to wilt and tolerant to *Botrytis* grey rot and salinity
4.	PCS-262 (Pragathi)	2015	80-160	15-18	Green stem and double bloom	High oil content and test weight Resistant to *Fusarium* wilt
Hybrids						
1	GCH-4 (SHB-18)	1986	90-180	25-28	Red stem, triple bloom, semi spiny	Highly tolerant to Leafhoppers and tolerant to wilt
2.	DCH-177 (Deepak)	2000	90-180	25-33	Red stem, single bloom, spiny	High yielding with resistance to wilt and whitefly but susceptible to leafhoppers
3.	DCH-519	2006	90-180	25-32	Green stem, triple bloom	High yielding with resistance to wilt and leafhoppers
4.	PCH-111	2010	90-180	25-32	Green stem, double bloom, spiny capsules	High yielding with resistance to wilt
5.	PCH-222	2012	90-180	25-32	Red stem, double bloom, spiny capsules	High yielding with resistance to wilt

Varieties are mostly preferred for rain fed environment and also under late sown conditions, while, hybrids are mostly preferred under high input/resource management conditions.

Cropping systems under which castor can be cultivated

Castor can be grown as sole crop or intercrop. The plant is characterised by wide plasticity in branching pattern and thus suitable for many intercropping systems. Castor + red gram (4:1 or 6:1) is the most predominant intercropping system among the farmers in the state of Telangana. However, for higher economic returns, Castor + red gram (1:1) should be followed. It can also be intercropped with green gram, cowpea, ground nut and black gram. Castor can be rotated with pigeon pea, cotton or groundnut also. As *rabi* castor can be taken up as an ID (irrigated dry) crop during first week of October, there are ample chances of growing a short duration and remunerative crops like maize or green gram during *kharif* season under rain fed conditions. Monocropping of castor cultivation in the same fields should be discouraged to avoid multiplication of *Fusarium* wilt pathogen.

Soils suitable for castor cultivation

Castor can be cultivated on all types of soils including light textured red soils; red chalka soils and low to medium depth heavy textured soils under rain fed as well as irrigated conditions. The crop can tolerate slight acidity (pH: 5.0-6.5) and moderate alkalinity (pH: 7.0-8.5). Adequate drainage or slope must be provided as castor is very sensitive to water logging and excessive moisture. Deep ploughing once in two to three seasons is advocated as it helps in breaking hard pan and conserving the moisture. Besides, it also exposes wilt causing pathogen and larvae and pupae of insects to the hot sun. Land should be ploughed thoroughly twice with cultivator, then clods should be crushed by using disc harrow followed by levelling the field using blade.

Seed treatment, spacing and sowing

Seeds need to be treated with Captan/Carbendazim @ 3.0 g/kg before sowing. It helps in protecting the plants from seed borne diseases. Alternately, seed can also be soaked overnight in carbendazim @ 2-3 g/lit. In wilt endemic areas, seed should be treated with *Trichoderma viride* @ 10 g/kg seed. Alternatively, 2.5 kg *Trichoderma viride* powder can be mixed in 125 kg of FYM/ha followed by decomposition and spreading in seed rows before sowing of castor. Depending on the desired spacing, the seed bed should be prepared by using suitable marker. Seed is sown by dibbling method at the intersecting point of the marker. Seeding depth could be 5cm. It can also be sown with the help of seed cum ferti drill in which case inter-row spacing can be maintained while intra row spacing can't be maintained perfectly. Optimum moisture must be ensured at or after sowing to get good germination in order to maintain optimum plant population. Soaking the seed in water for 24-48 hours before sowing ensures good germination, optimum plant stand and also imparts drought tolerance besides eliminating light weight and unfilled seeds.

Castor is mostly grown as a rain fed crop during *kharif* season from June 15th onwards. Optimum cut-off date is July 15th and last cut-off date is July 31st. Under the conditions of delayed onset of monsoon, the crop can be sown up to first week of August with some manipulation of agronomic practices. The agronomic practices like selection of varieties (Haritha and DCS-9), close spacing of 60 x 30 cm with 30% higher population and 25% more N than recommended dose need to be followed. Frequent intercultivation after each rain to create dust mulch so as to reduce evaporation has also to be adopted. During *rabi* season, castor has to be sown during first fortnight of October for higher yields. Nearly 80-100 mm rainfall is generally received during October month. The same will be useful for proper seed germination, establishment, and application of basal dose of fertilizer and thus, saves the cost of 2-3 irrigations. October rainfall also helps to avoid low temperature stress during November and December and high temperature stress during terminal stages. Delay in sowing from October 1st to November 15th results in decline in seed yield of castor by 21 to 44%.

Table 3.2: Seed rate and crop geometry for castor varieties and hybrids under different agronomic situations.

Condition	Soil type	Time of planting	Variety/ Hybrid	Seed rate (kg/ha)	Crop geometry (Cm)	Plant population ha⁻¹
Rain fed	Light textured	June-July	Variety	10.0	90x60	18,518
	Light textured	June-July	Hybrid	5.0	90x60	18,518
	Light textured	August	Variety	12.0	90x20-30 60x30	37,037 to 55,555
Late sown conditions rain fed (without irrigation)	Any soil	August first week	Variety	12.0	60x30	55,555

Table 3.2: *Contd...*

Condition	Soil type	Time of planting	Variety/ Hybrid	Seed rate (kg/ha)	Crop geometry (Cm)	Plant population ha^{-1}
Irrigated	Light textured	October	Variety	10.0	90x60	18,518
	Light textured	October	Hybrid	5.0	90x60	18,518
	Medium to heavy textured	October	Hybrid	5.0	120x60	13,888
Drip irrigation	Light textured	October	Hybrid	5.0	120x60	13,888
	Medium to heavy textured	October	Hybrid	5.0	150-180x60	9,259 to 11,111

Water management for castor cultivation

Castor crop needs 500-600 mm water for good growth and development. Moisture stress indicates excess or deficit moisture. Castor is highly sensitive to excess moisture. Hence, care should be taken to avoid water logging at any stage. Sufficient drainage or slope has to be provided to drain out excess water from the field. During *kharif* season, castor is mostly grown as rain fed crop. Further, it is a drought tolerant crop due to its' deep rooted nature and white waxy coating on different parts of the plant. Hence, castor does not need irrigation. However, it is not uncommon to experience long dry spells in rain fed agriculture which results in reduction in economic yields. Hence, the following drought proofing mechanisms should be adopted for reaping better yields.

* Frequent inter cultivation to create dust mulch thus reducing the loss of moisture through evaporation

* Ridge and furrow method of land configuration is better than flat bed method as former one helps in moisture conservation thus reduces the impact of dry spell. Making conservation furrows (dead furrow) at 2.4 to 3.6 m interval after 45 DAS is another option.

* However, in case of dry spell of more than 10 days in red soils and 15-20 days in black soils during mid/terminal stages, one or two life saving irrigations either from farm ponds or other sources will certainly increase the yield by 25-30%. Application of harvested water of farm ponds by sprinkler irrigation (20 mm) and drip irrigation (15-20 mm) once or twice at critical stages like flowering and primary and secondary spike development stages will increase the yield by 40-50%. Besides this practice also increases water use efficiency as compared to purely rain fed crop.

- Application of recommended dose of fertilizer in the presence of soil moisture enhances the seed and oil yield.

For *rabi* castor, irrigation has to be given after sowing in a dry soil. After establishment of crop with optimum plant population, irrigation has to be scheduled at Irrigation Water/ Cumulative Pan Evaporation (IW/CPE) ratio of 1.0 which means 12-15 days interval from October to December and 8 days interval from January onwards, for reaping higher yields. However, in the areas with less irrigation water, irrigation can be scheduled at IW/CPE ratio of 0.67 also (once in 15 days and 10 days, respectively). It is also always better to give irrigation by ridge and furrow method rather than flat bed method, in order to reduce water wastage. In the regions with meagre irrigation water availability, drip irrigation has to be scheduled at 0.6 pan evaporation i.e. for 1.5 to 2.0 hours from October to December and 2.5 to 3.0 hours from January onwards at an interval of three days. Drip irrigation helps to save nearly 30.0% of irrigation water and the water use efficiency can be doubled as compared to flood irrigation. Paddy needs 3500 mm water to produce one kg seed, while castor needs 250-300 mm water by conventional irrigation method and 100 mm water through drip irrigation to produce one kg seed. Nearly 2.5 times (conventional irrigation) and 3.5 to 4 times (drip irrigation) more castor area can be irrigated with the same quantity of water used for paddy cultivation.

Nutritional requirements of castor crop

Castor is mostly grown on marginal and sub marginal lands which are shallow in depth, poor in infiltration, low in water holding capacity and essential nutrients. Further, the crop management is neglected with regard to nutrition. However, it responds well to nutrient application and produce higher yields. Nearly 5 t ha^{-1} of farm yard manure or 1.25 to 1.5 t/ha of vermin compost has to be applied and incorporated into the soil well in advance to sowing for moisture conservation and nutrition. Nutrient recommendation must be based on soil test value which helps in avoiding excess fertilizer use thus reduces expenditure. Most of the castor growing soils in the state are mostly low in nitrogen, medium to high in phosphorus and potash. Castor crop yielding about 20.0 qt ha^{-1} removes 80 kg nitrogen, 20 kg phosphorus and 35 kg potassium. On an average, the nutrient application for castor grown under different environments is detailed below.

Table 3.3: Amount of Nutrients to be applied in castor grown under varied agronomic conditions.

Time of application	Nutrient levels (Kg t ha^{-1})					
	Varieties			Hybrids		
	N	P$_2$O$_5$	K$_2$O	N	P$_2$O$_5$	K$_2$O
Kharif castor						
Basal	30	40	30	40	60	30
30 DAS	15	-	-	14	-	-
60 DAS	15	-	-	13	-	-
90 DAS	-	-	-	13	-	-
Late sown *Kharif* castor						
Basal	35	60	30	-	-	-
30 DAS	20	-	-	-	-	-
60 DAS	20	-	-	-	-	-
Rabi castor under conventional irrigation						
Basal	-	-	-	40	60	30
30 DAS	-	-	-	14	-	-
60 DAS	-	-	-	13	-	-
90 DAS	-	-	-	13	-	-
Rabi castor under drip irrigation	-	-	-	120*	40	30

*To be given through fertigation

Seed treatment with biophos (Phosphate solubilising fungi) or PSB (Phosphate solubilising bacteria) before sowing along with application of 20 kg P$_2$O$_5$ ha^{-1} as basal is sufficient in case of low to medium P soils. It saves half of the expenditure on P fertilizer without affecting seed yield of castor. All the fertilizers have to be applied by pocketing method or in the furrow made with the help of plough near by the plant and followed by covering with soil, rather than broadcasting method or band placement. It helps in efficient utilization of applied nutrients by the crop besides avoiding fertilizer wastage and weed growth. In case of integrated nutrient management which envisages conjunctive use of organic, in organic and biological sources of nutrients for sustaining the soil health and crop productivity, application of 50% RDF (recommended dose of fertilizer) + Seed treatment with *Azospirillum*+25% N through FYM is advocated for castor under rain fed conditions.

In case of *rabi* castor grown under drip irrigation, 40 kg P_2O_5 and 30 kg K_2O ha^{-1} should be applied through SSP and MOP as basal by pocketing method at 15-20 DAS. On the other hand, N has to be applied in the form of urea through fertigation by following fertigation schedule. Nitrogenous fertilizers @ 20-50-30-20 kg N ha^{-1} has to be scheduled through fertigation during vegetative stage (0-45 DAS), primary spike development (46-90 DAS), secondary spike development (65-120 DAS) and tertiary spike development (85-150 DAS) stages, respectively.

Cultivation of castor in problematic soils

Castor can be grown on slightly acidic soils. It can tolerate salinity up to 8.0 dSm^{-1} beyond which germination, growth characters, yield attributes, yield and oil content besides relative water content are adversely affected. Hence, following management practices are to be adopted for achieving higher productivity

- Seed priming with 1% NaCl for three hours before sowing
- Sowing on the slope of ridges
- Spot application of 2 t ha^{-1} FYM and castor cake @ 1.0 t ha^{-1}
- Growing varieties/hybrids like 48-1, DCS-9 which can tolerate salinity upto 10 dSm^{-1}.

Adoption of these practices enhances the seed yield by 25% over farmers practice of sowing on flat bed, without FYM application and with no seed priming.

IPM (Pests, Diseases & Weed Control) of Castor

Castor crop is mainly affected by weeds, pests and diseases causing huge loss of economic yield if not controlled in time. All the crop husbandry practices will be futile if the crop is not timely and properly protected against crop pests including weeds, insects and diseases. So, they have to be managed/ controlled effectively to grow a good and healthy crop.

Integrated weed management in castor:

Weed menace is moderate to severe in castor due to slow initial growth up to 45 DAS, wider spacing (90x60 to 120x60 cm) and long duration nature of the crop (150-180 days). Thus, castor is sensitive to weed competition.

Weeding is the most important and critical agronomic practice. The critical period of crop weed competition for castor is 45-50 DAS. The predominant weed flora in castor field includes *Celosia argentia, Cyperus rotundus, Cynodon dactylon, Echnichloa sp., Dactyloctenia aegyptium, Digitaria sanguinalis and Digeria arvesnsis.* There are chances of failure of the crop under unweeded conditions. Hand weeding and intercultivation are the traditional methods for controlling the weeds. Single weed control method may not be possible and feasible under all circumstances. Hence, integrated weed management (IWM) which includes physical, mechanical and chemical weed control methods has to be adopted for timely weed control and achieving higher yields.

Pre-emergence application of Pendimethalin @ 1.0 kg a.i. ha^{-1} or alachlor @ 1.25 kg kg a.i. ha^{-1} (5.0 ml ltr^{-1}) within 2-3 DAS in the presence of soil moisture + two times intercultivation (IC) + one hand weeding is the best IWM module to be adopted for weed management in castor. Farmers' practice of 90x90 cm crop spacing + 3 IC+ 1 HW is also equally effective. Of late, power weeders, tractor drawn blades and mini tractors were tested for intercultivation and found to efficiently control the weeds. Nearly 10 labour and 35 hours time per hectare can be saved due to employment of suitable machinery for weed management and intercutlivation. In case of incessant rains where there is no scope for physical or mechanical methods of control, Paraquat (5 ml ltr^{-1}) or Glyphosate (10 ml ltr^{-1}) can be advocated through directed spray on weeds only. However, extreme care should be taken not to spray on the crop.

Integrated Pest Management (IPM) in castor crop

IPM utilizes all suitable techniques and methods in a compatible manner and maintains the pest populations at levels below economic injury. A number of non-chemical cultural practices form the core of IPM. However, it also advocates use of insecticides as a last weapon. More than 107 insect species are reported to feed on castor, but, only few are of economic importance. The details regarding nature of damage and IPM strategy for managing the important pests are detailed below.

Castor semilooper *(Achaea janata* L.):

It is a regular, serious and major defoliator pest on castor and its' activity is mostly seen during August to October. It causes defoliation leaving only stems and veins on the plant (Plate 9). A number of hymenopterous parasites attack the pest during the various stages of its life cycle. Castor + pigeonpea intercropping was found to have less infested by semilooper. The eggs are parasitised by releasing *Trichogramma evanescens minutum*

@ 1.25 lakh ha^{-1}. The larvae are heavily parasitised by a *braconid* parasite *Microplitis maculipennis* (Plate 10). Semilooper larva parasitized by *Microplitis maculipennis*) and a *chalcid - Euplectrus matternus* (Plate 11). A *braconid* larval parasite, *Rhogas sp.* have also been observed parasitizing on semilooper. Hand picking of older larvae in early stages, arranging bird perches @ 20-25 ha^{-1}, spraying of carbaryl @ 3 g lit^{-1} or novaluron @ 1.0 ml lit^{-1} or thiodicarb @ 1.5 g lit^{-1} if defoliation is above 25%.

Plate 9 Semilooper larva feeding on the castor leaf

Plate 10 Semilooper larva parasitized by *Microplitis maculipennis*

Plate 11 Semilooper larva parasitized by *Microplitis maculipennis* and *Euplectrus matternus.*

Tobacco caterpillar (*Spodoptera litura* Fabr)

This is the highly polyphagous pest and prefers castor as compared to other crops as it is more nutritious. It is active during August to November. The early instar larvae feed gregariously on under surface of the leaf leading to skeletonization of the leaves (Plates 12 and 13). Later they disperse and become solitary (Plate 14). They feed on the plant during nights and hide under soil crevices and debris during day. They also feed on the flower buds and flowers.

Plate 12 The gregarious patch of tobacco caterpillar feeding on under surface of the leaf

Plate 13 Completely skeletonised castor leaf due to infestation of *Spodoptera* larvae

Plate 14 Older solitary caterpillar of *Spodoptera*

It can be controlled by

- Summer ploughing thus exposing the pupae to predators and parasites
- Collect and destroy the egg masses and caterpillars in the early stages of infestation
- Arranging bird perches @ 20-25 ha^{-1}
- Monitoring the pest with pheromone traps (10-12 ha^{-1}) helps in timing of treatment

- The microbial agents like *Spodoptera* NPV, entomopathogenic fungi *Nomuraea rileyi,* pathogenic bacteria *Bacillus thuringiensis* and *B. cereus* are reported to exercise larval mortality.

- To control early stage larvae, spray neem seed kernel extract (NSKE) 5%. When ETL crosses 5-10 % infected plants, spray chlorpyriphos @ 2.5 ml lit^{-1} or acephate @ 1.5 g lit^{-1} or novoluron @ 1.0 ml ltr^{-1} or emamectin benzoate @ 0.5 g lit^{-1}.

- Placing the poison bait pellets at the base of the plant (12.5 kg of bran + 1.25 kg jaggery + 1.25 litre of monocrotophos or 1.25 kg carbaryl in little water to make the bait in to pellets for one hectare) helps in killing the grown up larvae.

Shoot and Capsule borer (*Conogethes punctiferalis* Guen)

The pest attacks the crop from flowering stage onwards and continues till maturity. It is serious during October-December. Larvae bore the capsules and webbing of the capsules along with excreta is seen (Plates 15 and 16). The pest can also attack inflorescence and terminal shoot. Collect and destroy the infested shoots and capsules. In the initial stages, spray acephate @ 1.5 g lit^{-1} and in case of severe infestation spray indoxacarb @ 1.0 ml ltr^{-1} or flubendiamide @ 0.2 g lit^{-1}.

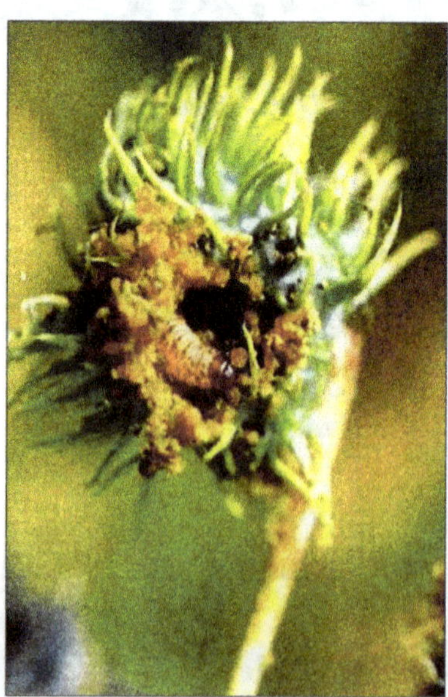

Plate 15 Capsule borer larva feeding inside the capsule

Plate 16 Webbings of the capsules, indicating the severity of infestation by the shoot and capsule borer larvae

Sucking pests

Leaf hoppers (*Empoasca flavescens* Fabr)

Peak infestation of leaf hoppers is seen during November to January. Nymphs and adults suck sap usually from the under surface of the leaves and inject toxin causing curling of leaf edges and leaves turn red or brown leading to appearance of hopper burn (Plate 17). The leaves dry up and shed (Plate 18). Grow double / triple bloom genotypes like PCH 111, DCS-9, 48-1, DCH-519, GCH 4 which are resistant / tolerant to leafhoppers. Monitor the pest incidence through light traps / coloured sticky traps. When ETL (2-3 adults leaf^{-1}) is crossed, spray Neem oil @ 5.0 ml lit^{-1} or monocrotophos @ 1.6 ml ltr^{-1} or dimethoate @ 2.0 ml ltr^{-1} or acephate @1.5 g ltr^{-1} in sequential manner depending on the need. Repeat the spray if required.

Plate 17 Leafhopper nymphs and adults sucking the sap from the under surface of the leaf

Plate 18 Typical hopperburn symptom caused by the infestation of leafhoppers

Integrated disease management (IDM) in castor crop

IDM advocates conjunctive use of cultural, mechanical, biological and chemicals methods to control the diseases. In Castor, wilt (20-50%) and *Botrytis* grey rot (5-85%) are the serious diseases causing economic yield losses. The details regarding nature of damage and IDM strategy for managing the pests are furnished below.

Seedling blight

This disease is caused by *Phytophthora parasitica* Dastur. The disease first appears on both the surfaces of the cotyledonary leaves as round patches of dull green colour and then spreads to the point of attachment causing the leaf to rot and fall (Plate 19). The infection then spreads to the stem due to which the seedling is killed following destruction of growing point or stem. Black lesions can be seen on the stem at base level. Young leaves curl inwards with black margins and drop off later and such branches die back. Care should be taken to avoid ill drained, damp and low-lying fields. Seed treatment with carbendazim (3 g kg^{-1}) or *Trichoderma viride* (10 g kg^{-1}) formulation can reduce disease incidence. Soil drenching with copper oxychloride @ 3g lit^{-1} or spraying with Dithane M-45 @ 3 g lit^{-1} will also control the disease.

Plate 19 Initial symptom of seedling blight on the cotyledonary leaves

Grow tolerant or resistant varieties like Jyothi, Jwala, GCH-4, DCH-30 and SHB-145. Avoid low lying patches which are prone to water logging. Destruction of crop debris, selection of healthy seed, providing irrigation at critical stages of the crop, seed treatment with thiram @ or carbendazim @ 3g kg^{-1} or *Trichoderma viride* @ 10 g kg^{-1} of seed are to be followed. Soil drenching with carbendazim @ 3 g lit^{-1} at 15 days interval will control the seedling blight disease.

Wilt

It is caused by *Fusarium oxysporum* f. sp. *ricini*. The disease incidence can be seen starting from seedling to maturity stage. Cotyledonary leaves turn to dull green colour, wither and die subsequently. Leaves will droop and drop off leaving behind only top leaves and diseased plants are sickly in appearance (Plate 20). Wilting of plants, root degeneration, collar rot, drooping of leaves and necrosis of affected tissue leading to the death of plants is observed. Necrosis of leaves starts from margins spreading to interveinal areas and finally to the whole leaf. Split open stem shows brownish discolouration with a prominent growth of white cottony mycelia in the pith of stem. The extent of seed yield loss due to wilt depends on the stage of incidence. For eg, 77% at flowering stage, 63% at 90 DAS and 39% at later stage on secondary branches has been reported.

Plate 20 Wilted plant showing drooping of leaves

Soil solarisation with linear transparent low density polyethylene sheets (25 microns) for six weeks during summer season reduces the wilt causing pathogen to a great extent. Deep ploughing, selection of disease free seed, growing tolerant and resistant varieties like Jyothi, Jwala, GCH-4, DCH-30 and SHB 145, avoiding water logging, burning of crop debris, green manuring and intercropping with red gram are some of the cultural/mechanical/low monetary practices to be followed. Seed treatment with Carbendazim @ 2-3 g kg^{-1} or *Trichoderma viride* 10 g kg^{-1} of seed is useful. Soil drenching with Carbendazim @ 2-3 g lit^{-1} at 15 days interval will control the root rot disease. Multiplication of 2 kg *T. viride* formulation by mixing in 50 kg FYM, sprinkling water and covering with polythene sheet for 15 days and then applying between rows of the crops is helpful in reducing the pathogen incidence.

Botrytis Grey rot

It is caused by *Botrytis ricini* Godfrey. Initial symptoms are small blackish spots on inflorescence from which drops of yellow liquid may exude. Fungal mycelia which grow from these spots spread the infection and produce characteristic appearance of affected raceme (Plate 21). The disease is problematic when rains occur during capsule formation and during prolonged weather. Total plant parts like leaves, stem, flower and capsules wither eventually. The affected flowers rot and are covered by grey coloured fungus. The disease spreads upwards infecting all flowers and capsules. Blue spots appear on the branches and laterals of the spike. Affected parts break off from the plant. Infection at the time of flowering results in flower rot and affects seed filling. Infected capsules are rot and drop from the plant resulting in heavy yield loss.

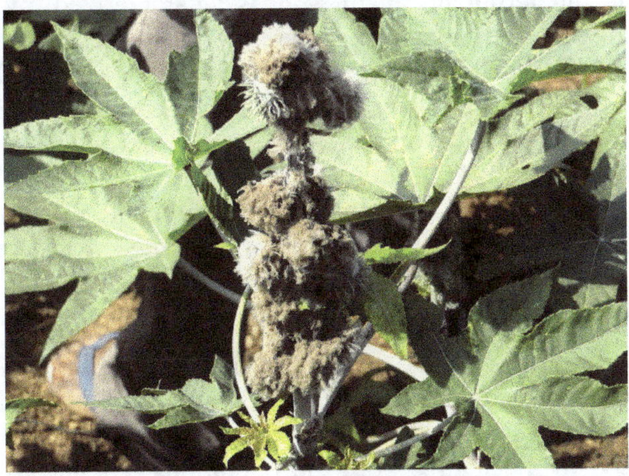

Plate 21 Inflorescence covered by greyish fungal growth

Night temperatures below 22°C, continuous drizzling and fog besides high relative humidity (>90%) are highly favourable for the development of disease. Seed treatment with carbendazim at 3g kg^{-1}, use of non spiny varieties (48-1), avoiding excess irrigation, close spacing and destruction of crop debris are some of the practices to be followed. Spraying of carbendazim (0.05%) or Thiophanate methyl (0.05%) at 15 days interval is to be done depending on weather forecast before and after rains. Application of 50 kg urea and 25 kg of MOP ha^{-1} after removal of diseased panicles is advocated as it helps in rejuvenation of crop and compensates the yield loss by producing fresh spikes.

Animals

Wild boars generally damage the food crops like groundnut, maize, sorghum, sunflower and paddy. However, it does n't attack castor crop due to its' non-edible nature. Hence, castor can be easily and safely grown in an area where wild boar is the problem. However, in the areas nearby forests like Amrabad mandal of Mahabubnagar district, wild boars remove the soil around the crop to feed on tubers of *Cyperus rotundus.* Neelgai (Blue bull) eats away entire castor spike which is a serious problem in North Indian castor growing states, however, it is absent in South India. The animals like cattle, sheep and goat feed on tender leaves. Monkeys though do not feed on castor but damage the crop by sitting on the plants.

Harvesting of castor

Castor is an indeterminate plant and is characterized by sequential flowering and spike formation. So, different order spikes come to maturity at different times. But, harvesting of matured spikes at right time before the seeds fall on the ground is very important. In case of *Kharif* crop, first spike matures by 90 DAS and thereafter, the subsequent spikes at an interval of 20-25 days. While in case of *rabi*, first spike matures by 110 DAS. The spikes can be harvested at physiological maturity or when colour of 60-70% of the capsules in a spike change from green to light yellow or brown. Yield loss will be significant if harvesting is delayed until all capsule in the spikes are fully dried.

Traditionally harvesting of spikes and threshing of capsules are done by hand and beating with wooden sticks or trampling with bullocks, respectively. However, in view of increasing labour shortage and wages, there is a need to mechanize both the operations. Of late, secatures for harvesting of spikes and motor operated castor threshers for threshing of dried capsules have come into existence. The use of secatures help to avoid damage to the palms and harvesting can be done even in hot summer which otherwise it is difficult for the labour to do manually. Threshing has to be done after drying the harvested spikes in the threshing yard. Finally winnowing is to be done to clean seeds and then dried for 2-3 days till they attain safe moisture content of 9-10 % and finally seeds are stored in gunny bags. Nearly 12 labour and 10 hours time per hectare can be saved due to mechanization of harvesting and threshing.

Yield levels and Economics of castor cultivation

A well-managed *kharif* crop yields 10-15 q ha^{-1}, while *Rabi* crop yields 23-25 q ha^{-1}. Adoption of drip irrigation and fertigation can increase the seed yield up to 35-38 q ha^{-1}. Full package of practices including improved variety/hybrid, thinning, timely weed control, proper nutrition, need based irrigation and plant health management will certainly improve the yields. An amount of Rs. 20,000 to 25,000 ha^{-1} and 25,000 to 30,000 ha^{-1} is generally incurred for raising a castor crop during *kharif* and *Rabi*, respectively. Based on the market rate prevailed in the last several years i.e. Rs. 3500 q^{-1}, it is estimated that, an amount of Rs. 5,000 to 27,500 ha^{-1} in *kharif* and Rs. 55,000 to 57,500 ha^{-1} in *rabi* season can be obtained as net returns.

Value addition to castor cultivation

Castor is known for its diversified uses in industrial, medical, Agriculture and domestic sectors. Its seeds contain 48-52% oil. The castor oil is unique among all vegetable oils owing to the presence of hydroxyl fatty acid (ricinoleic acid) up to 90%. Besides, it also maintains higher viscosity even at high temperatures and liquid form at low temperatures also. Further, it is considered as one of the best lubricants in the world owing to its' non-drying nature thus used in the manufacture of lubricants. It is also used in the preparation of Turkey oil which is used in cotton dyeing, printing and leather industries. Castor oil can be used in the manufacture of soaps, sebacic acid (required for the synthesis of Nylon fibre), production of hydraulic fluid, artificial leather, rubber, printing ink, recinol, lubricating and heavy duty automotive greases, telecom engineering plastics, production of nylon 11, nylon 6-10 and medicines etc. Castor oil can also be used as general laxative, in birth control, anti-partum treatment, conjunctivitis, skin ailments, inflammation, paralysis, piles and easing of stomach pain.

The castor cake can be applied as organic manure to improve the fertility of Agricultural fields as it contains 6.6% N, 2.6% P_2O_5 and 1.2% K_2O (cake from decorticated seed) and 4.5% N, 0.7% P_2O_5 and 1.9% K_2O (cake from un decorticated seed). It encourages soil microbial activity, promotes root development and winter cold hardiness. The castor leaf fall adds fertility to soil on decomposition. Castor stubbles and shelled capsules can be incorporated in to the soil or can be used in the preparation of vermin compost which in turn can be further applied to agricultural fields to improve the soil fertility.

Castor leaves can be fed to eri silkworms (Ericulture) to produce 'eri silk' which resembles cotton and is called as poor man's silk. It has been proved to be an ideal subsidiary occupation to a large number of rural and tribal populations as it provides ample scope for employment and income. Nearly 30% defoliation is permissible and the same can be used in ericulture without foregoing economic yield. Besides, the farmers can be benefitted with an additional income up to Rs. 6000 to 7500 ha^{-1} through ericulture which is a boon for a rain fed castor growers. Further, eri pupae are very much relished by tribal people and considered to be on par with mutton or chicken. The neutral lipid of silkworm pupae (*Bombyx mori* L.) is a good source of alpha linolenic acid (ALA), an essential fatty acid. Due to presence of linolenic acid to the tune of 43% in eri pupal oil, it is considered as a good source of omega 3 fatty acid.

Different types of food plants of eri silkworm

The food plants are categorized into three namely primary, secondary and tertiary hosts based on the preference by the silkworm and potentiality of silk yield. Primary host plants are mainly used for rearing of eri silkworm to ensure good cocoon harvest and more profit while the secondary hosts generally used as substitute on shortage of primary host leaf. However, eri silkworm could survive on the tertiary hosts but exhibit poor silk output or die without farming the cocoon.

Influence of Nutrient status and water content of Castor leaf on eri silk worm rearing

Castor (*Ricinus communis*) is the primary host plant of eri silkworm. The highly nutritious and nutrient balanced foods are the prime factor for healthy growth and development of any insect, as it provides the ultimate source of energy to the insects. The most important physiological factors in silkworm growth and silk productivity are nutrition. Almost all the nutrients required for its growth are derived from the host plant leaf. The nutritive value of host leaf depends on various factors such as storage duration, temperature, humidity, light, aeration, quality of leaves or shoots in a given space (density of storing) and many other conditions.

The eri silkworms exhibit differential response in the acceptance of castor varieties for feeding and show variation in the development. Castor being the primary host plant of eri silkworm, its leaf quality is of much importance to the healthy growth of worms and ultimately cocoon harvest.

In this regard, Govindan *et al.,* (2003) conducted an experiment on DCH-177 and local pink variety of castor in different locations of Karnataka. The results showed that the hybrid DCH-177 grown under recommended management practices recorded higher foliar nitrogen, phosphorous, potassium, calcium, magnesium and sulphur contents. This could be due to the difference in their inherent characters and better adaptability of the hybrid to rain fed condition. This could also be due to the application of recommended fertilizers and intercultural operations carried out as it has a favourable effect on the growth of castor. As per EL-Shaarawy *et al.,* (1975) red bloomy variety of castor was significantly superior with sugars, nitrogen, phosphorous and potassium than bloomy green variety. Kaleemurrahaman and Gowri, (1982) recorded 0.11%, 0.09% and 0.07% of phosphorous, 1.22%, 0.88% and 1.32% calcium and 0.53%, 0.83% and 0.74% magnesium in mature, middle and tender leaves respectively. Sannappa and Jayaramaiah (1999b) recorded higher phosphorous and potassium content in RC-8 (0.208 and 0.554%) and calcium, magnesium and sulphur content in Aruna genotype (1.67, 1.22 and 0.64%).

Sannappa and Jayaramaiah (1999a) analysed the leaves of different castor genotypes for the content of tannins and phenols in different castor genotypes in relation to economic characters of eri silkworm during 1996 at UAS, Bangalore. The total tannins and phenols were significantly higher in the PCS-121 genotype in tender, middle and mature leaves. The mean value of these groups was also higher. Significantly lower was in the Aruna genotype closely followed by the RC-8 genotype. The total tannins and phenols were found to have negative significant relationship with economic parameters of eri silkworm such as larval duration, larval weight, larval survival, ERR, cocoon weight, pupal weight, shell weight, shell percentage, moth emergence, fecundity and hatchability. The study clearly indicated that the level of tannins and phenols vary in different genotypes and they had a significant influence on the economic characters of eri silkworm.

The relationship between nutritive value of castor varieties and economic parameters of eri silkworm was investigated by Govindan *et al.,* (2006). The nutritive value of leaves viz., leaf moisture, total carbohydrates, crude protein, nitrogen, phosphorus, potassium, calcium, magnesium and sulphur showed significant positive correlation with rearing (mature larval weight, effective rate of rearing, cocoon yield and shell yield), cocoon (cocoon weight, shell weight, shell ratio, silk productivity and fibroin) and grainage parameters (pupal weight, rate of pupation, rate of moth emergence), while the trend was negative with larval duration, larval mortality, sericin and pupal duration. However, the fecundity of moths was found to have non-significant relationship with nutritive value of leaves.

Moisture content of the castor leaves plays a significant role on growth and development of eri silkworms. It is very well established that the young age silkworms require tender leaves with high moisture content, whereas, late age worms need coarse leaves with low moisture content. Kaleemurrahaman and Gowri, (1982) reported high moisture content of 72% in tender leaves, 69.4% in middle leaves and 66.7% in mature leaves. Basaiah (1988) recorded moisture content of 67.09% in the leaves of RC-8 variety of castor.

Sarmah *et al.,* (2011) stated that the ERR of eri silkworm was influenced by biochemical composition of the leaves of different castor accessions. The weight of larvae and cocoons were significantly influenced by nitrogen and crude protein content of the foliage. Silkworm rearing performance was better in Acc. 03 and Acc. 04 in terms of shell weight, shell ratio and effective rate of rearing.

In an experiment conducted by Chandrappa et *al.,* (2005) in Karnataka, India, during 2001 and 2002 the relationship between foliar constituents of 10 castor bean genotypes and economic traits of eri silkworm was established. The relationship between quality parameters of castor bean genotypes i.e. moisture, total chlorophyll, total carbohydrates and crude protein with economic traits of eri silkworm namely rearing (mature larval weight, total larval duration, ERR, cocoon and shell yield), cocoon (cocoon weight, shell weight, shell ratio, silk productivity, fibroin and sericin) and grainage traits (rate of pupation, pupal weight, pupal duration, rate of moth emergence, fecundity and egg hatching) was not significant. There was a significant positive relationship between cocoon traits (cocoon weight, shell weight, shell ratio, silk productivity and sericin) with elemental composition of castor bean genotypes i.e. N, P, K, Ca and S. Furthermore, the grainage traits, i.e. pupal duration and rate of moth emergence had significant positive correlation with N, P and K contents in the castor bean genotypes.

Effect of host plants on the Fibroin and Sericin content of eri silkworm cocoons:

Fibroin and sericin are the two main proteins of silk fibre. Sericin is the gummy substance around fibroin, the filament protein. The silk fibre of *Bombyx mori* L. consists of fibroin (70-80%), sericin (20-30%), waxy material (0.4-0.8%), carbohydrates (1.2-1.6%), colouring matter (0.2%)

and inorganic substances (0.7%). Though sericin is necessary for cohesion of the filaments, silkworms producing more sericin are not economical as it will be washed away during the process of cooking (Jolly and Krishnaswami, 1964). But the information about the sericin and fibroin of eri silk is lacking. So Munshibasaiah *et al.,* (1997) conducted an experiment to know the effect of host plants on the Fibroin and Sericin content of eri silkworm cocoons. As a part of the experiment, 10 eri cocoons raised on four hosts *viz.,* three varieties of castor (Local, Aruna and RC-8) and tapioca (Mangalore local) were collected and tested for fibroin and sericin after four to five days of drying as per the standard procedure.

The sericin and fibroin content of the eri cocoons were calculated by the formulae.

Sericin content: Dry weight of cocoon shell – Dry weight of cocoon shell after alkali treatment

Fibroin content: Dry weight of cocoon shell – Sericin content.

The results of the experiment show that the sericin content of eri silk fibre was significantly less on Aruna (31.06 mg/g shell) and RC-8 (31.58 mg/g shell). Whereas the fibroin content was significantly high on Aruna (68.94 mg/g shell) and RC-8 (6.45 mg/g shell) varieties of castor and lowest on tapioca (63.45 mg/g shell). The local variety of castor stands in intermediary position (Table 3.4).

Table 3.4: Sericin and Fibroin content (mg/g shell) in eri cocoons raised on different hosts.

Host	Sericin	Fibroin
Castor Local	33.70	66.30
Aruna	31.06	68.94
RC-8	31.58	68.42
Mangalore Local	36.54	63.45
CD (5%)	1.87	1.09

Nagaveni *et al.,* (2002) conducted an experiment to know the effect of feeding eri silkworms with diseased castor leaves on the economic parameters of cocoon. Healthy castor (*Ricinus communis*) leaves and those infected with spot and blight (*Cercospora ricinella* and *Alternaria ricini* respectively) were fed to eri silkworms. Cocoon assessment was done by random selection. The eri silkworms fed on infected leaves consumed more food and took a longer time to mature compared to those fed on healthy leaves, but suffered due to the non-availability of required nutrients in adequate quantities. Those fed on healthy leaves completed their larval period (23 days) faster with robust and uniform growth compared to silkworms fed on infected leaves. Feeding on infected leaves resulted in a significant reduction in larval weight, shell percentage and shell weight compared to feeding on healthy leaves. Sericin content was high in cocoons of silkworms fed on infected leaves compared to those fed on healthy leaves, while fibroin content was lower in cocoons of silkworms fed on infected leaves.

Effect of host plants on the biochemical aspects of eri silkworm in Nagaland University, Medziphema:

Since eri silkworm is polyphagous and feeds on different host plants, it is important to provide information about the amount of nutrient compositions especially protein. In this regard, Chang *et al.,* (2012) made an attempt to collect the information about the carbohydrate and protein contents in pupa of eri silkworm reared on different hosts like castor (*Ricinus communis* L.), tapioca (*Manihot utilissima* Pohl.), Gulancha (*Plumeria acutifolia* Poir), Gamari (*Gmelina arborea* Roxb.) and kesseru (*Heteropanax fragrans*). The total carbohydrate and protein contents were estimated in pre pupal stage by Anthrone and Lowry's method, respectively (Sadasivam and Manickam, 1996).

The carbohydrate and protein contents were the highest in larvae reared on kesseru (17.80 and 55.35%) followed by castor (16.65 and 51.55%), Gulancha (14.55 and 44.40%) and Gamari (13.60 and 39.00%), while the lowest values were recorded on tapioca (12.65 and 35.40%) (Table 3.5). Thus, kesseru is the most suitable host from nutritional point of view. However, due to faster growth rate of larvae feeding on castor makes it preferred choice.

Table 3.5: Carbohydrate and protein content of eri silkworm pupa

Host plants	Carbohydrate (%)	Protein (%)
Kesseru	17.80	55.35
Castor	16.65	51.55
Gulancha	14.55	44.40
Gamari	13.60	39.00
Tapioca	12.65	35.40

Comparative foliar constitutes of cultivars on rearing performance of varietal preference of castor cultivars on Eri silkworm:

A study was undertaken by Patel and Patel (2009a) to determine the effects of foliar constituents, moisture contents, crude protein, ether extracts (E. Ext), crude fiber, nitrogen free extracts (N.F.E.) and ash etc, of castor leaf of nine varieties on growth and development of eri worm. Leaf constituents were estimated following standard methods (AOAC, 1970). Based on the results it is revealed that higher moisture content (7.76%) and crude Protein (18.35%) was recorded in GCH-6. However, the least moisture contents and crude protein was noticed in GCH-2 (6.40%). Higher and lower crude fiber content of 15.87 and 7.87% was recorded in GCH-4 and Banaskantha Local respectively. Nitrogen free extracts (N.F.E.) was marked higher (47.59%) in GAUCH-1 and showed least content (39.35%) in GCH-4. Higher ether extracts (E.Ext) (10.61%) was recorded in GCH-2 (10.61%) and the least (3.93%) in GCH-6 genotype of castor.

Effect of leaf storage on nutritive value of castor, the primary food plant of eri silkworm:

The storing of leaf is important not only to produce nutritious and succulent leaf but also to preserve them fresh till they are consumed by silkworm. In this regard, Singha *et al.,* (2013) conducted an investigation to know the changes in nutritive components of castor leaves preserved in different duration for effective utilization in eri silkworm rearing at Assam Agricultural University, Jorhat, during autumn and spring season 2009-10. The tender, medium and mature leaves of castor were harvested separately in the morning hours which were preserved by heaped on mat and covered with wet gunny cloth and stored for 24 hours. The preserved leaves were subsequently used for chemical analysis to estimate moisture, crude protein, total soluble sugar, total mineral, crude fibre and total chlorophyll contents after 4 hour, 8 hour, 12 hour and 24 hours of storage duration. Fresh leaf samples of all the types without preservation (0 hour) were also subjected to chemical analysis for comparison.

Leaf storage had significant effect on nutrient contents of castor leaves (Table 3.6). Though moisture content was found in decreasing trend with the advancement of storage duration, 4-12 hour of leaf storage (76.53-75.09%) did not have significant effect in respect of moisture content. With the progress of leaf storage duration the amount of crude protein, total soluble sugar, crude fibre and total chlorophyll were decreased significantly while total mineral content increased. These changes in nutritional contents may be due to the breakdown of nutrients during storage of leaves. The nutrient contents of castor leaves have increasing trend with the advancement of leaf age from tender to mature except moisture (Pathak, 1988 and Dutta, 2000). Though moisture content has decreasing trend it was found at par in medium (75.42 %) and mature

(74.64%) leaves. The leaves of spring season was found to be better with significantly higher amount of moisture and total soluble sugar and lower amount of crude fibre while that of autumn season was registered better for crude protein, total mineral and total chlorophyll content. This variation in nutrient contents between the seasons may be due to the environmental factors like temperature, humidity, rainfall etc. Thus, it is imperative to conclude that castor leaves could be preserved covering with wet gunny cloth up to 12 hours without much deterioration of moisture for effective utilization in eri silkworm rearing. During preservation care should be taken to minimize moisture loss and chemical degradation to maintain quality of leaves for healthy growth and development of eri silkworm.

Table 3.6: Effect of storage duration, leaf type and season on nutritive value of castor leaves (Adopted from Singha *et al.,* 2013)

Factors	Moisture (%)	Crude protein (%)	Total soluble sugar (%)	Total mineral (%)	Crude fibre (%)	Total chlorophyll (mg/g fresh wt)
Storage duration						
0 hr	77.98	18.15	8.56	9.82	36.20	11.96
4 hrs	76.53	15.73	7.68	10.21	34.40	10.46
8 hrs	75.88	14.44	7.25	10.72	31.40	8.67
12 hrs	75.09	13.05	6.73	11.42	30.19	8.33
24 hrs	73.80	11.64	6.00	11.81	27.90	7.45
SEd±	0.88	0.34	0.20	0.26	0.40	0.31
CD (5%)	1.77	0.68	0.41	0.53	0.80	0.62
Leaf type						
Tender	77.52	13.68	6.60	8.95	25.31	8.48
Medium	75.42	14.22	7.18	10.20	33.52	9.16
Mature	74.64	15.92	7.96	13.21	37.23	10.50
SEd±	0.68	0.26	0.16	0.20	0.31	0.24
CD (5%)	1.37	0.52	0.32	0.41	0.62	0.48
Season						
Spring	78.57	13.44	7.74	9.57	31.65	8.11
Autumn	73.15	15.76	6.75	12.02	32.38	10.64
SEd±	0.56	0.21	0.13	0.17	0.25	0.20
CD (5%)	1.12	0.43	0.26	0.33	0.51	0.39
Interaction						
Storage duration x leaf type x season						
SEd±	NS	NS	NS	NS	0.98	0.76
CD (5%)	NS	NS	NS	NS	1.96	1.52

Nutritional content of healthy and diseased leaves of castor and papaya:

Castor and papaya leaves are the main food of eri silkworm. But both papaya and castor leaves are severely attacked by several fungi and viruses reducing the leaf yield. Even the quality of the leaves become inferior. Hence to assess the alterations in the nutritional content of healthy and diseased leaves of castor and papaya, Shree *et al.,* (2000) conducted an experiment by feeding the healthy and diseased leaves to eri silkworms. The fungal and viral affected leaves were analysed for major nutrients (nitrogen and potassium), secondary nutrients (calcium, magnesium and sodium), micro-nutrients (iron, manganese, copper, zinc, lead, silver, barium, molybdenum, chromium, nickel, cobalt, antimony, arsenic and bismuth). The results shows that eri silkworms dislike the diseased leaves and suffer when fed to them. All these elements either increase or decrease due to fungal/ viral infection. Some of them did not show any change in their content.

Suitability of castor varieties/Genotypes for eri silkworm rearing:

The nutrition that an insect derives during the course of development plays a significant role in its commercial exploitation (Slansky and Scriber, 1985). The suitability of host is determined through estimation of rate of ingestion, digestion, conversion efficiency of food and growth rate of the insect. The eri silkworm has a number of host plants but their acceptability and utilization by the larva varies. Though eri silkworm is hard enough to sustain the biotic and abiotic stresses, it relishes on the quality foliage for its nourishment and growth which is the least expected from dry land farming of host crops. Continuous dry spell, low soil profile and non adoptability of recommended practices might be the reason for low quality foliage output. Subsequently, undernourished silkworms are prone to diseases and crop losses. Lakshmanan and Geethadevi, (2005) also observed that farmers face low productivity under rain fed conditions of sericulture.

Among the various species of hosts, castor is considered as the most preferred food plant of eri silkworm and can be used for large scale rearings. This crop is widely grown in rain fed conditions of Telangana and Andhra Pradesh and under irrigated conditions in Gujarat, Rajasthan and Haryana (Chowdhury, 1982). There are many local and high yielding castor varieties and hybrids are available in these states. Hence there is a need for identification of suitable castor varieties / hybrids that meet the agro climatic needs of this area and yield better for successful rearing of eri silkworms to establish ericulture industry in the state. Hence identification of the factors that contribute to better quality leaves and the conditions under which those leaves are fed to eri silk worm larvae is highly essential (Sannappa and Jayaramaiah, 1999). However, very little information is available on eri silkworm rearing on castor in Andhra Pradesh and Telangana region which is a non traditional area for ericulture but where castor was grown extensively under rain fed conditions.

Eri silk production and productivity depends highly on feeds consumed by eri silkworms, which will be a function of feed sources. Preferred feed of eri silkworms is castor. However, there is an important difference in feed utilization efficiency when different castor genotypes were used as feed source for this silkworm (Kedir Shifa *et al.,* 2014). There are many reports available on the rearing performance of eri silkworm on various food plants under North Eastern conditions of the country (Saratchandra and Joshi, 1985; Biswas and Das, 2001; Hazarika *et al.,* 2003). It is important to find out the best suited food plant of castor with respect to growth and development of adult and ovipositional potency under different nutritional habits for further exploitation to increase the production of eri silk as a whole (Sarkar *et al.,* 2015). Extensive study has been done on the identification and suitability of different varieties/genotypes of castor.

Sarmah *et al.,* (2011) conducted an experiment in Assam to find the suitable castor variety for rearing of eri silkworm. They observed that among other castor varieties, Acc. 003 and Acc. 004 genotypes have an average of 344.70 and 334.50g leaf biomass per plant i.e. more than 13 MT/ha/year and the Acc.003 castor accession is the most suited host plant for eri silkworm in all aspects. The percentage of un-emerged cocoon was comparatively less in case of castor Acc.003 than that in other tested food plants. Both potential and effective fecundity in terms of eggs, egg size, fertilization of eggs and hatching % were found higher in case of castor Acc.003 (Sarkar *et al.,* 2015). Further, NBR-1 (a local non-bloomy red castor) was found as promising variety for eri silkworm rearing with potential leaf yield of 12 MT/ha/year (Sarmah *et al.,* 2002).

Bordoloi *et al.,* (2005) conducted an experiment in the department of sericulture, Assam Agricultural University, Jorhat during spring (March-April) and autumn (Nov – Dec), 2000-2001. They reared eri silkworms on different castor varieties *viz.,* local red petiole, DCS-9, DCH-32, DCH-177 and 48-1. Results revealed that among the castor varieties eri silkworm fed on 48-1 exhibited significantly highest mature larval weight with the shortest larval duration and comparatively better cocoon yield in both spring and autumn seasons. The variety was superior with respect to cocoon parameters like cocoon weight, shell weight and shell ratio in both the seasons.

Hazarika and Hazarika (1996) conducted an experiment to assess the preference of local (green, red, reed petiole and powdery varieties) and high yielding varieties (RC-8, Aruna, GCH-2, GCH-4, Co-1 and TNV-5) of castor for eri silkworm in Assam. The results show that red petiole varieties are superior for rearing eri silkworms. The larvae of eri silkworm fed on leaves of red petiole castor variety produced higher silk content and silk ratio.

Experiments conducted in Andhra Pradesh during 2011 to 2013 revealed that highest mature larval weight, ERR (%) and shortest larval duration was recorded in PCH-111. The genotype PCH-111 was superior by recording highest cocoon weight, shell weight and shell ratio closely followed by PCH-222 and GCH-4. Altogether the genotypes, PCH-111, PCH-222 and GCH-4 were found to be better performing genotypes for eri silkworm rearing in terms of yield attributing characters together with rearing performance (Lakshmi Narayanamma *et al.,* 2013; 2014; Lakshmi Narayanamma and Dharma Reddy, 2016). In another experiment, the castor genotypes PCH-111, PCS-262 and GCH-4 were found to be promising with respect to rearing performance indicators and cocoon traits of eri silkworm (Rajasri and Lakshmi Narayanamma, 2015). The castor genotype, PCH-111 was found to be superior to other genotypes in improving the rearing performance of eri silkworm with maximum records of mature larval weight, cocoon weight, pupal weight, shell weight and silk ratio and shorter larval period.

The experiments conducted during 2009 in Andhra Pradesh showed that the rearing performance of eri silkworm on DCH-519 genotype was found superior in terms of shell yield. This study reveals that castor farmers can switch to DCH- 519 hybrid for castor cultivation and ericulture also, as it yields more and would be double beneficial to the farmer in terms of castor oil and silk yield.

As per Kedir Shifa *et al.,* (2014), among the eight castor genotypes tested, Abaro and Acc 208624 expressed better performance in all the variables *viz.,* ingesta, digesta, approximate digestibility, reference ratio, relative consumption rate, efficiency to convert ingested leaves to larval biomass, efficiency to convert ingested food to cocoon and % efficiency to convert digested food to cocoon. The quality of leaf fed has a great influence on the amount of ingested food due to the physico-chemical characteristics in relation to the age of the leaf and the characteristics of the genotypes as stated by Legay (1958) and Friend (1958).

The experiments conducted during 2005-06 at UAS, Dharwad revealed that the eri silkworms nourished with the leaves of DCS-85 genotype registered significantly higher ERR (%), higher cocoon weight, cocoon yield, shell weight and shell yield. The eri worms fed on leaves of DCS-85 were found superior in respect of shell ratio, silk productivity, fibroin and sericin content. Eri pupae formed by the worms nourished with leaves of DCS-9 registered higher pupal weight. However, fecundity was more with DCS-85 genotype with highest egg hatching rate of 95-98% (Chandrashekhar and Govindan, 2010). In another experiment Chandrashekhar *et al.,* (2012) stated that there is lot of variation with respect to growth and development of eri silkworm on different castor genotypes.

Govindan *et al.* (2002) conducted the experiment in Karnataka to study the effect of castor genotypes on profitability of eri silkworm rearing and observed that eri silkworms fed on DCH- 177 recorded significantly higher larval weight, larval survival, effective rate of rearing, cocoon weight, shell weight, shell ratio, silk productivity, pupal weight, and rate of pupation, eclosion, fecundity and egg hatching with lower larval and pupal durations than those fed on local pink and raised under farmers' practices. In another experiment conducted by Sarmah *et al.,* (2012) eri larvae recorded significantly higher larval weight with DCH-177. Nokchensaba Kichu *et al.,* (2015) studied the effect of host plants on the rearing performance and cocoon characters of eri silkworm. Eri silkworm larvae were reared on the leaves of castor, kesseru, tapioca, white gulancha (*Plumeria acutifolia*) and red gulancha (*Plumeria frangipani*) under laboratory conditions. The castor plant recorded the shortest larval duration, highest larval weight, fecundity, hatching percentage, pupal weight, cocoon weight, shell weight, effective rate of rearing and growth index. In all aspects, white gulancha and kesseru were found to be equally suitable, while tapioca and red gulancha showed poor performance. However, these plants can be used as secondary hosts during the scarcity of the primary host.

Sachan and Bajpai (1973) observed superior larval growth, development and higher cocoon production when eri worms were reared on rosy castor varieties. Dookia (1980) recorded longest larval duration on EB-16 and minimum on J1 castor variety. Similarly larval duration was shorter on local variety and longer on RC-8 and Aruna varieties, while cocoon weight and shell ratio were more with EB-31 variety. Jayaramaiah and Sannappa, (2000a) reported that eri silkworms fed on RC-8 castor genotype recorded longer feeding duration, moulting and larval duration, while on SL-1, shorter feeding, larval durations. But the moulting durations were shorter on PCS-121 genotype. Basaiah (1988), Jayaramaiah and Sannappa, (2002b), Sannappa, (1997) and Misra (1999) recorded RC-8 and Aruna are the best hosts for eri silkworm. Patil *et al.,* (2000) reported that the castor genotype RG-323 recorded highest larval weight, cocoon weight, pupal weight, shell weight and fecundity. Reddy and Narayanaswamy (1999) reported that Bangalore local variety of castor is a suitable host for rearing eri silkworm, followed by Aruna and RC-8 varieties. But Misra (1999) stated that Aruna variety is a unique host for the rearing of eri silkworm.

The total food consumption by the eri silkworm larvae was lower when fed on GCH-5 (15.38 g) than those fed on GCH-2 (20.86 g). The larval weight of full grown larva was maximum when reared on leaves of Assam local (8.18 g), whereas it was minimum when reared on leaves of GCH-2 (6.56 g). The cocoon weight (3.328 g), as well as shell weight (0.410 g) and shell ratio (13.11 %) was maximum when eri worms were reared on GCH-2, GCH-4 and GCH-6 respectively, whereas the corresponding parameters were minimum (3.004 g, 0.37 g and 11.66 %) when they were reared on GC-2, GCH-2 and GAUCH-1, respectively (Patel and Patel, 2009b).

Rearing of eri silkworm on Tapioca

Castor is the primary food plant of eri silkworm and Tapioca/cassava (*Manihot esculenta* Crantz), is the next best host and secondary food plant of eri silkworm. Tapioca is a perennial shrub with variously branched stems growing to a height of 3-4 m, and produces enlarged tuberous roots. It is one of the most important tuber crops of the tropics widely cultivated in several countries like South America, Africa and South East Asia including India. It grows well in warm and humid climate, soils of all types having pH within 5.5 to 7.0. Tapioca cannot grow in saline or alkaline soils and cannot withstand water logging. The leaves of this plant are very large with long petioles having 5-10 lobes or sometimes more, dark green in colour but red, yellow and purple colour leaves are also found. Tapioca tuber is also used in sago industries. It is extensively grown in 0.50 lakh hectares in East Godavari district of Andhra Pradesh.

The experiment conducted by Hazarika *et al.,* (2015) with four different tapioca varieties *viz.,* H-97, Sree Vijaya, Balijan Local and Sree Jaya with different larval instars of eri silkworm revealed that the rate of food consumption, growth rate and digestibility of food decreases while conversion efficiency of ingested and digested food into body biomass increases with the advancement of larval stages. With the higher conversion efficiency of ingested as well as digested food into body biomass and lower rate of food consumption the silkworms grow at a much faster rate on H-97 followed by Sree Vijaya, Balijan Local and Sree Jaya varieties.

Reddy and Narayanaswamy, (1999) reared the eri silkworm on tapioca variety Mangalore local and inferred that Tapioca stood in serial sequence with regard to suitability of the silkworm after castor. Varieties of tapioca namely, Co-1, H-1423, M4 and H-2304 were found suitable for eri silkworm rearing under NE condition. The tapioca leaves can be utilized for eri silkworm rearing throughout the year excepting during the months of December to February (winter) (Chakravorthy and Neog, 2006). In Tapioca the variety KM-94 was most suitable for eri silkworm rearing (Kawabe *et al.,* 2014).

Sakthivel, (2016) evaluated seven popular Indian cassava varieties namely CO2, CO3, CO (TP) 4, H165, H226, Mulluvadi (MVD1) and Kunguma Rose for rearing of eri silkworm. The variety MVD1 recorded highest values of essential nutrients viz., protein, total carbohydrate, nitrogen, phosphorus, potassium and total minerals which were closely followed by H226. With respect to anti-nutrients, highest content of total tannins and HCN were recorded with the variety CO2. The variety MVD1 registered superior economic traits of eri silkworm followed by H226 while CO2 was noted as poor performer. All the nutrients exhibited positive correlation with economic traits except that of larval period which was decreased with increase in nutritional content of leaves while it was vice versa in case of anti nutrients and the order of merit of cassava varieties suitable for ericulture was recorded as MVD1 > H226 > CO (TP) 4 > CO3 > Kunguma Rose > H165 > CO2.

Comparative studies to utilize tapioca and castor for eri silkworm rearing:

According to Kapil, (1967) eri silkworms fed on leaves of Tapioca seldom exerted influence on larval growth and development. Further the age of the plant does not seem to influence the food value of castor leaves. Joshi and Misra, (1982) reported higher ERR on tapioca and castor-tapioca dietary regimes. Govindan *et al.,* (2002) found that substitution of tapioca in place of castor in later half of fifth instar did not deteriorate the larval and cocoon traits of eri silkworm. Rearing on castor leaves produced cocoons of higher quality in terms of silk % in almost all the seasons. However, during summer rearing on tapioca leaves gave more silk % (Rajadurai *et al.,* 2010). Among the tapioca varieties, Co-2 was superior to ME-1 and ME-120 for rearing and can be used singly or in combination with castor during shortage / non-availability of castor (Rajashekhargouda *et al.,* 2009).

Reddy and Narayanaswamy, (1999) reared the eri silkworm to study the consumption of food, leaf-cocoon and leaf-egg ratios on Bangalore local, Aruna and RC-8 varieties of castor and tapioca. The rate of leaf consumption by the larval instar varied with the varieties of the host plant. Leaf-cocoon ratios were higher in tapioca (19.17:1) and lower in Bangalore local variety (12.17:1) but, greater leaf-egg ratios were obtained in local castor (5.08:1) and lower in tapioca (3.40:1) leaves. Feeding of eri silkworm larvae on tapioca leaves for the first 15 days of larval period followed by castor was shown to be suitable for farmers to use in times of scarcity of either food plant. With this combination, duration and development of different instars of eri silkworm and adult emergence were normal (Hoque and Khan, 1990).

Patil *et al.,* (2000) reared the eri silkworm on 3 tapioca cultivars *viz.,* Co-2, ME-1 and ME-120 and on castor bean variety Aruna by keeping as control. Observations on survival and larval, pupal and cocoon characters were recorded. Castor bean resulted in the shortest larval duration. Among the tapioca cultivars, ME-1 registered the shortest larval duration, followed by Co-2 and ME-120. Castor bean was superior with regard to 5th instar larval, silk gland and cocoon weights, followed by Co-2, ME-1 and ME-120. However, the tapioca cultivars did not show significant differences for 5th instar larval and cocoon weights. It can be inferred that castor bean was the most suitable host plant for eri silkworm. Among the tapioca cultivars, Co-2 was superior to ME-1 and ME-120 for rearing, and can be used singly or in combination with castor bean.

Laboratory study was conducted in Gujarat to work out the effects of host plants *i.e.* castor, tapioca, papaya, *Jatropha, Champa,* and *Arduso* separately on biology and economic traits of eri silkworm, *S. cynthia ricini* Boisduval. The results revealed that the larvae when reared on castor, the egg period (15.0 ± 0.64 days), total larval period (25.20 ± 0.76 days), pre-pupal period (1.28 ± 0.45 days), pupal period (15.56 ± 0.71 days) and adult period (3.00 ± 0.40 days) in male and adult period (6.24 ± 0.92 days) in female were shorter than those reared on tapioca. The pre-oviposition period (1.60 ± 1.58 days), oviposition period (3.36 ± 0.70 days) and post oviposition period of female moth (1.28 ± 0.54 days), duration of total life cycle (63.28 ± 1.66 days) of female and duration of total life cycle of male (60.04 ± 1.42 days) of *S. c. ricini* were shorter when reared on castor than those reared on tapioca. The average egg laying capacity was greater (187.2 ± 43.17) with higher (90.44 ± 8.69%) hatching percent in castor reared eri silk worm than those reared on tapioca. The cocoon, shell and pupa of eri worms reared on castor were bigger and heavier and the shell ratio was greater than those reared on tapioca. The rearing rate (96 %) and growth index (3.08) of *S. ricini* were also greater when the *Samia* worms were reared on castor than those on tapioca (Patel and Patel, 2009b).

Host plants relationship in terms of cocoon colour and compactness of eri silkworm:

Preetirekha *et al.,* (2014) conducted a study to know the effect of host plants viz., castor, tapioca and kesseru on cocoon colour and compactness parameters of eri silkworm. It was revealed that host plants had significant effect on the compactness of cocoons. The cocoon spun by the eri silkworm fed on castor leaves were rated as more moderate followed by tapioca and Kesseru. Cocoons obtained from the eri silkworm fed on Kesseru leaves were rated as more hard followed by tapioca and castor (Tables 3.7 and 3.8).

Table 3.7: Effect of host plants on cocoon colour of eri silkworm

Host plants	Colour (% of respondents)			
	White (A)	Bright white (B)	Creamy white (C)	Dull white (D)
Castor	30	15	47	8
Tapioca	40	10	50	-
Kesseru	20	15	60	5

Table 3.8: Effect of host plants on cocoon compactness of eri silkworm

Host plants	Compactness (% of respondents)			
	Hard (A)	Moderate (B)	Soft (C)	Very soft (D)
Castor	10	75	15	-
Tapioca	20	60	20	-
Kesseru	40	45	35	-

Studies on defoliation of castor and cassava for eri silkworm rearing without adversely affecting crop yields

The size of land holdings possessed by small and marginal farmers in India is much below the minimum economically viable sizes, earning income almost equal to that of agricultural labourers. The income obtained from cultivation of agricultural crops in one hectare of land is not sufficient to cross the poverty line. Consequent to the increasing population and the decreased land man ratio in the country it calls for maximizing the productivity per unit area and time. Hence diversification of agriculture into various income fetching enterprises has become the order of the day. Rain fed castor and cassava farming in Telangana and Andhra Pradesh has been diversified into eri silkworm cocoon production which ensures additional income and family employment during crop leisure period which helps to sustain rural cottage industries as well as the farming community. Castor, the primary food plant of eri silkworm and cassava, the secondary food plant are extensively grown in 3.91 and 0.50 lakh hectares in Mahabubnagar in Telangana and East Godavari districts of Andhra Pradesh respectively. The concept to make use of foliage of castor and cassava crops, otherwise going waste, for eri silkworm rearing which assures additional income without affecting main crop yield has been popularised in the above districts since 2003. Further, it is an off-farm activity that ensures gainful family employment with 55% women participation (Jaya Prakash *et al.,* 2005; 2006). The net income from cocoon production is much higher in cassava crop than other crops (Jaya Prakash *et al.,* 2008). The two revenues thus generated (seed and cocoon), while keeping the oil interest still in the game, is a unique feature because none of the others has a host equivalent to castor which has its own product of great commercial value.

Castor genotypes will respond differently to defoliation which will affect ultimately the seed yield. Leaf defoliation studies will also be required to know the tolerance limit of castor genotypes to with-stand different levels of defoliation without affecting the seed yield. Therefore, attempts were made to find out promising castor genotypes for better eri silk productivity and which can withstand defoliation and also suitable for both the castor seed production and eri silk production.

Cultivating these crops solely for the purpose of eri silkworm rearing may not be economical and practical due to high cultivation cost. But when a portion of the total foliage can be harvested without affecting the main produce viz., castor seed and tapioca tuber, it may be economical and can contribute towards economic independence of farmers (Devaiah *et al.,* 1984). Raghavaiah, (2003) and Jayaraj, (2004) suggested that 25-40% leaves in tapioca can be harvested without having adverse effect on tuber yield, making it possible to rear 400-600 dfls/ha in 2-4 rearing seasons. This paves the way for raising the income level of tapioca farmers in dry land areas (Rajadurai *et al.,* 2010).

Castor can be exploited both for seed and eri cocoon production and provides additional farm income besides increasing non mulberry silk production in the country. Eight castor genotypes were subjected to four levels of defoliation *viz.,* no defoliation, 30%, 40% and 50% to know the effect of defoliation on seed yield and additional income generated by rearing eri silkworms at Regional Agricultural Research Station, Palem, Telangana during 2012-13. Significantly no deviation in seed yield was observed from non defoliation plot and 30% defoliated plants. When castor plants were defoliated in the vegetative stage, they were able to replace all the leaf area lost in a short period of time with no impact on the economic yield and defoliation beyond this level caused a reduction in the seed yield. That 30% defoliated leaf can be utilized for growing eri silkworms and can get an additional income of Rs. 2000 - 2500/- acre. However, by utilizing PCH-111, PCH-222 and GCH-4 genotypes farmers can reap higher gross and net returns with more CB ratio (Dharma Reddy and Lakshmi Narayanamma, 2016) (Table 3.9).

Table 3.9: Defoliation per cent vs. seed yield among selected castor genotypes during *Kharif* 2012-13

Genotype	Seed Yield (kg/ha)					
	No Defoliation	30% Defoliation	40% Defoliation	Deviation in Seed yield	50% Defoliation	Deviation in Seed yield
Haritha	1325	1320	1314	-0.83	1120	-15.47
Kranthi	1280	1272	1270	-0.78	1040	-18.75
Kiran	1252	1256	1230	-1.76	1060	-15.34
DPC-9	1204	1210	1180	-1.99	990	-17.77
PCH-111	1509	1520	1462	-3.11	1270	-15.84
PCH-222	1485	1462	1402	-5.59	1285	-13.47
GCH-4	1269	1260	1198	-5.59	1098	-15.6
DCH-177	1190	1182	1082	-9.08	950	-20.17
F-test	Sig	Sig	Sig	-	Sig	-
CD at 5%	142.21	139.46	106.21	-	94.26	-
SEM ±	74.24	85.65	92.41	-	64.25	-

Rajasri and Lakshmi Narayanamma, (2015) stated that the castor genotypes *viz.,* PCH-111, PCS-262 and GCH-4 can be grown for dual purpose i.e. for seed and silk. By utilizing the 30% foliage for eri worms, farmers can get an additional return of about Rs. 4000/- to 5000/- per ha.

As per Devaiah *et al.*, (1984) castor leaf can be harvested up to 25% and used for eri cocoon production without affecting the castor seed yield. Similarly, the research findings of Directorate of Oilseeds Research, Hyderabad revealed that defoliation to an extent of 25-30% does not affect the yield of castor as it has tremendous regenerating capacity (Teotia *et al.,* 2003). The South Indian States like Tamil Nadu, Karnataka, Telangana and Andhra Pradesh have been cultivating castor in large scale for production of seed. These states can successfully utilize 30-40% of castor leaves for obtaining additional income through eri silkworm rearing without impairing castor seed yield (Suryanarayana *et al.,* 2003). Misra (1999) stated that Aruna variety can withstand defoliation up to 40% without any significant effect on seed yield. At the point of D47-48% defoliation, a combined revenue of around Rs. 22, 227/ha, could be obtained as contributed by beans and cocoons equally.

The concept of eri silkworm rearing is to fetch additional income and generate gainful employment among rural populace by utilizing 25% of the unwanted foliage from host crops grown under rain fed conditions of Telangana and Andhra Pradesh. Eri silkworm rearing, being a subordinate enterprise, is considered as mere source of additional income although it is in no way comparable with the returns of mulberry sericulture. Lakshmanan *et al.,* (1996) observed that there were intangible returns from sericulture. However, the cost benefit ratio with rise in net returns was at higher side. The same trend may not last forever, as the cocoon price is totally determined by demand and supply of the product in the market.

It is reported that 25-30% of castor leaves can be utilized for eri silkworm rearing without affecting the seed production and as such ericulture can be a supportive economy for the poor, dry land cultivators and resource poor farmers of the southern Telangana zone and provide gainful employment to the women (Devaiah *et al.,* 1984; Raghavaiah, 2003 and Lakshmamma *et al.,* 2009). Misra, (2001) reported 16% net profit when castor was grown for seed production, whereas 34% profit was obtained when it was used for both seed and eri silk production.

Pandey (2003) realized the net returns of Rs. 3000 per acre during first year from ericulture and it was Rs. 13,256 from second year onwards. Chandrappa *et al.,* (2003) reported high CB ratio with JI-226 genotype when used both for castor seed and eri cocoon production. Singh *et al.,* (2014) realized high CB ratio, which revealed that ericulture is a profitable venture for the poor and marginal farmers of North Eastern India. However, there is an enormous scope for ericulture in castor growing areas without hampering castor seed production and it also provides a supportive economy for the small and marginal farmers. It is remarkable for its low investment, high returns which make it as a profitable venture and an ideal agro based industry for castor growers of Telangana region.

Lakshmamma *et al.,* (2009) carried out field investigation during 2005-07 to determine the yield loss to different levels of defoliation at different growth stages in 'DCS 9' castor variety. Treatments consisted of 4 defoliation levels (25, 50, 75 and 100%) at primary, secondary and tertiary spike initiation stages along with control. Per cent defoliation affected was not exactly at the prescribed levels, but nearer to them. Number of leaves removed at 25% defoliation ranged from 3 to 6 i.e. defoliation was 23-32% for 25% defoliation, 50-55% for 50%, 58-71% for 75% defoliation. With increase in per cent defoliation, there was significant reduction in seed yield. The reduction in seed yield ranged from 10 to 20%, 20 to 23%, 24 to 36% and 26 to 55% with 25, 50, 75 and 100 % defoliation, respectively. On an average, defoliation at primary stage showed 21% yield reduction and at secondary stage showed 33% yield reduction. Tertiary stage defoliation showed 26% yield reduction. Defoliation at secondary spike initiation stage was found to be the most sensitive stage which reduced seed yield of all order branches. In general, defoliation at any stage beyond 25% significantly reduced seed yield in castor. As there was only primary spike at primary defoliation stage, it affected only primary seed yield. But when defoliated at other two stages, the different branches were in different stages of growth and were affected. The data clearly showed the competition for assimilates from one order to next higher order.

Lakshmamma *et al.,* 2010 conducted another experiment for two consecutive years during 2005-07 using DCS-9 variety to test the effect of defoliation at different spike initiation stages on biomass partitioning. Treatments consisted of four defoliation levels (25%, 50%, 75% and 100%) at primary, secondary and tertiary spike initiation stages on entire plant along with control. With increase in defoliation at any stage there was significant reduction in stem, leaf, capsule dry weight and total dry matter. There was compensation in growth only up to 25% defoliation at any stage. Only primary seed yield was affected with defoliation at primary spike initiation stage, but with secondary stage defoliation yield of all order branches was affected. Yield of one order decreased with >25% defoliation at that order and >50% defoliation at next lower or higher order spikes.

Dinesh-Hans and Sundaramoorthy, (2002) also reported decrease in plant height, number of branches and spikes, and seed yield of castor with the increase in defoliation rate and frequency. In castor, removal of leaves proximal to the main spike significantly affected seed yield.

Ramesh and Prasad, (2001) conducted the defoliation studies in Aruna variety of castor. The results showed that removal of leaves proximal to main spike significantly affected seed yield. The top one, two and four leaves accounted to 10, 36 and 68 per cent leaf area. Clipping of the four leaves resulted in yield reduction of 33, 45 and 66 per cent respectively. On the other hand one, two and four leaves distal to main spike accounted to 6, 14 and 33 % leaf area. Upon defoliation the yield differences obtained were not significant. In the study of Ramesh and Prasad (2001), total defoliation of main spike resulted in yield reduction up to 68 %. The seed yield with the removal of four top leaves was on par with total defoliation, thereby revealing the importance of top leaves towards the sink development. Removal of all leaves, 4 leaves from top and bottom of the plant reduced yield by 46, 44 and 28% compared to 68, 66 and 40% of main spike yields. A maximum yield loss was noticed upon removal of top 4 leaves.

Defoliation and shading appear to decide the pattern of assimilate distribution in plants (King *et al.*, 1967). The reduction in yield due to defoliation implies that all leaves on the plant contribute to seed yield. Removal of any leaf decreases the yield. Reddy *et al.*, (1997) reported that maximum dry matter production in cultivar Aruna is achieved at optimum leaf area index of 0.85 and 0.62 at 90 DAS and at maturity respectively (Ramesh and Prasad, 2001).

Physiological relationships among leaf area, yield contributing characters and oil content in castor:

The weight of each individual seed plays an important but often neglected role in the determination of castor seed yield. Each individual seed has a large influence on its own fate, in opposition to the idea that they simply result from the assimilate supply from the mother plant (Egli, 2006). The variability in the individual seed weight has a potentially high impact on final seed production, but usually only the mean seed weight is considered. The mean seed weight is determined predominantly by the plants genetic potential (Egli *et al.*, 1987), but variability in seed weight can be caused by many physiological processes that are sensitive to environmental conditions. Variability in seed weight is inevitable even under controlled growing conditions (Hay *et al.*, 2010). Because phase I does not occur at the same time for all seeds in an indeterminate plant such as castor, high variability in seed characteristics is expected due to the environmental effect. Environmental factors can also create additional variability to seed weight by acting directly on the seed or indirectly when a change in the photosynthesis rate influences assimilates supply (Severino and Dick, 2013).

Severino *et al.,* (2010) conducted detailed studies on physiological relationships among levels of defoliation and yield contributing characters in Castor plant variety BRS Nordestina. The castor plants were subjected to bimonthly defoliations in order to evaluate the influence on development and yield components. Leaf area was calculated from the length of the main vein of each leaf. Defoliation treatments of 0, 15, 30, 45 and 60% of the total leaf area were applied. Castor plants could completely re-grow its leaf area when the defoliation occurred during the vegetative stage; however, this initial damage reduced the maximum leaf area obtained by the plant. Maturation of fruits occurred earlier when plants were subjected to defoliation. Defoliation treatments did not influence the number of fruits per raceme, but caused a steep reduction of the number of racemes per plant. Adaptations in the yield components were observed in the number of seeds per plant and in the seed oil content, but not in the mean seed weight. Any level of defoliation repeated along the plant growth cycle was found to cause a reduction of the oil yield. The loss of $1m^2$ of leaf area caused the reduction of 37.8 g of castor seed or 24.4 g of castor oil.

Leaf photosynthesis depends on the intensity of the intercepted solar radiation. As the plant canopy consists of several layers that partially overlap and shade one another, when light is attenuated, less photosynthesis can be realized in the inner layers (Larcher, 2003). However, leaf function is not only photosynthesis, but it also act as transitory storage tissue. Primary accumulation of photosynthates and absorbed nutrients in castor can occur in leaves, to be translocated afterwards to seeds. Assimilated carbon can also be stored early in the growth cycle in organs like stems and roots, in such a way that losing part of the leaves would not directly affect seed fill (Ahmadi and Joudi, 2007). For this reason, effects are not the same when defoliation occurs in vegetative or reproductive stages, or when the defoliation is limited to leaves of the upper or lower part of the plant.

Utilization of senescent tapioca leaves for the rearing of eri silkworm:

Cassava/Tapioca (*Manihot esculenta*), is a secondary host plant of eri silkworm. In general tapioca farmers can divert a portion of foliage (25-40%) to rear eri silkworm commercially and get additional income without affecting the tuber quality and yield (Krishna Rao, 2003). Moreover, a huge amount of senescent foliage is available at the time of tuber harvest and is wasted by the farmers (Sakthivel *et al.,* 2010). These senescent leaves generally have poor nutritive values which affect the growth and development of eri silkworm and cocoon yield. In order to utilise those leaves, Sakthivel and Qadri, (2013) have undertaken an experiment to know the effect of feeding senescent tapioca leaves on economic traits of eri silkworm. The results of the study suggests that, senescent tapioca leaves wasted at the time of tuber harvest could be successfully utilized by supplementing 10% jaggery solution and soya powder @ 10g/100g of leaves for rearing of eri silkworm which ensured considerable improvement in economic traits of the worms and silk yield. Sugarcane jaggery and soya flour are cheap sources of supplement and hence could be utilized to a greater extent for enhancing the eri silk productivity besides supporting the livelihood of tapioca growers with assured additional income through ericulture.

Feasibility of defoliation of cassava for eri silkworm rearing in Cambodia without adverse effect on tuber yields:

Cassava is one of the main crops in Cambodia, and its harvested areas and yields have been increasing in the last five years and it is estimated to expand year by year. Ericulture, the raising of eri silkworm, a kind of wild silkworm originated in Assam province, India, was introduced to Cambodia in October, 2010. Since then, farmers face the difficulty to ensure fresh leaves for feeding eri silkworm. So trimming methods of leaves was studied. Accordingly, attention has been paid to evaluate the suitable amounts of leaves trimmed for ericulture without damaging cassava yields.

KM 94, one of cassava variety that is widely adopted by local farmers in Cambodia, was cultivated from November 2011 to July 2012, a period of 8 months, at the experimental field in Royal University of Agriculture located in Phnom Penh. After 4 months passed from planting, the leaves were trimmed at 0%, 20%, 40%, 60%, 80% and 100%, respectively. At 8 months passed from planting, the yields were compared among plots of different trimming percentage and discussed the suitable amounts of leaves trimmed for ericulture without damaging cassava tubers. The results showed that the yields of 100% trimmed were significantly smaller than that of 0% to 80% trimmed. Moreover, there was no significant difference in yields amongst plots trimmed 0% to 80%. Accordingly, it was concluded that cassava leaves can be trimmed up to 80% without adversely affecting cassava yields (Kawabe *et al.,* 2014).

Other Alternate Hosts for rearing eri silkworm:

Though castor is the main host plant of eri silkworm, its leaves are not available throughout the year and the crop has to be sown after every six months. The castor is a warm season crop, the ideal season for plantation of castor is March to April and September to October in the northeast region of India. During winter season (long dry spell), its leaf yield gets considerably reduced. Moreover, its cultivation practices depend on rainfall conditions resulting in uncertainty in raising the castor crop and in turn ericulture. Further, several pests and diseases affect the castor crop and in controlling them large quantity of chemicals are used frequently. The chemical sprayed leaves are harmful to eri silkworm. In the event of harsh climatic conditions tree based ericulture is the need of hour (Mitalee Baruah, 2012a and 2012b). In this direction, search for alternate perennial tree host plants available in the north-eastern region of India for continuous eri silkworm rearing is started (Sarmah *et al.,* 2015).

At the time of scarcity of castor leaf, the use of secondary food plants of eri silk worm has become a regular practice amongst the rearer. Consequently, this has led to the process of adopting the practice of curtailment of eri silkworm rearing during the period of food scarcity, which ultimately results in low productivity of the crop. In the periods where there is scarcity of the primary host plant for eri silkworm rearing, the secondary host plants meet the requirement of the food to some extent. Thus, the use of appropriate secondary host plant becomes absolutely necessary in pulling off some of the loss created during food scarcity and maintains a steady rate of production of crop.

Eri silkworm is a polyphagous species, and feeds on host leaves of about fifty kinds of plants belonging to Euphorbiaceae, Araliaceae, Apocynaceae, Simaroubiceae etc. However, all the food plants are not equally good for eri silkworm rearing and eri silkworm shows differential behaviour, when reared on different food plants. Based on number of larvae feeding on the host leaves, the descending order of food preference of eri silkworm larva was castor (7.50 larvae) > papaya (6.25 larvae) > tapioca (3.25 larvae) > Jatropha (1.50 larvae) > *Ailanthus* (1.25 larvae) and > *Plumeria* (0.25 larva) (Patel and Patel, 2010).

Patil *et al.,* (2009) conducted an experiment to test the suitability of two *Jatropha* sps *viz., Jatropha curcas* and *J. gossypifolia* as feed for eri silkworm. Known number of eri silkworms brushed on castor bean immediately after hatching, and after the 1st, 2nd, 3rd and 4th moults were released on the leaves of *Jatropha curcas* and *J. gossypifolia.* Observations on survival, larval, pupal and cocoon characters were recorded. The results showed that irrespective of the larval stage released on *J. gossypifolia*, the larvae scraped the leaves and survived for 2-3 days, but could not complete even a single instar. The larvae reared on *J. curcas* scraped the leaves, but survived for only 3-4 days, indicating their non-preference. However, 32% of the 3rd instar larvae fed on *J. curcas* reached the 4th instar with prolonged larval duration of 8 days but failed to undergo the 4th moult. Rajashekhargouda *et al.,* (2009) stated that eri silkworms failed to complete life cycle on *J. curcas* as well as *J. gossypifolia* and do not serve as viable alternate hosts.

Few authors reported the possibility of utilization of *Ailanthus* species in eri silkworm rearing. Four genotypes of *Ailanthus* are reported in India viz., *A. excelsa* (Borkesseru)*, A. grandis* (Borpat), *A. altissima* and *A. malabarica*. In Chinese literature, reports were available on the use of *Ailanthus* leaves for the silk industry in Shantung Province (Ahmed *et al.,* 2015). Out of these four species, two members of *Ailanthus* species *viz., Ailanthus grandis* (Barpat) and *Ailanthus excelsa* (Barkesseru) are advantageous because they are perennial, evergreen, quick growing tall trees and not generally browsed by the cattle (Deori and Khanikor, 2015). As *Ailanthus* leaves are found abundantly throughout the year, they can be successfully utilized as a substitute food for eri silkworm rearing during the scarcity of castor leaves. An investigation was carried out to evaluate the proximate composition of eri silkworm pupae fed on the leaves of Barpat and Barkesseru on moisture, crude fat, crude protein, total minerals and total carbohydrate contents along with castor leaves. Pupae obtained from castor fed silkworm showed significantly higher amount of crude fat and crude protein. No significant differences were observed in moisture content, crude fibre, and total mineral and total carbohydrate in the pupae obtained from silkworms fed on Barpat and Barkesseru leaves than that of pupae obtained from castor fed silkworms. So the pupae of eri silkworm fed on Barpat and Barkesseru leaves also could be easily utilized as a source of proteinaceous food as good as pupae obtained from castor fed silkworms for human consumption as well as feed for fish, poultry, piggery, cattle and other purposes (Deori *et al.,* 2016).

Studies revealed the superiority of the combination of Castor (*Ricinus communis*) with Borpat (*Ailanthus grandis*) leaves feeding in early and late stages of silkworm, respectively. The lowest larval duration, higher mature larval weight, single cocoon weight, single shell weight, cocoon yield per disease free laying (dfl) in numbers, effective rate of rearing and cocoon shell yield per 100 dfls were found the highest in treatment of Castor in combination with Borpat, which was at par with the treatment

Borpat alone i.e. feeding from brushing (1st instar) till spinning (5th instar). The lowest performances in all economic characters such as higher larval duration and the lowest mature larval weight, single cocoon weight, single shell weight, cocoon yield per dfl in number and cocoon shell yield per 100 dfls were recorded in the treatment of eri silkworm feeding from 1st instar till spinning on Kesseru plant. The highest fecundity was recorded in the treatment of Castor in combination with Borpat, which was at par with the treatment of Castor in combination with Borkesseru. Similarly, the higher hatchability, moth emergence and lower cocoon: dfl ratio was recorded in the treatment of Castor in combination with Borpat. Adoption of Borpat tree for rearing of eri silkworm among eri rearers will help in commercialization of ericulture and will overcome the problem of non-availability of sufficient leaves throughout the year besides mitigation of deforestation and effects of climate changes (Ahmed *et al.*, 2015).

The comparative fresh leaf biomass productivity (metric tonnes per ha) found to be highest in Borpat (33.45) followed by Borkesseru (28.24), Kesseru (21.24) and Castor (11.45). It revealed that Borpat produces higher fresh leaf biomass of 57.48% and 192.14% in comparison with Kesseru and Castor, respectively. In case of Borkesseru, the annual biomass production is more by 33.15% and 146.64%.

An experiment on effect of two *Ailanthus* species *viz.*, *A. grandis* and *A. excelsa* and *Ricinius communis* on cocoon and yarn characteristics of eri silkworm was carried out at Assam Agricultural University, Jorhat. Results revealed that castor is better for rearing of eri silkworm considering the economic cocoon characters *viz.*, cocoon weight, shell weight, pupal weight and size of the cocoons. Borpat and Borkesseru fed cocoons appeared to be better for qualitative physical characters of the yarn *viz.*, tensile strength, tenacity, elongation and imperfection. Though with significantly lower degumming loss and shorter degumming period, the yarn yield was registered higher for castor fed cocoons than both the *Ailanthus* species, there was no variation in size of the spun yarn produced from the cocoons feeding on the food plants tested (Saikia and Dutta, 2013). Thus *Ailanthus* leaves can also be best utilized for rearing of the silkworm during scarcity of castor leaves (Chowdhury, 1982; Khanikor *et al.*, 1997 and Shaw, 1998).

Eri silkworm, *Samia c. ricini* feeds primarily on castor (*R. communis*). However, Kesseru (*Heteropanax fragrans* Seem) is considered as another major perennial food plant. Besides these two, eri silkworm being polyphagous, feeds on several alternative host plants, *viz.*, Payam (*Evodia fraxinifolia*), Tapioca (*Manihot esculanta*), Barkesseru (*Ailanthus excelsa*) Barpat (*A. grandis*), Gulancha (*Plumeria acutifolia*), Gamari (*Gmelina arborea*), etc. Bindroo *et al.,* (2007) reported 24 plant species as hosts of eri silkworm. The tribals and several weaker sections of society of the region depend mostly on forest based eri food plants for silkworm rearing as sustainable source of livelihood. Hence, a study has been conducted to screen out best food plants for rearing of eri silkworm (Naik and Murthy, 2013) at UAS, Dharwad, Karnataka. Among 23 plant species tried for ericulture (Table 3.10), five plant species have been accepted by the eri silkworm as food plants. Out of five plant species, eri larvae had good feeding response and survivability on fountain tree (*Spathodea companulata* Pal.), banyan tree (*Ficus bengalensis* L.) and Indian almond (*Terminalia catappa* Linn.), moderate response on carrot leaves (*Daucus carota* Linn.). The reason may be due to stimulant action (gustatory stimulants) and nutritional content of leaf. But the eri larvae recorded slight feeding response and survivability when fed on jack fruit leaves (*Artocarpus heterophyllus* Lam.). Non- acceptance of this host may be due to anti feedant (feeding deterrent), poor phagostimulant activity and the presence of alkaloids in the leaf. For no feeding, it may also be due to the odour (olfactory stimulants) which repelled the worms from feeding site and morphological feature of leaves. The perusal of literature revealed that these hosts except fountain tree (Patil, 2004) were recorded for the first time on eri silkworm. Patil (2004) reported fountain tree as the new host plant of eri silkworm. The new host plants of eri silkworm recorded first time may be the substitute host plants of eri silkworm.

In a laboratory study, Subramanianan *et al.,* (2013) used 10 plant species, *viz., Calotropis gigantean, Nerium odorum,* subabul (*Leucaena leucocephala*), *Parthenium hysterophorum, Annona squamosa, Pongamia pinnata,* Coconut leaf, banana leaf, *Sesbania grandiflora* and country almond *Terminalia catappa* for their suitability for feeding by the 5th instar larvae of eri silk worm after pre-starvation period of 30 min. In case of *Calotropis gigantea,* the larvae started feeding initially, then stopped in a short period and showed 100% larval death within an hour. In all other cases except the country almond *Terminalia catappa* and subabul, the worms were not feeding at all, or they just nibbled and started wandering. On the other hand, the feeding was moderate and continuous on subabul and country almond. Hence, these two plants also can be utilised during the off season.

Table 3.10: New host plants selected for the study (Naik and Murthy, 2013)

S.No.	Common name	Botanical name	Family	Leaf character
1.	Jack fruit	*Artocarpus heterophyllus* Lam.	Moraceae	Glossy, leathery, thick
2.	Kar-Khair- (Banni)	*Acacia ferruginea* DC.	Mimosaceae	Bipinnate, thickness
3.	Bahunia	*Bauhinia purpurea* L.	Caesalpiniaceae	Cleft, oblong, papery
4.	Bamboo	*Bambusa arundinaceae* Retz.	Bambusaceae	Distichous, hairiness
5.	Yellow bamboo	*Bambusa vulgaris* L.	Bambusaceae	Leaf sheath pubescent
6.	Banyan tree	*Ficus bengalensis* L.	Moraceae	Alternate, latex content
7.	Soapnut tree	*Sapindus emarginatus*	Sapinidaceae	Oblong, small leaves
8.	Copper pod	*Peltophorum ferrugineum* Decne	Caesalpiniaceae	Bipinnate, small, thick leaves
9.	Pongamia	*Pongamia pinnata* Linn.	Fabaceae	Imparipinnate, hard
10.	Raintree	*Samanea saman* Jacq.	Mimosaceae	Pinnate, thickness
11.	Fountain tree	*Spathodea companulata* Pal.	Bignoniaceae	Broad, dark green
12.	Flamboyant tree	*Swietenia mahogoni* Marophylla	Meliaceae	Paripinate, thick
13.	Tabubia	*Tabubia argentina*	Bignoniaceae	Pubescence, thick
14.	Indian almond	*Terminalia catappa* Linn.	Combertaceae	Glabrous leaves
15.	Cassia	*Cassia auriculata* Linn.	Caesalpiniaceae	Elliptic, hard
16.	Peepal tree	*Ficus religiosa* L.	Moraceae	Broadly oblong, less latex
17.	Anjan	*Hardwickia binata* Roxb.	Fabaceae	Paripinate, hard
18.	Khirani	*Manilkara hexandra* Roxb.	Sapotaceae	Branchlets, hard
19.	Paradise tree	*Simarouba glauca* DC.	Fabaceae	Pinnate, small
20.	Carrot	*Daucus carota* Linn.	Umbelliferae	Pinnetely, succulent
21.	Rye grass	*Acalypha gracilens* A.	Euphorbiaceae	Oblong, red colour
22.	Lantana	*Lantana camara* L.	Verbenaceae	Pinnately compound
23.	Golden shower tree	*Cassia fistula* Linn.	Caesalpiniaceae	Paripinnate, hard

According to Dayashankar (1982) and Devaiah *et al.*, (1985) out of four different host plants (Castor, Tapioca, White plumeria (*Jatropa curcas*) and Red plumeria (*Jatropa gossypifolia*) tested, castor was the most preferred and the best host plant for eri silkworm followed by Tapioca. Among the tapioca varieties, Co-2 was superior to ME-1 and ME-120 for rearing and can be used singly or in combination with castor during shortage / non-availability of castor.

Neelu Nangia *et al.*, (2000) reported the host plant preference of eri silkworm in the order of merit was castor > tapioca > papaya > barkesseru > gulancha. Castor and tapioca are the two most important host plants for eri silkworm though there are certain perennial tree species like *kesseru* and *payam* in North Eastern states may serve to provide supplementary food in the off season. Kumar *et al.*, (1993) reported that leaves of *kesseru* were next to the castor leaves in terms of cocoon harvest and other economic parameters in South India, the match wood tree species, *Ailanthus excelsa* is common in the plains and hills is also an alternative host for eri silkworm. Sachan and Bajpai, (1973) observed superior larval growth, development and higher cocoon production when eri silkworms were fed with leaves of Rosy castor variety followed by S- 30, EB-31 and EB-16 varieties. Papaya plantation is also coming up on a large scale in many parts of South India and the leaves can be used productively. The progressive growth of eri silkworm was superior when fed on castor. Further, the larvae receiving the castor leaves during fifth instar had better growth irrespective of the diet used earlier, i.e. whether tapioca/castor (Joshi, 1987 and Subramanianan *et al.*, 2013).

Kesseru is widely distributed in the North Eastern States of India, both in wild and cultivated condition. Taking the advantage of its perennial nature, Kesseru is being utilized in various developmental schemes on eri food plants like Augmentation of Eri food plant (AEFP), Catalytic Development Programme (CDP) and Cluster Promotion Programme (CPP) *etc*. Besides, most of Government farms of Assam and Meghalaya planted kesseru for eri silkworm rearing. *Ailanthus* belongs to family Simaroubaceae is a perennial host plant for rearing of eri silkworm. Hence, the performance of a high yielding eri silkworm breed C_2 developed by Central Silk Board, CMER&TI, Lahdoigarh was studied in different food plants *viz.*, Castor, Kesseru, Borpat and Borkesseru. Castor could be judged as the best food plant of eri silkworm. Kesseru ranks second among all the food plants of eri silkworm and utilized as ruling perennial food plant. Other two perennial food plants *viz.*, Borpat and

Borkesseru are rarely used but those are potential eri food plants. During the study all food plants showed positive economic character in rearing performance (Sarmah *et al.,* 2015).

An attempt was made to rear the eri silkworm on the leaves of *Terminalia catappa* (Badam) for the first time under Western Maharashtra conditions in India and reared under laboratory condition. It was observed that the badam is a potential new host of eri silkworm. The growth parameters and cocoon characters of the eri silkworm reared on badam leaves were normal. The eri silkworm successfully lasted its life cycle ranged from 47 to 53 days. The effective rate of rearing, weight of pre-spinning larvae, cocoon, shell and pupae ranged from 90 to 95 per cent, 7 to 9 g, 3.02 to 3.10 g, 0.45 to 0.50 g, 2.60 to 2.90g respectively. Parameters such as length of cocoon shell, width of shell, peduncle, shell thickness etc were also normal. The fecundity ranged from 400 to 425 eggs (Kavane, 2014a).

Kavane, in the year (2015) made another attempt to rear the eri silkworm on the leaves of *Terminalia arjuna* (Arjun) for the first time under Western Maharashtra conditions in India and reared under laboratory condition. It was observed that the arjun is also a potential new host of eri silkworm. The growth parameters and cocoon characters of the eri silkworm reared on arjun leaves were normal. The eri silkworm successfully lasted its life cycle ranged from 52 to 55 days. The effective rate of rearing, weight of pre-spinning larvae, cocoon, shell and pupae ranged from 40 to 45 per cent, 6 to 7 g, 2.80 to 2.90 g, 0.40 to 0.42 g, 2.35 to 2.40g respectively. The fecundity ranged from 400 to 425 eggs.

Kavane, (2014b) attempted and reared the eri silkworm under laboratory conditions on the leaves of Papaya, *Carica papaya* as a primary food for the first time under Western Maharashtra region in India. It was observed that the papaya is a potential host of eri silkworm as the growth parameters and cocoon characters of the eri silkworm reared on papaya leaves were normal. The eri silkworm successfully lasted its life cycle ranged from 47 to 53 days. The effective rate of rearing, weight of pre-spinning larvae, cocoon, shell and pupae ranged from 45 to 55 per cent, 6.10 to 7 g, 2.49 to 2.70 g, 0.30 to 0.32 g, 2.10 to 2.60g, respectively. The fecundity ranged from 350 to 415 eggs. The finding of arjun, badam and papaya as a potential new host for eri silkworm has opened new vistas in promoting vanya silk industry in Maharashtra state.

Narayanaswamy *et al.,* (2006) studied the activity of amylase in digestive juice and haemolymph in relation to consumption indices and economic parameters in eri silkworm as influenced by feeding on castor, tapioca and barkesseru during fifth instar. The activity of digestive and haemolymph amylases increased steeply from first day to fourth day of fifth instar and thereafter their activity declined towards the onset of spinning. The activity of both the enzymes had established positive correlation with consumption, digestion, approximate digestibility, mature larval weight, cocoon weight, pupal weight, shell ratio but negatively correlated with consumption index, growth rate, efficiency of conversion of ingested food, efficiency of conversion of digested food and larval duration. It was observed that the relationship was stronger for both castor and barkesseru than tapioca. The amylase activity of eri silkworm fed on barkesseru was at par with those fed on castor, indicating that barkesseru could be a good substitute for castor to rear fifth instar eri silk worm. Hence, barkesseru could be exploited as a suitable substitute for castor to rear eri silkworms in commercial production of eri silk (Kumari *et al.,* 2009).

Deka *et al.,* (2011) conducted the studies on larval growth and spinning parameters of eri silkworm on castor (red variety), kesseru and tapioca. The evaluation of data revealed that castor (red variety) has shown supremacy over the other food plants i.e. kesseru and tapioca for larval growth as well as spinning characters. The data showed that the minimum larval duration (days), fecundity, effective rate of rearing percentage and other spinning characters e.g. silk ratio percentage were better in castor compared to other two host plants.

Grimm and Somarriba, (2001) conducted an experiment by feeding the larvae of eri silkworm with the leaves of physic nut, *Jatropha curcas* from ten different provenances of Africa, Asia and the America. By rearing the eri silkworm on Jatropha leaves collected from Mexico and Nicaragua enabled the insects to complete its life cycle. Feeding on these leaves during the first four larval stages and castor during fifth instar showed a significant improvement on cocoon characters, chrysalis and cocoon cortex weight, larval and pupal development time as well as oviposition frequency and number of eggs. The same combination in reverse order did not improve the performance of eri silkworm. Hence the physic nut leaves can be utilised for rearing the silkworm up to fourth instar only.

Devi and Kalita, (2009) investigated on the impact of feeding of Japanese weed leaves (*Mikania micrantha*) on the larval duration, larval growth, cocoon parameters and fecundity of eri silk worm. The rearing was conducted at laboratory condition. The larvae were reared on three experimental treatments *viz.,* rearing of eri silkworm on *Mikania micrantha* from 1st instar till maturity; rearing on *Ricinus communis* from 1st to 4th instar larva and the 5th instar larva fed with *Mikania micrantha* and *Ricinus communis* in equal proportion till maturity; and rearing of eri silkworm on *Ricinus communis*- the control. In spite of heavy mortality (90%) of 1st instar larva shown by Ex tr-I, the survivors (10 %) were highly adapted to the new environment of food plant *Mikania micrantha*. The larval and cocoon parameters studied in the first set of experiment were superior over the other two sets of experiment. The larval, pupal, cocoon, shell weight and fecundity were recorded to be highest in the first treatment (88.54 g/10, 24.57 g/10, 20.50 g/10, 3.63 g/10 and 250 eggs) respectively followed by second and control treatments (77.39 g/10, 24.52 g/10, 27.53 g/10, 2.71 g/10 and 248 eggs; 77.58 g/10, 23.89 g/10, 27.68 g/10, 3.48 g/10 and 248 eggs respectively). Further, the highest shell ratio (12.63%) was recorded in the larva treated with Japanese weed from 1st instar till maturity. Hence the Japanese weed leaves can also serve as alternate host during the scarcity of the primary host.

Suitability of host plants for Eri silkworm in Tamil Nadu

Subramanianan *et al.,* (2013) studied on the suitability of different host plants in Tamil Nadu region during September 2003 to January 2005.

Feeding response of *S. cynthia ricini* was poor in the host plants viz., *Thespesia populnea, Antigonon leptopus, Erythrina indica, Vitex negundo, Ipomea carnea, Albezia lebbeck, Calotropis gigantea* and the mortality rate was high in *Azadirachta indica, Jatropha glandulifera*. Low mortality and moderate feeding was observed in the case of *Glyricidia maculate* and *Carica papaya*. In the case of *Calotropis gigantea*, the larvae started feeding immediately when fed but stopped within short period and resulted in 100% larval death within an hour. In all other cases the larvae were not shown any growth and the larval stage existed for long time, except in country almond (*Terminalia catapa*) and subabul (*Leucaena leucocephala*) the worms were fed moderate and continued the rearing. Hence *Terminalia catapa* and subabul were recorded as alternative host plants in Tamil Nadu for mature worms and worms were shown good feeding and no mortality was observed (Table 3.11).

Table 3.11: Suitability of host plants for eri silkworm
(Subramanianan *et al.,* 2013)

Plants tested	Feeding response	Remarks
Albezia lebbeck	No feeding	High mortality
Antigonon leptopus	No feeding	High mortality
Azadirachta indica	Slight feeding	High mortality
Coffea robusta	Nibbling/Low feeding	High mortality
Coffea Arabica	Nibbling/Low feeding	High mortality
Calotropis gigantean	No feeding	High mortality
Carica papaya	Moderate feeding	Low mortality
Erythrina indica	No feeding	High mortality
Glyricidia maculate	Moderate feeding	Low mortality
Ipomea carnea	No feeding	High mortality
Jatropha glandulifera	Moderate feeding	Low mortality
Terminalia catapa	Good feeding	No mortality
Thespesia populnea	No feeding	High mortality
Vitex negundo	No feeding	High mortality
Leucaena leucocephala	Good feeding	No mortality

Earlier reports mentioned that *S. c. ricini* was unable to survive beyond the 1st instar on sweet potato or papaya (Huq *et al.,* 1991). Tender leaves were chosen for the chawkie rearing in case of castor, tapioca and country almond. On contrast *Ailanthus, Jatropha* and papaya gave 100% mortality at chawkie stage, when the same host plants were tried in the 5th instar the survival percentage was very poor. Among the different host plants the castor and tapioca gave better performance in respective with larval weight, cocoon weight and ERR.

As stated by many workers, eri silkworm mostly prefers castor as a primary host plant for feeding. In Punjab state except castor, other known hosts of eri silkworm are not much prevalent. So to find out other alternate hosts, Rabinder Kaur and Virk, (2012) conducted an experiment in the Sericulture Laboratory, Punjab Agricultural University, Ludhiana to determine the possibility of rearing of eri silkworm on papaya, cauliflower, brinjal, potato and castor in the state. The results showed that there was 100 per cent mortality of first instar larvae on papaya and cauliflower leaves in 7.55±1.99 days and 6.10±0.31 days respectively. Only 20 and 10 per cent third instar larvae were able to form the cocoon in 14.0±2.67 and 11.67±0.58 days, respectively. No third instar larvae completed the larval period on brinjal and potato leaves and died after 9.55±4.19 and 5.50±1.28 days, respectively. All the larvae successfully completed the larval period in 25.4±1.80 days on castor leaves with 100 per cent cocoon formation. Hence it is concluded that castor was the only host plant for the successful commercial rearing of eri silkworm in Punjab. The list of the food plants that serve as alternate hosts for eri silkworm is listed in the Table 3.12. These host plants can be utilised under the scarcity of the primary host, castor.

Table 3.12: Different plant species tested for their suitability for rearing eri silkworm, *S. c. ricini*

S.No	Common Name	Scientific Name
1	Castor	*Ricinus communis* Linn
2	Barkesseru	*Ailanthus excelsa; A. glandulosa; A. altissima; A. malabarica, A. triphysa*
3	Hollynock	*Althea rosea* Cav.
4	Kesseru	*Heteropanax fragrans* Seem.
5	Celery	*Apium graveolans* L.
6	Neem	*Azadirachta indica* Juss.
7	Barberru	*Berberis* sp.
8	Payam	*Evodia fraxinifolia* Hock.
9	Tapioca/ Cassava	*Manihot utilissima; M. esculenta* Crantz.
10	Papaya	*Carica papaya* L.
11	Horn bean	*Carpinus betulus* L.
12	Laburnum	*Cassia Fistula* L.
13	Gulancha phool/ Champa	*Plumeria acutifolia; P. acuminate* Ait.; *Plumeria rubra* Linn.
14	Bhotera/ Gulancha	*Jatropa curcas; J. multifidea; J. gossipifolia*
15	Masuri	*Coriaria nepalensis* Wall.
16	Gamari	*Gmelina arborea* L.
17	Walnut	*Juglans regia* L.
18	Indian privet	*Lawsonia alba* Lamk.
19	Privet	*Ligustrum robustum* Blume
20	Cherry	*Prunus avium* L.
21	Plum	*P. domestica* L.
22	Apple	*Pyrus malus* L.
23	Ber	*Zyziphus jujuba* Lamk.; *Z. alatium; Z. mauritiana* Lamk.
24	Korha	*Sapium ellgeniafolium; S. tenuifolium* Buch-ham
25	Bazraman	*Zanthorylum yhesta*
26	Thebow	*Hodgsonia heteroclita*

Table 3.12: *Contd…*

S.No	Common Name	Scientific Name
27	Cinnamomum	*Cinnamomum glanduliferum; C. cecidodaphne*
28	Phootkhol	*Micromulum minutum*
29	Jack fruit	*Artocarpus heterophyllus* Lam.
30	Kar-Khair (Banni)	*Acacia ferruginea* DC.
31	Bahunia	*Bauhinia purpurea* L.
32	Bamboo	*Bambusa arundinaceae* Retz.
33	Yellow bamboo	*Bambusa vulgaris* L.
34	Banyan tree	*Ficus bengalensis* L.
35	Soapnut tree	*Sapindus emarginatus*
36	Copper pod	*Peltophorum ferrugineum* Decne
37	Pongamia	*Pongamia pinnata* Linn.
38	Raintree	*Samanea saman* Jacq.
39	Fountain tree	*Spathodea companulata* Pal.
40	Flamboyant tree	*Swietenia mahogoni* Marophylla
41	Tabubia	*Tabubia argentina*
42	Indian almond	*Terminalia catappa* Linn.
43	Cassia	*Cassia auriculata* Linn.
44	Peepal tree	*Ficus religiosa* L.
45	Anjan	*Hardwickia binata* Roxb.
46	Khirani	*Manilkara hexandra* Roxb.
47	Paradise tree	*Simarouba glauca* DC.
48	Carrot	*Daucus carota* Linn.
49	Rye grass	*Acalypha gracilens* A.
50	Lantana	*Lantana camara* L.
51	Golden shower tree	*Cassia fistula* Linn.
52	Guava	*Psidium guajava*
53	Cauliflower	*Brassica oleracea* var. *botrytis*
54	Potato	*Solanum tuberosum*
55	Brinjal	*Solanum melongena*
56	Swallow wort	*Calotropis gigantea*
57	Nerium	*Nerium odorum*
58	Subabul	*Leucaena leucocephala*
59	Congress weed	*Pathenium hysterophorum*
60	Indian beech	*Pongamia pinnata*
61	Seetaphal	*Annona squamosa*
62	Agati/Humming bird tree	*Sesbania grandiflora*
63	Prickly ash	*Zanthoxylum armatum, Z. limonella*
64	Japanese weed	*Mikania micrantha*

Feasibility of Sequential feeding of eri silkworm by different host plants:

The Eri silkworm, *S. cynthia ricini,* Biosduval can be reared indoors throughout the year to produce silk depending on the availability of suitable host plants (Debraj *et al.,* 2003). However, rearing performance of eri silkworm by using the leaves of different food plants produced varied results such as prolonged larval duration, reduced larval weight, cocoon weight, shell weight, shell ratio and pupal weight. On the other hand, eri host plants are interchangeable at 3rd instar to 5th instar especially during the scarcity of castor leaves (Dutta and Khanikor, 2005 and Mukul Deka *et al.,* 2011). Hence, the new practice of utilizing sequential use of food plants for commercial rearing is worked out by Venu and Munirajappa, (2013).

In the study of Venu and Munirajappa, (2013) it was clearly proved that, the rearing of eri larvae by feeding the castor leaves from 1st to 3rd instar and interchanging with tapioca leaves during 4th and 5th instars improved the commercial characteristic features such as larval weight, larval duration, ERR, cocoon weight and shell weight. Therefore, they recommended the use of castor and tapioca leaves sequentially for successful production of eri silk. In eri silkworm rearing, more than 80-85% of food is consumed by 4th and 5th instar larvae. Hence, substitution of Castor with Borpat leaves from 3rd instar onwards (late-stage rearing) or rearing with Borpat alone from brushing till spinning would make the sector sustainable and economically vibrant among the primary stakeholders. Further, Borpat is a forest-based tree, conservation and utilization of same in ericulture will also help in mitigation of deforestation and ill effects of climate changes.

Joshi, (1992) reared the eri silkworm on four experimental diets *viz.,* TT = First to fifth instar on tapioca, TC = First to third instar on tapioca and fourth to fifth instar on castor, CT = First to third instar on castor and fourth to fifth instar on tapioca and CC = First to fifth instar on castor. These studies were conducted in the laboratory with $26\pm2^{o}C$ temperature and $90\pm5\%$ relative humidity up to fourth instar and continuing rest of the development with the same temperature but with $60\pm5\%$ relative humidity. It was observed that an average larva ingested more amounts of dietary constituents (total sugars, total proteins and total lipids) on CC and TC dietary regime than on TT and CT. Similar pattern can be seen for co-efficient of apparent digestibility (CAD) values except total sugars, where the value is higher on TT (86.6 ± 0.02) than other three.

Scope of enhancing eri silk production by incorporating Phagostimulants in the diet of eri slik worm

The insects like all other heterotrophs must consume other organisms either plant or animal in order to acquire energy-rich molecules required by them. Efforts of a good number of researchers unravelled insect's dietary requirements as carbohydrates, proteins, lipids, nucleotides, vitamins and minerals besides water as essential component. Carbohydrate requirements include polysaccharides such as starch, glycogen, chitin, celluloses; oligosaccharides including sucrose, trehalose and simple sugars like glucose, galactose and fructose. Complex carbohydrates like chitin, celluloses are digested by enzymes secreted by symbiotic bacteria and protozoa living inside their digestive tract. Phytophagous insects particularly those of Lepidoptera must consume large volumes of plant material in order to extract enough proteins to meet their metabolic needs. Protease enzymes in insects' digestive system break down proteins into constituent amino acids which are in turn used by cells to build enzymes and hormones as well as proteins needed for muscle, egg yolk, ribosome, cuticle and many other purposes. As ten essential amino acids like lysine, tryptophan, histidine, phenylalanine, leucine, isoleucine, threonine, methionine, valine, and arginine that cannot be synthesised from other amino acids or similar chemical building blocks, must be supplemented in insects' feed. Proteins required by insects are actins, myosin, resilin, arthropodin, enzymes, hormones, ribosome and peptides like brain hormone, bursicon.

Lipids are important in the later larval stages for survival, reproduction and development (Nation, 2001 and Chapman, 1998). Lipid compounds for insect growth and development are fats, fatty acids and steroids. The fat molecules through enzymatic breakdown by lipase in the mid gut releases energy required by insect body for various life processes. Fatty acids also serve as building blocks for cuticle's waxes and the glandular synthesis of certain pheromones and defensive compounds. Insects appear to lack enzymes for metabolic pathway of steroid synthesis and they must obtain steroid compounds especially cholesterol, directly from their diet. Steroid building blocks are used to make hormones like ecdysteroids and growth factors. Insects feeding on steroid-deficient diets generally survive as immature but fail to moult properly into the adult stage. Lipids, phospholipids, triglycerides, sterols, sterol esters and fatty acids are usually important phagostimulants for many different insects (Nation, 2001). For nutrition, insect body also have the requirement of nucleic acids, purines, pyrimidines, mono-nucleotides and di-nucleotides.

In addition insects also acquire water, vitamins, and minerals from their food. Most insects are highly-adapted for water conservation and get most of the water they need directly from their food. Insects usually require seven vitamins like thiamine, riboflavin, nicotinic acid, pyridoxine, pantathenic acid, folic acid, and biotin, because they can't synthesize the vitamins. Insects have the dietary requirements of minerals in very small amount such as iron, copper, calcium, sodium, potassium ions, phosphorus, and sulphur. Given the known composition of insects, it is reasonable to assume that sodium, potassium, calcium, magnesium, chloride and phosphate are essential minerals for the function of insects (Nation, 2001). Many phytophagous insects need quite large amount of potassium and only trace amounts of sodium. Balance of nutrients may be important due to small body size and the stress placed on several physiological systems by having to deal with excesses while attempting to accumulate suboptimal levels of the major nutrients (Genc, 2006). Most phytophagous insects strongly regulate their nutrient uptake and in case of imbalanced diets, they employ regulatory rules that govern the extent to which nutrients occurring in excess or deficit are eaten.

It has been found by several researchers that some nutrients can have phagostimulatory effect themselves. Some nutrient chemicals can be synergised by other constituents that enhance their phagostimulatory effect well above that of nutrient phagostimulants. Perhaps the combined phagostimulatory effect is greater than that of nutrients in the plant. This effect could arise through the interaction of chemicals at the insects sensory receptors so that the sign stimulus dominates the information that the insect receives. So in the present context of studies, certain nutrient supplements fortified with insect phagostimulants (phagostimulant formulation H1) were attempted in growth and development of Muga silkworm, *Anthereae assamensis* Helfer.

There are no studies available on eri silkworm which could demonstrate the role of phagostimulants in enhanced food consumption and higher silk production. However, some information is available in case of Muga Silkworm (*Antheraea assamensis* Helfer). Researches conducted by Barman and Rama Krishna Rajan, (2011) demonstrated that supplemented nutrients with phagostimulants works synergistically. Muga silkworm larvae responded positively in their growth and development. All the supplemented nutrients ascorbic acid, proline and Na cyclamate responded positively and may be the constituents in a nutrient supplemented phagostimulant spray formulation in their best combination for Muga silkworm. The nutrient combination ascorbic acid-proline has strong synergistic response in larval tissue growth while ascorbic acid-Na cyclamate combination has synergistic effect on cocoon production e.g. ERR. Thus, ascorbic acid is an essential nutrient supplement for both body tissue growth and survivability. Motoyuki Sumida and Hiroki Ueda, (2007) found that dietary sucrose functions as an effecter molecule to midgut cells in the 5th instar *B. mori* larvae and as an output response, produces suppressed sucrease protein. The highest activity was observed with 3% sucrose diet and it was higher than the control indicating apparent activation. Thus, there is need for studies in case of eri silk worm to evaluate different nutrients as well as different plant extracts as phago stimulants which enhances food consumption and enhanced production of better quality silk.

Feeding deterrents

Very few researchers have worked on this aspect pertaining to eri silkworm. Dutta *et al.,* (1986) worked on *Tithonia diversifolia,* commonly called as tree marigold or Mexican sunflower. Mexican sunflower is a woody herb or succulent shrub. The nutrients that were isolated from this wild sunflower plant like Tagithinin A and C and hispodolin were acted as potent feeding deterrents when evaluated against fourth instar larvae of eri silkworm. As per the reports of Bhattacharya *et al.,* (1995) Cynaropicrin isolated from the annual herb, *Tricholepis glaberrima* proved to be a potential feeding deterrent when tested against IV instar larvae of eri silkworm. This compound restrained appreciable feeding up to 40% up to two days from treatment. As per the laboratory results of Dutta *et al.,* (1989) 7-methonyl-coumarin isolated from *Eupatorium ayapana,* a tropical American shrub acts as a potential deterrent on *Samia c. ricini.* Latestly, Grimm and Somarriba, (2001) in their studies to test the suitability of physic nut, *Jatropha curcas* for rearing eri silkworm found that the seeds of Jatropha contain phorbol ester content that inhibit the growth of the eri silkworm during fifth instar stage.

Summary and Conclusions

➢ The word "Eri" is derived from the Sanskrit term "Eranda" which refers to the castor, the primary host plant of eri silkworm.

➢ Eri silkworm, *Samia cynthia ricini* Boisduval (Lepidoptera: Saturniidae) is a polyphagous insect and feeds on the leaves of several plants. Among various food plants, castor (*Ricinus communis* L, Euphorbiaceae), is the primary and most preferred host plant by this insect. Hence, it has been accepted for commercial rearing since time immemorial.

➤ Ericulture is more popular and being practiced in North-Eastern states of India. Of late it has spread to other non-traditional states of India like Andhra Pradesh, Telangana, Madhya Pradesh, Tamil Nadu, Karnataka, etc.

➤ Ericulture broadly comprises two inter-linked activities i.e. food plant cultivation and maintenance to feed eri silkworm (rearing).

➤ Castor, the major food plant of eri silk worm, can be grown on semi-arable, degraded and wastelands under rain fed condition. Castor can be grown as sole crop and it also fits in intercropping system.

➤ Though castor is cultivated in 13 states in India, only three states *viz.,* Gujarat, Rajasthan and Andhra Pradesh occupy the lions share.

➤ During *kharif* season, castor is mostly grown as rain fed crop. Further, it is a drought tolerant crop due to its' deep rooted nature and white waxy coating on different parts of the plant. In *Rabi* season, irrigation has to be given at an interval of 12-15 days from October to December and 8 days from January onwards, for reaping higher yields of castor.

➤ On an average, castor crop yielding about 20.0 qts/ha removes 80 kg nitrogen, 20 kg phosphorus and 35 kg potassium. Following full package of practices including improved variety/hybrid, thinning, timely weed control, proper nutrition, need-based irrigation and plant health management is a prerequisite for enhanced leaf production for eri silk worm rearing as much for obtaining higher seed yields. Usually by adopting recommended package of practices farmer can get 10-15 q of seeds/ha during *Kharif* and 23-25 q/ha during *Rabi* season.

➤ The eri silkworms exhibit differential response in the acceptance of castor varieties for feeding and show variation in the development. The nutritive value of host leaf depends on various factors such as storage duration, temperature, humidity, light, aeration, quality of leaves or shoots in a given space (density of storing) and many other conditions.

➤ Selection and cultivation of suitable castor variety is a prerequisite for a successful ericulture. With the advancement in varietal improvement, a number of varieties have been developed and tested for commercial cultivation and rearing of eri silkworm. The castor genotypes like PCH-111, PCS-262, PCH-222, GCH-4, DCS-85, DCS-9, Aruna, DCH-177, DCH-519, NBR-1, Acc.003, Acc.004, Abaro, Acc.208624, DCS-9, PCS-121, EB-16, J1, SL-1, 48-1, Bangalore local and red petiole variety were found highly suitable by various workers throughout the country for eri silkworm rearing. The eri silkworm reared on these varieties/hybrids registered significantly higher effective rate of rearing % (ERR), cocoon weight, cocoon yield, shell weight, shell yield. The larvae reared on these genotypes, recorded the highest larval weight and took the shortest duration to complete its larval period.

> In castor, the eri worms that were fed on healthy leaves completed their larval period (23 days) faster with robust and uniform growth compared to silkworms fed on leaves infected with spots and blight.

> Leaf storage had significant effect on nutrient contents of castor leaves. In an experiment conducted to know the effect of leaf storage on nutritive value of castor leaves inferred that they could be preserved covering with wet gunny cloth up to 12 hours without much deterioration of moisture for effective utilization in eri silkworm rearing.

> Tapioca, another important host of eri silkworm, is also an important tuber crop widely cultivated in several countries in South America, Africa and South East Asia including India. Tapioca is a perennial shrub with variously branched stems growing to a height of 3-4 m with very large leaves and long petioles. Tapioca varieties H-97 Co-1, H-1423, M4, H-2304, KM-94 and MVD_1 were better for eri silkworm rearing.

> Despite the fact that castor leaves serve as chief food for rearing of eri silk worms, the combination of castor and tapioca leaves could also be beneficially used for commercial rearing of eri silkworm.

> Host plants had significant effect on the compactness of cocoons. The cocoons spun by the eri silkworm fed on castor leaves were rated as more moderate followed by tapioca and Kesseru.

> At the time of Tapioca tuber harvesting, the senescent tapioca leaves could be successfully utilized by supplementing 10% jaggery solution and soya powder @ 10g/100g of leaves for rearing of eri silkworm which ensured considerable improvement in economic traits of the worms and silk yield besides supporting the livelihood of tapioca growers with assured additional income through ericulture

> Cultivating these crops solely for the purpose of eri silkworm rearing may not be economical and practical due to high cultivation cost. But when a portion of the total foliage can be harvested without affecting the main produce *viz.,* castor seed and tapioca tuber, it may be economical and can contribute towards gainful employment of women and economic independence of farmers.

➢ Experiments conducted in Punjab to test the feasibility of papaya, cauliflower, brinjal and potato for eri silkworm rearing shows that castor was the only host plant for the successful commercial rearing of eri silkworm.

➢ Physiological relationships among leaf area, yield contributing characters and oil content in castor revealed that defoliation influenced the number of seeds per plant and the seed oil content, but not the mean seed weight of a particular genotype. The loss of $1m^2$ of leaf area caused the reduction of 37.8 g of castor seed or 24.4 g of castor oil. However, it could be deduced that 25-30% of castor leaves can be utilized for eri silkworm rearing without affecting the seed production.

➢ Defoliation studies on cassava in Cambodia revealed that up to 80% trimming did not significantly lower tuber yields revealing vast scope in utilizing cassava leaves for eri silkworm rearing.

➢ Apart from castor and cassava, search for alternate perennial tree host plants was continued. Among sixty four plant species belonging to Euphorbiaceae, Araliaceae, Apocynaceae Simaroubiceae etc. families tested by various workers for their suitability for eri silkworm rearing, on twenty five plant species *viz.*, *Ailanthus altissima*, *Ailanthus excelsa*, *Ailanthus grandis*, *Ailanthus triphysa*, *Carica papaya*, *Cinnamomum cecidodaphne*, *Coriaria nepalensis*, *Evodia fraxinifolia*, *Gmelina arborea*, *Heteropanax fragrans*, *Terminalia catappa*, *Jatropha curcas*, *Jatropha multifida*, *Mannihot esculenta*, *Terminali arjuna*, *Sapium eugeniifolium*, *Sapium sebiferum*, *Spathodea companulata*, *Ficus bengalensis*, *Daucus* carota, *Leucaena leucocephala*, *Micania micranta*, *Zanthoxylum armatum*, *Zanthoxylum limonella*, *Zizyphus mauritiana* eri silkworm could complete its life cycle. Hence these can be considered as secondary hosts for eri silkworm.

➢ In an experiment conducted to find the suitable alternate hosts for eri silkworm stated that *Terminalia catappa* and *Leucaena leucocephala* were recorded as alternative host plants in Tamil Nadu for mature worms.

➢ Feeding of fifth instar eri silk worm on castor, tapioca and barkesseru during fifth instar increased the activity of digestive and haemolymph amylases steeply from first day to fourth day of fifth instar. Thereafter their activity declined towards the onset of spinning. The activity of both the enzymes had been positively correlated with food consumption, digestion, approximate digestibility, mature larval weight, cocoon weight, pupal weight, shell ratio of eri silk worm.

➤ Digestive and haemolymph amylases established negative correlation with consumption index, growth rate, efficiency of conversion of ingested food, efficiency of conversion of digested food and larval duration.

➤ Sequential feeding of eri silkworm on different host plants revealed that when larvae were fed with castor leaves during early larval stages (1st to 3rd instar) followed by tapioca leaves during late larval stages (4th and 5th instar) led to higher ingestion of dietary constituents (total sugars, total proteins and total lipids) and resulted in desirable rearing parameters like larval weight, larval duration, ERR, cocoon weight and shell weight.

➤ Studies on exploiting Insect Growth stimulants in other plants to enhance silk production by eri silkworm showed that Japanese weed *Mikania micrantha* either alone or as fortification to castor leaves as 40% aqueous extracts increased the appetite of the larvae leading to better consumption and good growth of the eri silkworm. Dusting of 20% *Cassia sericea* on 48 hr old fifth instar eri silkworm larvae or castor leaves extra foliated with 10% aqueous extracts of *Amaranthus spinosus* significantly improved the growth characters of eri silk worm.

➤ Some nutrients can have phagostimulatory effect themselves while some nutrient chemicals can be synergised by other constituents to enhance their phagostimulatory effect. Supplemented nutrients ascorbic acid, proline and Na cyclamate in case of Muga Silkworm (*Antheraea assamensis* Helfer) and 3% sucrose solution in case of *Bombyx mori* have been observed to have phagostimulatory effect. Castor leaves extra foliated with 10% aqueous extracts of *Amaranthus spinosus* significantly improved larval and pupal weight of eri silk worm *Samia cynthia ricini* by 6.7 and 6.4% respectively over absolute control. Since the studies available on eri silkworm on this aspect are scanty, this can be a future research aspect on eri silk worm rearing.

➤ As per the few records available on feeding deterrents, the nutrients that were isolated from the wild sunflower plant, *Tithonia diversifolia viz.*, Tagithinin A and C and hispodolin and Cynaropicrin isolated from the annual herb, *Tricholepis glaberrima* were acted as potent feeding deterrents against fourth instar larvae of eri silkworm. 7-methonyl-coumarin isolated from *Eupatorium ayapana* and phorbol ester content of *Jatropha curcas* inhibited the growth of the eri silkworm during fifth instar stage.

References

Ahmadi A and Joudi M. 2007. Effects of timing and defoliation intensity on growth, yield and gas exchange rate of wheat grown under well-watered and drought conditions. *Pakistan Journal of Biological Sciences* **10**: 3794–3800.

Ahmed S A, Sarkar C R, Sarmah M C, Ahmed M and Singh N I. 2015. Rearing Performance and Reproductive Biology of Eri Silkworm, *Samia ricini* (Donovan) Feeding on *Ailanthus* Species and Other Promising Food Plants. *Advances in Biological Research* **9**(1): 07-14.

AOAC 1970. Methods of Analysis. Association of Official Agricultural Chemists. 9th Ed. Washington, D.C.

Arora G S and Gupta I J. 1979. Taxonomic studies of some of the Indian non-mulberry silk moths (Lepidoptera: Saturniidae). *Meories of Zoological Survey of India* **16**(1): 1-64.

Barman and Rama Krishna Rajan. 2011. Studies on Effects of Nutrient Supplements Fortified with Phagostimulants Formulation H1 on Growth and Development of indoor Reared *Antheraea assamensis* Helfer (Lepidoptera: Saturniidae). *International Journal of Biology* **3**(1): 167-173.

Basaiah J M M. 1988. Consumption and utilization of castor and tapioca by the eri silkworm. M.Sc. (Seri) Thesis, UAS, Bangalore Pp. 119.

Bhattacharya P R, Baria N C and Ghosh A C. 1995. Cynaropicrin from *Tricholepis glaberrima*: A potential insect feeding deterrent compound. *Industrial Crop and Products* **4**(4): 291-294.

Bindroo B B, Singh, Tiken N, Sahu A K and Chakravorty R. 2007. Eri silkworm host plants. *Indian Silk* **2**: 13-16.

Biswas N and Das P K. 2001. Effect of food plant species on rearing performance of eri silkworm, *Samia ricini* Donovan. *Bulletin of Indian Academy of Sericulture* **5**: 36-38.

Bordoloi S, Dutta P P and Handique P. 2005. Evaluation of castor varieties through bioassay and cocoon characters of eri silkworm (*Samia cynthia ricini* Boisduval). *New Agriculturist* **16**(1/2): 109-111.

Chakravorty R and Neog K. 2006. Food plants of eri silkworm (*Samia ricini* Donovan) their rearing performance and prospects for exploitation. Abstract of *National workshop on eri food plants* Oct 11-12 pp 1-27.

Chandrappa D, Sannappa B, Ramakrishna Naika, Manjunath K G and Govindan R. 2003. Differential performance of castor genotypes on seed yield and eri cocoon production and their economic analysis under rain fed conditions. *Global Journal of Biology, Agriculture and Health Sciences* **1**(2): 1-3.

Chandrappa D, Govindan R, Sannappa B and Venkataravana P. 2005.Correlation between foliar constituents of castor genotypes and economic traits of eri silkworm *Samia cynthia ricini* Boisduval. *Environment and Ecology* **23**(2): 356-359.

Chandrappa D. 2003. Performance of Eri silkworm on some castor genotypes and economics of Ericulture-cum-castor seed production. Ph D Thesis, UAS, Bangalore p. 146.

Chandrashekhar S and Govindan R. 2010. Studies on the performance of castor genotypes on rearing cocoon of eri silkworm. *International Journal of Plant Protection* **3**(2): 219-224.

Chandrashekhar S, Sannappa B, Manjunath K G and Govindan R. 2012. Efficacy of castor genotypes on bio-assay, cocoon and grainage traits of the domesticated vanya silkworm *Samia cynthia ricini* Boisduval. *Journal of Pharmacy Research* **5**(3): 1346-1349.

Chapman R F. 1998. The Insects, *Structure and Function*. 4th edition. Cambridge University Press, UK. 770pp.

Chowdhury S N. 1982. Eri silk industry. Directorate of Sericulture and Weaving, Govt. of Assam, Guwahati.

Damodaram T and Hegde D M. 2011. Oilseed Situation: A Statistical Compendium. Directorate of Oilseeds Research, Hyderabad.

Dasari Prasanna, Gurajala Bhargavo and Manjula. 2013. Evaluation of castor varieties based on the performance of eri silkworm *Samia cynthia ricini*. *International Journal of Biological & Pharmaceutical Research* **4**(12): 835-839.

Dutta P, Barua N C, Chaudhuri R P and Sharma R P. 1989. Insect feeding deterrent principle from *Eupatorium ayapana* Vent. *Journal of Research, Assam Agricultural University* **1**(2): 74-77.

Dayashankar K N. 1982. Performance of eri silkworm, *Samia cynthia ricini* Boisduval on different host plants and economics of rearing on castor under Dharwad conditions. M Sc. (Ag) Thesis, UAS, Bangalore P. 86.

Debaraj Y, Datta R N, Das P K and Benchamin K V. 2003. Eri silkworm crop improvement – A review. *Indian Journal of Sericulture* **41**: 100-105.

Deka M, Dutta S and Devi D. 2011. Impact of Feeding of *Samia cynthia ricini* Boisduval (red variety) (Lepidoptera: Saturniidae) in respect of larval growth and spinning. *International Journal of Pure and Applied Sciences and Technology* **5**(2): 131-140.

Deori G and Khanikor D P. 2015. Rearing performance of Eri silkworm (*Samia ricini* Boisd.) on *Ailanthus excelsa* and *Ailanthus grandis*. *Journal of Experimental Zoology* **18**(2): 839-841.

Deori G, Khanikor D P and Sarkar C R. 2016. Evaluation of Proximate composition of eri silkworm pupae on *Ailanthus grandis* and *Ailanthus excelsa* leaves. *Journal of Experimental Zoology* **19**(1): 515-516.

Devaiah M C, Govindan R and Thippeswamy C. 1984. Economics of rearing eri silkworm *Samia cynthia ricini* Boisduval on castor, *Ricinus communis* L. *Indian Journal of Sericulture* **23**: 117-120.

Devaiah M C, Rajashekhar Gouda R, Suhas Y and Govindan R. 1985. Growth and silk production in *Samia cynthia ricini* Boisduval fed on four different host plants. *Indian Journal of Sericulture* **24**: 33-35.

Devi M and Kalita J. 2009. Use of Japanese weed (*Mikania micrantha*) as food plant of eri silkworm (*Samia ricini* Donovan). *Environment and Ecology* **27**(4): 1635-1637.

Dharma Reddy K and Lakshmi Narayanamma V. 2016. Ericulture – An additional income generation source for the resource poor farmers of South India. *Journal of Insect Science* **29**(1): 88-92.

Dinesh-Hans and Sundaramoorthy. 2002. Optimization and defoliation tolerance evaluation of four released castor varieties in semi-arid farming. *Plant-Archives* **2**(2): 285-288.

Dookia B R. 1980. Varied silk ratio in cocoons of eri silkworm (*Philosamia ricini* Hutt.) reared on different castor varieties in Rajasthan. *Indian Journal of Sericulture* **19**: 38-40.

Dutta P, Bhattacharya P R, Radha L C, Bordoloi D N, Barua N C, Chowdhury P K, Sharma R P and Barua J N. 1986. Feeding deterents of *Philosamia ricini* from *Tithonia diversifolia*. *Phytoparasitica* **14**(1): 77-80.

Dutta L C, Kalita P and Phukan S N. 2015. Effect of leaf storage on nutritional changes in food plants and its impact on larval growth and cocoon quality of eri silkworm, *Samia ricini* Boisd. *Journal of Applied Zoological Researches* **26**(1): 57-61.

Dutta P P and Khanikor D P. 2005. Rearing performance of eri silkworm (*Samia ricini*) with interchange of food plants in different seasons. *Bulletin of Indian Academy of Sericulture* **9**(2): 31-35.

Dutta L C. 2000. Effcet of castor varieties on growth, nutrition and cocoon characters of eri silkworm, *Samia cynthia ricini* Boisduval. *Ph D Thesis*, Department of Sericulture, Assam Agricultural University, Jorhat.

Egli D B, Wiralaga R A and Ramseur E L. 1987. Variation in seed size in soybean. *Agronomy Journal* **79**: 463–467.

Egli D B. 2006. The role of seed in the determination of yield of grain crops. *Australian Journal of Agricultural Research* **57**: 1237–1247.

El-Shaarawy M F, Gomma A A and El-Garthy A T 1975 Chemical determination and utilization of dietary constituents of two castor bean varieties by larvae of eri silkworm. *Zeitschrift fuer Angewandte Entomologie* **78**: 171-176.

FAO 1987. Manual on Sericulture. Central Silk Board, Bangalore.

Friend W G. 1958. Nutritional requirements of phytophagous insects. *Annual Review of Entomology* **3**: 57-74.

Fukuda T V, Higushi Y and Mastuda M 1961. Artificial food for eri silkworm raising. *Agricultural and Biological Chemistry* 25: 417-420.

Fukuda T. 1963. A Semi-synthetic diet for Eri-silkworm Raising. *Agricultural and Biological Chemistry* 27(9): 601-609.

Genc H. 2006. General principles of insect nutritional ecology. *Trakya University Journal Science* **7**(1): 53-57.

Ghose C C. 1949. Silk production and weaving in India. Council of Scientific and Industrial Research, Memoir.

Gogoi 1998. Studies on certain aspects of wild Eri silkworm with special reference to its rearing performance. *Sericologia* **39**(3): 463-468.

Govindan R, Sannappa B, Bharathi V P, Singh M P and Hegde D M. 2006. Correlation coefficients between nutritive value of castor varieties and bioassay parameters of eri silkworm. *Bulletin of Indian Academy of Sericulture* **10**(2): 15-19.

Govindan R, Devaiah M C and Rangaswamy H R. 1978. Effect of different food plants on the growth of fifth instar larvae and economic traits of cocoons of *Philosamia ricini* Hutt. *Proceedings of All India Symposium on Sericulture* October 23-26, Bangalore.

Govindan R, Sannappa B, Bharathi V P and Ramakrishna Naika. 2002. Influence of castor genotypes on economic traits of eri silkworm. *Insect Environment* **8**: 72-73.

Govindan R, Sannappa B, Bharathi V P, Singh M P and Hegde D M. 2003. Status of major and secondary nutrients in leaves of castor varieties at varied cultivation practices in Karnataka. *Karnataka Journal of Agricultural Sciences* **16**(4): 565-569.

Grimm C and Somarriba A 2001. Suitability of physic nut, *Jatropha curcas* L., leaves for rearing eri silkworm, *Samia cynthia ricini* (Boisd.). *Forests, Trees and Livelihoods* **11**(1): 3-12.

Hay F R, Smith R D, Ellis R H and Butler L H. 2010. Developmental changes in the germinability, desiccation tolerance, hard seeded ness, and longevity of individual seeds of *Trifolium ambiguum*. *Annals of Botany* **105**: 1035–1052.

Hazarika M, Dutta L C, Bhattacharjee J and Patgiri P. 2015. Nutritional efficiency of eri silkworm, *Samia ricini* Boisd. reared on four different varieties of Tapioca. *Journal of Experimental Zoology* **18**(2): 831-833.

Hazarika P K and Hazarika L K. 1996. Effcet of castor varieties on performance of eri silkworm. *Indian Journal of Entomology* **58**(4): 284-290.

Hazarika U, Barah A, Phukan J D and Benchamin K V. 2003. Studies on the effect of different food plants and seasons on the larval development and cocoon characteristics of silkworm *Samia cynthia ricini* Boisduval. *Bulletin of Indian Academy of Sericulture* **7**(1): 77-85.

Hoque A A M M and Khan A R.1990. Development of the eri silkworm, *Samia cynthia ricini* (Boisd) on castor-papaya leaves combinations. *Annals of Entomology* **8**(1): 15-18.

Huq S B, Haque M A and Khan A B. 1991. Rearing Performances of Eri-silkworm, *Philosamia ricini* Hutt. (Lepidoptera: Saturniidae) on Five Different Host Plants. *Bangladesh Journal of Agricultural Sciences* **18**(2): 169-171.

Jaya Prakash P, Jaikishan Singh R S, Sanjeeva Rao B V and Suryanarayana N. 2006. Rearing performance of Eri silkworm (*Samia cynthia ricini* Donovan) on castor and tapioca in non-traditional areas. Paper presented in the National workshop on Eri food plant, Central Muga and Eri Research & Training Institute, Central Silk Board, Lahdoigarh (Assam), Guwahati, October 11-12, 2006, Pp. 38-49.

Jaya Prakash P, Jaikishan Singh R S, Sanjeeva Rao B V, Vijay Kumar M and Suryanarayana N. 2008. Economic viability of Eri silkworm rearing on rain fed castor and cassava crops in Andhra Pradesh. *Indian Journal of Sericulture* **47**(1): 7-11.

Jaya Prakash P, Sasikala A and Suryanarayana N. 2005. Ericulture – A tool for rural women development in Andhra Pradesh, India. Proceedings of XX Congress of International Sericulture Commission, Vol. II(3), International Sericulture Commission, France and Central Silk Board, Bangalore, India, December 15-18, 2005, pp. 106-113.

Jayaraj S. 2004, Eri culture in Tamil Nadu, Kerala and Pondichery states: Some experiences. In: *Proce. Workshop Prospects Dev. Ericulture,* Karnataka, UAS, Dharwad, pp: 10-29.

Jayaramaiah M and Sannappa B. 2000a. Influence of castor genotypes on rearing performance of different eri silkworm breeds. *International Journal on Wild Silkmoth & Silk* **5**: 29-31.

Jayaramaiah M and Sannappa B. 2000b. Correlation co-efficients between foliar constituents of castor genotypes and economic parameters of the eri silkworm, *Samia cynthia ricini* Boisduval (Lepidoptera: Saturniidae). *International Journal on Wild Silkmoth & Silk* **5**: 162-164.

Jolly M S and Krishnaswami S 1964. Sericin content in the cocoons of indigenous silkworm races (*Bombyx mori*). *Indian Journal of Sericulture* **3**: 17-18.

Joshi K L and Misra S D. 1982. Silk percentage and effective rate of rearing of eri silkworm, *Philosamia ricini* Hutt. (Lepidoptera: Saturniidae) reared on two host plants and their combinations. *Entomon* **7**: 107-110.

Joshi K L. 1987. Progression factor for growth in eri silkworm in relation to diet. *Indian Journal of Sericulture* **26**: 98-99.

Joshi K L. 1992. Utilization of dietary constituents (total sugars, total proteins and total lipids) by larvae of the Eri silkmoth, *Philosamia ricini* Hutt. (Lepidoptera: Saturniidae). *Indian Journal of Entomology* **54**(1): 62-65.

Kaleemurrahaman M and Gowri C. 1982. Foliar constituents of the food plants of eri silkworm (*Philosamia cynthia ricini*). *Proceedings of the National Academy of Sciences* India **48**: 349-353.

Kapil R P. 1967. Effects of feeding different host plants on the growth of larvae and weight of cocoons of *Philosamia ricini* Hutt. *Indian Journal of Entomology* **29**: 295-296.

Kavane R P. 2014a. *Carica Papaya* – A Potential host of *Philosamia ricini* eri silkworm under Western Maharashtra condition. *Indian Journal of Applied Research* **4**(11): 472-474.

Kavane R P. 2014b. *Terminalia catappa* – A Potential new host of *Philosamia ricini* eri silkworm under Western Maharashtra condition. *International Journal of Advanced Research* **2**(11): 433-438.

Kavane R P. 2015. *Terminalia arjuna* – A new host of *Philosamia ricini* eri silkworm under Western Maharashtra condition. *Indian Journal of Pharma and Bio Sciences* **6**(1): 787-792.

Kawabe K, Mihara M and Itagaki K. 2014. Changes in Cassava yields with Trimmed leaves for Eri-culture in Kampong Cham Province, Cambodia. *International Journal of Environmental and Rural Development* **5**(1): 182-187.

Kedir Shifa, Waktole Sori and Emana Getu. 2014. Feed utilization efficiency of eri-silkworm (*Samia cynthia ricini* Boisduval) (Lepidoptera: Saturniidae) on eight castor (*Ricinus communis* L.) genotypes. *International Journal of Innovative and Applied Research* **2**(4): 26-33.

Khanikor D P, Dutta S K and Khound J. 1997. Exploitation of Barkesseru, *A. excelsa* leaves as a substitute to castor leaves for rearing eri silkworm. *New Agriculturist* **8**(1): 7-12.

King R W, Wardlaw I F and Evans L T. 1967. Effect of assimilate utilization on photosynthetic rate in wheat. *Planta* **77**: 261-276.

Krishna Rao J V. 2003. Large scale development of ericulture in India. *Indian Silk* **42**(1): 27-29.

Krishnaswami S, Madhava Rao N R, Suryanarayana S K and Sundaramurthy T S. 1971. *Sericulture Manual-3* Silk Reeling, FAO, Rome Pp. 112.

Krishnaswami S. 1978. New Technology of Silkworm Rearing. *Bulletin of Centre for Sericulture Research and Training Institute* Mysore, pp. 1-24.

Kumar R, Gargi, Prasad D N and Saha L M. 1993. A Note on secondary food plants of Eri silkworm. *Indian Silk* **31**(9): 20-21.

Kumari S S, Narayanaswamy K C and Manjunath Gowda. 2009. Amylase activity in eri silkworm, *Samia cynthia ricini* Boisduval as influenced by host plants. *Indian Journal of Ecology* **36**(1): 71-74.

Lakshmamma P, Lakshminarayana M, Lakshmi Prayaga, Alivelu K and Lavanya C. 2009. Effect of defoliation on seed yield of castor (*Ricinus communis*). *Indian Journal of Agricultural Sciences* **79**(8): 620-623.

Lakshmamma P, Lakshmi Prayaga, Lakshminarayana M and Alivelu K. 2010. Source manipulation induced variation in dry matter partitioning to reproductive sink in Castor. *Indian Journal of Plant Physiology* **15**(3): 213-218.

Lakshmanan S and Geethadevi R G. 2005. Studies on economics of sericulture under dry farming conditions in Chamarajnagar district of Karnataka. *Indian Journal of Sericulture* **44**(2): 183-185.

Lakshmanan S, Mallikarjna B, Jayaram H, Ganapathy Rao R, Subramaniam M R, Geetha Devi R G and Datta R K. 1996. Economic issues of production of mulberry cocoon in Tamil Nadu – A Micro Economic Study. *Indian Journal of Sericulture* **35**(2): 128-131.

Lakshmi Narayanamma V and Dharma Reddy K. 2016. Evaluation of promising castor genotypes for eri silk culture. *Journal of Insect Science* **29**(1): 172-178.

Lakshmi Narayanamma V, Dharma Reddy K and Vishnuvardhan Reddy A. 2013. Castor genotypes on rearing and cocoon parameters of eri silkworm, *Samia cynthia ricini*. *Indian Journal of Plant Protection* **41**(2): 127-131.

Lakshmi Narayanamma V, Dharma Reddy K and Vishnuvardhan Reddy A. 2014. Identification of promising castor genotypes for rearing eri silkworm, *Samia cynthia ricini*. *Indian Journal of Plant Protection* **42**(2): 135-140.

Larcher W. 2003. Physiological Plant Ecology, 4th ed. Springer-Verlag, Berlin, Pp. 513.

Legay J M. 1958. Recent advances in silkworm nutrition. *Annual Review of Entomology* **3**: 75-86.

Maynard A L and Loosli L. 1962. Animal Nutrition, 5th edn. *McGraw Hill Book Co.*, Inc., New York, pp: 533.

Misra S D. 1999. ARUNA variety, a unique host for the eri silkworm. Part I. Two revenues from the same resource. *Indian Journal of Sericulture* **38**(2): 95-101.

Misra S D. 2001. Aruna variety, a unique host for the eri silkworm. Part 2. Ericulture for stressed ecosystem. *Indian Journal of Sericulture* **40**(1): 21-26.

Mitalee Baruah 2012a. Improvement of cocoon parameters of eri silkworm (*Philosamia ricini*) in their nutritional level during different rearing seasons. *Research Analysis and Evaluation* **5**(36): 38-39.

Mitalee Baruah 2012b. Studies on larval weight and shell ratio of Eri silkworm (*Philosamia ricini*) on castor, kesseru and treated kesseru by foliar spray. *IJCAES Special issue on basic, Applied & Social Sciences* **2**: 133-137.

Motoyuki Sumida and Hiroki Ueda. 2007. Dietary sucrose suppresses mid gut sucrease activity in Germfree, fifth instar larvae of the silkworm, *Bombyx mori*. *Journal of Insect Biotechnology and Sericology* **27**(1): 131-137.

Mukul Deka, Saranga Dutta and Dipali Devi. 2011. Impact of feeding of *Samia cynthia ricini* Boisduval (red variety) (Lepidoptera: Saturniidae) in respect of larval growth and spinning. *International Journal of Pure and Applied Sciences and Technology* **5**(2): 131-140.

Munshibasaiah J M, Reddy D N R, Vijayendra M and Narayanaswamy K C. 1997. Effect of host on the Fibroin and Sericin content of Eri silkworm cocoons. *Bulletin on Sericulture Research* **8**: 89-91.

Nagaveni V, Shree M P and Kumar K R. 2002. Effect of feeding eri silkworms with diseased castor leaves on the economic parameters of cocoon. *Indian Journal of Sericulture* **41**(2): 155-156.

Naik M and Murthy C. 2013. Evaluation of new host plant species for ericulture. *International Journal of Plant protection* **6**(2): 444-448.

Narayanaswamy K C, Saritha Kumari K, Manjunath Gowda and Amarnatha N. 2006. Barkesseru (*Ailanthus excelsa* Roxb.), an ideal substitute for castor to rear Eri silkworm, *Samia cynthia ricini* Boisduval. *Environment & Ecology* **24**(4): 720-725.

Nation J L. 2001. Insect Physiology and Biochemistry. *Boca Raton, Fla.*, CRC Press 485 pp.

Neelu Nangia, Jagadish P S and Nageshchandra B K. 2000. Evaluation of the volumetric attributes of the Eri silkworm reared on various host plants. *International Journal of Wild Silkmoth & Silk* **5**: 36-38.

Nokchensaba Kichu, Imtinaro L and Pankaj Neog 2015. Influence of host plants on rearing performance and cocoon characters of eri silkworm, *Samia cynthia ricini* Boisduval. *Indian Journal of Entomology* **77**(1): 88-91.

Pandey R K. 2003. Ericulture in Uttar Pradesh. *Indian Silk* **41**(12): 21-24.

Patel B S and Patel G M. 2009a. Effects of host plants on biology and economic traits of Eri silkworm, *Samia cynthia ricini* Boisduval. *Insect Environment* **14**(4): 172.

Patel B S and Patel G M. 2009b. Effects of rearing conditions on rearing of Eri silkworm, *Samia cynthia ricini* Boisduval. *Insect Environment* **14**(4): 171.

Patel B S and Patel G M. 2010 Food Preference of eri silkworm, *Samia cynthia ricini* Boisduval. *Insect Environment* **16**(1): 32.

Pathak A K. 1988. Studies on nutrition, growth and cocoon characters of eri silkworm, *Philosamia ricini* Hutt. fed on different varieties of leaves. *M. Sc Thesis*, Department of Sericulture, Assam Agricultural University, Jorhat.

Patil G M and Savanurmath C J. 1994. Eri silkworm – The poorman's friend. *Indian Silk* **33** (4): 41-45.

Patil R R, Sunita Kusugal and Ganga Ankad. 2009. Performance of eri silkworm, *Samia cynthia ricini* Boisd on few food plants. *Karnataka Journal of Agricultural Sciences* **22**(1): 220-221.

Patil G M, Kulkarni K A, Patil R K and Badiger K S. 2000. Performance of eri silkworm, *Samia cynthia ricini* Boisd. on different castor genotypes. *International Journal of Wild Silkmoth & Silk* **5**: 193-195.

Patil G M. 2004. Organically raised *Michelia champaca* L. a potential new host plant of eri silkworm, *Samia cynthia ricini* Biosduval. Abstract of the *National Seminar on prospects of organic sericulture and seri-byproduct utilization* pp. 39-52.

Preetirekha C, Rajesh Kumar and Khanikar D P. 2014. Host Plants relationship in terms of cocoon colour and compactness of Eri silkworm (*Samia ricini*). *Biological Forum – An International Journal* **6**(2): 340-343.

Rabinder Kaur and Virk J S. 2012. Evaluation of different plants for rearing of eri silkworm, *Samia cynthia ricini* under Punjab conditions. *Journal of Insect Science* (Ludhiana) **25**(3): 270-271.

Raghavaiah C V. 2003. Strategy for promotion of eri silk through utilization of castor (*Ricinus communis* L.) for foraging. *Indian Silk* **42**(1): 33-35.

Rajadurai S, Philip T and Shekhar M A. 2010. Seasonal Rearing performance of eri silkworm, *Samia cynthia ricini* (Boisduval) on castor and Tapioca under South Karnataka conditions. *Indian Journal of Sericulture* **49**(2): 134-137.

Rajashekhargouda R Patil, Sunita K and Ganga A. 2009. Performance of eri silkworm, *Samia cynthia ricini* Boisd on few food plants. *Karnataka Journal of Agricultural Sciences* **22**(1): 220-221.

Rajasri M and Lakshmi narayanamma V. 2015. Eri silkworm rearing on different castor genotypes and their economic analysis. *The Bioscan* **10**(2): 567-571

Rajesh Kumar, and Gangwar, S. K. 2010. Impact of varietal feeding on *Samia ricini* Donovan in spring and autumn season of Uttar Pradesh. *ARPN Journal of Agricultural Biological Sciences* **5**(3): 46-51.

Ramanjaneyulu A V, Vishnuvardhan Reddy A and Madhavi A. 2013. The impact of sowing date and irrigation regime on castor (*Ricinus communis* L.) seed yield, oil quality characteristics and fatty acid composition during post rainy season in South India. *Industrial Crops and Products* **44**: 25-31.

Ramesh T and Prasad M M K D. 2001. Effect of source manipulation on yield of castor *(Ricinus communis* L.) *Madras Agricultural Journal* **87**(10/12): 661-663.

Reddy D N R and Narayanaswamy K C. 1999. Effect of host on the consumption rate, leaf-cocoon and leaf egg ratio of Eri silkworm, *Samia cynthia ricini* Boisduval. *Entomon* **24**(1): 67-70.

Reddy D N R, Kotikal Y K and Vijayendra M. 1989. Development and silk yield of Eri silkworm, *Samia cynthia ricini* (Lepidoptera: Saturniidae) as influenced by the food plants. *Mysore Journal of Agricultural Sciences* **23**: 506-508.

Reddy Y V, Singh B G and Reddy S N. 1997. Growth analysis in castor. *Indian Journal of Plant Physiology* **2**: 87-89.

Sachan J N and Bajpai S P. 1973. Response of castor (*Ricinus communis* L.) varieties on growth and silk production of Eri silkworm *Philosamia ricini* Hutt. (Lepidoptera: Saturniidae). *Annals of Arid Zone* **11**: 112-115.

Sahu M N, Bhuyan and Das P K. 2006. Eri silkworm, *Samia ricini* (Lepidoptera: Saturniidae) Donovan, Seed production during summer in Assam. In: Proceeding of Regional seminar on *"Prospects and problems of sericulture as an economic enterprise in North West India"*, Dehradune, India, pp: 490-493.

Saikia P and Dutta L C.2013. Evaluation of *Ailanthus* species in relation to cocoon and yarn characters of eri silkworm, *Samia ricini* Boisd. *Journal of Applied Zoological Researches* **24**(2): 173-175.

Sakthivel N and Qadri S M H, Anbazhagan R, Krishnamoorthi T S and Jayaraj S. 2010. Cropping system of eri food plants for economically viable ericulture in Tamil Nadu. *Asian Textile Journal* **19**(4): 70-76.

Sakthivel N and Qadri S M H. 2013. Effect of fortification of senescent tapioca leaves on economic traits of Eri silkworm. *Indian Journal of Sericulture* **52**(1): 44-47.

Sakthivel N. 2016. Evaluation of cassava varieties for eri silkworm, *Samia cynthia ricini* Boisduval. *Munis Entomology & Zoology* **11**(1): 165-168.

Sannappa B and Jayaramaiah M. 1999a. Interaction of Castor genotypes and breeds of Eri silkworm in relation to cocoon and grainage parameters. *Mysore Journal of Agricultural Sciences* **33**: 214-223.

Sannappa B and Jayaramaiah M. 1999b. Mineral constituents of selected genotypes of castor, *Ricinus communis* L. *Mysore Journal of Agricultural Sciences* **33**: 297-300.

Sannappa B, Ramakrishna Naika, Govindan R and Subramanya G. 2007. Influence of some castor genotypes on larval, cocoon and grainage traits of eri silkworm (*Samia cynthia ricini* Biosduval). *International Journal of Agricultural Sciences* **3**(1): 139-141.

Sannappa B. 1997. Evaluation of castor genotypes for ericulture. *M. Sc. (Seri) Thesis*, University of Agricultural Sciences, Bangalore 167 pages.

Saratchandra B and Joshi K L. 1985. A comparative study of bunch and tray feeding in Ericulture using castor (*Ricinus communis*) and Kesseru (*Heteropanax fragrans*) as food plant. *Sericologia* **25**(1): 11

Sarkar B N, Sarmah M C and Giridhar K. 2015. Grainage performance of eri silkworm *Samia ricini* (Donovan) fed on different accession of castor food Plants. *International Journal of Ecology and Ecosolution* **2**(2): 17-21.

Sarmah M C, Ahmed S A, Sarkar B N, Debaraj Y and Singh L S. 2012. Seasonal variation in the commercial and economic characters of eri silkworm, *Samia ricini* (Donovan). *Munis Entomology & Zoology* **7**(2): 1268-1271.

Sarmah M C, Chutia M, Neog K. Das R, Rajkhowa G and Gogoi S N. 2011. Evaluation of promising castor genotype in terms of agronomical and yield attributing traits, biochemical properties, and rearing performance of eri silkworm, *Samia ricini* (Donovan). *Industrial Crops and Products* **34**(3): 1439-1446.

Sarmah M C, Datta R N, Das P K and Benchamin K V. 2002. Evaluation of certain castor genotypes for improving ericulture. *Indian Journal of Sericulture* **41**(1): 62-63.

Sarmah M C, Sarkar B N, Ahmed S A and Giridhar K. 2015. Performance of C_2 breed of eri silkworm, *Samia ricini* (Donovan) in different food plants. *Entomology and Applied Science Letters* **2**(1): 47-49.

Severino L S and Dick L A. 2013. Seed abortion and the individual weight of castor seed. *Industrial Crops and products* **49**: 890-896.

Severino L S, Maria A O Freire, Amanda M A Lucena, Leandro S Vale. 2010. Sequential defoliations influencing the development and yield components of castor plants (*Ricinus communis* L.). *Industrial Crops and products* **32**: 400-404.

Shaw C. 1998. Evaluation of *Ailanthus* species in relation to nutrition, growth and cocoon characters of eri silkworm, *Philosomia ricini* Hutt. *M.Sc. Thesis*, AAU, Jorhat.

Shree M P, Chandra M and Ravi Kumar K. 2000. Nutritional content of healthy and diseases leaves of castor and papaya, the food plants of the eri silk moth *Samia ricini* Donovan. *Bulletin of Indian Academy of Sericulture* **4**(2): 84-86.

Singh R, Handique P K and Sonowal P. 2014. Techno economic feasibility of ericulture in the plains of North East India. *Biological Forum- An International Journal* **6**(1): 44-47.

Singha T A, Dutta L C and Kalita P. 2013. Effect of leaf storage on nutritive value of castor, *Ricinus communis* Linn: The primary food of eri silkworm, *Samia ricini* Boisd. *Journal of Applied Zoological Researches* **24**(1): 55-57.

Sinha A K, Choudhury S K, Brahmachari B N and Sengupta K. 1986. Foliar constituents of the food plants of temperate tasar silkworm *Antheraea proylei*. *Indian Journal of Sericulture* **25**(1): 42-43.

Slansky F Jr and Scriber J M. 1984. Food consumption and utilization. In comprehensive Insect Physiology, Biochemistry and Pharmacology (eds. G A Kerkut and L I Gilbert) **4**: 87-16.

Subramanianan K, Sakthivel N and Qadri S M H. 2013. Rearing technology of eri silkworm (*Samia cynthia ricini*) under varied seasonal and host pant conditions in Tamil Nadu. *International Journal of Life Sciences Biotechnology and Pharma Research* **2**(2): 130-141.

Suryanarayana N, Das P K, Sahu a K, Sarma M C and Phukan J D. 2003. Recent advances in ericulture. *Indian Silk* **4**: 5-12.

Teotia R S, Sathyanarayana K, Rajashekar K, Goel R K and Krishna Rao J V. 2003. UNDP assistance in assistance of ericulture. *Indian Silk* **41**(14): 13-18.

Thangavelu K and Phukon J C. 1983. Food Preference of Eri Silkworm *Philosamia ricini* Hutt (Saturnidae: Lepidoptera). *Entomon* **8**: 311-315.

Venu N and Munirajappa. 2013. Impact of Independent and sequential feeding of different host pants on economic traits of Eri silkworm, *Philosamia ricini* Hutt. *International Journal of Science and Nature* **4**(1): 51-56.

Waldbauer G P. 1968. The consumption and utilization of food by insects. *Advances in Insect Physiology* **5**: 229-288.

CHAPTER 4

Artificial Diet for Rearing Eri Silkworm and Other Related Silkworms

Among the commercially exploited non-mulberry silkworms, the eri silkworm, Samia cynthia ricini Boisduval is the only species domesticated completely and adapted to indoor rearing all through the year. Rearing silkworms on leaves of host plants require more labour and the leaves used must be fresh enough to meet the preferences of silkworm. If low quality leaf is fed to early instars of silkworms there is risk of infection in chawki worms. The year-round rearing of eri silkworm on fresh leaves is not convenient, because majority of the food plants shed their leaves. To solve this problem, artificial diet was prepared for the silkworms. Rearing on artificial diet allows complete prevention of various pathogenic microorganisms infecting through food source. Farmers who rear silkworms can buy the diet instead of cultivating and need not have a field for themselves. Raising of insects on artificial food has been successful in a few cases, but raising of insects, on artificial food is not so easy and there are very few successful cases. This chapter discusses in detail about the Historical perspective of rearing eri silk worm on artificial diet, different methods of preparation of artificial diet, difficulties faced while preparation, modifications and consumption indices of eri silkworm on artificial diet etc.

Eri silkworm, *Samia cynthia ricini* Boisduval is a polyphagous species and eats fresh leaves of about fifty kinds of plants including the castor oil plant (Fukuda, 1963). Castor (*Ricinus communis* L.) is considered as a primary host of eri silkworm. The rearing of this insect became a problem under the scarcity of these host plants and further rearing silkworms on leaves require more labour. The leaves used must be fresh enough to meet the preferences of silkworm, and the feed has to be given to the silkworm three to four times a day. The year-round rearing of eri silkworm on fresh leaves is not possible, because the food plants like *Ricinus communis, Manihot esculenta* and *Ailanthus glandulosa* shed their leaves. If low quality leaf is fed to early instars of silkworms there is risk of infection in chawki worms. This is also the major constraint for eri cocoon productivity. Maximization of cocoon crop by improving chawki (early instars) rearing on a nutritive and balanced diet to reduce the mortality rate, *etc.*, is therefore necessary to develop an artificial diet for chawki rearing, *i.e.*, starter.

This diet should contain essential nutrients required for healthy and uniform growth of silkworms, as production of good quality cocoons is prompted by good and perfect chawki rearing. Rearing of eri silkworm on artificial diet allows complete prevention of various pathogenic microorganisms infecting through food source. Farmers who rear silkworms can buy the diet instead of cultivating host plants and need not have a field for themselves (Mangammal and Sri Devi, 2012). The main intention in the preparation of the semi-synthetic diet is to make it possible to rear eri silkworms in all the seasons with the omission of the cultivation of food plants such as *Ricinus communis* etc. One of the other purposes is to find a clue in preparation of a synthetic diet consisting of chemically defined nutrients such as amino acids, sugars, vitamins, mineral substances etc. which play an important role in studies on nutrition and silk production of the eri-silkworm (Fukuda *et al.*, 1961).

In Japan, the artificial diets are being used commercially for young age silkworm rearing in order to stabilize the crops and reduce the cost of rearing. In China, Lu and Chien (1979), improved the artificial diet and are using for mass rearing of *S. c. ricini*. However, the studies on artificial diet under tropical conditions are scanty. The awareness on the importance of artificial diet and its use is increasing of late even in tropics. Rearing of insects on artificial food has been successful in few cases, but the surviving of insects that eat usually green leaves, on the artificial food is not so easy and there are only few successful cases. However, Fukuda *et al.* (1960a and 1960b) have succeeded in rearing silkworm (*Bombyx mori* L.) on artificial diet (Mangammal *et al.*, 2013).

Historical perspective of rearing eri silk worm on artificial diet

There were no reports regarding the rearing of eri silkworm on artificial diet till 1961, when Fukuda and his co-workers prepared a combined diet, consisting of powder of dry leaves of castor-oil plant (*Ricinus communis*), sucrose, 'kinako' (a powder of parched soy bean), agar-agar, sodium dehydroacetate as an aseptic and water. They showed that this diet enabled hatchlings of the eri-silkworm (*Samia cynthia ricini*), one of the leaf-feeding insects, to grow to adult stage. The raising of eri silkworms on the combined diet was by no means inferior to the raising of the insect on fresh leaves of castor-oil plant, one of the proper foods for this insect, and was rather better in result on cocoon fibres produced and on eggs laid. An almost same result was also obtained by using a combined diet whose main constituents were the powder of dry leaves of *Ailanthus glandulosa*, one of the other host food plant for this insect (unpublished data) (Fukuda, 1963). Later in 1963 the same author modified the diet. It consists of kinako', sucrose, dry brewer yeast, β-sitosterol, L-ascorbic acid, agar-agar, sodium dehydroacetate as an aseptic and water. It does not contain any extracts from leaves of food plants such as *Ricinus communis* or *Ailanthus glandulosa*, not to mention the powder of dry leaves of these plants and named it as semi-synthetic diet.

Preparation of Semi-synthetic diet for rearing eri silk worm-the process and difficulties

Kinako', dry brewer yeast, ascorbic acid and cellulose powder were weighed out in a mortar, and the mixture was stirred well with a glass rod. Weighed agar agar and sucrose were dissolved by heating in the appropriate amount of the solution containing sodium dehydroacetate. As the agar-agar solution cooled below 65°C the 'kinako'-yeast-ascorbic acid-cellulose powder mixture was added, thoroughly stirred in, hardened at room temperature, and stored in a refrigerator at 2°C. Stored diets were renewed after every 4 days. The sodium dehydroacetate solution was prepared by dissolving 2g of sodium dehydroacetate in 1lit of water. One ml of this solution contains 2mg of sodium dehydroacetate. For the β-sitosterol, 9g of cellulose powder was added to the ether solution containing 1g of β-sitosterol, the mixture was stirred well with a glass rod, and the ether was absolutely removed from the mixture. One g of this dry mixture contains 100mg of β-sitosterol. Powdered agar-agar, β-sitosterol, dry brewer yeast, ascorbic acid and the 'kinako' were commercial ones respectively. The eri silkworms were raised on this diet as per the regular procedure. The raising of the eri-silkworm on the semi-synthetic diet was somewhat inferior in the points of larval raising, cocoon fibres produced and eggs laid down as compared to that on fresh leaves of *Ricinus communis* and *Ailanthus glandulosa* or on the combined diets whose main constituents were the powder of dry leaves of these plants. Therefore the authors thought it appropriate to improve the present semi-synthetic diet in future by altering its composition.

Trials to rear eri silk worm on semisynthetic diet without extracts from leaves of food plants

In attempting to formulate a semi-synthetic diet which does not contain any extracts from leaves of food plants such as *Ricinus communis*, *Ailanthus glandulosa* etc., it seemed that the initiation of feeding would involve very specific stimuli, as eri silk worm ate food plants of the limited range. However, a finding that eri-silkworms willingly ate the agar-based diet, named as basal diet, consisting of only 5 components of agar-agar, sucrose, β-sitosterol, cellulose powder and water gave an important suggestion not only for the host plant selection of this insect, but also in attempting to formulate a semi-synthetic diet. Namely, it seemed that host selection of this insect might be determined not only by the existence of specific stimuli in the plants, but also largely by the absence of repellent substances in plants. Thus agar-based diet stated above might prove a satisfactory starting point in attempting to formulate a semi-synthetic diet. No.5 diet (Table 4.1), a modification of the basal diet in whose composition appreciable amount of the 'kinako' and the brewer yeast were used as a protein and vitamin source to replace the removed cellulose powder, was at first prepared by Fukuda. This diet enabled isolated fifth instar moulted newly to grow to snatuied larva, but attempts to rear hatchlings on this diet were unsuccessful, all dying by the 4th instar. By supplying newly appreciable amount of ascorbic acid and much more quantities of 'kinako' to the No. 5 diet, based upon knowledge obtained in chemical analyses on the combined diet which enabled hatchlings to grow to adults, the No.11 diet (Table 4.1) was prepared. This diet enabled at last hatchlings to grow to adults. This No.11 diet was then named as a semi-synthetic diet for eri-silkworm raising. However, even this No11 diet had many faults which must be removed. With this No11 diet, the time from hatching to spinning a cocoon is prolonged, the weight of the cocoon fibre produced by one worm is less, the number of eggs laid by one moth is fewer, etc.

Table 4.1: The composition of artificial diets prepared for the rearing of Eri silkworm, *S. c. ricini* (Fukuda, 1963).

Substance	No. of diets											Combined diet
	1	2	3	4	5	6	7	8	9	10	11	
Agar-agar (g)	1.5	1.5	1.5	1.5	1.5	1.5	1.5	1.5	1.5	1.5	1.5	1.5
Kikano (g)	-	-	-	-	1.0	-	1.0	1.0	1.0	2.5	2.5	1.0
Sucrose (g)	1.0	-	1.0	-	1.0	1.0	-	1.0	1.0	1.0	1.0	1.0

Table 4.1: *Contd...*

Substance	No. of diets											Combined diet
	1	2	3	4	5	6	7	8	9	10	11	
β-Sitosterol (mg)	50	50	-	-	50	50	50	50	-	50	50	-
Dry brewer yeast (g)	-	-	-	-	1.0	1.0	1.0	-	1.0	1.0	1.0	-
L-ascorbic acid (mg)	-	-	-	-	-	-	-	-	-	-	25	-
Cellulose powder (g)	6.5	7.5	6.5	7.5	4.5	5.5	5.5	5.5	4.5	3.0	3.0	-
Sodium dehydroacetate (mg)	-	-	-	-	54	54	54	54	54	54	54	54
Water (ml)	27	27	27	27	27	27	27	27	27	27	27	27
Powder of dry leaves of *R. communis* or *A. glandulosa* (g)	-	-	-	-	-	-	-	-	-	-	-	5.5

Importance of palatability of the diet in rearing eri silk worm

In attempting to formulate a semi-synthetic diet for eri -silkworm raising, we must take also especially into consideration the palatability of the insect for diets prepared. As understood from some experimental results of Fukuda that eri-silkworms are not fond of the diets in which, equal weight of other protein source such as egg albumin etc. is used to replace 'kinako' in the diet. The reason is such replacement reduced the feeding of eri silkworm. Further, eri-silkworms lose their palatabilities for diets with higher content of the ascorbic acid (e.g. more than 5 times) or of the brewer yeast (e.g. more than twice) than present semi-synthetic diet. These results prove that the appetite of eri silkworm for nutrients is sensitive and peculiar. Therefore, the quality and the quantity of nutrients adopted in preparation of a semi-synthetic diet must naturally be limited in a narrow range. When we make every possible attempt to formulate a semi-synthetic diet, it needs to satisfy completely two demands: the diets prepared are nourishing, and they do not lose the appetite of the insect. The present semi-synthetic diet consists of 8 components of agar-agar, 'kinako', sucrose, brewer yeast, ascorbic acid, β-sitosterol, cellulose powder, and water. Seven components except β-sitosterol are necessary at least as constituents of the semi-synthetic diet which enables hatchlings to grow to adult stage. For the addition of β-sitosterol no definite conclusion could be arrived at as to whether it is necessary to add β-sitosterol to the diet or not. The reason isthe content of the sterols contained in the 'kinako' and the brewer yeast might be insufficient to the insect needs. But Fukuda could not venture into eliminating the β-sitosterol from the composition of the present semi-synthetic diet. The reason is that in successive investigations, β-sitosterol-added diet made sometimes better record than β-sitosterol-less diet in the eri-silkworm raising. Further, in another experiment using a chemically defined diet, β- sitosterol was shown to be essential for eri-silkworms.

Nutritional requirement of ascorbic acid to eri silk worm that could be revealed by rearing on semi synthetic diet

On the nutritional requirements of this insect, the present semi-synthetic diet could scarcely reveal except a few points, because the diet is not yet defined chemically. However, as ascorbic acid is scarcely contained in constituents such as the 'kinako', the brewer yeast, the agar-agar etc. of the semi synthetic diet prepared by Fukuda, there is no doubt about that this insect requires ascorbic acid. On the ascorbic acid requirement of insects, it seemed that hardly any of the insects studied until very recently was said to require ascorbic acid, because their food did not contain any. However, as artificial diets which support growth of phytophagous insects have developed, the importance of ascorbic acid which distributes universally among green plants was recognized especially in leaf-eating insects such as locusts, mulberry silkworm etc. It is at present postulated by many scientists that ascorbic acid will induce feeding by insects, that it prevents the oxidation of unprotected unsaturated fatty acids in diets, and that it has some nutritive importance. In mulberry silkworms that have close relations with eri-silkworms, it has recently been postulated that ascorbic acid not only has some nutritive importance, but also induces remarkably feeding by this insect. However, in the case of eri-silkworms the effects of ascorbic acid to induce feeding was scarcely recognized, and the vitamin seemed only to be metabolic importance. Furthermore, some experiments using the basal diet and the others indicated that sucrose stimulated strongly feeding by the eri-silkworm, no less than other many insects, in addition to its metabolic importance.

Successful attempt by Fukuda in rearing eri silk worm on a practical scale on semi synthetic diet without any extracts from leaves of food plants

The current study by Fukuda emphasizes a fact that it is possible to rear eri-silkworms on an artificial food, named as a semi-synthetic diet, which does not contain any extracts from leaves of the food plants, such as

Ricinus communis and *Ailanthus glandulosa*, proper foods for this insect, not to mention the dry leaf powder of these plants. The composition of the semi-synthetic diet devised was as follows: Agar-agar 1.5g, β-Sitosterol 50mg, 'Kinako' (a powder of parched soy bean), 2.5g, Ascorbic acid 25mg, Sucrose 1.0g, Sodium dehydroacetate as an aseptic 54mg, Dry brewer yeast 1.0g, Water 27ml, Cellulose powder 3.0g. The present semi-synthetic diet enabled 48 out of 100 hatchlings to spin their cocoons, but it has yet many faults which should in future be improved; i.e., the time from hatching to spinning a cocoon is prolonged, the weight of the cocoon fibre produced by one worm is less, and the number of eggs laid down by one moth is fewer. In developing a semi-synthetic diet ascorbic acid was shown to be essential for eri silkworms, and it was also found that sucrose stimulated feeding by this insect in addition to its metabolic importance. The improved semi-synthetic diet paves way for making it possible to rear eri-silkworms in all the seasons with the omission of the cultivation of food plants, such as *Ricinus communis* etc.

Attempts by other workers to rear eri silk worm on artificial diets

Later Kaleemurrahman and Muthukrishnan, (1981) prepared the artificial diet for *Samia cynthia ricini* (Boisd.) from 6 g cassava (tapioca) leaf powder, 1 g parched soya bean powder, 1 g agar agar, traces of streptomycin sulfate and 50 ml distilled water. The results shows that the larvae fed on the diet from the 1st instar onwards or from the 3rd instar onwards did not differ much from larvae fed on tapioca leaves in body size, cocoon weight or duration of larval development, but their body colour remained white throughout all stages. Adults emerging from the cocoons were normal in size and laid viable eggs, although one set of females (treated in the 3rd instar) laid only 85.8 eggs on average as compared with over 120 eggs for larvae that were reared on tapioca leaves or on the diet from the 1st instar onwards.

Hosny *et al.,* (1986) evaluated different varieties of castor, for their usage in semi-synthetic diet for the eri silkworm, *S. c. ricini* (Boisd). Laboratory experiments were carried out to determine the most favourable variety of castor leaves (*Ricinus communis*) for use as a powder in synthetic diets for eri silk worm. Larval duration, mortality and the weights of larvae, pupae, fresh cocoons and cocoon cortices for both sexes and the number of eggs deposited per female were evaluated and found that the most favourable source of castor leaf powder in the diets were those from the red variety.

Abdel-Khalek, (2002a) investigated the effect of artificial diet on certain biological aspects, fecundity and silk production of the eri silkworm. The highest larval and pupal weights, heaviest silk gland, maximum weights of fresh and dry cocoons and cocoon shell as well as high cocoon ratio were recorded after rearing the larvae on the artificial diet suggested by Kaleemurrahman and Gouri, (1982) after replacing tapioca leaf powder with castor leaf powder and adding sucrose and ascorbic acid (D5), followed by the same diet after replacing soya bean powder with bran powder (D1), and the same diet after adding casein, sucrose, and ascorbic acid (D6).

In another experiment, Abdel-Khalek, (2002b) fed the artificial diet to fourth and fifth larval instars of eri silk worm containing different components. The control diet was that recommended by Kaleemurrahman and Gouri, (1982) but the tapioca component was replaced with castor leaf powder. The larvae consumed 3.78 and 18.41 g of the control artificial diet during the fourth and fifth instars, respectively. The corresponding values for fresh castor bean leaves were 3.16 and 18.08 g, with no significant difference between the values for both types of food. Replacing the soya bean in the control diet with the same amount of bran caused an increase in the weight of food ingested by fifth instar larvae. The greatest amount of food consumed was recorded in the control diet after replacing soya bean with bran, the control diet after adding sucrose and ascorbic acid, and the control diet after adding sucrose, ascorbic acid, and casein. The same trend occurred for approximate digestibility also. The highest efficiency of converted food was recorded after feeding the larvae on fresh leaves; 19.20 and 27.93% were recorded for efficiency of conversion of ingested food to body matters (ECI) and efficiency of conversion of digested food to body matters (ECD) of the fifth instar larvae, respectively. Larvae fed on the control diet gave the lowest values of both ECI and ECD (Abdel-Khalek, 2002b).

Later in the year 1993, the diet was modified by Bhattacharya. The basic composition of this diet was commodity (soaked form) – 17.67g, yeast powder – 3.07g, sodium ascorbate – 0.31g, methyl-p-hydroxybenzoate- 0.31g, sorbic acid – 0.15g, agar-1.54g, formaldehyde (10%) – 0.15 ml and water – 76.80 ml. Parul Chaudhary and Neeta Gaur, (2012) tried to rear the eri silkworm on this diet. For this diet the 21 grains *viz.,* Bajra, Jau, Jowar, Maize, Barnyard millet, Rice, Wheat, Bengal gram, Black gram, Green garm, Lentil, Lobia, Pea, French bean, Red gram, Soybean, Castor, groundnut, Linseed, Mustard, Niger were added after soaking for 24 hours in water along with castor leaf. The results showed that out of the 21 diets tested, pupation was observed only in 10 diets (Table 4.2). On castor leaf pupation started at 24 days after feeding (DAF), while in rice, barnyard millet, green gram, black gram and Bengal gram at 26 days and in jowar it started at 28 DAF. Highest larval survival of 60.0% was recorded on green gram followed by barnyard millet (53.6%). Per cent pupation was highest on green gram-based diet (60.0%), while lowest was recorded on jowar based diet (13.33%). Mean cocoon weight was highest in green gram-based diet (2.89g), while it was lowest in red gram-based diet (1.3g). Mean shell weight and shell ratio was highest on green gram-based diet indicating its usefulness for the growth and development of this insect. Adult emergence did not occur in any of the diets. So, these diets are not suitable for breeding purpose of this insect. However, these diets can be used for rearing the larvae up to pupation under adverse conditions or scarcity of food and thus silk can be obtained from the cocoons.

Table 4.2: Development indices and economic parameters of eri silkworm on diets prepared with yeast powder by using different soaked stored commodities (Parul Chaudhary and Neeta Gaur, (2012).

Treatments	Larval period (Days)	Larval survival (%)	Pupation (%)	Mean cocoon weight (g)	Mean shell weight (g)	Shell ratio (%)
Jowar	28.00	23.30(28.78)*	13.33(21.14)	1.50	0.50	20.07(26.18)
Maize	27.33	43.30(40.71)	43.33(40.77)	2.22	1.13	49.34(44.68)
Rice	26.00	26.60(30.78)	26.60(30.78)	1.86	0.48	27.86(31.17)
Barnyard millet	26.00	53.60(53.06)	16.66(23.36)	2.01	0.48	17.44(24.24)
Bengal gram	26.33	33.30(35.21)	33.30(35.21)	2.16	0.38	35.82(36.43)
Black gram	26.00	40.00(39.14)	40.00(39.14)	2.20	0.67	32.53(33.75)
Green gram	26.00	60.00(51.14)	60.00(51.14)	2.89	0.90	71.24(59.86)
French bean	26.00	20.00 (22.14)	16.66(23.85)	1.35	0.64	52.37(46.67)
Red gram	26.00	16.60(19.92)	16.66(23.85)	1.30	0.40	31.37(33.82)
Linseed	26.00	56.60(49.22)	56.66(48.92)	2.47	1.14	45.00(41.53)
Castor leaf	24.00	86.60(72.28)	93.33(77.70)	2.55	1.20	46.03(42.13)
CD at 5%	0.97(0.63)	12.08(9.81)	18.17(12.73)	NS	NS	NS

Further modifications in the artificial diet for eri silk worm rearing:

So the artificial diet was further modified by Mangammal and Narayanaswamy, (2010). Leaves picked from castor plant were dried at not higher than 40°C under air-blast and then powdered to pass through 80- mesh sieve. The inhibitor solution was prepared by dissolving 0.1g of Vitamin K_3, 0.2g of Sodium Dehydroacetate and 0.2g Sodium Sorbate in 100ml of distilled water by heat. The inhibitor solution was used to prevent the artificial diet against the infection of micro-organisms. To the hot mixture containing the agar-agar, the sucrose, the inhibitor solution and the water, powdered leaves and the parched soybean were added. The mixture was stirred well and was hardened at room temperature. The composition of artificial diet is given in Table 4.3.

Table 4.3: Composition of Artificial Diet (Mangammal and Narayanaswamy, 2010)

Castor leaf powder	11.5g
Powder of parched soybean	1.0g
Sucrose	1.0g
Agar-agar	1.5g
Inhibitor solution	5ml
Distilled water	20ml

Consumption indices of eri silk worm reared on artificial diets

Quantitative work on artificial diets has usually involved only measurements of the amount of a particular nutrient required per unit of diet. This defines the relationship between the requirements for particular nutrients, but says nothing of food intake, absolute requirements or the efficiency of food utilization (House, 1962). Most measurements of intake and utilization have been made with insects feeding on natural foods. In this context, Mangammal and Narayanaswamy, (2010) conducted an experiment by rearing the eri silkworms with different treatments. The treatments were feeding the eri silkworm on artificial diet up to first instar + feeding on castor from second to fifth instar (T_1), feeding the eri silkworm on artificial diet up to second instar + feeding on castor from third to fifth instar (T_2), feeding the eri silkworm on artificial diet up to third instar + feeding on castor from fourth to fifth instar (T_3) and control i.e. feeding of eri silkworms on castor throughout the larval period by taking 50 larvae per replication.

The results showed that the food consumption and digestion were maximum when eri silkworms were reared on artificial diet up to first instar + castor from second to fifth instar, followed by the eri silkworms which were fed on artificial diet up to second and third instars + castor during remaining instars. However, the growth rate (GR) (growth rate explains how much weight has increased per day per gram of mean body weight during the stage), consumption index (CI), approximate digestibility (AD) (It is a precise measure of digestibility to evaluate the best food compared to the amount of food digested. It measures the digestible portion of the food that is ingested), efficiency of conversion of ingested food (ECI) (It is an overall measure of an insects ability to utilize for growth, the food which is digested. It mainly depend on approximate digestibility, which on one hand, converted into body substance and on the other hand metabolized for energy to maintain life) and efficiency of conversion of digested food (ECD) (It explains the ability of insect to utilize the portion of food that is digested for growth and development) of eri silkworms reared on artificial diet were almost similar to those of eri silkworms reared on castor leaves.

It is obvious from the results of Mangammal and Narayanaswamy (2010) that the eri silkworms fed on artificial diet during first instar + castor from second to fifth instar were able to consume and digest more food which in turn led to increased larval growth (Table 4.4). The consumption indices were better among the eri silkworms which derive more nutrition from artificial food during the early instars. Hence, artificial diet can be used for rearing of early stage eri silkworms for better productivity.

Table 4.4: Consumption indices of Eri silkworm, *Samia cynthia ricini* Boisduval as influenced by feeding on artificial diet (Mangammal and Narayanaswamy, 2010)

Treatments	Total during Larval period (g/50 larvae)		Mean of five instars				
	Consumption	Digestion	CI	GR	AD (%)	ECI (%)	ECD (%)
T_1	852.96	624.18	2.00	0.352	79.31	17.24	32.43
T_2	847.64	620.25	1.99	0.339	78.44	16.78	32.32
T_3	815.78	604.96	1.95	0.309	80.72	15.88	31.74
Control	853.18	626.30	2.14	0.375	79.01	17.37	33.23
CD at 5%	3.012	0.730	0.003	0.003	0.306	0.386	0.182

Influence of artificial diet on economic and grainage parameters of eri silk worm

Mangammal *et al.,* (2013) investigated on the influence of artificial diet on economic and grainage parameters of *Samia caynthia ricini* along with natural diet for comparison in the department of sericulture, UAS, GKVK, Bangalore with consecutive rearings conducted during September 2007 to October 2007, November 2007 to December 2007 and January 2008 to February 2008, respectively. The treatments were same as that of Mangammal and Narayanaswamy, (2010) experiment. In case of control, the worms were fed with castor leaves five times a day during first three instars at 6.00 AM, 11.00 AM, 2.00 PM, 6.00 PM and 10.00 PM with tender leaves. For the larvae of fourth instar onwards, feeding was given four times a day at 6.00 AM, 1.00 PM, 5.00 PM and 10.00 PM with coarse leaves. The eri silkworms fed with castor leaves were reared using plastic trays, while those on artificial diet were reared in vials. A small amount of diet was placed on the inner wall of the vial and ten newly emerged larvae (0-12h old) obtained from stock culture were placed with the help of a camel hair brush. Five replications were used in each treatment. Initially larvae were reared in groups but after 10 days, they were reared individually and diets were provided ad-libitum.

Observations on economic parameters such as cocoon weight, shell weight, cocoon shell ratio and silk productivity were recorded (Table 4.5). Grainage parameters like rate of pupation, pupal weight, rate of moth emergence, fecundity, weight of eggs and hatchability were also recorded (Table 4.6). The fecundity of the individual female was recorded by allowing five gravid female moths in four replications from each treatment to lay eggs in bamboo trays provided with a paper at the bottom, which were kept in the dark room. Up to two days of oviposition were considered for the calculation of fecundity. The economic parameters such as rate of cocooning, cocoon weight, shell weight, shell ratio, productivity were significantly maximum in eri silkworms fed with artificial diet up to first instar + castor up to fifth instar, which was followed by feeding of eri silkworms on artificial diet up to second instar + castor from up to fifth instar and feeding of eri silkworm on artificial diet up to third instar + castor during remaining instars. The grainage parameters like rate of moth emergence, fecundity, weight of eggs and hatchability were also maximum among the eri silkworms reared on artificial diet during first instar + castor leaves up to fifth instar compared to other treatments. The quality of cocoons obtained from eri silkworms fed on artificial diet was normal as those of castor fed eri silkworms. Hence, the artificial diet can be used for rearing eri silkworm up to third instar, as balanced nutrition for production of quality cocoons and eggs.

Table 4.5: Cocoon parameters of Eri silkworm, *Samia cynthia ricini* Boisduval as influenced by feeding on artificial diet (Mangammal *et al.,* 2013)

Treatments	Cocoon weight (g)	Rate of cocooning (%)	Shell weight (g)	Pupal weight (g)	Shell ratio (%)	Silk productivity (cg/day)
T_1	2.51	95.00	0.29	2.17	11.81	4.84
T_2	2.48	94.00	0.27	2.19	11.29	4.55
T_3	2.24	91.75	0.25	1.79	10.64	4.13
Control	2.62	96.00	0.32	2.28	12.02	5.25
CD at 5%	0.109	1.237	0.021	0.301	0.361	0.363

Table 4.6: Grainage parameters of Eri silkworm, *Samia cynthia ricini* Boisduval as influenced by feeding on artificial diet (Mangammal *et al.,* 2013)

Treatments	Rate of pupation (%)	Rate of moth emergence (%)	Fecundity (eggs / laying)	Weight of eggs (g/100 eggs)	Hatchability (%)
T_1	94.38	97.25	165.75	0.157	84.50
T_2	93.75	96.00	160.00	0.154	81.50
T_3	90.45	94.00	156.00	0.147	77.50
Control	95.60	98.25	171.50	0.159	88.00
CD at 5%	0.295	0.959	2.035	0.002	1.444

Further studies on the economic parameters and consumption indices of eri silkworm, on artificial diet

Mangammal and Sridevi, (2012) conducted an experiment to study the economic parameters and consumption indices of eri silkworm, *Samia c. ricini* on artificial diet in the Department of Sericulture, University of Agricultural Sciences, Gandhi Krishi Vignana Kendra, Bangalore. The treatments included are T_1- Feeding eri silkworms on artificial diet up to first instar + further feeding on castor leaves from second to fifth instar. T_2- Feeding eri silkworms on artificial diet up to second instar + further feeding on castor leaves from third to fifth instar. T_3- Feeding eri silkworms on artificial diet up to third instar + further feeding on castor leaves from fourth to fifth instar. T_4- Feeding eri silkworms on artificial diet up to fourth instar + castor during fifth instar. T_5- Feeding eri silkworms on artificial diet up to spinning. T_6 (Control)- Feeding eri silkworms on castor leaves throughout the larval period. The observations on larval weight, larval duration, moulting duration and larval mortality were recorded.

Larval weight: The results showed that the larval weight was significantly maximum in eri silkworms fed with artificial diet up to first instar which was followed by feeding of eri silkworms on artificial diet up to second instar + castor leaves up to fifth instar and feeding of eri silkworm on artificial diet up to third instar + castor during remaining instars. The variation in larval weight may be due to the difference in nutritional composition of the artificial diet and leaf. There was a gradual increase in larval weight from first to fourth instar and steep increase during fifth instar. This may be due to maximum food consumption and growth rate during fifth instar (Table 4.7).

Larval duration: The total larval duration was significantly shorter (20.78 days) when eri silkworms were reared on castor leaves during the entire larval period. The total larval duration was longer in case of artificial diet fed eri silkworms up to third instar + castor from fourth to fifth instar (22.51 days) followed by the larval duration of 21.76 and 21.34 days in eri silkworms provided with artificial diet during first and second instars, respectively + castor during remaining instars. There was no significant difference between the treatments and control with respect to larval duration during first three instars, but significant differences were noticed during latter two instars (Table 4.7). These results are in close agreement with the findings of Choudhuri, (2000) who reported that the eri silkworms reared on artificial diet containing castor leaf powder recorded longer larval duration over natural food plant, where the duration was slightly less.

Moulting duration: There was no significant difference between the treatments and control in respect of first moult duration. It was minimum in eri silkworms which received artificial diet up to third instar + castor from fourth to fifth instar (48.31 hours) followed by the moulting duration of 48.63 and 48.69 hours in eri silkworms reared on artificial diet up to second and first instars, respectively + castor during remaining instars (Table 4.8). In general, moulting duration was found to be higher when eri silkworms were reared on artificial diet and transferred to castor leaves. This may be due to sudden change in feeding habit, which may take some time for acclimatization for the larvae when they were transferred from artificial diet to leaves. The larval period was maximum in artificial diet fed eri silkworms compared to those fed on castor leaves because of increased moulting duration, so they were unable to moult simultaneously and were lagging behind. These results conform to the results of Okauchi, (1969) who reported that the larvae reared on artificial diet were lagging behind in growth, which can be made uniform by adding few drops of ecdysone to the diet. Choudhuri, (2000) also reported that eri silkworms reared on artificial diet containing castor leaf powder recorded longer larval duration over eri silkworms reared on natural food plant, where the duration was slightly less.

Larval mortality: The larval mortality was significantly low (2.56 and 1.88 per cent during first and second instar, respectively) in eri silkworms fed on artificial diet up to first instar + castor from second to fifth instar followed by eri silkworms fed on artificial diet up to second instar + castor from third to fifth instar (2.50, 1.93 and 1.38 per cent during first, second and third instar, respectively). The eri silkworms fed on artificial diet up to third instar + castor up to fifth instar had significantly higher mortality (2.43, 2.81, 4.88 and 1.55 per cent during first, second, third and fourth instar, respectively) (Table 4.9). However, there was no mortality in castor fed eri silkworms during all the instars. There was maximum mortality when eri silkworms were provided with artificial diet and when they were transferred to castor leaves the mortality was reduced and there was no mortality during subsequent instars. But the mortality was maximum when eri silkworms were fed on artificial diet upto fourth instar. There was severe mortality during fourth instar i.e., immediately after third moult. So the worms in the treatments provided with artificial diet up to fourth instar + castor during fifth instar and artificial diet throughout the larval period were unable to reach fifth instar and spin cocoons. Yanagawa *et al.,* (1991) also reported that the rate of survival was slightly lower when larvae were reared on artificial diet. The reason for the mortality of the larvae during first instar (when reared on artificial diet) is attributed to the composition of the diet. In most of the cases it is due to the factor preventing the feeding or the presence of growth inhibitors in the diet.

Table 4.7: Mature larval weight and total larval duration of Eri silkworm, *Samia cynthia ricini* Boisduval as influenced by feeding on artificial diet (Mangammal and Sridevi (2012).

Treatments	Mature Larval weight (g/50 larvae)	Total larval duration (days)
T_1	370.97	21.76
T_2	369.97	21.34
T_3	360.86	22.51
T_4	-	10.67
T_5	-	11.01
T_6 (Control)	375.75	20.78
CD at 5%	0.302	0.060

Table 4.8: Larval duration of different moults in Eri silkworm, *Samia cynthia ricini* Boisduval as influenced by feeding on artificial diet (Mangammal and Sridevi (2012).

Treatments	Moulting duration (hours)			
	I Moult	II Moult	III Moult	IV Moult
T_1	48.69	120.69	24.31	24.38
T_2	48.63	96.63	72.38	24.19
T_3	48.31	96.63	120.31	24.63
T_4	48.25	96.81	48.75	-
T_5	48.50	96.63	48.38	-
T_6 (Control)	48.25	24.25	24.25	24.19
CD at 5%	-	0.467	0.442	0.345

Table 4.9: Larval mortality in different instars of Eri silkworm, *Samia cynthia ricini* Boisduval as influenced by feeding on artificial diet (Mangammal and Sridevi (2012).

Treatments	Larval mortality (%)				
	I Instar	II Instar	III Instar	IV Instar	V Instar
T_1	2.56	1.18	-	-	-
T_2	2.50	1.93	1.38	-	-
T_3	2.43	2.81	4.88	1.55	-
T_4	2.63	2.38	4.88	89.76	-
T_5	2.49	2.69	4.48	90.00	-
T_6 (Control)	-	-	-	-	-
CD at 5%	0.556	0.697	0.522	0.474	-

As per the results of these experiments, feeding of eri silkworms on artificial diet should be restricted to the first and second instars only, covering chawki rearing. Usage of the diet in the later stages has certain limitations like difficulty in preservation and use i.e. maintenance of aseptic condition, besides the correct temperature and humidity to prevent the physical and chemical degradation of the diet (Sengupta, 1990).

Rearing performance of Muga silkworm on semi-synthetic diet:

Muga silkworm, *Antheraea assama* Westwood (Lepidoptera: Saturniidae) is semi-domesticated in nature and reared outdoor resulting in heavy crop loss right from brushing to spinning of cocoons especially during seed crop of summer and winter seasons. In case of exothermic insects, the environmental factors such as temperature, humidity, rainfall etc. play a commanding role in every stage. The attack of pests and parasites, diseases, abnormal changes of temperature, humidity, rainfall, wind in different seasons, agricultural chemicals and other pollutants like dust, smokes spilling from oil fields declines the production of muga silk. Shortage of food plants, non-availability of good seeds, lack of personnel to take up farming operations also reduce the production of muga silk. Rearing of insects on their natural hosts is time consuming due to regular change of food material and also involves man power (Siddiqui and Dey, 2002). Insects can often be reared throughout the year on artificial diet irrespective of food source and season economically in a limited space giving uniform specimens of known age whose nutrition and metabolism can be regulated and manipulated. Thus, insects with special traits can be developed (Singh, 1977). Indoor rearing on semi synthetic diet may contribute immensely in solving the problems of outdoor muga silkworm rearing and help in the process of domestication of the insect like that of mulberry silkworm, *B. mori*.

Preparation of semi-synthetic diet of muga silkworm

The ingredients taken for the preparation of semi-synthetic diet of muga silkworm were Som (*Persea bombycina* King, Lauraceae) leaf powder, soybean powder, sucrose, agar, cellulose, β-sitosterol, vitamins and antibiotics (Saikia and Hazarika, 2015). This diet was developed by Chatterjee, (D$_3$, 1994) (Table 4.9). The experiment was conducted with seven different semi-synthetic diets viz., MD$_1$, MD$_2$, MD$_3$, MD$_4$, MD$_5$, MD$_6$, MD$_7$ in the Entomology laboratory of Assam Agricultural University, Jorhat, Assam. These diets were formulated by modifying the diet D$_3$.

MD$_1$ diet: The MD$_1$ diet was prepared by adding Vitamin E (Tocopherol), 0.2 mg streptomycin, 125 mg β-sitosterol to D3. In this formulation amount of propionic acid is increased from 75 µl to 100 µl to reduce the fungal growth (Table 4.10).

At first, exact quantity of sucrose, soybean powder, streptomycin, propionic acid, β-sitosterol were weighed and then dissolved in 5.5 ml hot water and mixed thoroughly. Later, thiamine, pyridoxine, vitamin E, ascorbic acid, leaf powder and agar were added one after another along with 21.5 ml water. All these mixing were done in a 100 ml beaker. The lower portion of the beaker was submerged into hot water to prevent the agar from quick solidification. When all the ingredients were mixed thoroughly, the mixture was kept under room temperature for some time for cooling. The pH of the freshly prepared diet was measured by using a pH meter. The pH of MD$_1$ diet was recorded to be 4.20. The diet was kept at least for 24 hrs in refrigerator at 5-8°C before providing to silkworm. The remaining diets were prepared in the same procedure.

Table 4.10: Ingredients used to prepare different semi-synthetic diets of Muga silkworm, *A. assama* (Saikia and Hazarika, 2015).

Ingredients	Quantity							
	D_3 (Chatterjee, 1994)	MD_1	MD_2	MD_3	MD_4	MD_5	MD_6	MD_7
1. Leaf powder (g)	5	5	5	5	5	5	5	5
2. Soybean powder (g)	2.5	2.5	2.5	2.5	2.5	2.5	2.5	2.5
3. Sucrose (g)	1.5	1.5	1.5	1.5	1.5	1.5	1.5	1.5
4. Agar (g)	1.0	1.0	1.0	1.0	1.0	1.0	1.0	1.0
5. Cellulose powder (g)	1.0	1.0	1.0	1.0	1.0	1.0	1.0	1.0
6. Ascorbic acid (g)	0.15	0.15	0.15	0.15	0.15	0.10	0.20	0.25
7. Thiamine (mg)	0.2	0.2	0.2	0.2	0.2	0.2	0.2	0.2
8. Pyridoxine (mg)	0.2	0.2	0.2	0.2	0.2	0.2	0.2	0.2
9. Propionic acid (µl)	75	100	100	100	100	100	100	100
10. Vitamin E (capsule)	-	1/4	-	1/4	-	1/4	1/4	1/4
11. Multivitamin multimineral capsule	-	-	1	1	-	1	1	1
12. Streptomycin (mg)								
13. β-sitosterol (mg)	-	0.2	0.2	0.2	0.2	0.2	0.2	0.2
14. 30g methyl parahydroxy benzoate and 40 g sorbic	-	125	125	125	125	120	130	135
acid dissolved in 340 ml of 95% ethanol	-	-	-	1	-	1	1	1
15. Water (ml)	27	27	27	27	27	27	27	27

MD_2 diet: In this formulation all the ingredients were same with MD_1 diet, except in place of vitamin E, one capsule of multivitamin was added as an additional ingredient. Here the pH was found to be 4.21.

MD_3 diet: Into the MD_2 diet, one capsule of multivitamin along with vitamin E and 1 ml of methyl parahydroxy benzoate and sorbic acid solution were added. pH of the diet was 4.22.

MD_4 diet: In this diet, quantity of propionic acid and β-sitosterol added was different from that of D_3 diet formulation. In MD_4 diet pH was recorded as 4.24.

Preparation of MD_5, MD_6, MD_7 diets: These three diets were modified from that of MD_3 diet. Quantity of ascorbic acid and β-sitosterol added were different. These three diets were also prepared by following MD_3 diet preparation procedure. pH of these three diets *viz.*, MD_5, MD_6, MD_7 was recorded as 4.18, 4.29 and 4.35 respectively.

Rearing of Muga silkworm: In order to study the growth and development of muga silkworm on semi-synthetic diet, the newly hatched larvae were selected randomly and transferred separately on the semi-synthetic diet inside the culture tube. One drop of fresh Som leaf juice is placed in each diet except that of D_3 diet and provided with a sterile bamboo stick to help the insect during moulting and to give comfort while feeding. Mouths of these tubes were covered with a muslin cloth and tied by elastic. This was done to prevent the insect going outside. After this they were kept inside the incubator. The temperature and relative humidity of the incubator were maintained at 24°C and 85% respectively. The diet was replaced every alternate day by fresh diet to avoid any contamination. The rearings were carried out in three seasons *viz.*, post monsoon (October-November), pre monsoon (March-April) and monsoon season (April-May).

Larval duration: The results of the experiment show that, out of the seven semi-synthetic diets formulated, MD_3 diet supported the larval growth as well as pupation of *A. assama*. Those larvae reared on MD_3 completed 1st, 2nd and 3rd instar with larval duration of 7.02, 7.38 and 8.05 days respectively whereas, longer larval duration was observed in MD_7 diet (9.20 days for 1st instar) in D_3 and MD_2 diet (9.20 days for 2nd instar) and in D_3 diet (11.00 days for 3rd instar). In MD_1, MD_2, MD_4, MD_5, MD_6, MD_7 and D_3 larvae did not survive beyond the 3rd instar (Table 4.11). It may be the result of inadequacy of nutrients present there. An innovation was made to raise 1st and 2nd instars indoor on MD_3 diet and release them into the twigs and also to study their survivability etc. The third instar larvae when transferred from MD_3 diet to twigs of *P. bombycina* took 7.45 days, 8.21 days and 16.14 days to complete 3rd, 4th and 5th instar respectively. This attempt showed great promise and may contribute immensely for the growth of the industry.

In terms of nutritional requirements, MD_3 diet may be somewhat suitable and could support the growth of the muga silkworm. One of the constituent of the diet was Vitamin E. Vit. E is necessary for fertility, normal functioning of muscles and to prevent oxidation of vitamin A, fatty acids and sulphur (Rajendiran *et al.,* 1993). Streptomycin was used against bacterial contamination as the D_3 diet decomposes in a very short period. Sterols cannot be synthesized by insects having feeding stimulating action and required for normal growth, structural component of cell and for the synthesis of ecdysteroid, growth regulations and to improve dietary efficiency (Yungen, 2000). Larval duration was prolonged in the present experiment. Chatterjee, (1994) and Hazarika, (1995) also recorded prolonged larval duration of *A. assama* on artificial diet up to the fourth instar only, while Hazarika and Deka, (2002) could able to rear the silkworm up to pupal stage. The observed prolongation of larval period of *A. assama* was by 15 days on semi-synthetic diet than rearing on detached leaves.

Larval weight: The highest larval weight (0.020, 0.037, 0.346, 1.583, 4.164g in 1st, 2nd, 3rd, 4th and 5th instar respectively) was recorded in MD_3 diet whereas, the lowest larval weight of 0.012g in 1st instar was recorded in MD_4 and D_3 diet, 0.026g in second and 0.276g in 3rd instar in D_3 Diet while in larvae transferred from MD_3 to *P. bombycina* it was 0.378, 2.291 and 7.530g in 3rd, 4th and 5th instar respectively (Table 4.12). Chatterjee, (1994) and Hazarika, (1995) also found lower larval weight of *A. assama* on semi-synthetic diet. Thangavelu *et al.,* (1991) found difference in larval weight of 5th instar larva, while reared indoor compared to outdoor rearing. The decreased larval weight in indoor worms may be due to less accumulation of tissue matter in room condition, leaf moisture (Kataky and Hazarika, 1994; Bordoloi and Hazarika, 1998 and Hazarika *et al.,* 2006).

Effective rate of rearing (%): The ERR on MD_3 diet was recorded to be 3.16 % (Table 4.13). Chatterjee, (1994) and Hazarika, (1995) recorded 0% ERR of *A. assama* reared in artificial diet. Hazarika and Deka, (2002) stated that 43.5% *A. assama* larva survived on the semi-synthetic diet up to fifth instar. Trivedy *et al.,* (2003) found 98% survival of *B. mori* larvae upto second instar on semi-synthetic diet (Nutrid).

Table 4.11: Larval duration of *Antheraea assama* reared on semi-synthetic diet (Saikia and Hazarika, 2015).

Diet	Instars (Days)					Total larval duration (days)
	1st	2nd	3rd	4th	5th	
D_3	8.60	9.20	11.00	-	-	-
MD_1	8.60	8.80	9.00	-	-	-
MD_2	8.50	9.20	9.60	-	-	-
MD_3	7.02	7.38	8.05	11.70	17.61	51.32
MD_4	8.30	9.00	8.50	-	-	-
MD_5	7.50	7.80	8.33	-	-	-
MD_6	7.84	8.25	8.40	-	-	-
MD_7	9.20	8.60	10.00	-	-	-
CD at 5%	0.808	1.165	-	-	-	-

Table 4.12: Larval weight of *Antheraea assama* reared on semi-synthetic diet (Saikia and Hazarika, 2015).

Diet	Larval weight (g)				
	1st	2nd	3rd	4th	5th
D_3	0.012	0.024	0.271	-	-
MD_1	0.013	0.027	0.292	-	-
MD_2	0.015	0.029	0.309	-	-
MD_3	0.020	0.037	0.346	1.583	4.164
MD_4	0.012	0.026	0.276	-	-
MD_5	0.017	0.032	0.326	-	-
MD_6	0.019	0.031	0.304	-	-
MD_7	0.015	0.029	0.291	-	-
CD at 5%	0.001	0.0002	-	-	-

Table 4.13: Effective rate of rearing (%) of *Antheraea assama* reared on semi-synthetic diet (Saikia and Hazarika, 2015).

Diet	Instar wise survival (%)				
	1st	2nd	3rd	4th	5th
D_3	24.00	8.00	4.00	-	-
MD_1	32.00	12.00	8.00	-	-
MD_2	36.00	16.00	12.00	-	-
MD_3	69.97	53.33	41.00	35.33	3.16
MD_4	28.00	8.00	6.00	-	-
MD_5	40.00	28.00	12.00	-	-
MD_6	48.00	32.00	20.00	-	-
MD_7	36.00	20.00	10.00	-	-
MD_3 (larvae transferred from MD_3 diet to leaf after 2nd instar	-	-	55.00	50.00	5.00

The results of the Saikia and Hazarika (2015) attempt showed that, other than MD_3, no diet could support the larval growth of *A. assama*. All the nutrients like β-sitosterol and multivitamin-multimineral capsule in optimum quantity in the MD_3 diet that played a crucial role. Though the diets MD_5, MD_6, MD_7 were supplemented with multivitamin-multimineral capsule also, but may be due to balance of ascorbic acid (0.15 g) and β-sitosterol (125 mg) led MD_3 diet to complete larval development with shorter duration although other diets failed to complete larval development. The feeding rate on the semi-synthetic diets by the larvae was low possibly due to rapid loss of moisture from the diet which may affect the growth and development of silkworm. In MD_3 diet growth and development of *A. assama* was found somewhat better. In the study of Saikia and Hazarika (2015), D_3 (Chatterjee diet), MD_1, MD_2 and MD_4 diets showed fungal growth though 75 ml propionic acid was added in D_3 and 100 µl propionic acid was used in other three diets. These diets showed lower survival percentage compared to MD_3, MD_5, MD_6 and MD_7 diet. Lower survival may be due to the occasional unequal growth resulting from artificial diet. MD_3 diet was added with multivitamin, multimineral capsule along with 1 ml solution of 30 g methyl parahydroxy benzoate and 40g sorbic acid dissolved in 340 ml of 95 percent ethanol. Hence MD_3 diet recorded better performance compared to all other diets. So the results of the Saikia and Hazarika (2015) study showed that MD_3 diet is apparently promising and with further modification may help in the domestication of muga silkworm.

Neog *et al.,* (2011) studied the feeding habits of *Antheraea assamensis* and influence of four host plants *viz., Persea bombycina* King, *Litsea polhantha* Jussieu, *L. salicifolia* Roxburgh and *L. citrata* Blume and effect of chemical stimulants on the feeding behaviour in the muga silkworm. The results showed the nutritional superiority of young and medium leaves of *P. bombycina* with respect to soluble protein, total phenol and phenylalanine ammonia lyase activity was observed in the leaves compared to other host plants. Attraction and feeding tests with detached leaves and artificial diet with different chemical stimulants revealed that a mixture of the flavonoids, myrcetin, and trimethoxy dihydroxy flavone with sterol compound β-sitosterol elicited the most biting behaviour by *A. assamensis* larvae. While linalyl acetate alone attracted larvae towards the leaves of the host plants, a mixture of caryophyllene, decyl aldehyde and dodecyl aldehyde was found to both attract them to the host leaves and cause biting behavior. Azaindole was found to deter them from the host plants.

Rearing of Oak Tasar silkworm, *Antheraea proylei* J. on artificial diet:

Oak Tasar silkworm, *Antheraea proylei* J feeds naturally on *Quercus serrata* plant. However, Reeta Luikham *et al.,* (2010) made an attempt to formulate the artificial diet for *Antheraea proylei* by using the base diet of *A. yamamai*. The base diet was modified by using different stages of food plant (*Quercus serrata*) *viz.,* tender, medium, mature and mixed leaves and tested by feeding to *A. proylei* larvae in the laboratory maintaining the temperature at 28±1°C and 70-80% R.H. The result revealed that first instar larvae with 67.66% survived on the diet prepared with medium oak leaf as compared to 97.68% survived in control (fed on *Q. serrata* leaf) within five days. The second instar larvae fed with artificial diet survived with prolonged larval duration without entering to moult and ultimately died. However, when fed on natural *Q. serrata* from 2nd instar onwards, it spins cocoon which is at par with the control. The larvae fed on synthetic diet (without *Q. serrata* leaf) could not survive beyond 24 h of feeding. Hence, the artificial diet prepared with medium leaf (Apd-3) is found suitable for rearing *A. proylei* first instar larvae when there is scarcity of food plant leaves.

Rearing of Monovoltine strains of *B. mori* by altering artificial diet and mulberry leaf selections for higher silk productivity

One of the most important characteristics of the silkworm (*Bombyx mori* L.) (Lepidoptera: Bombycidae) is its ability to convert plant proteins to produce silk. The unique natural nourishment of *B. mori* is the mulberry leaf while, under controlled laboratory conditions, larvae can be reared also on artificial diets. In 2004 an original diet recipe was developed and patented by CRA-API (Cappellozza *et al.*, 2005). In contrast to mulberry leaf, artificial diet does not encounter any seasonal changing in its quality (Scriber and Slansky, 1981) and it is also exploitable in germ-free rearing systems (Sumida and Ueda, 2007). However, diet employment is currently limited to high-tech applications because it is quite expensive in comparison to mulberry leaf. Artificial diet, which represents pathogen-free pabulum, is preferred when the silkworm is used as a bioreactor to obtain recombinant proteins (Kato *et al.*, 2010 and Tatemastu *et al.*, 2012) or as a biological model (Hamamoto *et al.*, 2005 and 2009 and Kaito and Sekimizu, 2007). Therefore, in order to widen artificial diet utilization, it is necessary to obtain silkworm strains which are able to produce a large amount of silk with a high efficiency of food transformation into textile fibre, so that diet rearing can be cheap enough to be affordable. Further more this insect's characteristic should be retained when the selected *B. mori* strains are reared again on the mulberry leaf, so that multiple-purpose silkworms can be developed through one effort of selection only. To date selection processes based on nutritional indices have been carried out on tropical polyvoltine strains (Ramesha *et al.*, 2010 and 2012), while attempts in this direction have not been recently devoted to monovoltine strains, due to the progressive loss of the importance of sericulture in temperate countries in the last century.

Another reason for the scarce application of this technique of selection is because it is a very time demanding process which employs a lot of manual labour and is necessary to weigh the larvae, the administered leaf or the diet, the remaining food, and the excreta individually and on a daily basis. In addition, low heritability of nutritional indexes and lack of application of appropriate statistical tools for analysis of phenotypic data are further constraints (Seidavi, 2009). In fact, sericulture key-characters are under the concerted action of several genes (polygenic or quantitative traits) and non hereditary factors. Up-to-date advances in genetic marker-assisted selection have been made mostly for characters ruled by single genes, while for quantitative traits results are still at a very early stage (Esfandarani *et al.*, 2002). Therefore, a new attempt was planned by Saviane *et al.*, (2014) to rear the monovoltine strains of *B. mori* by altering artificial diet and mulberry leaf accelerates. Experiment was conducted with two strains viz., 129 and 124 reared on leaf and 129T10 and 124T10 on artificial diet. For each strain (129T10, 129, 124T10 and 124) 8 males and 8 females were reared individually both on mulberry leaf and artificial diet until cocoon spinning. The fresh weight of larvae, newly added food (leaf or diet), faeces and left-over food were recorded on a daily basis and used to calculate nutritional indices.

After three years of evaluation, the results showed that 129T10 expressed the best performance. On the whole, it showed the lowest ingesta (I) and consumption index (CI) values on both foods consistently for males and females. The conversion efficiency of ingesta to larva, cocoon and shell were the highest and the grams of ingesta necessary to produce one gram of cocoon and silk shell were the lowest, indicating that the larval enzymatic apparatus used to convert nutrients to body matter and, more specifically, to silk proteins was finely tuned (Rahmathulla *et al.*, 2004). On the other hand, 124T10 showed contrasting results on the two different foods. The ingesta value was the highest on artificial diet. However, when switched to mulberry leaves, its performances clearly changed and 124T10 showed the best I value and results related to CI were very similar to those of 129T10. Moreover, ECI to shell had also the same pattern, similar to 129 and 124 on the diet, to 129T10 on the mulberry leaf. Results suggest that this strategy can be used to select highly performing strains adapted to both foods and that selection on artificial diet indirectly ameliorates food conversion efficiency by larvae. Obtained pure lines can also be used to produce hybrids suitable for rearing on both leaves and diet (Saviane *et al.,* 2014).

Physiological comparison between mulberry leaves diet and artificial diet on growth and development of Thai silkworm, *Bombyx mori* L. var. *nangtui*:

Pimporn *et al.,* (2013) planned an experiment to explore the effect of a diet of mulberry leaves as compared to an artificial diet on Thai silkworm, *Bombyx mori* L. var. *nangtui* growth and development as well as the antibiotic therapeutic responses of bacterial infected silkworms. The 4th-instar silkworms with the same body weight were selected for the study. They were divided into two groups, each group containing 100 silkworms. One group was fed with mulberry leaves whereas the other was fed with the artificial diet. The silkworms fed with mulberry leaves as well as those with artificial diets were weighed and the numbers of survival were counted. The mulberry leaves- fed silkworms (MFS) showed faster growth rate (0.2 g body weight day^{-1}) than the artificial diet-fed silkworms (AFS). The MFS exhibited bigger body size and greater body weight than the AFS. It was found that the median lethal dose (LD50) of *Staphylococcus aureus* cells needed to kill 1-g body weight silkworms was 108 cells ml^{-1} for both MFS and AFS. However, the median effective dose (ED50) of the antibiotics (ampicillin, ceftriaxone, cephalexin, and gentamicin) for AFS was two times higher than those for MFS. The results indicate that some elements in mulberry leaf enhance the therapeutic effect of the antibiotics in the silkworm model. This study suggests that high attention on silkworm diet should be paid to the research relevant to drug therapeutic evaluation using silkworms as an animal model.

In this study, a diet of mulberry leaves showed many different effects on the silkworms as compared to an artificial diet. The first experiment demonstrates the effects on the silkworm growth, development and on the starving resistance. The silkworms fed with mulberry leaves showed faster growth than those fed with the artificial diet. Mulberry leaves have high (75 to 85%) *in vivo* digestible dried matters (Kandylis *et al.,* 2009). The leaves also contain sugars, proteins, fats, moisture, fibres and organic acids. The results of this experiment indicated that mulberry leaves contain more essential compounds suitable for the growth of silkworm as compared to an artificial diet. Bosquet, (1983) reported that food intake is not only related to the growth of the animal, but also acts as a developmental signal. From the results of the present experiment, Pimporn *et al.,* (2013) inferred that some compounds in mulberry leaves might act as developmental signals for the silkworms. During starvation, the survival numbers of those silkworms fed with artificial diet were greater than those of the ones fed with mulberry leaves. Silkworms store energy as glycogen, lipid and storage protein (Janarthanan *et al.,* 1999; Nagata and Kobayashi, 1990 and Horie and Inokuchi, 1978), which are obtained by conversion from food ingredients. The ingredients of mulberry leaves and those of the artificial diet are different. The artificial diet contains soybean protein as the amino acid source (Ito *et al.,* 1975; Escaffre and Kaushik, 1995), glucose as the sugar source (Kim and Jang, 2011) and soybean oil as the fatty acid source (Kandylis *et al.,* 2009). These differences might explain the higher growth rate of the AFS in comparison with the MFS. The ED 50/MIC (Minimum Inhibitory Concentration) ratio is well correlated to the pharmacokinetic feature of antibiotics. The ratio values of the silkworms fed with the artificial diet were more than two times greater than those fed with the mulberry leaves, suggesting that the ingredients in mulberry leaves affect the pharmacokinetics of the antibiotics in silkworms. Sun *et al.,* (2011) reported that mulberry leaves contain various compounds that can inhibit the activity of cytochrome P450, a major enzyme involved in drug metabolism. It is possible that the degradation of antibiotics by this enzyme in the MFS silkworm body was inhibited by the compounds in the mulberry leaves. This might have led to the sustained high drug concentration in the haemolymph of the silkworms and supported the survival of the silkworms against the injected bacteria, resulting in the reduction of their corresponding ED50 values. Zhou *et al.,* (2008) reported that different diets could alter the expression of proteins in relation to the immune system, digestion and absorption of nutrients, and energy metabolism in silkworms. The report suggests that the different diets, mulberry leaves and artificial diet, may affect the immunological resistance of silkworms to bacterial infection. Further studies of the chemical compounds in mulberry leaves responsible for lowering the effect of the ED50 of antibiotics are needed to establish Thai silkworm as a model animal for the evaluation of the therapeutic effects of antibiotics.

Summary and Conclusions

➤ The rearing of eri silkworm, *Samia cynthia ricini* Boisduval becomes a problem under the scarcity of host plants and requires more labour. Hence, to develop an artificial diet for rearing of at least early instars or "chawki rearing" is important. In Japan, the artificial diets are being used commercially for young age silkworm rearing in order to stabilize the crops and reduce the cost of rearing.

➢ Fukuda and his co-workers were first to prepare a combined diet, consisting of powder of dry leaves of castor-oil plant (*Ricinus communis*), sucrose, 'kinako' (a powder of parched soy bean), agar-agar, sodium dehydroacetate as an aseptic and water that enabled hatchlings of the eri-silkworm to grow to adult stage.

➢ Later they prepared a semi synthetic diet consisting of Kinako, dry brewer yeast, ascorbic acid, cellulose, agar agar, sucrose, β-sitosterol and sodium dehydroacetate. The raising of the eri-silkworm on the semi-synthetic diet was somewhat inferior to that on fresh leaves of *Ricinus communis* and *Ailanthus glandulosa* or on the combined diets whose main constituents were the powder of dry leaves of these plants

➢ By varying different constituents of the semi synthetic diet the palatability and nutritional requirements of eri silk worm could be assessed. Finally, Fukuda and his co-workers were successful in rearing eri silkworm on semi synthetic diet without any extracts from leaves of food plants from first instar larva up to adult stage.

➢ Among the attempts by other workers to rear eri silk worm on artificial diets, on the diets with cassava (tapioca) leaf powder, parched soya bean powder, agar agar, traces of streptomycin sulfate and distilled water the first instar larvae could be reared up to adult stage.

➢ In later attempts, tapioca leaf powder in the above diet was replaced with castor leaf powder along with addition of sucrose and ascorbic acid; or in the same diet soya bean powder was replaced with bran powder; and to the same diet casein, sucrose, and ascorbic acid were added. In all these cases first instar larvae could be reared up to adult stage.

➢ Other workers tried to rear eri silk worm on diet with basic composition as commodity (soaked form) – 17.67g, yeast powder – 3.07g, sodium ascorbate – 0.31g, methyl-p-hydroxybenzoate- 0.31g, sorbic acid – 0.15g, agar-1.54g, formaldehyde (10%) – 0.15 ml and water – 76.80 ml. By varying the commodity with 21 types of grains viz., Bajra, Jau, Jowar, Maize, Barnyard millet, Rice, Wheat, Bengal gram, Black gram, Green garm, Lentil, Lobia, Pea, French bean, Red gram, Soybean, Castor, groundnut, Linseed, Mustard, Niger, there was success up to pupation but not in adult emergence.

➢ Measurements of food intake and utilization of eri silk worm when reared on artificial diets revealed that food consumption and digestion were maximum when eri silkworms reared on artificial diet up to first instar + castor leaves from second to fifth instar.

➢ Influence of artificial diet on economic and grainage parameters of eri silk worm revealed that the economic parameters such as rate of cocooning, cocoon weight, shell weight, shell ratio, productivity were significantly maximum in eri silkworms fed with artificial diet up to first instar + castor up to fifth instar.

➢ Further studies on the economic parameters and consumption indices of eri silkworm, on artificial diet showed that feeding of eri silkworms on artificial diet should be restricted to the first and second instars only, covering chawki rearing. Later on the larvae should be shifted to natural diet consisting of leaves of host plants like castor.

➢ Rearing performance of Muga silkworm *Antheraea assamensis* Westwood on semi-synthetic diet consisting of Som (*Persea bombycina* King, Lauraceae) leaf powder, soybean powder, sucrose, agar, cellulose, β-sitosterol, vitamins and antibiotics showed that MD_3 diet recorded better performance compared to all other diets. In this diet (MD_3) multivitamin, multimineral capsule was added along with 1 ml solution of 30 g methyl parahydroxy benzoate and 40g sorbic acid dissolved in 340 ml of 95 per cent ethanol.

➢ Rearing of Oak Tasar silkworm, *Antheraea proylei* J on the artificial diet modified by using different stages of food plant (*Quercus serrata*) *viz.,* tender, medium, mature and mixed leaves, revealed that the larvae fed on synthetic diet (without *Q. serrata* leaf) could not survive beyond 24 h of feeding but when fed on natural *Q. serrata* from 2nd instar onwards, it spins cocoon which is at par with the control.

➢ Rearing of Monovoltine strains of *Bombyx mori* by altering artificial diet and mulberry leaf selections for higher silk productivity was conducted with two strains viz., 129 and 124 reared on leaf and 129T10 and 124T10 on artificial diet. After three years of evaluation, the results showed that 129T10 expressed the best performances.

➢ Physiological comparison of growth and development of Thai silkworm, *Bombyx mori* L. var. *nangtui* between mulberry leaves diet and artificial diet revealed that the silkworms fed with mulberry leaves showed faster growth than those fed with the artificial diet. Further, mulberry leaves and artificial diet, can influence the immunological resistance of silkworms to bacterial infection and curative efficiency of antibiotics used against bacterial diseases.

References

Abdel-Khalek A A. 2002a. Evaluation of some dietary components on certain biological aspects and digestive efficiency of eri silkworm I. *Biological studies. Annals of Agricultural Science (Cairo)* **47**(3): 1045-1054.

Abdel-Khalek A A. 2002b. Evaluation of some dietary components on certain biological aspects and digestive efficiency of eri silkworm II. Consumption and utilization of food. *Annals of Agricultural Science (Cairo)* **47**(3): 1055-1062.

Bordoloi S and Hazarika L K. 1998. Response of muga silkworm, *Antheraea assama* to host quality. *Entomon* **23**: 111-115.

Bosquet G. 1983. Nutritional and non-nutritional stimulation of protein synthesis in the fat body of *Bombyx mori. Insect Biochemistry* **13**: 281-288.

Cappellozza L, Cappellozza S, Saviane A and Sbrenna G. 2005. Artificial diet rearing system for the silkworm *Bombyx mori* (Lepidoptera: Bombycidae): Effect of vitamin C deprivation on larval growth and cocoon production. *Applied Entomology and Zoology* **40**(3): 405-412.

Chatterjee J. 1994. Formulation of artificial diet for muga silkworm (*Antheraea assama* Westwood). M.Sc. (Agri) Thesis, Assam Agricultural University, Jorhat.

Choudhuri C C. 2000. Rearing of eri silkworm, *Philosamia ricini* and *Philosamia cynthia ricini* on artificial diet. In: *National Conference on Strategies of Sericulture Research and Development* CSR&TI, Mysore, p. 94-95.

Escaffre A M and Kaushik S J. 1995. Survival and growth of first-feeding common carp larvae fed artificial diets containing soybean protein concentrate. *Aquaculture* **129**: 253.

Esfandarani M T, Bahreini R and Tajabadi N. 2002. Effect of mulberry leaves moisture on some traits of the silkworm (*Bombyx mori* L.). *Sericologia* **42**(2): 285-289.

Fukuda T V, Higushi Y and Mastuda M 1961. Artificial food for eri silkworm raising. *Agricultural and Biological Chemistry* **25**: 417-420.

Fukuda T, Higuchi Y and Matsuda M. 1960a. Artificial food for eri silkworm. *Indian Journal of Sericulture* **1**(1): 12-16.

Fukuda T, Suto M and Higuchi 1960b. Silkworm rearing on the artificial food. *Journal of Sericulture Science of Japan* **29**: 1-3.

Fukuda T. 1963. A Semi-synthetic diet for Eri-silkworm Raising. *Agricultural and Biological Cemistry* **27**(9): 601-609.

Hamamoto H, Kamura K, Razanajatovo I M, Murakami K, Santa T and Sekimizu K. 2005. Effects of molecular mass and hydrophobicity on transport rates through nonspecific pathways of the silkworm larva midgut. *International Journal of Antimicrobial Agents* **26**(1): 38-42.

Hamamoto H, Tonoike A, Narushima K, Horie R and Sekimizu K. 2009. Silkworm as a model animal to evaluate drug candidate toxicity and metabolism. *Comparative Biochemistry and Physiology C* **149**(3): 334-339.

Hazarika L K and Deka P C. 2002. Indoor rearing of muga silkworm, *Antheraea assama* on detached leaves and artificial diet. In: *Fourth International Conference on wild silk moths at Yogyakarta*, Indonesia, April 23-27.

Hazarika L K, Kataky A and Bhuyan M. 2006. A note on wild muga silkworm and indoor rearing of its counterpart. In: Muga (*Antheraea assama*) silkworm Biochemistry, Molecular Biology and Biotechnology to improve silk production (Ed. Unni B G), RRL, Jorhat, Assam, P.75-78.

Hazarika S. 1995. Effect of relative humidity on the survivability of muga silkworm (*Antheraea assama* Westwood) on artificial diet. M.Sc. (Agri) Thesis, Assam Agricultural University, Jorhat, Assam.

Horie Y and Inokuchi T. 1978. Protein synthesis and uric acid excretion in the absence of essential amino acids in the silkworm, *Bombyx mori*. *Insect Biochemistry* **8**: 251-254.

Hosny A, Mariy F M and Megalla A H. 1986.Evaluation of using different varieties of castor leaves in a semi-synthetic diet for the eri silkworm *Philosamia cynthia ricini* (Boisd). *Annals of Agricultural Science, Moshtohor* **24**(4): 2237-2247.

House H L. 1962. Insect Nutrition. *Annual Review of Biochemistry* **31**: 653-672.

Ito T, Horie Y and Nakasone S. 1975. Deterrent effect of soybean meal on feeding of the silkworm, *Bombyx mori*. *Journal of Insect Physiology* **21**: 995-1006.

Janarthanan S, Sankar S, Krisgnan M and Ignacimuthu S. 1999. Nutritional and hormonal effects on storage proteins in the silkworm, *Bombyx mori* L. *Journal of Bioscience* **24**: 53-57.

Kaito C and Sekimizu K. 2007. A silkworm model of pathogenic bacterial infection. *Drug discoveries & therapeutics* **1**(2): 89-93.

Kaleemurrahaman M and Gowri C. 1982. Foliar constituents of the food plants of eri silkworm (*Philosamia cynthia ricini*). *Proceedings of the National Academy of Sciences India* **48**: 349-353.

Kaleemurrahman, M and Muthukrishnan, T. S. 1981. Artificial diet for eri silkworm. *Science and Culture* **47**(11): 402-403.

Kandylis K, Hadjigeorgiou I and Harizanis P. 2009. The nutritive value of mulberry leaves (*Morus alba*) as a feed supplement for sheep. *Tropical Animal Health and Production* **41**: 17-24.

Kataky A and Hazarika L K. 1994. Indoor rearing of muga silkworm. *Annual Technical session of Assam Science Society*, Tejpur University, December, 24 Pp. 59-60.

Kato T, Kajikawa M, Maenaka K and Park E Y. 2010. Silkworm expression system as a platform technology in life science. *Applied microbiology and biotechnology* **85**(3): 459-470.

Kim G N and Jang H D. 2011. Flavonol content in the water extract of the mulberry (*Morus alba* L.) leaf and their antioxidant capacities. *Journal of Food Science* **76**: C869-C873.

Lu H S and Chien J F. 1979. Advances in research on the artificial diet for rearing eri-silkworm, *Philosamia cynthia ricini. Scientia Agricultura Sinica* **1**: 84-90.

Mangammal P and Sri Devi G 2012. Influence of artificial diet on larvae of eri silkworm, *Samia cynthia ricini* Boisduval. *Madras Agricultural Journal* **99**(4-6): 390-393.

Mangammal P, Narayanaswamy K C and Manjunatha Gowda. 2013. Influence artificial diet on economic parameters of the eri silkworm *Samia cynthia ricini* Boisduval. *Mysore Journal of Agricultural Sciences* **47**(3): 532-538.

Mangammal, P. Narayanaswamy, K. C. 2010. Consumption indices in eri silkworm, *Samia cynthia ricini* Boisduval as influenced by feeding on artificial diet. *Indian Journal of Ecology* **37**(1): 81-85.

Nagata M and Kobayashi J. 1990. Effects of nutrition on storage protein concentrations in the larval haemolymph of the silkworm, *Bombyx mori. Journal of Sericultural Science, Japan* **59**: 469-474.

Neog K, Unni B and Ahmed G. 2011. Studies on the influence of host plants and effect of chemical stimulants on the feeding behavior in the muga silkworm, *Antheraea assamensis. Journal of Insect Science (Madison)* **11**: 133.

Okauchi T. 1969. Recent study of phytoecdysone. *Insect Repellents* **38**: 140-156.

Parul Chaudhary and Neeta Gaur. 2012. Rearing of eri silkworm on artificial diets. *Pantnagar Journal of Research* **10**(2): 253-255.

Pimporn A, Hiroshi H, Kazuhisa S and Siriporn O. 2013. Physiological comparison between mulberry (*Morus alba* L.) leaves diet and artificial diet on growth development as well as antibiotic therapeutic response of silkworms. *Australian Journal of Crop Science* **7**(13): 2029-2035.

Rahmathulla V, Mathur V and Geetha Devi R. 2004. Growth and dietary efficiency of mulberry silkworm (*Bombyx mori* L.) under various nutritional and environmental stress conditions. *Philippine Journal of Science* **133**(1): 39-43.

Rajendiran P M, Himantharaj M T, Rajan R K, Veeranna G N, Meenal A and Datta R K. 1993. Role of vitamins in silkworm nutrition. *Indian Silk* **31**(11): 50-51.

Ramesha C, Anuradha C M, Lakshmi H, Sugnana K S, Seshagiri S V, Goel A K, Suresh K C and Kumari S S. 2010. Nutrigenetic traits analysis for identification of nutritionally efficient silkworm germplasm breeds. *Biotechnology* **9**(2): 131-140.

Ramesha C, Lakshmi H, Kumari S S, Anuradha C M and Kumar C S. 2012. Nutrigenetic screening strains of the mulberry silkworm, *Bombyx mori*, for nutritional efficiency. *Journal of Insect Science* **12**(15): 1-17.

Reeta Luikham, Keisa T J, Singh K C and Singh N I. 2010. Rearing of oak tasar silkworm, *Antheraea proylei* J. on artificial diet. *Bulletin of Indian Academy of Sericulture* **14**(1): 81-89.

Saikia M and Hazarika L K. 2015. Rearing performance of Muga silkworm *Antheraea assama* Westwood (Lepidoptera: Saturniidae) on semi-synthetic diet. *Journal of Experimental Zoology, India* **18**(2): 825-829.

Saviane A, Toso L, Righi C, Pavanello C, Crivellaro V and Cappellozza S. 2014. Rearing of monovoltine strains of *Bombyx mori* by alternating artificial diet and mulberry leaf accelerates selection for higher food conversion efficiency and silk productivity. *Bulletin of Insectology* **67**(2): 167-174.

Scriber J M and Slansky Jr F. 1981. The nutritional ecology of immature insects. *Annual Review of Entomology* **26**(1): 183-211.

Seidavi A. 2009. Determination and comparison of nutritional indices in commercial silkworm hybrids during various instars. *Asian Journal of Animal and Veterinary Advances* **4** (3): 104-113.

Sengupta K. 1990. Artificial diet for silkworm: Are we nearer the goal? *Indian Silk* **29**: 16-18.

Siddique K H and Dey D. 2002. Artificial diets – A tool for insects Mass rearing. Jyoti Publishers, EG-102. Inderpuri, New Delhi P. 209.

Singh P. 1977. Artificial diets for insects, Mites and Spiders. IFI/ Plenum Data Company, New York P. 594.

Sumida M and Ueda H. 2007. Dietary sucrose suppresses midgut sucrase activity in germfree, fifth instar larvae of the silkworm, *Bombyx mori*. *Journal of Insect Biotechnology and Sericology* **37**: 31-37.

Sun F, Shen L and Ma Z. 2011. Screening for ligands of human aromatase from mulberry (*Morus alba* L.) leaf by using high-performance liquid chromatography/tandem mass spectrometry. *Food Chemistry* **126**: 1337-1343.

Tatemastu K, Sezutsu H and Tamura T. 2012. Utilization of transgenic silkworms for recombinant protein production. *Journal of Biotechnology & Biomaterials* **9**:004.

Thangavelu K, Bajpai C M and Bania H R. 1991. Indoor rearing of tropical silkworm. *Indian Silk* **30**(6): 19-20.

Trivedy K, Nair K S, Ramesh M, Gopal N and Kumar S N. 2003. New semi-synthetic diet, "Nutrid"–A technology for rearing young instar silkworms in India. *Indian Journal of Sericulture* **42**: 158-161.

Yanagawa H, Watanabe K and Nakamura M. 1991. Application of feed ingredients for livestock diet by using polyphagous strains of the silkworm *Journal of Sericulture Science of Japan* **58**: 401-406.

Yungen M. 2000. Artificial diet rearing for the silkworm II. Nutritive components of artificial diets. *Bulletin of Indian Academy of Sericulture* **4**(2): 10-17.

Zhou Z H, Yang H J, Chen M, Lou C F, Zhang Y Z, Chen K P, Wang Y, Yu M L, Yu F, Li J Y and Zhong B X. 2008. Comparative Proteomic Analysis between the Domesticated Silkworm (*Bombyx mori*) reared on fresh mulberry leaves and on artificial diet. *Journal of Proteome Research* **7**: 5103-5111.

CHAPTER 5

Life Cycle and Rearing Technology of Eri Silkworm

Among the different vanya silk worms, the eri silkworm, Samia cynthia ricini (Boisduval) is highly domesticated and suitable for indoor rearing. Rearing of eri silkworm is an important and essential part of ericulture. Eri silkworm being a poikilothermic insect is very sensitive to the environmental conditions. For successful rearing of silkworm, rearing house, rearing stands, rearing trays (bamboo and plastic), chandrikas, leaf basket etc are required. Eri silkworm is a hardy race and less susceptible to diseases. Quality seed is the backbone of sericulture industry. The grainage technique of eri silkworm is quite different from that of other non-mulberry silkworms and domestication renders the grainage operation systematic and easy. This chapter discusses in detail about the eri silkworm rearing technology, methods of eri silkworm rearing, eri silkworm seed production, techniques of seed production, rearing performance of eri silkworm, progressive embryonic development of eri silkworm egg, evolving rearing technology for eri silkworm etc.

Ericulture, rearing of eri silkworm is a labour oriented, less investment, agrarian small-scale industry which suits both marginal and small land holders because of its high returns, short gestation period. Further it also creates opportunity for own family employment round the year. Ericulture serves as an important tool for rural reconstruction, benefiting the weaker sections of the society (Lakshmanan *et al.,* 1998). It also provides direct and indirect employment to a large number of families in rural areas and helps to up lift the socio-economic status of small and marginal farmers. Among the different vanya silk worms, the eri silkworm, *Samia cynthia ricini* (Boisduval) is highly domesticated and suitable for indoor rearing. So to rear the eri silkworms, we need a rearing house. A silkworm rearing house is a place where the silkworms are reared to produce the cocoons. Eri silkworm being a poikilothermic insect is very sensitive to the environmental conditions. The cocoon quality and yield are adversely affected if optimum environmental conditions i.e. temperature, relative humidity, ventilation, illumination, hygiene etc. are not provided to the silkworms. The rearing house should be selected such that it should have facilities for creation and maintenance of the optimal environmental conditions inside the house. Other equipment includes rearing stands, rearing trays (bamboo and plastic), chandrikas, leaf basket etc.

The various operations involved in silkworm rearing are leaf production, leaf harvest, disinfection, silkworm rearing, mounting, harvesting of cocoons, transportation etc. Unlike mulberry silkworm, eri silkworm is a hardy race, less susceptible to diseases, but it is inevitable to disinfect the rearing room and other appliances used for eri silkworm rearing, because the success purely depends on the effective disinfection (Plates 22-25).

Plate 22 Disinfected rearing room ready for rearing

Plate 23 Disinfected rearing stands and trays

Plate 24 Disinfected leaf basket ready to store the leaves

Plate 25 Disinfected chandrikas

Many disinfectants are in use *viz.,* formalin, bleaching powder, chlorine dioxide, decol etc (Venugopal *et al.,* 2012). Disinfection means destruction of disease causing microbes. Disinfection of rearing house, rearing appliances, silkworm rearing bed and maintenance of hygiene during rearing are the most essential activities for successful cocoon production. Traditional sericulturists use cow dung as paste on bamboo rearing trays and on mud floor of the silkworm rearing house. It enhances the durability of the bamboo tray and gives smoother surface to the floor. It was observed that the application of the cow dung paste assists in persistence of pathogens. The persistent pathogen forms secondary source of infection and contamination during silkworm rearing (Selvakumar *et al.,* 2007).

Disinfection and maintenance of hygienic condition during rearing are essential factors for preventing occurrence of diseases. Sodium hypochlorite acts as a leaf surface disinfectant. Sodium hypochlorite is effective against bacteria, viruses and fungi and it disinfects the same way as chlorine does. When sodium hypochlorite dissolves in water, two substances *viz.*, hypochlorous (HOCl) and the less active hypochlorite ion (OCl-) are formed, which play a role for oxidation and disinfection. The efficacy of Sodium hypochlorite in controlling mortality due to bacterial and viral diseases has been reported in tasar and muga silkworms (Sahay *et al.,* 2008; Singh *et al.,* 2014). The application of sodium hypochlorite solution to the foliage before the rearing and during the rearing reduces the mortality of eri silk worm due to bacterial and viral diseases thereby contributing in increasing the cocoon productivity (Goswami *et al.,* 2015).

Eri Silkworm Rearing Technology

Rearing of eri silkworm is an important and essential part of ericulture. It is generally reared indoor using either castor or kesseru leaves. Between the leaf fed to and the fabric produced by eri silkworm, there is a big interlinking of complex processes. In general silkworm rearing is divided into two main phases *viz.,* young age silkworm rearing and late age silkworm rearing depending upon the nutritional requirements and environmental conditions to be maintained during rearing. In both the phases of rearing, all the requirements are different and the techniques of rearing are also quite distinct. The first to third instar worms form the young age and the remaining two instars (fourth and fifth) form the late age worms.

Young Age Silkworm Rearing

Young age silkworm rearing is also known as "Chawki rearing." It generally starts from hatching of crawling insects called "larvae" from the eggs (Plates 26 and 27). The period of eating food prior to first moult is called first instar or stadium. Then the larva casts off its skin (ecdysis) and grows to the second stage. Again, it starts eating food leaves and enters in to the stage of second moult. In this way, it crosses second, third and fourth moulting stages (Plates 28 and 29). Thus the larva keeps on growing through five successive larval instars to reach the stage of cocoon formation. A good cocoon crop at the end of successful rearing is greatly influenced by the conditions of young worm rearing. Therefore, young age worms should be reared under good rearing conditions providing them with suitable and good quality leaves. Temperature from 26-28°C and relative humidity ranging 85-90 % are ideal conditions for young age rearing. Sufficient tender (glossy) and fresh leaves should be provided to young age larvae. Maximum care should be taken not to expose young age worms to extreme heat and cold. The young age worms are very delicate and should be handled with utmost care. In spite of normal care during young age worms rearing, the chances of loss of worms, particularly first and second instar are more than the grown up worms.

Plate 26 Eggs laid by the female moth

Plate 27 Chawki larvae hatched from the eggs

Plate 28 Process of moulting

Plate 29 Third instar larvae of eri silkworm

Late Age Silkworm Rearing

Late age worms consist of fourth and fifth instars which need preferably lower temperature and humidity during rearing conditions (Plates 30 and 31). These stages consume more quantum of food than the young age worms because the worm has not only to develop silk glands and to increase growth rate, but also has to store up the reserve food materials for the future stages like pupa and imago. Therefore, these stages should be provided with as much quality food as they require. The late age worms consume 80-85 % leaves supplied during the entire larval period. They attain significant growth during this stage. If the chawki rearing is conducted perfectly, it results in healthy and robust worms with less mortality and the late age worm rearing will be easy with more chances of a successful crop. But proper care by providing with sufficient food is essential to obtain the full potential of larval growth, maximum yield and best cocoon quality. The environmental and nutritional conditions required in late rearing are different from that of young age. The ideal conditions of temperature ranged 24-26°C and relative humidity ranged 70-80 % should be maintained during the rearing of late age worms. These stages should be provided with semi-matured leaves. The feeding of dried, yellow and diseased leaves deteriorates the health of the worms and even results in death due to diseases. Since 80-85 % of leaves are consumed by these stages, continuous supply of quality leaves after preservation in the leaf chamber needs to be ensured.

Plate 30 Fourth instar larvae of eri silkworm

Plate 31 Fifth instar larvae ready to spin cocoons

Maintenance of larvae

During the entire rearing period, the stage wise general maintenance of the leaves is required considering the following aspects a) Feeding and its frequency b) Bed cleaning c) Spacing of worms d) Handling of moulting worms and care e) Collection and destruction of weak, diseased and undernourished larvae.

Feeding and its frequency

The suitability of food plant leaves differs according to the period of larval growth. Without physical and biochemical knowledge, the leaf quality cannot be judged simply based on the position of the leaves on the plants. While plucking leaf from the same shoot, the softness and the degree of maturity may vary widely according to the position of the leaves. Therefore, it is desirable to pluck more than one leaf from one shoot which appeared to be suitable for the worms' particularly young ages. After collection, the leaves should be washed in water and preserved in the leaf preservation chamber covering with wet gunny cloth / bag all around. It is better to provide castor leaves without petiole in tray rearing. At least 4-5 feedings should be given per day including in the night at regular intervals during the young age rearing. In late age worms due to increase of feeding capacity, 5 feedings per day are essential. In the night time more than sufficient leaves should be provided to fulfil the required consumption throughout the night. It is advisable to prepare a routine feeding schedule and follow till the end of the rearing strictly.

Bed cleaning

As soon as the larvae grow-up, the unconsumed leaves and litter increase in the rearing bed which ultimately causes changing of atmosphere and favours multiplication of pathogenic organisms. Hence, timely bed cleaning is essential to keep the worms healthy. Frequent cleaning is better but it involves more labour and ultimately rendering silkworm rearing uneconomical. Therefore, it is necessary to prepare stage wise cleaning schedule. Only one cleaning is sufficient during first stage worms, otherwise the loss of worms will be more. In second stage, two times bed cleaning is required, after first moult and before second moult. It should be done after one or two feedings. Three bed cleanings should be resorted to the third and fourth stages, first after second moult, second in the middle and the third before third moult. But in fifth instar stage, the rate of food consumption increases substantially compared to previous instars. Hence, the bed becomes thick and damp soon. Therefore, it is necessary for daily bed cleaning in this stage, preferably in the morning after one or two feedings. The method of bed cleaning practiced in eri culture is simple and easy. Prior to bed cleaning, feeding should be given with new foliage. Then the worms along with the new foliage should be transferred to a new rearing tray carefully. Maximum care should be taken not to harm the worms during handling. This method of bed cleaning is therefore to be accomplished through experience.

Spacing of worms

The maintenance of optimum number of worms per unit area according to the size or stage of the worms during rearing is called spacing of worms. Proper spacing and good aeration keep the worms healthier. Overcrowding of the worms in the tray leads to competition for food and space and ultimately leads to undernourishment and unhealthy growth of the larvae often resulting in crop loss. Optimum spacing is, therefore, to be accomplished through experience. Spacing of the worms should be made combining with cleaning. Two times rearing space is sufficient from first stage to third stage. In fourth stage, it may be necessary to increase the space by two or three times and again in the fifth stage, two times space is required to increase than the previous stages. In a standard size rearing tray (1.0 m dia), 300 number of fifth stage worms can be reared conveniently.

Handling of moulting worms and care

A good rearing is judged by the uniformity of the larvae entering into moulting and emerging from moulting. The brushing and feeding of the worms play key role for uniform moulting, if handled properly. As the worms are entering to moult, they stop feeding, become lethargic with less movement. If 75-80 % of the worms enter into the moult, there is no need to feed the rest of the worms. But first feeding has to be provided only when 80 per cent of the worms emerge out of moult. The worms moult four times during its larval life span. Moulting is a very sensitive period during which the worms cast off its old skin and the body is soft and delicate. The larvae take 12-36 hours to complete the moulting process during the different instars and different seasons. It is important to keep the rearing bed dry when the worms are in moult. No bed cleaning should be done during moult.

Collection and destruction of weak, diseased, and undernourished larvae

During rearing, the weak, injured, diseased and irregular worms should be collected immediately and put in 2 per cent formalin solution. Such worms should be buried or burnt carefully to prevent spread of diseases. If rearing continues mixing with these worms, the losses due to contamination and disease spread will be high and even crop failure may take place.

Matured worm collection and mounting

After completion of larval life span, the matured 5th stage larvae discard their complete excreta consisting of liquid and semi-solid substances. Now the worms are ready for spinning cocoons. Before spinning, the worms stop feeding and become restlessly moving here and there to search a suitable place for cocooning. The matured worm produces a hollow sound when it is rubbed gently between fingers. This is the time for picking the ripe worms and putting them on mountages. Before mounting process, the required number of mountages should be kept ready well in time. The worms should be collected carefully for mounting. While cocooning, it is observed that the worms require at least two supporting sides. The size of the cocoon depends upon the size of the space available for cocooning. The rearing of the worms takes place during the day time till midday. The commonly used mountages are chandrikas, basket filled with dry leaves, jail (a bundle of dry leaves like mango, jack fruit, some ornamental plants etc.) and gunny bag filled with dry leaves. The leaves should not be completely dried, semi-dried leaves are suitable for easy spinning. After keeping the optimum number of worms in the respective mountages, it is covered by newspaper or cloth to make support and to create calm and semi-dark environment, a suitable condition for cocooning until it is complete. However, if disturbance occurs larvae stop spinning for a short period. Due to unavailability of suitable place and/or due to disturbance, the larvae spin defective cocoons and even fail to spin cocoons. In some cases, if the physiological condition inside the body is abnormal, the larvae often die without pupating. Mounting of immatured and over matured larvae results in poor cocoon quality. The quality of cocoon also depends upon the type of mountages, density of worms in mounting and different mounting methods / models. During spinning, temperature, relative humidity and aeration influence cocoon quality. The ideal condition for spinning is around 24-25°C temperature and 60-70 % relative humidity. While mounting, the optimum number of worms should be maintained per mountage is 300 worms per chandrika of 1.0 m diameter size.

Harvesting and assessment of cocoons

After completion of spinning, the larval skin is cast off and pupation takes place. The pupa has a thin cuticular skin which is soft and may get ruptured easily, if disturbed. The last and important step is harvesting of cocoons from the mountage in time. Cocoons should be harvested after 5-6 days of spinning in summer and 8-9 days in winter (Plate 32). The harvesting process is the best time for sorting of cocoons according to the quality. The cocoons should be sorted out into good, double, melted, stained, dead or inferior, cut or pierced cocoons. Good commercial cocoons should be shifted and dried perfectly after the harvest. Cocoons should be preserved carefully to protect from fungal infestation and attack from pest and predators. Cocoons should be assessed on the basis of cocoon weight, shell weight and silk ratio.

Plate 32 Cocoons spin by the eri silkworm larvae

Methods of Eri Silkworm Rearing

There are three methods of eri silkworm rearing, i.e. Bunch rearing, Tray rearing and platform rearing. The farmers of the North Eastern region generally employ bunch and tray rearing methods either singly or in combination together. The new recommendation of the R&D institutions is that the first three instars should be reared on trays, and the fourth and fifth instars on the bunch.

Bunch Rearing

In bunch rearing method, about 10-12 leaves (castor) or branches (of kesseru) are tied together to make a bundle and hung vertically on a horizontal bamboo/wire/string support. Then the worms are allowed to feed on the tied leaves. The foliage is changed by keeping fresh bunch near the exhausted one and the worms crawl over the new one. Just below the hanging bunches bamboo mat or tray is kept on the floor so that the worms which fall down are not contaminated with dust on the floor and can be picked-up and put on the bunches. Bunch rearing is simple and easy with minimum cost and also yields a better crop due to more hygienic condition. In this method, minimum manpower is utilized for bed cleaning, etc, but strict maintenance like timely replacement of old bunches is required. Besides, there is no soiling of the leaves due to excreta of the worms as these are fallen down directly beneath the bunch.

However, more worms cannot be accommodated on a bunch and a greater space is required for large scale rearing.

This bunch method of rearing is much cleaner, and can accommodate more worms in less space. In this method, minimum man power is utilized for bed cleaning and soiling of leaves due to mixing with excreta is prevented. However, strict requirements like timely replenishment of leaves, and frequent exposure of the worms leading to mechanical injury due to falling from the bunch, are the drawbacks of bunch method of rearing.

Tray Rearing

In tray rearing method, the worms are reared providing the leaves on the tray. Trays are made up of either bamboo or wood in different shapes and sizes. The shapes are round (bamboo made), square and rectangular (wooden) (Plate 33). The young age (I-III instars) silkworm rearing is conducted either in the wooden trays of 50 cm x 60 cm x 5 cm size or in bamboo tray (of 70 cm dia.). However, bamboo tray of size 1.0 m diameter is more convenient to rear 10-15 dfls (disease free layings) until 2nd / 3rd instar, while 600-700 worms can be reared up to 4th instar and 300 worms in the final instar which also provides sufficient space. Although the bunch rearing method has more advantages with more hygienic condition and less cost, the tray rearing facilitates more capacity of rearing by using rearing stand of 6-7 tier system with two trays per tier.

Plate 33 Rearing of eri silkworms by tray rearing method

This tray rearing method is generally adopted in most of the experimental rearings to maintain constant number of worms and also to facilitate sorting out of weak, irregular and deformed/ diseased worms.

Mechanical injuries are by and large avoided in this method of rearing. However, this method requires maximum space and appliances. Further, beds are often soiled with the excreta of worms if proper and timely bed cleaning is not done. Such conditions often attract fungal/ bacterial/ viral infections leading to loss of crops due to outbreak of diseases.

Platform Rearing

In addition to the above two methods of traditional eri silkworm rearing, platform rearing for the last two stages is done by giving individual leaf of castor or kesseru shoots. It is generally recommended for mass scale rearing (not less than 100 dfls) of late age eri silkworms. The platform can be prepared using wood or bamboo and its shelves can be arranged in two tiers with an interval of 30 inches or in three tiers with an interval of 27 inches in between the tiers. The rearing seat of the shelves is prepared by using nylon mesh or bamboo mats. The ideal size of the platform is 5feet width and 7feet length. Length can be extended with the availability of the space and rearing capacity.

This method has many advantages. Platform rearing saves labour up to 50% when compared to traditional methods like bunch rearing or tray rearing. Labour dependent risk is reduced and handling of silkworm is minimized. Hence, contamination and spreading of diseases is reduced. Better hygienic conditions can be maintained. Better aeration in the bed is ensured. Cost of production is reduced. But the only disadvantage is, this method occupies larger space.

Eri Silkworm Seed Production

Production of absolutely disease free, particularly the pebrine (protozoan disease of silkworm) free layings is highly essential for harvesting successful cocoon crops. Hence, strict hygienic methods should be maintained in its production centres of grainages. This paves the way for a sustainable eri silk industry. Quite a substantial supply of eri silkworm seed comes from the rearers themselves. The eri rearers usually prepare their own seed. Since the private rearers lack hygienic grainage facilities, the layings prepared by them are quite often not disease free due to non-examination of mother moths during seed preparation.

However, a few farmers follow traditional methods to maintain eri silkworms for seed purpose during off-season. They keep a little quantity of selected seed cocoons wrapping in a thin cloth and put in a corner of a room. They know that when it will emerge. Exactly by that time they open and allow emerging of the adult moths freely. In the morning hours the newly emerged adult moths crawl on the cloth and dry the wings by sitting on a suitable place. The moths couple in the evening. The couples are left undisturbed till the next morning. They are decoupled in the morning and put on the kharika for egg laying. The eggs are then collected. In this way, the farmers of Assam prepare hardly 5 to 10 layings during the off seasons. Thus farmers maintain a few generations of eri silkworms. These eggs are reared for their own. Then these farmers start preparation of commercial eggs and sell to other farmers.

Eri silkworm is multivoltine in nature and hence, there are 4-6 overlapping generations in a year. The farmers of the North Eastern region do continuous rearing of the eri silkworms throughout the year. At the same time, eri silkworm is hardy and resistant to diseases yet it suffers from pebrine disease occasionally. Rearers do not stick much importance to this fact. Therefore, management for quality seed production is very important by adopting scientific methods of egg production right from seed crop rearing to egg production. This will enable to fulfil the demand of the farmers' good quality seeds.

Pre-requisites of seed preparation

Grainage room: A grainage room having working area of 34' x 18' x 12' bamboo made house with thatched roof and mud plastered wall is preferable. The site of the grainage room should be elevated, dry and damp free with good ventilation. The ground surface should be such that it is easy to clean and drain water and disinfect without damaging the room. Further, the egg processing room should have the provision for moth emergence, hanging of kharikas (small bundles of straw or a split bamboo stick of 30 cm long with a hook or cut branches of trees having one hook like side projection used for egg laying by eri silkworm), egg laying, microscopic examination etc.

Disinfection of grainage room and appliances

- Close the door and windows of the grainage room and make it air tight for 5-7 days before grainage operation.
- Spray 2% formalin solution or 5% bleaching powder solution on the walls and large appliances for disinfection. Dip the smaller appliances in 5 % bleaching powder solution.
- Use bleaching powder with 30% chlorine content.
- Spray 2% formalin + slaked lime mixture @ 1 lit/2.5 m^2 area and close the grainage room.

- Under high humid condition, fumigate the grainage room with 10% formalin solution and leave the room closed for 24 hours.
- Use disinfection mask, hand gloves etc. while disinfecting.
- Formaldehyde is more effective at a temperature of > 20°C. Therefore, disinfection should preferably be done during sunny days.
- Complete the whole process of disinfection at least 3-4 days before arrival of cocoon consignment and leave the door and windows open for proper aeration.

Seed cocoon collection and transportation

After harvesting, the farmers remove the cocoons from the jail or chandrika. Then those are transported in loose condition in bamboo baskets to the grainage. Long distance transportation is done by truck from far off places to the grainage houses. While transporting, seed cocoons should not be transported in sunny hours and exposure to direct sunlight should be avoided. They should not be transported during rains. Dropping/ falling of cocoons over the hard surface and vigorous shaking of seed cocoons should be avoided.

Selection and preservation of seed cocoons: Before consigning the seed cocoons in the grainage room, seed cocoons are subjected to visual examination. Stained, deformed or thin layered, pierced, uzifly infested and dead cocoons are rejected. Only healthy, well-built and neat cocoons are considered as seed cocoons. Seed cocoons should be stored in a single layer on round bamboo trays on wooden racks or moth cages. Maintain 22-24°C temperature and 70-80% relative humidity inside the grainage room.

Technique of Seed Production

Moth emergence: Generally, eri moths start emerging early in the morning (3-5 AM) and continues till mid-day. The emergence of moths from cocoons takes place after about two weeks of cocoon formation at 20-22°C. However, in natural conditions it takes 15-18 days in summer and 35-40 days in winter. The freshly emerged moths are moist, with scaly hairs on the soft body flexible, wings small and curled. These moths stay motionless near the cocoon from which the moths emerge.

Moth picking: Since moth emergence and wing spreading are all fast, the male and female moths must be gathered separately immediately after the emergence itself to prevent injuries to each other. At that time deformed and weak moths should be rejected. Within a batch of cocoons, a small quantity of exceptionally early or late emerging moths should be eliminated and not included for mating.

Mating: Eri moths have good coupling aptitude. The optimum temperature for mating is about 20°C. Generally, mating takes place at 16-18 hours on the day of emergence. Then equal number of male and female moths is put on a tray and covered by another tray or inside the moth cages for about 16 hours and allow the moths to mate (Plate 34). Females that emerge in the evening usually do not mate the first night.

Plate 34 Coupled male and female eri silkworm moths

Picking up and tying of mated moths: Within an hour, more than 2/3rd moths mate. When the sound of moth flying subsides, the mated moths can be picked up. While picking up the mated moths, the wing base of mated female moths is held with thumb and fore fingers and tied it on the kharikas by a thread passing under the shoulder joint of the right wing. The kharikas should be hanged evenly on the string, not to be too close preventing separation of mated moths. There should be no sound, strenuous vibrations and bright light in the room. The room should have proper ventilation.

Decoupling: Mating for 12 hours is sufficient enough to ensure full fertility. The mated moths are separated during the next day between 8 to 9 hours. The separation of mated moths should be done gently. By one hand the abdomen of the female moth is gently held and then the male moth is pulled down without causing injury to reproductive organs. The separated female moths are put on the kharika without removing the thread while male moths are rejected. Then the kharika is shaken gently to let the female moth pass urine fully. This is very important because, if the mother moths have not urinated or have not thoroughly urinated then those would normally have a late egg deposition. Further, such moths would also contaminate their eggs when they urinate during the course of egg laying. In case of shortage of male moths good male moths are selected, put into another tray and preserved at low temperature for second time use.

Moth sorting: The purpose of moth sorting is to eliminate diseased moths and unhealthy moths, as well as the moths with abnormal or underdeveloped wings. The moths whose scaly hairs are dropped and the moths which are inert, inactive and incapable of copulation are also to be rejected to improve the quality of eggs. Generally, a moth whose body is fat and large, abdomen is soft, segment is relaxed and the amount of eggs is less is considered as bad one.

Egg laying: The decoupled females are allowed to lay eggs on kharikas in a vertical position (Plate 35). Optimum temperature in all the seasons for egg laying is below 22-24°C with high humidity. The female moths deposit the eggs in clusters glued together in rows projecting from the substrate. They are allowed to lay eggs for a maximum period of two days. The number of eggs laid by one female moth is usually 200-250, but sometimes the number can be as high as 400 to 600 eggs.

Plate 35 Female moth laying eggs on kharika

Collection of female moths after egg laying: After the egg laying, female moths are collected on the next day early morning and are microscopically examined.

Mother moth examination: In order to produce disease free layings / seeds and to determine the presence of pebrine spores, mother moths are subjected to microscopic examination in the following way.

- Crush the individual moth in porcelain crushing set.

- Homogenize it in 2 ml of 0.6% K_2CO_3 or 0.5% KOH solution using the same mortar and pestle.

- Prepare five fields of the homogenate in a smear on a glass slide and cover them with cover slips.

- Then examine them under the compound microscope or phase contrast microscope with 675x magnification.

The individual moth homogenate is certified as free of infection if pebrine spore is not detected. The mother moths infected with pebrine should immediately be rejected along with the kharika, cotton thread and eggs laid by that particular moth or moths. It is advisable to burn them to destroy the pathogen spores.

Disinfection of eggs: After ascertaining the disease freeness of layings, the dirty materials such as scales and urine stuck to the eggs are washed away with clean water and immersed in 2% formalin solution for 10-15 minutes. The eggs are again washed in pure water until no smell of formalin is detected. This helps in removal of pathogens adhering to the egg shell and further prevents secondary contamination. The eggs are then dried in the shade spreading in single layer on blotting paper and packed in laying boxes / bags (Plate 36).

Plate 36 Disinfected eggs ready for incubation

They are now ready for incubation or delivering to the farmers for rearing. Packing should be accompanied with the following details,

- Name of the grainage

- Name of the race / lot number
- Weight in grams or number of eggs
- Date of egg laying
- Probable date of hatching

Transportation of eggs: Eggs are silkworm seeds in which active development of embryo is in progress. While delivering to the farmers, if care is not taken in handling them, harmful effects may result. Hence, the following measures should be followed carefully.

- Do not expose to direct sunlight, high temperature and dry air
- Physical shock on unpaved road should be avoided
- There should not be direct or indirect contact with tobacco, chemical fumes, fertilizer, petroleum, insecticides etc.
- Do not contact with hands contaminated with pathogenic microorganisms
- Perforated egg boxes should be used
- Do not store eggs in enclosed places with poor ventilation.

Rearing Performance of Eri Silkworm

Egg laying pattern in *Samia cynthia ricini* Boisduval

The Sericigenous Saturniids like *Antheraea mylitta* Drury and *Antherarea assamensis* West wood deposit eggs in clusters of 5 to 10 and 8 to 10 respectively over a period of 6 to 7 days. But in eri silkworm the literature about the egg laying aspect is meagre. So Shailaja *et al.,* (2000) conducted an experiment to observe the clumping pattern of eri silkworm. As a part of this study, the female and male moths of *S. c. ricini* were allowed to mate for 3 hours and gravid females were placed separately on bamboo trays to achieve oviposition. Observations were made on the number of egg clumps or clusters deposited by each moth and number of eggs per clump, at an interval of 24 hrs till the death of moths. The results showed that a female moth on an average deposited two to three egg clumps per day over an oviposition period of six days. The size of the clump varied among moths and also among days of oviposition (Plate 37). In a clump the eggs were piled one over the other in 3 to 5 layers placing anterior ends in one direction. The size of the clump-I and clump-II, which were deposited on first day was 68.75 and 130.00 eggs respectively. The number of eggs per clump was observed to be maximum on second day of oviposition in both clump-I (148.50) and clump-II (58.00). From third day onwards, the size of the clump-I was higher than that of clump-II and the size of both the clumps was gradually reduced reaching a minimum on sixth day. However, the moth laid three clumps on third day with 52.60, 29.60 and 11.50 eggs per clump. So from the observations of Shailaja *et al.,* (2000), it is clear that the eri silk moth lays eggs in clusters and the number of clusters and number of eggs per cluster varies according to days of oviposition. In scientific egg production technology, it has been emphasized to select eggs laid by eri moths within 48hrs, because such eggs are best in terms of hatchability and rearing performance (Reddy *et al.,* 2000).

Plate 37 Egg clumps laid by the female moth

Performance of eri silkworm as influenced by eggs obtained from different days of oviposition

In *Samia c. ricini* the egg laying starts one to two hours after mating and continues for three to four days. However, eggs laid on first two days are collected for rearing (Devaiah *et al.,* 1978). Similarly, in *Antheraea mylitta*, the moths lay eggs over a period of six to seven days, but only the eggs deposited in the first 72 hours are considered for rearing (Jolly *et al.,* 1979). The rearing performance of the silkworms obtained from eggs laid on different days has been investigated to certain extent by Vijayaraghavan *et al.,* (1983). In continuation, Ramakrishna Naika *et al.,* (2006) attempted a study to know the rearing performance of eri silkworms obtained from eggs laid on different days of oviposition. The eggs were collected from twenty copulated moths for five successive days separately and split into four replications. Observations were recorded on fecundity at different days of oviposition, single egg weight, egg hatchability, weight of ten fifth instar larvae, larval duration, effective rate of rearing, single cocoon weight, shell weight, pupal weight, shell ratio, moth emergence and fecundity.

The results showed that the fecundity (193.25 eggs), single cocoon weight (0.0019 g) and hatchability (97.84%) were highest in the first day laid eggs and least in the eggs laid on fifth day (Table 1). Similar trend was observed for fifth instar larval weight (38.08 g/10 larvae) and effective rate of rearing (94.39%), but the larval duration was the same (636 hrs) among the batches of silkworms obtained from eggs laid on first three days. The cocoon weight, shell weight, pupal weight and moth emergence showed significantly favourable values in case of eggs laid during first two days (Table 5.2). However, the shell ratio did not differ significantly, though the eggs laid on first two days showed marginally higher values. The fecundity of the moths emerging from first day laid eggs was significantly higher than those emerging from fourth and fifth day laid eggs.

From the findings of Ramakrishna Naika *et al.,* (2006), it is obvious that the eggs laid on first day are superior with respect to almost all the parameters studied (Table 5.1 & 5.2). However, the eggs laid on second day did show almost similar performance. Mohanty (1998) has reported that egg laying in case of tropical tasar silk worm continues up to seventh day, but the eggs laid up to third day are collected to avoid wide disparity during rearing. It can therefore be concluded that eggs of eri silkworm collected batch wise from first to second day or even up to third day can be utilized for economical rearing, since they contribute nearly 90 per cent of the total eggs laid by eri silk moths.

Table 5.1: Egg and growth parameters of eri silkworm as influenced by different days of oviposition (Adopted from Ramakrishna Naika *et al.,* 2006)

Days of oviposition	Fecundity (No.)	Single egg weight (g)	Hatchability (%)	Fifth instar larval weight (g/10 larvae)	Larval duration (Hrs.)	ERR (%)
First day	193.25	0.0019	97.84	38.08	636	94.39
Second day	72.50	0.0017	93.66	37.44	636	92.37
Third day	45.50	0.0012	91.76	32.18	636	86.13
Fourth day	20.25	0.0010	81.99	28.89	720	81.01
Fifth day	14.50	0.0011	60.14	28.14	756	78.36
LSD (5%)	21.76	0.0002	14.15	2.92	12.01	4.02

Table 5.2: Cocoon and grainage parameters of eri silkworm as influenced by different days of oviposition (Adopted from Ramakrishna Naika *et al.,* 2006)

Days of oviposition	Single cocoon weight (g)	Shell weight (g)	Pupal weight (g)	Cocoon shell ratio (%)	Moth emergence (%)	Fecundity (No.)
First day	2.77	0.3855	2.3740	13.92	92.45	317.25
Second day	2.73	0.3638	2.3452	13.29	91.42	303.75
Third day	2.32	0.2962	2.0195	12.72	89.30	298.00
Fourth day	2.15	0.2780	1.8610	12.94	88.63	285.00
Fifth day	2.13	0.2707	1.8390	12.66	87.14	290.00
LSD (5%)	0.203	0.038	0.177	-	1.486	19.266

Effect of refrigeration on the hatchability of eri silkworm eggs at Central Sericultural Research and Training Institute, Mysore

Quality seed is the backbone of sericulture industry. A shortfall in the availability of disease free silkworm seed will definitely hit the industry in the country. The grainage technique of eri silkworm is quite different from that of other non-mulberry silkworms and domestication renders the grainage operation systematic and easy. The present grainage technology remained obsolete and could not make a breakthrough in production of industrial layings. For commercial grainage operations, sometimes it becomes necessary to postpone the hatching to overcome the problems such as synchronization in brushing, distribution etc. Under such circumstances, refrigeration becomes an ideal way to delay hatching or synchronization of hatching of silkworm eggs, keeping in view the large scale requirement of layings, availability of leaf etc. Though there is ample information on the refrigeration of eggs of *Bombyx mori* to postpone hatching, with regard to refrigeration of eri silkworm eggs it is scanty. Govindan *et al.,* (1980) reported adverse effect on hatching when 5 days old eggs of *S. cynthia ricini* were refrigerated beyond 48 hours. Hence an attempt has been made by Somaprakash and Sathya Prasad, (2009) to study the effect of refrigeration of different aged eggs of *S. c. ricini* for various durations on hatching. The experiment was conducted during the year 2006-07 at the Eri Research Laboratory of Central Sericultural Research and Training Institute, Mysore. The eri silkworm rearing was conducted by utilizing the local green castor variety leaf. The male and female moths of *S. cynthia ricini* were allowed to mate at room temperature of 24 to 27^0C and relative humidity of 60 to 75 % for 4 hours. The eggs collected on first day were used for the experiment. The eggs after attaining an age of 24 and 48 hrs were collected and refrigerated at 5, 10 and 15^0C for 5, 10 and 15 days. After the stipulated duration of refrigeration, the eggs were released and kept for incubation at 24 ± 2^0C and 65 ± 5 per cent RH. At pinhead stage of embryonic development, the eggs were black boxed for 48 hrs followed by exposure to light for 4 hrs. A control was also kept at room temperature of 24 to 27^0C with relative humidity of 60 to 75 per cent. Observations on incubation period and egg hatchability were recorded.

The results indicated that when 24 hrs old eggs were refrigerated at 5, 10 and 15^0C for 5, 10 and 15 days, the incubation period decreased with increase in duration of refrigeration at 5 and 15^0C. The egg hatchability was significantly low at all the temperatures compared to control. The maximum (83.39%) and minimum (7.90%) hatchability was recorded at 15^0C (5 days) and 5^0C (15 days) respectively (Table 5.3). When 48 hrs old eggs were refrigerated at 5, 10 and 15^0C for 5, 10 and 15 days, incubation period did not vary significantly at 5 (9.4 ±0.24) and 10^0C (9.2±0.49). Consistently higher percentage of egg hatching was observed when the eggs were refrigerated for 5 days, irrespective of different temperature regimes and at 15^0C for 10 days (Table 5.4).

When 24 hrs old eggs were refrigerated at 5^0C for 5, 10 and 15 days, the incubation period of 24 hrs old eggs decreased significantly as against 48 hrs aged eggs. The incubation period of 48 hrs old eggs was on par with control, when the eggs were incubated at different temperatures, for different durations except at 5^0C for 15 days, 10^0C for 10 days, 15^0C for 5, 10 and 15 days where both the treatments differed significantly. Similarly, 24 hr old eggs recorded significantly lower hatching percentage (7.90 to 83.39) compared to 48 hr old eggs (10.32 to 92.57). Further, irrespective of age of the eggs, the hatching percentage decreased drastically with the increase in duration of consignment period compared to control (Table 5.5).

Table 5.3: Effects of consignment duration and temperature on incubation and hatching of 24 hrs old eggs of *S. cynthia ricini* (Adopted from Somaprakash and Sathya Prasad, 2009)

Consignment duration (days)	Incubation period (days) at different temperatures			Hatching (%) at different temperatures		
	5^0C	10^0C	15^0C	5^0C	10^0C	15^0C
5	8.4±0.24[b]	8.2±0.49[b]	8.2±0.0[b]	71.96±0.71[b]	72.06±1.05[b]	83.39±2.16[b]
10	8±0.00[c]	7.6±0.24[b]	7.0±0.0[c]	61.54±0.72[c]	61.01±0.84[c]	63.82±1.23[c]
15	8±0.00[c]	7.8±0.49[b]	6.0±0.0[d]	7.90±0.31[d]	11.52±0.21[d]	18.96±0.55[d]
Control	10±0.00[a]	10±0.00[a]	10.0±0.0[a]	90.79±0.50[a]	90.79±0.50[a]	90.79±0.50[a]
F value	Sig	Sig	Sig	Sig	Sig	Sig
Cd at 5%	0.125	1.232	0.00	1.967	2.482	4.346

Table 5.4: Effects of consignment duration and temperature on incubation and hatching of 48 hrs old eggs of *S. cynthia ricini* (Adopted from Somaprakash and Sathya Prasad, 2009)

Consignment duration (days)	Incubation period (days) at different temperatures			Hatching (%) at different temperatures		
	5^0C	10^0C	15^0C	5^0C	10^0C	15^0C
5	9.4±0.24[b]	9.2±0.49[ab]	9.0±0.0[b]	90.77±1.12[a]	91.67±0.55[a]	92.6±0.5[a]
10	9.0±0.00[b]	8.6±0.24[b]	7.4±0.2[c]	75.92±0.97[b]	86.73±1.91[b]	91.5±0.4[a]
15	9.4±0.24[b]	8.8±0.49[b]	7.0±0.0[d]	10.32±0.92[c]	18.35±0.75[c]	69.7±2.4[b]
Control	10±0.00[a]	10±0.00[a]	10.0±0.0[a]	90.79±0.50[a]	90.79±0.50[a]	90.8±0.5[a]
F value	Sig	Sig	Sig	Sig	Sig	Sig
Cd at 5%	0.580	1.232	0.125	3.044	3.662	4.221

Table 5.5: Effects of temperature and consignment period on incubation and hatching of eri silkworm eggs (Adopted from Somaprakash and Sathya Prasad, 2009)

Egg age	5^0C 5 days		5^0C 10 days		5^0C 15 days	
	Incubation (days)	Hatching (%)	Incubation (days)	Hatching (%)	Incubation (days)	Hatching (%)
24 hrs	8.4±0.24[b]	71.96±0.71[b]	8.00±0.00[b]	61.54±0.72[c]	8.00±0.00[c]	7.90±0.31[c]
48 hrs	9.4±0.24[a]	90.77±1.12[a]	9.00±0.00[a]	75.92±0.97[b]	9.40±0.24[b]	10.32±0.92[b]
Control	10±0.00[ab]	90.79±0.50[a]	10.00±0.00[a]	90.79±0.50[a]	10.00±0.00[a]	90.79±0.50[a]
F Value	Sig	Sig	Sig	Sig	Sig	Sig
CD at 5%	1.360	5.565	0.0	5.145	0.304	4.284

Egg age	10^0C 5 days		10^0C 10 days		10^0C 15 days	
	Incubation (days)	Hatching (%)	Incubation (days)	Hatching (%)	Incubation (days)	Hatching (%)
24 hrs	8.20±0.49[b]	72.06±1.05[b]	7.60±0.24[b]	61.01±0.84[c]	7.80±0.49[b]	11.52±0.21[c]
48 hrs	9.20±0.49[ab]	91.67±0.55[a]	8.60±0.24[b]	8.73±1.91[b]	8.80±0.49[ab]	18.35±0.75[b]
Control	10.00±0.00[a]	90.79±0.50[a]	10.00±0.00[a]	90.79±0.50[a]	10.00±0.00[a]	90.79±0.50[a]
F Value	NS	Sig	Sig	Sig	Sig	Sig
CD at 5%	2.721	5.040	1.360	8.439	1.924	3.630

Egg age	15^0C 5 days		15^0C 10 days		15^0C 15 days	
	Incubation (days)	Hatching (%)	Incubation (days)	Hatching (%)	Incubation (days)	Hatching (%)
24 hrs	8.00±0.00[c]	83.39±2.16[b]	7.00±0.00[b]	63.82±1.23[b]	6.00±0.49[b]	18.96±0.55[c]
48 hrs	9.00±0.00[b]	91.57±0.50[a]	7.40±0.24[b]	91.51±0.36[b]	7.00±0.00[b]	69.72±0.39[b]
Control	10.00±0.00[a]	90.79±0.50[a]	10.00±0.00[a]	90.79±0.50[a]	10.00±0.00[a]	90.79±0.50[a]
F Value	Sig	Sig	Sig	Sig	Sig	Sig
CD at 5%	0.0	8.916	0.304	5.404	0.0	9.826

Govindan *et al.,* (1980) observed adverse effect of refrigeration of 1 to 2 day old eggs at 7 ± 1^0C beyond 48 hrs, but in the study by Somaprakash and Sathya Prasad (2009), consignment of eggs at 5^0C for 5 days did not show any significant variation in the hatching percentage over the control, thus implying that 24 hr and 48 hr old eggs can be conveniently consigned for 5 days. However, the findings of the present study showed that low temperature had an adverse effect on hatching percentage. Sannappa *et al.,* (2002), also reported an adverse effect on hatching percentage with increase in age of the eggs and refrigeration duration at low temperature. Vishwakarma (1983) made similar observations in case of *Philosamia ricini* Hutt. However, these authors have not noticed any significant impact of refrigeration on incubation period which is not in agreement with the observations made in the study by Somaprakash and Sathya Prasad (2009), wherein significant decline in incubation period was observed. The reason for reduction in incubation period may be due to heat shock effect, as the eggs refrigerated at 5^0C were brought to room temperature without an intermediate temperature. However, the heat shock did not have any adverse effect on the hatchability, thus suggesting that the eggs incubated at 5^0C for 5 days cannot be brought to the room temperature without passing through the intermediate temperature.

From the study of Somaprakash and Sathya Prasad (2009), it is inferrred that 1-2 day-old eri silkworm eggs can be consigned at a temperature of 5^0C for 5 days and 2 day old eggs at 15^0C for 5 to 10 days without affecting egg hatchability. These findings are also helpful to tide over the problems in commercial grainages in times of exigency such as synchronization and/or postponement of brushing due to shortage of leaf at farmer's conditions.

Changes in carbohydrate, lipid and moisture contents during the embryonic development in *Samia c. ricini*:

In the eri silkworm, *Samia cynthia ricini* Boisduval, the embryonic development is important in egg production. As the metabolic activities will be intense during different stages of embryogenesis, it requires constant energy for growth, differentiation and to maintain the cells in living condition. Among the major food reserves, glycogen is the main source of energy during embryogenesis in *Antheraea mylitta* (Pant and Sharma, 1976) and *Bombyx mori* (Chino, 1958). To gather such information during embryonic development of *S. c. ricini*, Krishnappa *et al.,* (2001) made an attempt to study the changes in the levels of carbohydrate, lipid and moisture contents during the embryogenesis of *S. c. ricini*. The eggs of plain white race of eri silkworm in five samples of 100 each were subjected for the estimation of carbohydrate, lipid and moisture contents from the day of oviposition till hatching at an interval of 24 hrs.

The results of Krishnappa *et al.,* (2001) revealed that the total carbohydrate content showed a slight increase on the first day (25.1 µg/mg egg), and gradually decreased towards eighth day (4.0 µg/mg egg). The total free sugars were high initially (19.5- 20.2µg/mg egg) up to 1st day followed by rapid depletion on 2nd and 3rd days (1.8 to 2.8µg/mg egg). However, there was abrupt increase on fourth day (16.0µg/mg eggs) followed by gradual decline reaching to a minimum of 1.3 µg/mg eggs by 8th day. The rapid depletion in free sugar content on the second and third days, in the study of Krishnappa *et al.,* (2001) probably coincides with the maximum embryonic length stage. The depletion has been attributed to its requirement for the formation of rudiments and segmentation of the embryo. The gradual depletion of both carbohydrates and sugars till eighth day suggests their utilization for embryonic process, metabolism and chitin synthesis. The lipid content of eggs remained almost similar up to 2nd day (10.7 to 10.8%), followed by gradual decline throughout the development reaching to 5.1% by10th day (Table 5.6). A gradual decrease in the level of lipids during embryogenesis was reported in normal and diapausing eggs of *B. mori* (Nakasone, 1979). The moisture content of the eggs of *S. c. ricini* in the study of Krishnappa *et al.,* (2001) varied from 65 to 71% during different embryonic developmental stages. However, it is very difficult to pin point the reasons for such variations from their data.

Table 5.6: Changes in the carbohydrates, free sugars, lipids and moisture content during embryonic development of *Samia cynthia ricini* Boisduval (Adopted from Krishnappa *et al.,* 2001)

Days of development	Total carbohydrate (µg/mg egg)	Free sugars (µg/mg egg)	Non-sugar carbohydrates (µg/mg egg)	Lipid content (%)	Moisture content (%)
0	22.4	19.5	2.8	10.8	68.8
1	25.1	20.2	4.8	10.8	65.9
2	20.3	1.8	6.5	10.7	70.5
3	19.5	2.8	16.6	9.3	71.9
4	16.2	16.0	0.13	8.8	67.1
5	13.0	12.3	0.76	8.5	69.8
6	10.6	10.5	0.18	7.7	71.0
7	6.1	3.84	2.3	7.2	69.1
8	4.0	1.3	2.7	6.2	69.8
9	4.6	3.0	2.6	5.6	70.7
10	4.4	4.2	0.15	5.1	71.7

Nitrogen metabolism during the progressive embryonic development of *S. cynthia ricini*:

During the process of embryogenesis, metabolic activities occur for the growth and differentiation of the cells. The eggs of *Samia c ricini* being cleidoic type cannot draw nutrients from outside or cannot throw off from inside the egg, except for oxygen and carbon dioxide. Among the major food reserves, protein accounts for only a small fraction of the total materials utilised for development in cleidoic eggs and there is only a slight decrease in the original supply of proteins. According to Pant and Sharma, (1968) the levels of total egg nitrogen and shell nitrogen remain constant during the embryogenesis of *Samia cynthia ricini*. With advancement of age, the protein nitrogen and non-protein nitrogen decrease. However, the amino nitrogen decreases till the third day but increases subsequently. Krishnappa *et al.,* (1999) made studies on the variation of the nitrogen and proteins in *S. c ricini* during embryogenesis. The eggs of known age were obtained by rearing the plain white race of eri silkworm on castor leaves and were subjected to nitrogen and protein analysis from the day of oviposition till egg hatching at 24 hours interval.

Results showed that the total nitrogen content in *Samia cynthia ricini* eggs remained almost same (61.3 to 61.8 µg/mg egg) during the entire development of the embryo. However, fluctuations were noticed in protein nitrogen (between 40.8 and 55.4µg/mg egg) and non-protein nitrogen (between 5.9 and 20.7 µg/mg egg) (Table 5.7). Among different fractions of proteins, total crude protein varied between 383.5 to 386.4 µg/mg egg while, total protein varied between 225.0 to 346.6 µg/mg egg during different stages of embryonic development of eri silk worm eggs. Insoluble protein showed a variability of 190.7 to 290.0 µg/mg egg during different stages of embryonic development. However, it is interesting to note that the soluble protein fraction exhibited considerable variation from 19.2 to 65.8 µg/mg egg during different stages of embryonic development, while collagenous protein also showed a remarkable variability between 42.0 to 87.3 µg/mg egg during different stages of embryonic development of eri silk worm eggs. Generally, changes in the protein levels are associated with their utilization for the organogenesis and chitin biosynthesis during the progressive embryonic development (Table 5.7).

Table 5.7: Changes in the levels of nitrogen and different proteins during embryogenesis of *S. c. ricini* (Adopted from Krishnappa *et al.*, 1999)

Days of development of egg	Total Nitrogen (µg/mg egg)	Protein nitrogen (µg/mg egg)	Non-protein (µg/mg egg)	Total crude protein (µg/mg egg)	Total protein (µg/mg egg)	Insoluble protein (µg/mg egg)	Soluble protein (µg/mg egg)	Collagenous protein (µg/mg egg)
0	61.3	50.5	10.8	383.5	315.9	252.4	63.4	81.1
1	61.6	48.7	12.9	385.4	305.4	243.6	60.8	87.3
2	61.7	41.8	19.8	385.6	261.8	198.3	63.4	85.1
3	61.5	40.8	20.7	385.7	225.0	190.7	64.2	80.2
4	61.5	44.6	16.8	384.4	279.1	213.2	65.8	77.9
5	61.4	54.8	6.5	384.2	343.0	280.2	62.8	74.4
6	61.4	55.4	5.9	384.0	346.6	290.0	56.6	57.7
7	61.4	46.1	15.3	384.2	288.2	268.9	19.2	48.0
8	61.5	41.8	19.6	384.7	261.6	240.6	21.0	42.0
9	61.5	47.5	13.9	384.6	297.3	278.6	23.7	62.7
10	61.8	43.2	18.4	386.4	270.0	246.0	24.1	68.4

Table 5.8: Suitability of low-cost materials for eri silkworm rearing (Adopted from Subramanianan *et al.*, 2013)

Treatments	5th day 5th instar wt (g)	Larval mortality (%)	Unequal Larvae/sq.ft (%)	Good cocoons (%)	Flimsy cocoons	Cocoon weight (g)	ERR (%)	Silk ratio (%)
T1. GB + NP	6.23	2.25[a]	3.50[a]	79.60[b]	20.40[b]	2.31[a]	84.80[c]	13.43[c]
T2. CL + NP	6.40	1.00[a]	2.30[a]	85.80[c]	14.20[a]	2.51[ab]	90.95[d]	14.19[d]
T3. NP	6.20	3.70[ab]	4.33[a]	75.40[b]	24.60[b]	2.28[a]	77.05[b]	12.18[b]
T4. UB + NP	5.90	9.00[c]	9.30[b]	65.70[a]	34.30[c]	2.17[a]	71.70[a]	11.69[a]

NB: The values in a column followed by the same letter are not significantly different at P = 0.05

Progressive Embryonic Development of Eri Silkworm Egg

Generally, the development of embryo from oviposition till hatching is mainly dependent on the temperature during incubation. Krishnappa (1989) and Reddy *et al.,* (2000) reported that the eri embryo requires 10 days (240 hrs) to complete development at a constant temperature of $23.5 \pm 1.0^{\circ}C$ in BOD incubator. The sequence of development according to the age has been described as below.

Day '0' (2 hours): The embryogenesis of the insect starts with mitotic division of the fertilised egg nucleus (zygote) by cleavage. On the day of oviposition the division of the cleavage nucleus occurs, followed by its movement towards the periphery of the egg. In most of the insects, the type of cleavage noticed is meroblastic. In eri egg, a superficial cleavage seems to be more common with its occurrence at the centre.

First Day (24 hours): At 24 hours after oviposition, the germ band formation was complete followed by its invagination indicating the signs of gastrulation. Towards the end of gastrulation, formation of serosal layer was observed. Vitellophages were clearly noticed in large number in the yolk.

Second Day (48 hours): Amnion and serosal layers were clearly visible. Invagination of stomodaeum was observed. Vitellophages were found in large numbers throughout the yolk. Initiation of segmentation was noticed and the embryo to reach the maximum length stage. Amniotic fold was clear, covering the embryo. Serosa was completely covered with the yolk.

Third Day (72 hours): The embryo just occupied about 60 per cent of the total egg space, the maximum length stage. Eighteen segments were noticed of which the four anterior segments constitute the head, successively three form the thorax and the remaining eleven, the abdomen. Thoracic leg rudiments were observed. Serosa covering the entire yolk and amniotic layer surrounding the embryo become evident. Invagination of the ectoderm was noticed to form stomodaeum. Endodermal cells, believed to have originated from yolk cells, but difficult to differentiate them to form midgut, started accumulating during this stage. Yolk cells or vitellophages were persistent. The maximum length stage appeared at 72 hours, after oviposition, which is much delayed, might be due to the incubation temperature.

Fourth Day (96 hours): The embryo was shortened with increased girth. Initiation of involution was noticed, which may be designated as involution pre period. Invagination of the proctodaeum was extended further to sixteenth segment. The initiation of blastokinesis was noticed at 96 hours. Endodermal cells were clearly noticed to give rise to the midgut epithelium.

Fifth day (120 hours): The development comprises of involution of blastokinesis mid stage. The stage is characterised by its 'S' shaped embryo in the middle region of the yolk. Amniotic and serosal layers become elongated and increase in girth. The stage was noticed between 4 to 6 days, which may be due to low incubating temperature, in contrast to 3 to 4 days at 25°C. Blastokinesis helps the embryo turn its head towards the micropyle, enabling the larva to hatch out of the egg after the embryonic development. Any increase in temperature during this stage may adversely affect the embryo. This point would help in improving the grainage technology.

Many events takes place in this stage, notable of them include the visibility of rudimentary brain. Gnathal segments fused to give rise to head appendages. Thoracic leg rudiments were distinct. Invagination of stomodaeum and proctodaeum become prominent. The midgut formation was clear. Generally, it is believed that the embryonic membranes are usually ruptured during involution. But, in eri embryo, only a little rupture occurred to the serosa which was not complete. Involution can also occur without rupture of the membranes.

Sixth Day (144 hours): During the period, the embryo can be considered as an involuted structure of the post period stage, where the involution is completed and the embryo moved to the periphery. The foregut formation was complete and the midgut cells were distinguishable. Development of the brain can be seen. External processes start developing from this stage. The stage occurs in less than 5 days in *S. c. ricini.*

Seventh Day (168 hours): The embryo elongated and thickened. Head was almost completely formed. Rudiments of abdominal legs were clearly discernible. Alimentary canal was almost formed. Silk glands were clearly noticed. The yolk content diminished, embryo occupied most of the egg volume. The stage was noticed on the sixth day in *S. c. ricini.*

Eighth Day (192 hours): The embryo almost occupied the egg. Pigmentation, chitinisation of the head, prothoracic hood, setae on dorsal side of embryo and thoracic legs were seen. Mouth parts were completely formed. Ocelli are clearly visible.

Nineth Day (216 hours): Further growth of the embryo with notable increase in girth. The tail end curved to accommodate itself in the egg space. Darkening of the chitinised mouth parts, head capsule and setae were clearly visible. Six ocelli and spiracles were visible. Head orientation was towards the broader (micropylar) end.

Tenth Day (240 hours): The fully developed embryo hatched out on the tenth day morning. Increase in the embryonic development period might be due to lower temperature (23.50 ± 1^{0}C) during incubation, which was nine days at 25^{0}C and seven days at 27^{0}C (Reddy *et al.,* 2002).

Evolving Rearing Technology for Eri Silkworm

Suitability of low-cost materials for Eri silkworm rearing in Tamil Nadu

In eri silkworm rearing technology, feeding must be in harmony with the eating habit and appetite of larvae and must be done economically. Next to feeding, cleaning is an equally important factor. It is necessary for the health and progress of the worms, because pilling of litter makes bed moist. The moist condition favours multiplication of pathogenic microorganisms which in turn affects the health of worms and culture. Considering all these factors, Subramanianan *et al.,* (2013) planned an experiment to test the low cost materials for eri silkworm rearing in Tamil Nadu, the following types of bed materials (locally available low cost or waste materials) were used for preparing rearing bed and compared their suitability for ericulture using local castor and tapioca varieties. The treatments included are

   T1 Gunny bag (GB) + Newspaper (NP)

   T2 Coconut leaf mat (CL) + Newspaper (NP)

   T3 Newspaper alone (NP)

   T4 Urea bag (UB) + Newspaper (NP)

Observations were recorded on larval weight, larval mortality, cocoon weight, ERR (%) and silk ratio (%). Among the different types of bed materials used in this experiment, coconut leaf mat with newspaper and gunny bag with newspaper were better in all the parameters and statistically significant compared to other types of bed materials recording lower number of unequal larvae, lower mortality per square foot and lower number of flimsy cocoons together with higher larval weight, higher silk ratio and higher ERR% (Table 5.8). However, the cocoon weight was not statistically significant among the treatments.

Impact of larval population density on rearing parameters of Eri silkworm in Tamil Nadu

With a view to optimize the space and maximize productivity, two higher levels of larval population were tested in comparison with the standard 50 larvae per sq. ft. The study was conducted by Subramanianan *et al.,* (2013) in a farmer's rearing house with the following treatments:

T1 50 worms per square foot (the standard)

T2 60 worms per square foot

T3 70 worms per square foot

Spacing experiment revealed that the treatment one with 50 larvae/sq. ft. was statistically superior in terms of minimum larval mortality, minimum number of unequal larvae (2%), maximum larval weight (7.27 g), highest ERR% (2.65%), higher cocoon weight and maximum silk ratio% (13.57%) (Table 5.9). But when the larval populations were increased per square foot, the rearing parameters observed were below optimum. Huq *et al.,* (1991) reported that the highest density desirable for eri silkworm was found to be 150 larvae/tray (0.9 x 6 m^2) when reared on castor and cassava in the laboratory at 27°C and 80% RH. The larvae/tray attempted varied between 50-275 under the same environmental conditions.

The study conducted by Saranga Dutta *et al.,* (2013) on the impact of crowding of eri silkworm larvae showed that crowding affected the growth of larvae because as the densities increased the growth rate decreased and the duration of larval development was lengthened. The larval weight decreased and the size of the cocoons was also affected, showing a positive correlation with the other parameters. The emerged moths were smaller in the denser lot than in the less dense lots. The emergence of the moths was delayed in the denser regime. As the density increased the fecundity decreased. Fiber quality studied by Instron Tensile Tester revealed that the larvae reared in small numbers showed high tensile properties. This study suggests that during rearing of eri silkworm crowding should be avoided by providing an optimum condition as it affects the silkworm ecologically and economically.

Evaluation of suitable feeding schedule for Eri silkworm rearing

Growth and development of eri silkworm depend on the food availability or the frequency of feeding. Hence, Subramanianan *et al.,* (2013) conducted an experiment to find out the interval of feeding using castor Pink Double Bloom Variety during the fifth instar larval stage.

T1 3 times feeding per day

T2 4 times feeding per day

T3 5 times feeding per day

Table 5.9: Impact of Population Density on Rearing Parameters of Eri Silkworm (Adopted from Subramanianan *et al.*, 2013).

Treatments	5th day 5th instar wt (g)	Larval mortality (%)	Unequal Larvae/sq. ft (%)	Good cocoons (%)	Flimsy cocoons	Cocoon weight (g)	ERR (%)	Silk ratio (%)
T1	7.27[c]	0.67[a]	2.00[a]	88.9[c]	95.2[c]	6.2[a]	2.65[b]	13.57[b]
T2	6.73[b]	2.33[b]	4.33[b]	81.6[b]	87.4[b]	5.8[a]	2.23[a]	13.44[b]
T3	6.00[a]	4.00[c]	8.00[c]	67.6[a]	76.3[a]	8.7[b]	2.04[a]	12.53[a]

NB: The values in a column followed by the same letter are not significantly different at P= 0.05

Table 5.10: Suitable feeding schedule for Eri silkworm rearing (Adopted from Subramanianan *et al.*, (2013).

Treatments	5th day 5th instar wt (g)	Unequal Larvae/sq.ft (%)	Good cocoons (%)	Flimsy cocoons	Cocoon weight (g)	ERR (%)	Silk ratio (%)
T1	5.90[a]	4.00[b]	90.5[a]	89.4[a]	9.5[b]	2.44[a]	12.13[a]
T2	6.57[b]	2.67[a]	97.0[b]	96.6[b]	3.0[a]	2.68[b]	14.23[c]
T3	6.87[b]	3.00[a]	96.0[b]	95.7[b]	4.0[a]	2.65[b]	13.88[b]

NB: The values in a column followed by the same letter are not significantly different at P= 0.05

Table 5.11: Bed cleaning schedule in ericulture (Adopted from Subramanianan *et al.,* (2013).

Treatments	5th day 5th instar wt (g)	Larval mortality (%)	Unequal Larvae/sq.ft (%)	Good cocoons (%)	Flimsy cocoons	Naked pupae (%)	Cocoon weight (g)	ERR (%)	Silk ratio (%)
T1	6.80	2.20	11.65	80.20[b]	19.80[a]	90.10[b]	1.90	90.43	13.42
T2	6.40	4.37	11.20	79.00[b]	21.00[a]	87.45[ab]	2.1 0	90.27	12.79
T3	6.30	6.05	12.80	71.93[a]	28.07[b]	83.70[a]	1.30	90.17	13.30
T4	6.30	6.88	15.40	69.78[a]	30.22[b]	79.45[a]	4.30	90.09	12.40

NB: The values in a column followed by the same letter are not significantly different at P = 0.05

Table 5.12: Quantitative and qualitative characters of Eri cocoons harvested from different mountages (Adopted from Narzary *et al.,* 2014)

Mountages	Cocoon harvest (%)	Good cocoon (%)	Cocoon weight (g)	Shell weight (g)	Pupal weight (g)	Shell ratio (%)	Defective cocoon (%)	Size (cm)		Shape
								Length	Breadth	
Chandrika	94.00	97.87	3.16	0.43	2.73	13.61	2.13	4.95	1.92	Normal
Hingori leaf jail	91.33	96.34	2.33	0.27	2.06	11.59	3.66	4.54	1.66	Normal
Banana leaf jail	90.00	91.10	2.32	0.27	2.05	11.64	8.90	4.41	1.65	Normal
Jackfruit leaf jail	76.00	84.20	2.29	0.26	2.03	11.35	15.80	4.13	1.55	Normal
Straw mountage	90.67	89.62	2.80	0.36	2.44	12.86	10.38	4.55	1.69	Normal
Folded newspaper	78.00	82.88	2.35	0.26	2.09	11.06	17.12	4.28	1061	Normal
Paper mash egg tray	75.33	84.02	2.27	0.27	2.00	11.89	15.98	3.39	1.54	Irregular
S.ED ±	2.05	2.01	0.15	0.02	0.13	0.50	2.01	0.06	0.02	
CD (5%)	4.39	4.30	0.31	0.04	0.28	1.08	4.30	0.14	0.05	

The frequency of feeding studies revealed that when larvae were fed for 3 times per day more number of unequal larvae, lower rate of good cocoons and lower ERR were observed compared to 4 and 5 times feeding per day (Table 5.10). The cocoon characters like cocoon weight, ERR (%) and silk ratio were not greatly influenced by the frequency of feeding (Table 5.9). Hence the study indicated that the number of feedings is not having much influence if sufficient food is provided during three times per day. However, Huq *et al.,* (1991) reported that the optimum number of feeds was 4/day at an interval of 6 hours.

Evaluation of bed cleaning schedule in Ericulture

During eri silkworm rearing, at least 4-5 feedings should be given per day including in the night at regular intervals during the young age rearing. In late age worms due to increase of feeding capacity, 5 feedings per day are essential and hence to remove the unconsumed leaves and litter, bed cleaning is essential. Because the pilling of litter makes beds moist. Such moist condition favours multiplication of pathogenic microorganisms and affects the health of worms and culture. In this regard, Subramanianan *et al.,* (2013) conducted an experiment for schedule of cleaning with the following treatments during fifth instar.

T1 Regular bed cleaning

T2 Cleaning once in two days

T3 Cleaning once in three days

T4 Cleaning once in fifth instar at 6th day

The results of the bed cleaning experiment revealed that the regular bed cleaning every day recorded the highest larval weight (6.80g), lowest mortality (2.20%), more good cocoons (80.20%), higher ERR % (90.43%) and highest silk ratio (13.42%). Lowest number of flimsy cocoons (19.80) were recorded in the treatment I, while the highest number of 28.07 and 30.22 number of flimsy cocoons were recorded in the treatments 3 and 4 respectively (Table 5.11). Hence it is advisable to clean the bed regularly to get higher cocoon yield.

Effect of mountages on cocoon and reproductive parameters of eri silkworm in the Department of Sericulture, Assam agricultural university, Jorhat

Most of the Sericigenous lepidopteran insects secrete silk after larval maturity and spin cocoons. Cocoon spinning stage is the productive phase and the cocoons are commercially exploited for silk production and egg production for further rearing. Silkworms are reared and allowed to spin cocoons on an artificial arrangement called mountage. The basic function of the mountage is to provide a specific angular space to each silkworm to spin its cocoons and to minimize the loss of silk during spinning. The process of transferring mature silkworm larvae to a suitable mountage to provide proper anchorage for cocoon formation is called mounting (Chandrakanth *et al.,* 2004). Different types of mountages are used in different countries which are generally made from locally available cheap and simple material. In India, mountages used for spinning of cocoons are bamboo spiral mountage (chandrika), plastic collapsible mountage, bottle brush mountage, rotary card board mountage and dried grass/straw/twigs etc (Rajan *et al.,* 1996). Sound rearing practices during the larval period and quality of mountage provided to the mature larvae play an equal part in deciding the quality of the cocoon. Hence Narzary *et al.,* (2014) made an effort to test and compare the quality of cocoon and reproductive parameters of eri silkworm by using different types of mountages in the department of Sericulture, Assam agricultural university, Jorhat, during 2010-11.

The eri silkworm larvae were reared on castor leaves in the laboratory under natural room temperature and humidity condition from first to fifth instar. When silkworm larvae get mature, they were collected daily by hand and transferred to seven different types of mountages *viz.,* chandrika, dry hingori (*Castanopsis indica*) leaf jail, dry banana (*Musa* spp) leaf jail, jack fruit (*Artocarpus heterophyllus*) leaf jail, straw (*Oryza* spp) mountage, folded newspaper and paper mash egg tray. Observations were recorded on cocoon parameters *viz.,* cocoon harvest (%), good cocoon (%), cocoon weight (g), shell weight (g), pupal weight (g), shell ratio (%), defective cocoon (%), size of the cocoons (cm) and shape of the cocoons. Other observations on reproductive parameters like moth emergence (%) and fecundity (no. of eggs laid/female) after harvesting of cocoons were also recorded.

The results of the experiment showed that the qualitative and quantitative characters of eri cocoon such as cocoon harvest (94.00%), good cocoons (97.87%), cocoon weight (3.16g), shell weight (0.43g), pupal weight (2.73g), shell ratio (13.61%) were found significantly highest in the chandrika followed by hingori leaf jail, banana leaf jail and straw mountage, but lowest was observed in folded news paper and paper mash egg tray (Table 5.12). The significant effect on defective cocoon percentage was recorded highest in the folded newspaper (17.12%) and lowest in chandrika (2.13%). The size and shape of cocoons were seen to be almost similar in all the mountages. Irregular shaped cocoons were found to be maximum in paper mash egg tray. The moth emergence (95%) and fecundity (294.13) were found significantly highest in chandrika (Table 5.13). The results indicated that the increase or decrease in the number of defective cocoons vary depending on structure and the material used for construction of the cocooning frame (mountage) which can also significantly affect the moth emergence and fecundity. The usage of chandrika, hingori leaf jail, banana leaf jail and straw mountage would improve cocoon weight, shell ratio, moth emergence and fecundity which are the need of the hour for farmers. These are quite economic, easily available with the rearers and also require very less cost as compared to chandrika. These are biodegradable and can be prepared every time freshly and need not be stored which also help in preventing contamination of pathogens to the next generation. Hence the rearers can adopt mounting of eri silkworm on these mountages conveniently.

Table 5.13: Reproductive characters of eri cocoons harvested from different mountages (Adopted from Narzary *et al.,* 2014)

Mountages	Reproductive parameter	
	Moth emergence (%)	Fecundity (No. of eggs laid/female)
Chandrika	95.00	294.13
Hingori leaf jail	90.00	273.30
Banana leaf jail	86.67	272.10
Jackfruit leaf jail	73.33	221.73
Straw mountage	85.00	272.57
Folded newspaper	75.00	193.07
Paper mash egg tray	70.00	209.47
S.Ed ±	4.27	9.56
CD (5%)	9.16	20.51

Evaluation of different mountage materials locally available in Tamil Nadu for Eri silkworm rearing:

In order to study the spinning ability and quality of cocoon of eri silkworm on different locally available mountage materials, Subramanianan *et al.,* (2013) tried the suitability of different mountages as follows:

T1 Rotary mountage

T2 Dried Sapota leaves

T3 Dried Coconut leaves

T4 Green Coconut leaves

T5 Plastic mountage

T6 Dried banana leaves

T7 Dried sugarcane leaves

T8 Dried palm leaf

T9 Dried sugarcane sheath

T10 Dried guava leaves

T11 Dried mango leaves

The ripened worms were collected from the rearing bed and allowed for spinning on the above mentioned materials during September 2003 – January 2005 in Tamil Nadu. The present study revealed that the worms allowed to spin on the banana, rotary, green coconut, mango leaves and plastic mountage were recorded statistically significant maximum shell weight (0.35-0.33 g) compared to the dried coconut, sapota, sugarcane leaf and sheath, palm and guava leaves (0.32-0.30 g). But, the silk ratio of the cocoons obtained from the different mountage materials used in this study were statistically not significant and ranged from 12.87 to 13.97% (Table 5.14). The length of the cocoon was maximum in worms allowed to spin on the banana, sapota, coconut, guava and mango leaves and ranged from 6.1 to 5.82 cm. The length of the cocoon was minimum when worms were allowed to spin on rotary mountage. The breadth of cocoon ranged from 2.09 to 2.58 cm and it was maximum in worms which spun on sapota (2.58) and minimum on palm leaf (2.09). Hence, it is clear that the mountage materials have influenced the cocoon and shell weight of eri silkworm and the mountages made with banana leaves, green coconut leaves, dried mango leaves, rotary and plastic mounatges can be used for spinning of cocoons of eri silkworm.

Table 5.14: Suitability of mountage materials locally available in Tamil Nadu for Ericulture (Adopted from Subramanianan *et al.,* (2013).

Materials	Cocoon weight	Shell weight	SR%	Cocoon Length	Cocoon Breadth
Rotary mountage	2.53[ab]	0.34[ab]	13.71	4.48[a]	2.43[b]
Dried Sapota leaves	2.46[a]	0.32[a]	13.17	5.83[c]	2.58[bc]
Dried Coconut leaves	2.47[a]	0.32[a]	12.90	6.00[c]	2.34[b]
Green Coconut leaves	2.67[ab]	0.34[ab]	12.87	5.74[bc]	2.39[b]
Plastic mountage	2.36[a]	0.33[a]	13.97	5.52[b]	2.51[b]

Table 5.14: *Contd...*

Dried banana leaves	2.46^a	0.35^{ab}	14.33	6.10^c	2.51^b
Dried sugarcane leaves	2.38^a	0.31^a	13.10	5.60^b	2.22^a
Dried palm leaf	2.35^a	0.31^a	13.08	5.48^b	2.09^a
Dried sugarcane sheath	2.25^a	0.31^a	13.86	5.33^b	2.42^b
Dried guava leaves	2.32^a	0.30^a	13.05	5.83^c	2.40^b
Dried mango leaves	2.51^{ab}	0.33^a	13.14	5.82^c	2.46^b
F-Value	2.39	2.12	NS	14.07	3.32

Effect of cocoon mountage substrates on cocoon and grainage parameters of eri silkworm in the Department of Entomology, Central Agricultural University, Imphal

Production of eri cocoon depends on many factors such as the food plants used for rearing silkworm, agro-techniques adopted for raising the plantation of food plant, strain/ eco race of the silkworm reared, rearing condition, feeding frequency, care taken during the mounting period and hygienic conditions maintained during rearing and the post cocoon harvesting activities etc. It is felt that cocoon mountage, the substrate on which the mature silkworms are allowed to form cocoons, can affect the quantitative and qualitative aspects of the cocoons. Dried leaves and twigs (called as jail) of several plants have been utilized traditionally as cocoon mountages in NE India, however, the comparative efficacy of these substrates is scanty. In this context, Sarma and Singh, (2007) made a comparative study to know the effect of different cocoon mountage substrates locally available in Manipur on cocoon parameters and grainage behaviour of eri silkworm.

The experiment was conducted during Autumn and Spring rearing of 2002 and 2003 in the Department of Entomology, Central Agricultural University, Imphal. The fully mature larvae were subjected to six treatments *viz.,* Banana jail, Eucalyptus jail, Mango jail, Guava jail, Chandrika and Bi-layered net. Every treatment was covered with nylon net except bi-layered net in order to restrict the movement of the silkworm within the treatment. Harvested cocoons were sorted out as normal and abnormal cocoons, and then the abnormal cocoons were categorized as double cocoon, stained cocoon, thin-shelled cocoon and smaller cocoon. Cocoons with 20% less weight than that of normal cocoons were selected as smaller cocoons. The cocoons of each replication under observation were sorted in single layer on the tray till the emergence of moth. The results of the experiment revealed that the cocoon mountage substrates had variable impact on some of the important ericultural parameters. Chandrika and banana jail emerged to be the best cocoon mountage with a better record of normal cocoon formation (85.00% and 82.50%), cocoon weight (20.20 g and 19.70 g per ten cocoons), moth emergence (88.50% and 88.10%), fecundity (275.61 and 282.19 per laying) and shell-percent to an extent of 13.13% and 12.10% respectively. From ericulture point of view the mountage substrates had a descending order of efficiency as follows: Chandrika = Banana jail > Eucalyptus jail > Bi-layered net > Guava jail = Mango jail (Table 5.15).

A simple score method has been adopted for ranking the mountage substrates based on their relative performance in relation to those parameters for which significant difference among the mountage substrates was noticed. A score of 1/2/3/4/5/6 (1 to 6) was given to a treatment against its rank of 6th /5th /4th /3rd /2nd /1st (6 to 1) in relation to a parameter, and thus, the total score of a treatment for five parameters has been considered to determine their overall rank. An equal score was given to all the treatments which were statistically at par for a particular parameter (Table 5.16).

Studies on the impact of substrata on ovipositional behaviour of eri silkworm

Quality seed is the backbone of sericulture industry. According to Subba Rao and Kakoty (1975), the present grainage technology in eri silkworm is handicapped by the lack of precise ecological knowledge on moth emergence, egg production and hatching percentage. Timely supply of superior quality eggs helps to sustain ericulture as a commercial crop. For expanding this industry, there is a dire need for organized seed production and supply source to the farmers. Though literature is available on some aspects of rearing technology, host plant cultivation and grainage aspects of this insect, there is a dearth of information on many aspects of ericulture in general and grainage activities such as synchronization of moth emergence and production of disease free and viable quality eggs in particular which are of paramount importance for timely supply of layings for successful ericulture practice. Although studies of earlier workers indicated increased ovipositional potential and egg recovery on Kharika, there are certain controversies and constraints encountered in *S. cynthia ricini*.

Table 5.15: Relative effects of cocoon-mountage-substrates in relation to ericultural parameters
(Adopted from Sarma and Singh, 2007)

Treatment Substrate	Cocoon parameters							Grainage behaviour			
	Normal cocoon (%)	Abnormal cocoon (%)					Cocoon weight (g/10 nos)	Shell (%)	% Moth emergence	Fecundity (eggs/ laying)	Hatching rate (%)
		Double	Stained	Thin-shelled	Small	Total					
Chandrika	82.50 (65.27)	1.50 (7.04)	6.00 (14.18)	4.50 (12.25)	5.55 (13.63)	17.50	19.70	13.13 (21.22)	88.50 (70.18)	275.61	87.03 (68.87)
Bi-layered net	79.50 (63.08)	3.00 (9.78)	7.00 (15.34)	4.00 (11.54)	6.55 (14.83)	20.50	18.50	12.89 (21.05)	83.30 (65.88)	230.83	86.89 (68.78)
Banana jail	85.00 (67.21)	0.00 (0.00)	5.00 (12.92)	6.00 (14.18)	4.00 (11.54)	15.00	20.20	12.10 (20.36)	88.10 (69.82)	282.19	87.11 (68.95)
Guava jail	76.00 (60.67)	0.00 (0.00)	15.00 (22.79)	5.00 (1.92)	4.00 (11.54)	24.00	18.80	12.41 (20.62)	81.67 (64.65)	215.09	86.48 (68.44)
Eucalyptus jail	84.50 (66.81)	0.00 (0.00)	10.00 (18.44)	3.00 (9.98)	3.50 (10.18)	16.50	19.50	12.33 (20.93)	84.33 (66.66)	200.38	87.02 (68.87)
Mango jail	77.00 (61.34)	0.00 (0.00)	12.50 (20.70)	4.50 (12.25)	6.00 (14.18)	23.00	18.20	11.88 (20.18)	82.77 (65.50)	288.49	86.77 (68.70)
CD (5%)	2.10	0.12	1.19	NS	0.34	-	1.41	0.44	3.32	26.17	NS

NB: The values in parentheses are angular transformed values

Table 5.16: Score table of cocoon-mountage-substrates based on relative performance in ericultural operations (Adopted from Sarma and Singh, 2007)

Treatment	Normal cocoons		Cocoon weight		Shell percent		Moth emergence (%)		Fecundity		Overall	
	Rank	Score	Rank	Score	Rank	Score	Rank	Score	Rank	Score	Rank	Score
Chandrika	2	5	1	6	1	6	1	6	1	6	29	1
Bi-layered net	3	4	2	5	1	6	2	5	2	5	25	3
Banana jail	1	6	1	6	2	5	1	6	1	6	29	1
Guava jail	4	3	2	5	2	5	2	5	2	5	23	4
Eucalyptus jail	1	6	1	6	2	5	2	5	3	4	26	2
Mango jail	4	3	2	5	3	4	2	5	1	6	23	4

Studies conducted at Central Sericultural Research and Training Institute, Mysore

Somaprakash and Sathya Prasad (2008) made an attempt to systematically collect the information on the egg recovery on different oviposition substrata in order to recommend a cheap, highly efficient and eco-friendly substratum for oviposition of the eri moth. The experiment was conducted in the eri rearing house of Central Sericultural Research and Training Institute, Mysore during 2005-07.

The cocoons were sorted out and placed in a single layer in 2 x 3 inch plastic trays covered with wire mesh for emergence. The moths were collected and allowed to mate for 4-5 hrs. After mating these were de paired and the females were kept on different substrata *viz.,* nylon bag, bamboo strips, plastic rope, coir rope and mulberry twigs and kharika which were positioned vertically by tying one end to the stand. The kharika served as control. The observations were recorded on day wise fecundity up to three days.

The results of the experiment revealed that higher egg recovery was observed on bamboo strips (264.75±5.00) followed by plastic rope (235.50±8.68) over kharika (220.75±9.55) and significantly lower egg recovery was noticed in nylon bag and mulberry twigs. This implies that bamboo strips are more preferred over the other substrata (Table 5.17). Irrespective of the month of rearing and oviposition, maximum total egg recovery was observed on bamboo strips (386.80±15.23) followed by plastic rope (368.50±14.10) and mulberry twigs (337.80±12.93) over the control, that is kharika (330.55±10.17), suggesting that the physical factors did have impact on the oviposition. It was also observed that irrespective of oviposition, maximum eggs were laid on the Ist day and the egg laying was continued for three days. Significantly higher egg recovery was observed during the month of September, followed by May, August and March suggesting that the temperature and relative humidity prevalent during September month under Mysore conditions are more congenial for higher egg productivity (Table 5.18).

Hence it is concluded that bamboo strips are more convenient and preferred over kharika for egg laying, while the other substrata such as plastic rope and mulberry twigs did not show any significant variation when compared. The higher preference of bamboo strips is attributed to availability of more surface area for the moths to alight and oviposit.

Table 5.17: Effect of substrata on day-wise egg recovery in *S. cynthia ricini* (Adopted from Somaprakash and Sathya Prasad, 2008)

Substratum	Day-wise fecundity (number)			
	Day 1	Day 2	Day 3	Total
Nylon bag	172.75 ± 4.88^d	104.25 ± 11.64^a	48.00 ± 4.23^{ab}	325.25 ± 15.23^c
Bamboo strips	264.75 ± 5.00^a	81.20 ± 5.26^{bc}	41.50 ± 1.78^b	386.80 ± 9.26^a
Plastic rope	235.50 ± 8.68^b	96.50 ± 7.57^{ab}	36.75 ± 2.72^c	368.50 ± 14.10^{ab}
Coir rope	221.00 ± 8.63^b	65.50 ± 5.60^c	33.50 ± 1.70^c	320.00 ± 8.44^c
Mulberry twigs	195.25 ± 9.13^c	88.15 ± 8.17^{ab}	54.50 ± 5.72^a	337.80 ± 12.93^{bc}
Kharika	220.75 ± 9.55^b	78.00 ± 2.58^{bc}	32.00 ± 2.86^c	330.55 ± 10.17^c
F value	Sig	Sig	Sig	Sig
CD at 5%	36.27	33.85	15.97	55.02

Table 5.18: Impact of grainage operation period on egg recovery of *S. cynthia ricini* (Adopted from Somaprakash and Sathya Prasad, 2008)

Substratum	Day-wise fecundity (number)			
	September	March	August	May
Nylon bag	383.20 ± 4.00^b	221.00 ± 2.25^c	372.80 ± 1.71^b	324.00 ± 16.12^b
Bamboo strips	426.00 ± 4.36^a	326.00 ± 5.82^a	384.20 ± 3.53^a	411.00 ± 10.44^a
Plastic rope	388.20 ± 5.64^b	274.00 ± 4.47^b	370.80 ± 2.87^b	441.00 ± 8.00^a
Coir rope	361.00 ± 6.52^c	267.00 ± 8.45^b	321.00 ± 5.15^b	331.00 ± 9.7^b
Mulberry twigs	379.40 ± 3.15^b	250.00 ± 20.02^b	369.00 ± 1.58^b	352.80 ± 10.12^b
Kharika	375.20 ± 5.64^b	260.00 ± 3.32^b	331.00 ± 2.00^c	356.0 ± 3.74^b
Mean	385.50 ± 4.10	266.33 ± 6.85	358.13 ± 4.50	369.30 ± 8.78
F value	Sig	Sig	Sig	Sig
CD at 5%	10.73	21.89	7.06	23.81

Studies conducted at Nagaland University, Medziphema

Similar experiment was conducted by Chang *et al.,* (2012) in the Department of Entomology, School of Agricultural Sciences and Rural Development, Nagaland University, Medziphema during 2010 with different treatments like twigs, paper box, coir rope and nylon bag by keeping the kharika (thatch stick) as control. In this experiment the egg recovery was significantly highest on kharika (13.75 and 10.75 respectively) on 1st and 2nd day followed by twigs, paper box, coir rope and nylon bag (Table 5.19). However, there was no significant difference among treatments in egg recovery on 3rd day. Moreover, maximum egg recovery was recorded on 1st day in all the substrata and it was decreasing as the days increased. The highest egg laying in kharika is attributed due to its smooth surface which encourages the female to deposit all the eggs formed in the ovarian follicules. Earlier, Gupta *et al.,* (1990) recorded the highest oviposition in smooth surface i.e. glasses. Debaraj *et al.,* (2003) also reported the highest egg recovery on 1st day and it decreased on 2nd and 3rd day. It can be concluded that kharika is the most suitable substrata for egg laying and can be utilized at commercial scale. Twigs, paper box, coir rope and nylon bag showed almost equal performance next to kharika, so these will be good tools to get maximum egg recovery with low cost and also affordable by farmers.

Table 5.19: Effect of substrata on egg recovery in eri silkworm (Adopted from Chang *et al.,* (2012)

Substratum	*Day-wise fecundity (number)		
	1st day	2nd day	3rd day
Kharika	13.75	10.75	9.99
Wigs	10.91	10.20	9.88
Paper box	10.62	10.19	9.96
Coir rope	10.30	09.95	9.70
Nylon bag	09.86	09.85	9.56
SEm±	0.272	0.236	0.081
CD at 5%	0.61	0.53	0.18

*Values are mean of ten observations with three replications

Hence from the studies of Somaprakash and Sathya Prasad, 2008 and Chang *et al.,* (2012), it can be inferred that preference of eri silkworm oviposition substrata differs depending upon the place and season of experimentation.

Different Host Plants and Different Methods of Rearing Eri Silkworm and their Impact on Rearing Parameters

Subramanianan *et al.,* (2013) studied different methods of rearing and their impact on rearing parameters of eri silkworm with 11 treatments listed below. Same size of the platform (6 x 2 ft.) was used for all the treatments with a larval population of 50 per sq. ft. All the developmental parameters were observed on 50 larvae per replication in three places selected at random on each bed. The treatments include

T1 Bed method of rearing on double bloom green castor variety

T2 Bunch method of rearing on double bloom green castor variety

T3 Bed + Bunch method of rearing on double bloom green castor variety

T4 Bed method of rearing on Kumkum Rose variety of tapioca

T5 Bunch method of rearing on Kumkum Rose variety of tapioca

T6 Bed + bunch method of rearing on Kumkum Rose variety of tapioca

T7 Bed method of rearing on Burma White variety of tapioca

T8 Bunch method of rearing on Burma White variety of tapioca

T9 Bed + bunch method of rearing on Burma White variety of tapioca

T10 Floor rearing on Kumkum Rose variety of tapioca

T11 Tray rearing on Kumkum Rose variety of tapioca

The observation on different host plants and methods of rearing experiment indicated that the bunch and bunch + bed methods of rearing on castor caused less mortality, lower unequal cocoons and greater ERR (90.00 and 88.30%) compared to other methods and these were statistically significant. Though the bunch method of rearing either for castor or tapioca was consuming more time, more labour and more space, the bunch method provides good aeration which results in less mortality lower number of unequal larvae and maximum ERR% (Table 5.20). The results of the experiment indicate that the maximum larval weight (7.6 g) has been obtained in bed + bunch method of rearing on double bloom green castor, highest cocoon weight (2.96 g) in bed method of rearing on Kumkum Rose variety, maximum effective rate of rearing (90.0%) in bunch method of rearing on double bloom green castor. Further, maximum shell weight (0.44 g) and highest silk ratio (14.9 g) were recorded in bed method of rearing on Kumkum Rose variety.

Table 5.20: Influence of combinations of methods of rearing and varieties of host plants on larval and grainage parameters of eri silkworm (Adopted from Subramanianan *et al.,* 2013)

Treatment	Mature larval wt. (g)	3 days old cocoon wt. (g)	Shell wt. (g)	Silk ratio (%)	ERR (%)
T1	7.2	2.62	0.35	13.48	82.70
T2	6.8	2.48	0.32	12.94	90.00
T3	7.6	2.76	0.40	14.13	88.30
T4	7.5	2.96	0.44	14.90	84.00
T5	6.7	2.85	0.42	14.70	82.00
T6	6.9	2.93	0.43	14.70	83.00
T7	6.2	2.85	0.39	13.70	75.00
T8	6.4	2.69	0.37	13.70	77.00
T9	6.0	2.57	0.35	13.60	76.00
T10	6.6	2.88	0.41	14.20	83.00
T11	6.2	2.95	0.43	14.60	84.00
F-value	10.60	11.70	9.80	10.60	5.79

Effect of feeding methods of tapioca leaves on economic traits of eri silkworm in Tamil Nadu

The method and form of providing leaves to silkworms along with the frequency and the quantity of leaf provided are known to influence the performance of eri silkworm (Sannappa *et al.,* 2002). Ravishankar *et al.,* (2000) recorded higher larval weight, ERR, cocoon weight, pupal weight, shell weight, shell ratio, fecundity and hatching percentage when the worms were fed on whole shoot throughout the larval period. Similarly, Reddy *et al.,* (1999) found that feeding on castor leaves in shoots (4-5 leaves) was better than other methods. According to Saratchandra and Joshi (1984), the mature larval weight was lower (4.68 g) in tray feeding method throughout the larval feeding against bunch feeding method which was higher (4.8g) during December-January in Assam. Further Saratchandra and Joshi (1984) also reported moth emergence of 75.15% and 80.07% in tray feeding methods compared to 81.98% and 82.59% in bunch feeding method on castor and kesseru host plants respectively. The fecundity was 265 eggs in tray rearing method against 275 eggs in bunch method of feeding. However, the authors suggested that supply of castor leaves on trays up to third instar, a branch with 10-12 castor leaves for fourth and fifth instars can be adopted to get better results. Jayaraj *et al.,* (2004) identified that leaf feeding on rearing bed (bed method), hanging in bunches (bunch method) and bed cum bunch methods were suitable for rearing of eri silkworm on castor but observed significantly improved cocoon traits on bed cum bunch method. Large scale rearing of eri silkworm on tapioca was successfully employed using 3-5 tier shelves either by feeding the leaves with petiole or leaves with whole shoot (Sakthivel, 2004).

By evolving eri silk worm technology on tapioca leaves, there is feasibility of additional remuneration to about 1.6 lakh tapioca growers in India mainly in Southern states of Tamil Nadu, Kerala and Andhra Pradesh. In this regard, Sakthivel, (2014) conducted a study to find out the suitable method of feeding tapioca leaves in different seasons for better growth of eri silkworm and silk yield. Eri silkworms were reared in different seasons during 2009 to 2012 on the ruling variety of tapioca in Tamil Nadu, MVD1. A total of six rearings per year (@ two rearings per season) were conducted. The larvae were mass reared up to second instar and then the treatments were imposed from 3rd instar till maturity of worms. The treatments comprised of T1= Leaf (L) (offering leaves along with petiole), T2= Shoot (S) (offering leaves with entire shoot), T3=

Bunch feeding (tying 20-25 leaves in bunch and hanging over the rearing bed), T4= Leaf cum bunch feeding (L+B) and T5= Shoot cum bunch (S+B) feeding. Observations were recorded on economic traits of silkworms.

Results of the experiment revealed that the shoot cum bunch feeding method registered significantly higher cocoon yield (50.13, 61.57 and 75.16) in summer, rainy and winter seasons respectively. Similarly higher shell yield (6.97, 9.64 and 12.13 kg/100 dfls) and higher SR % (13.90, 15.65 and 16.36) in summer, rainy and winter seasons respectively were also recorded in shoot cum bunch feeding method. The next best treatment was shoot feeding (48.85, 60.39 and 73.74 and 6.47, 8.90, 11.88 kg/100 dfls with SR% of 13.48, 14.73 and 16.10) followed by leaf cum bunch feeding (47.95, 58.87 and 72.36 and 6.39, 8.70 and 11.80 with SR% of 13.08, 14.77 and 16.30) and bunch feeding (47.11, 57.45 and 71.42 and 6.04, 8.13 and 11.48 with SR% of 12.82, 14.15 and 16.07) irrespective of season. Leaf feeding method recorded least yield except that of winter season in which it was on par with bunch feeding recording 46.32, 55.81 and 71.56 and 5.81, 7.85 and 11.44 kg/100 dfls and SR% of 12.54, 14.07 and 15.99 in summer, rainy and winter seasons respectively (Table 5.21)

In the investigation by Sakthivel (2014), shoot cum bunch feeding method registered highest cocoon and shell yield and also SR% in all the seasons. The next best treatment was shoot feeding followed by leaf cum bunch feeding irrespective of seasons.

Evaluation of different methods of laying preparation of eri silkworm in Karnataka

Devaiah *et al.,* (1981) conducted a laboratory experiment in Karnataka with 3 methods of preparing females of the silk moth *Samia cynthia ricini* (Boisd.) for oviposition *viz.,* females were tied by a thread to twigs of castor suspended in a rearing room, females were enclosed in a bamboo basket and females were enclosed in a cellule on an egg card by clipping their wings. Observations were recorded on the number of eggs laid per female. The results showed that, when the mated females were tied to the twigs of castor, the females laid the highest number of eggs i.e. 240.08, followed by 229.33 eggs in bamboo basket and 188.08 eggs in a cellule. Hence it can be inferred that out of the three methods, the bamboo basket was more convenient for laying preparation than the other two methods.

Table 5.21: Effect of different feeding methods of tapioca leaves on the economic traits of eri silkworm
(Adopted from Sakthivel, 2014)

Methods (M)	Season (S)	Larval period D:H	Mature larval weight (g)	ERR %	Coocon yield (kg/100 dfls)	Shell yield (kg/100 dfls)	SCW (g)	SSW (g)	SR (%)	Fecundity (no)	Hatching (%)
T1 Leaf (L)	S	21:19	5.66	75.26	46.32	5.81	2.14	0.27	12.54	329.13	75.56
	R	23.06	6.18	83.39	55.81	7.85	2.32	0.33	14.07	344.38	88.30
	W	30.10	6.69	93.06	71.56	11.44	2.67	0.43	15.99	361.07	95.26
T2 Shoot (S)	S	21:19	5.86	79.01	48.85	6.47	2.16	0.29	13.48	335.96	80.35
	R	23.06	6.34	89.53	60.39	8.90	2.34	0.35	14.73	349.18	92.80
	W	30.10	6.92	95.47	73.74	11.88	2.68	0.43	16.10	364.33	96.90
T3 Bunch (B)	S	21:19	5.72	76.29	47.11	6.04	2.14	0.28	12.82	333.19	78.85
	R	23.06	6.28	85.57	57.45	8.13	2.33	0.33	14.15	345.68	88.88
	W	30.10	6.78	92.70	71.42	11.48	2.68	0.43	16.07	361.33	95.39
T4 L+B	S	21:19	5.74	77.15	47.95	6.39	2.15	0.28	13.08	332.66	80.15
	R	23.06	6.30	87.06	58.87	8.70	2.35	0.35	14.77	346.00	92.04
	W	30.10	6.88	93.33	72.36	11.80	2.69	0.44	16.30	364.88	96.10
T5 S+B	S	21:19	5.87	80.40	50.13	6.97	2.17	0.30	13.90	338.35	81.45
	R	23.06	6.42	90.47	61.57	9.64	2.36	0.37	15.65	350.27	92.10
	W	30.10	6.99	95.97	74.16	12.13	2.68	0.44	16.36	365.00	96.67
CD (5%)	M		0.471	2.349	1.503	0.189	0.0042	0.0035	0.17	3.149	2.158
	SN	-	0.413	1.761	1.206	0.193	0.0040	0.0029	0.14	3.273	1.515
	M x SN		0.088	3.979	2.663	0.400	0.0085	0.0064	0.31	6.749	3.505
	SN x M		0.092	3.939	2.697	0.432	0.0090	0.0066	0.32	7.319	3.389

Effect of Leaf Storage on Nutritional Changes in Food Plants and its Impact on Larval Growth and Cocoon Quality of Eri Silkworm

The eri silkworms are reared indoor on trays or on hanging leaf bundles providing 4-5 times feedings per day at regular interval depending upon the age of the larvae. Leaves harvested from the plant lose water rapidly due to withering and considerably affects the leaf quality and also the edibility or palatability of leaves to silkworms. Duration of storage, temperature, humidity, light, aeration, initial quality of leaves at harvest time and many other conditions greatly influence the final quality of leaves at the time of feeding. Therefore, storage of leaf is important to preserve them fresh as well as succulent and nutritious till they are consumed by the silkworm. Different approaches of leaf storage methods are advocated during mulberry silkworm rearing in different regions and situations, but these are lacking in eri silkworm. Hence Dutta *et al.,* (2015) conducted a study with an aim to develop a suitable method for preservation of eri silkworm food plant leaves considering the nutritional changes in different durations and its impact on larval growth and cocoon quality of the silkworm.

The experiment was carried out in two commercial seasons *viz.,* spring and autumn during the years 2012-13. The leaves of three food plants i.e. castor, kesseru and borpat were harvested in the morning hours and preserved separately in three different methods *viz.,* open floor (heaped on mat and covered with wet gunny cloth), bamboo basket (preserved in bamboo made leaf chamber and covered with wet gunny cloth) and poly bag (preserved in poly bag and covered with wet gunny cloth) and stored for 48 hours. The preserved leaves were subsequently used for rearing of eri silkworm and subjected to chemical analysis to estimate moisture, total soluble sugar, crude protein, total mineral, crude fibre and total chlorophyll after 12, 24, 36 & 48 h of storage durations. Fresh samples (0 h) of all the food plant leaves were also used for chemical analysis and a separate batch of silkworm were reared on food plant-wise with fresh (unpreserved) leaves for comparison. Observations were recorded on larval, growth and cocoon parameters *viz.,* larval duration, larval weight, cocoon yield, cocoon weight, shell weight and pupal weight.

The results revealed that the leaf preservation duration and leaf storage methods had significant effect on the nutritional value of the food plant leaves of eri silk worm. The leaves could be preserved up to 48 hours without much deterioration of moisture, protein and total soluble sugar for rearing of eri silkworm. Leaf preservation in Bamboo basket covered with wet gunny cloth is better than in poly bag and open floor considering the retaining capacity of nutrient contents in the food plant leaves and larval growth and cocoon characters of the silkworm (Table 5.22).

Leaf storage method has significant effect on larval growth and cocoon quality of eri silkworm. Leaf preservation in bamboo basket was found to be at par with fresh leaves and better than poly bag and open floor considering the larval growth, cocoon yield and cocoon quality (Table 5.23) (Plate 38). After preservation, feeding of castor leaves though registered better for larval growth parameters with higher cocoon yield, Kesseru and borpat leaves appeared to be better for economic cocoon characters with higher silk content on the cocoons. These variations among the food plants might be due to variations of nutrient retaining capacity among the food plants after storage. It was reported that the environmental condition during rearing period also play a significant role in the growth and development of the silkworm and ultimately the cocoon crop. Hence, the investigations of Dutta *et al.,* (2015) revealed that the leaves of spring flash were better than autumn flash for rearing of eri silkworm after storage considering the economic larval growth and cocoon quality.

Plate 38 Leaf storage in Bamboo basket

Table 5.22: Effect of host plant leaves, storage durations and methods on nutrient contents of eri silkworm (Adopted from Dutta *et al.*, 2015)

Factors	Moisture (%)	Protein (%)	Total soluble sugar (%)	Total mineral (%)	Crude fibre (%)	Chlorophyll (mg/g)
Storage duration (hours)						
Fresh (0 hrs)	74.24	13.47	10.32	8.95	46.42	57.70
12 hrs	73.25	12.37	9.89	9.21	45.79	57.35
24 hrs	71.50	11.09	8.90	9.66	44.93	54.86
36 hrs	70.43	10.13	7.84	9.93	43.30	52.79
48 hrs	69.31	9.57	6.47	10.30	41.88	50.85
CD at 5%	1.32	0.50	1.83	0.25	1.13	2.47
Leaf storage method						
Bamboo basket	73.30	12.57	9.25	8.93	42.35	49.27
	(98.78)	(93.32)	(89.63)	(99.66)	(91.23)	(86.29)
Poly bag	71.65	9.08	8.07	10.04	44.81	53.30
	(96.51)	(67.41)	(78.20)	(+1.12)	(96.53)	(93.02)
Open floor	68.43	10.73	7.51	10.37	45.05	59.31
	(92.17)	(79.66)	(72.77)	(+1.16)	(97.05)	(97.28)
CD at 5%	2.28	0.77	0.81	1.80	1.87	3.28
Food Plant						
Castor	79.47	12.68	6.75	11.77	43.32	46.84
	(91.70)	(71.22)	(69.52)	(+14.55)	(91.98)	(89.32)
Kesseru	68.93	10.70	9.28	8.70	45.65	59.94
	(93.98)	(72.25)	(43.84)	(+14.67)	(87.16)	(92.02)
Borpat	66.83	9.00	10.02	8.36	44.42	57.34
	(94.75)	(69.43)	(78.23)	(+16.28)	(93.22)	(88.27)
CD at 5%	2.28	0.77	0.81	1.80	1.87	3.28
Interactions	NS	NS	NS	NS	NS	NS

NB: The values in parentheses are percentage of retaining capacity of nutrients after 48 hrs of storage

+ indicates increase in percentage of nutrients after 48 hrs of storage

Table 5.23: Effect of storage of food plants leaves on larval growth, cocoon yield and cocoon characters of eri silkworm (Adopted from Dutta *et al.,* 2015)

Factors	Larval duration (days)	Larval weight (g)	Effective rate of rearing* (%)	Cocoon weight (g)	Shell weight (g)	Shell ratio (%)	Pupal weight (g)
Leaf storage method							
Fresh leaves	22.33	5.01	83.02	3.12	0.38	12.16	2.75
Bamboo basket	22.58	4.84	82.79	3.08	0.37	12.09	2.70
Poly bag	22.71	4.58	80.17	2.81	0.30	10.80	2.51
Open floor	22.75	4.81	81.80	3.03	0.34	11.42	2.69
CD at 5%	NS	0.14	NS	0.07	0.01	0.64	0.07
Food Plant							
Castor	21.88	5.10	84.63	2.75	0.32	11.64	2.46
Kesseru	23.13	4.58	80.83	3.19	0.38	11.96	2.82
Borpat	22.79	4.72	79.92	2.98	0.35	11.76	2.64
CD at 5%	0.68	0.20	2.34	0.15	0.01	NS	0.15
Season							
Autumn	23.58	4.56	80.83	2.86	0.35	11.52	2.53
Spring	21.61	5.04	82.75	3.09	0.33	11.35	2.74
CD at 5%	0.56	0.16	NS	0.12	0.01	NS	0.13
Interactions	NS	NS	NS	NS	NS	NS	NS

*cocoon yield per 100 larvae

Effect of cocoon refrigeration in eri silkworm, *Samia c. ricini* on eclosion, fecundity and hatching

Swamy *et al.,* (1988) conducted a laboratory experiment to study the effect of cocoon refrigeration on eclosion, fecundity and hatching of eri silkworm. Pupae of 9, 10 and 11 day old were stored for 2, 4 and 6 days at 5°C. The results revealed that there was no significant difference in adult emergence between the pupae refrigerated for different times or between pupae of different ages. Refrigeration of 10 day old pupae for 4 days resulted in lower fecundity than refrigeration of 2-day-old pupae for the same time. Similarly, refrigeration of 11 day old pupae for 6 days resulted in lower fecundity than refrigeration of pupae of the same age for 2 days. Pupal age and duration of refrigeration did not have significant effects on egg viability, except that refrigeration of 9 day old pupae for 6 days and 10 day old pupae for 4 days resulted in lower percentage hatching.

Embryo isolation and egg preservation technology of eri silkworm *Samia c. ricini*:

Embryo isolation and egg preservation is one of the important aspects for synchronizing the eri silkworm rearing activities. Sarkar *et al.,* (2012) studied the effect of egg preservation on embryo isolation by using 36-40 hrs old eggs. The results of the experiment revealed that the 36-40 hrs old age group eggs can be preserved maximum days without any adverse effect. Bioassay conducted on the same group of eggs hatched after 25 days of oviposition recorded cocoon weight (3.2 gm), shell weight (0.42 gm) and ERR (68%), which were on par with the fresh eggs. From the study, it can be inferred that multivoltine eri eggs can be preserved in the particular embryonic stage and ages.

Summary and Conclusions

➢ Among the different vanya silk worms, the eri silkworm, *Samia cynthia ricini* (Boisduval) is highly domesticated and suitable for indoor rearing.

➢ Ericulture, rearing of eri silkworm is a labour oriented, low investment, agrarian small scale industry which suits both marginal and small land holders because of its high returns, short gestation period, and also it creates opportunity for own family employment round the year.

➢ The rearing house should be selected such that it should have facilities for creation and maintenance of the optimal environmental conditions inside the house.

- ➢ Eri silkworm is a hardy race, less susceptible to diseases. However, disinfection of the rearing house is inevitable activity before conducting silkworm rearing.

- ➢ Disinfection means destruction of disease causing microbes. Disinfection of rearing house, rearing appliances, silkworm rearing bed and maintenance of hygiene during rearing are the most essential activities for successful cocoon production.

- ➢ The silkworm rearing is divided into two main phases *viz.,* Young age silkworm rearing and late age silkworm rearing depending upon the nutritional requirements and environmental conditions to be maintained during rearing.

- ➢ Young age silkworm rearing is also known as "Chawki rearing." It generally starts from hatching of crawling insects called larva from the egg.

- ➢ Temperature from 26-28°C and relative humidity ranging 85-90 % are ideal conditions for young age rearing.

- ➢ Late age worms consist of fourth and fifth instars which need preferably lower temperature and humidity during rearing conditions.

- ➢ The ideal conditions of temperature ranged 24-26°C and relative humidity ranged 70-80 % should be maintained during the rearing of late age worms.

- ➢ During rearing, as soon as the larvae grow-up, the unconsumed leaves and litter increase in the rearing bed favouring multiplication of pathogenic organisms. Hence, timely bed cleaning is essential to keep the worms healthy.

- ➢ Moulting is a very sensitive period during which the worm casts-off its old skin and the body is soft and delicate. The larvae take 12-36 hours to complete the moulting process during the different instars and different seasons.

- ➢ Matured fifth instar larvae stop feeding and become restlessly moving here and there to search a suitable place for cocooning.

- ➢ Bunch method of rearing is much cleaner, can accommodate more worms in less space with minimum man power. However, the necessity of timely replenishment of leaves and falling the worms from the bunch often resulting in mechanical injury are the drawbacks of bunch method.

- ➢ Tray rearing method is generally adopted in most of the experimental rearings to maintain constant number of worms and also to facilitate sorting out of weak, irregular and deformed/ diseased worms. Mechanical injuries are by and large minimum in this method of rearing.

➢ Beds are often soiled with the excreta of worms, favouring fungal/ bacterial/ viral infections there by leading to loss of crops due to outbreak of diseases.

➢ Platform rearing is generally recommended for mass scale rearing (not less than 100 dfls) of late age eri silkworms. The advantages of this method are labour dependent risk is reduced and handling of silkworm is minimized. Hence, contamination and spreading of diseases is reduced. Better hygienic conditions, aeration in the bed is ensured with less cost. But, occupies larger space.

➢ Production of absolutely disease free, particularly the pebrine (protozoan disease of silkworm) free layings is highly essential for harvesting successful cocoon crops.

➢ A grainage room having working area of 34' x 18' x 12' bamboo made house with thatched roof and mud plastered wall is preferable for seed production.

➢ To get disease freeness of layings, the eggs are washed with clean water and immersed in 2% formalin solution for 10-15 minutes and again washed with pure water until no smell of formalin is detected.

➢ The eri silk moth lays eggs in clusters and the number of clusters and number of eggs per cluster varies according to days of oviposition. Eggs laid within 48 hours are best in terms of hatchability and rearing performance.

➢ Eggs laid on first day are superior with respect to cocoon and grainage parameters studied. However, the eggs laid on second day did show almost similar performance.

➢ Eri silkworm eggs of 1-2 day-old can be consigned at a temperature of 5^0C for 5 days and 2 day old eggs at 15^0C for 5 to 10 days. This will not affect the egg hatchability.

➢ The moisture content of the eggs of *S. c. ricini* varied from 65 to 71% during different embryonic developmental stages.

➢ The eggs of *Samia c ricini* being cleidoic type cannot draw nutrients from outside or cannot throw off from inside the egg, except for oxygen and carbon dioxide.

➢ The levels of total egg nitrogen and shell nitrogen remain constant during the embryogenesis of *Samia cynthia ricini*. With advancement of age, the protein nitrogen and non-protein nitrogen decrease.

- Among different fractions of proteins, total crude protein varied between 383.5 to 386.4 µg/mg egg while, total protein varied between 225.0 to346.6 µg/mg egg during different stages of embryonic development of eri silk worm eggs.

- The embryo requires 10 days (240 hrs) for complete development at a constant temperature of $23.5 \pm 1.0°C$ in BOD incubator.

- In an experiment conducted to test the suitability of low-cost bed material used for eri silkworm rearing, coconut leaf mat with newspaper and gunny bag with newspaper were better in cocoon and grainage parameters.

- During rearing of eri silkworm crowding should be avoided by providing an optimum condition and fifty worms per square foot was the optimum population density for better eri silk productivity.

- In an evaluation of suitable feeding schedule for Eri silkworm rearing, the number of feedings is not having much influence. Further, three times feeding per day is good enough if sufficient food is provided.

- The results of the bed cleaning experiment revealed that the regular bed cleaning every day recorded the highest larval weight (6.80g), lowest mortality (2.20%), more good cocoons (80.20%), higher ERR % (90.43%) and highest silk ratio (13.42%).

- The basic function of the mountage is to provide a specific angular space to each silkworm to spin its cocoons and to minimize the loss of silk during spinning.

- Effect of mountages on cocoon and reproductive parameters of eri silkworm showed that, highest qualitative and quantitative characters of eri cocoons were recorded in chandrika followed by hingori leaf jail, banana leaf jail and straw mountage.

- The mountages made with banana leaves, green coconut leaves, dried mango leaves and rotary and plastic mountages can be used for spinning of cocoons of eri silkworm in addition to chandrika.

- From ericulture point of view the mountage substrates in Imphal had a descending order of efficiency as follows: Chandrika = Banana jail > Eucalyptus jail > Bi-layered net > Guava jail = Mango jail.

- Studies conducted at CSRTI, Mysore, on the impact of substrata on ovipositional behaviour of eri silkworm showed that higher egg recovery was observed on bamboo strips (264.75±5.00) followed by plastic rope (235.50±8.68) over kharika (220.75±9.55).

> In Nagaland University, Medziphema, kharika is the most suitable substrata for egg laying and can be utilized at commercial scale followed by Twigs, paper box, coir rope and nylon bag.

> The observation on different host plants and methods of rearing experiment indicated that the bunch and bunch + bed methods of rearing on castor caused less mortality, lower unequal cocoons and greater ERR (90.00 and 88.30%) compared to other methods.

> In an experiment conducted to know the effect of feeding methods of tapioca leaves on economic traits of eri silkworm in Tamil Nadu revealed that the shoot cum bunch feeding method registered highest cocoon and shell yield and also SR% in all the seasons.

> In Karnataka, enclosing the females in bamboo basket was more convenient for laying preparation of eri silkworm, than they were tied to twigs of castor and females enclosed in a cellule on an egg card.

> Duration of storage, temperature, humidity, light, aeration, quality of leaves and many other conditions greatly influence the nutritive value and quality of leaves meant for feeding eri silkworm.

> The leaves could be preserved up to 48 hours without much deterioration of moisture, protein and total soluble sugar for rearing of eri silkworm. Leaf preservation in Bamboo basket covered with wet gunny cloth is better than in poly bag and open floor considering the retaining capacity of nutrient contents in the food plants leaves and larval growth and cocoon characters of the silkworm.

> Embryo isolation and egg preservation is one of the important aspects for synchronizing the eri silkworm rearing activities. The results revealed that 36-40 hrs old age eggs can be preserved maximum days without any adverse effect.

References

Chandrakanth K S, Srinivasa Babu G K, Dandin S B, Mathur V B and Mahadevamurthy T S. 2004. Development of improved mountages. *Indian Silk* **43**(5): 7-12.

Chang B, Imtinaro L and Neog P. 2012. Effect of ovipositional substrata on egg recovery and biochemical aspects of eri silkworm, *Samia cynthia ricini* (Boisduval). *Indian Journal of Entomology* **74**(4): 398-406.

Debaraj Y, Sarmah M C and Suryanarayana N. 2003. Seed technology in Eri silkmoth-experimenting with other oviposition devices. *Indian Journal of Sericulture* **42**(2): 118-121.

Devaiah M C, Govindan R and Rangaswamy H R. 1978. Performance of eri silkworm *Philosamia ricini* on castor leaves under Karnataka conditions. *All India Symposium on Sericultural Sciences, Bangalore* Pp.48.

Devaiah M C, Govindan R, Thippeswamy C and Rangaswamy H R. 1981. Evaluation of different methods of laying preparation of eri silkworm, *Samia cynthia ricini* Boisduval. *Current Research* **10**(7): 120-122.

Dutta L C, Kalita P and Phukan S N. 2015. Effect of leaf storage on nutritional changes in food plants and its impact on larval growth and cocoon quality of eri silkworm, *Samia ricini* Boisd. *Journal of Applied Zoological Researches* **26**(1): 57-61.

Goswami D, Singh N I, Mustaq Ahmed, Rajesh Kumar, Mech D and Giridhar K. 2015. Impact of Integrated Chawki Rearing Technology on cocoon production of Muga Silkworm, *Antheraea assamensis* Helfer. *Biological Forum – An International Journal* **7**(1): 146-151.

Govindan R, Devaiah M C, Rangaswamy H R and Thippeswamy C. 1980. Effect of refrigeration of the eggs of eri silkworm, *Samia cynthia ricini* Boisduval on hatching. *Indian Journal of Sericulture* **19**: 13-15.

Gupta B K, Kharoo V K and Sahani N K. 1990. Effect of different laying sheets on number of eggs laid by silkworm *Bombyx mori* L. *Biove* **1**(1): 79-80.

Huq S B, Haque M A and Khan A B. 1991. Rearing performances of Eri-silkworm, *Philosamia ricini* Hutt. (Lepidoptera:Saturniidae) on five different host plants. *Bangladesh Journal of Agricultural Sciences* **18**(2): 169-171.

Jayaraj S, Amuthamurugan D, Subramanian K, Kumar M M, Madesu C, Qadri S M H, Sakthivel N, Anbalagan R and Mathew P J. 2004. Ericulture in Tamil Nadu, Pondicherry and Kerala states in India: A study on feasibility, technology development, refinement and transfer. *Half Yearly progress Report for the period ending 31 December* 2004, Central Silk Board, Bangalore, India pp. 40-45.

Jolly M S, Sen S K, Sonwalkar T N and Prasad G K. 1979. Non Mulberry Silks. *FAO Agricultural Services Bulletin, Rome* Pp.178.

Krishnappa B L, Reddy D N R and Narayanaswamy K C. 1999. Nitrogen metabolism during the progressive embryonic development of *Sami cynthia ricini* Boisduval. *Sericologia* **39**(4): 647-651.

Krishnappa B L, Reddy D N R and Narayanaswamy K C. 2001. Changes in carbohydrate, lipid and moisture contents during the embryonic development in *Samia cynthia ricini* Boisduval. *Indian Journal of Sericulture* **40**(2): 180-181.

Krishnappa B L. 1989. Embryology of *Samia cynthia ricini* Boisduval (Lepidoptera: Saturniidae). *M. Sc. (Agri) Thesis*, University of Agricultural Sciences, Bangalore p.116.

Lakshmanan S, Mallikarjuna B, Ganapathi Rao R, Jayaram H and Geetha Devi R G. 1998. Studies on adoption of Sericultural innovations at farmers level in Tamil Nadu. *Indian Journal of Sericulture* **17**(1): 44-47.

Mohanty P K. 1998. Tropical Tasar Culture in India. *Daya Publishing House, New Delhi* Pp.112-113.

Nakasone S. 1979. Changes in lipid components during embryonic development of the silkworm, *Bombyx mori. Journal Sericultural Scienec, Japan* **48**: 526-532.

Narzary N, Bhattacharjee J, Khanikor D P, Patgiri P and Singha T A. 2014. Effect of mountages on cocoon and reproductive parameters of eri silkworm (*Samia ricini* Boisduval). *Journal of Experimental Zoology, India* **17**(2): 609-611.

Pant R and Sharma K K. 1968. Changes in nitrogenous compounds during embryonic development of *Philosamia ricini. Indian Journal of Experimental Biology* **6**: 110-112.

Pant R and Sharma K K. 1976. Variation in glycogen, total carbohydrates, free sugars, proteins, total free amino acids and some enzymes (active phosphorylase acid and alkaline phosphatises) during embryonic development of *Antheraea mylitta. Current Science* **45**: 125-128.

Rajan R K, Tamio Inokuchi and Datta R K. 1996. Manual on mounting and harvesting technology. *JICA Bivoltine Sericulture Technology Development Project*, CSR &TI, Mysore.

Ramakrishna Naika, Manjunath Gowda and Narayanaswamy K C. 2006. Performance of eri silkworm *Samia cynthia ricini* Boisduval as influenced by eggs obtained from different days of oviposition. *International Journal of Agricultural Sciences* **2**(2): 414-415.

Ravishankar H M, Reddy D N R, Reddy R N and Baruah A M. 2000. Evaluation of different methods of application of castor leaves to eri silkworm in relation to cocoon and egg production. *International Journal of Wild Silkmoth & Silk* **5**: 118-121.

Reddy D N R, Baruah A M and Reddy R N. 1999. Ericulture – An overview. *Bulletin of Indian Academy of Sericulture* **3**: 91-95.

Reddy D N R, Manjunath Gowda and Narayanaswamy K C. 2002. Ericulture-An Insight. *Zen Publishers*, Bangalore 82 Pages.

Reddy D N R, Narayanaswamy K C and Devaiah M C. 2000. The eri silkworm egg. *Zen Publishers*, Bangalore P. 41.

Sahay A, Ojha N G, Kulshrestha V, Singh G P, Roy D K and Suryanarayana N. 2008. Application of sodium hypochlorite for controlling virosis and bacteriosis in tasar silkworm, *Antheraea mylitta* D. *Sericologia* **48**(1): 71-79.

Sakthivel N. 2004. Ericulture on castor and tapioca in Tamil Nadu. In: *Proceedings of workshop on Prospects for development of ericulture in Karnataka.* 12th June, 2004, Central Silk Board, Bangalore. P. 78-81.

Sakthivel N. 2014. Effect of feeding methods of tapioca leaves and seasons on economic traits of eri silkworm, *Samia cynthia ricini* Boisduval. *Indian Journal of Sericulture* **53**(1): 63-67.

Sannappa B, Jayaramaiah M, Govindan R and Chinnaswamy K P. 2002. Advances in Ericulture. *Seri Scientific Publishers, Bangalore, India* p 3.

Saranga Dutta, Mukul Deka, Dipali Devi. 2013. Impact of crowding on larval traits and silk synthesis in Eri silkworm. *Entomological News* **123**(1): 49-58.

Saratchandra B and Joshi K L. 1984. Studies on bunch and tray feeding methods of rearing for eri silkworm *Philosamia ricini*. *Annual Report, Regional Sericultural Research Station* (RSRS), Central Silk Board, Titabar (Assam), pp. 96-100.

Sarkar B N, Sarmah M C, Dutta P and Dutta K. 2012. Embryo isolation and egg preservation technology of eri silkworm *Samia ricini* (Donovan) (Lepidoptera: Saturniidae). *Munis Entomology and Zoology* **7**(2): 792-797.

Sarma A K and Singh M P. 2007. Do cocoon mountage substrates affect the cocoon parameter and grainage behaviour of eri silkworm, *Samia cynthia ricini* Boisduval differentially. *Indian Journal of Entomology* **69**(2): 185-188.

Selvakumar T, Nataraju B and Sivaprasad V. 2007. Impact of disinfection on silkworm bamboo rearing trays smeared with cow dung paste and soil. *Indian Journal of Sericulture* **46**(2): 194-186.

Shailaja T A, Reddy D N R and Narayanaswamy K C. 2000. Egg laying pattern in *Samia cynthia ricini* Boisduval. *Insect Environment* **6**(2): July-September 53-54.

Singh N I, Goswami D, Mustaque A and Giridhar K. 2014. Efficacy of sodium hypochlorite in controlling viral and bacterial diseases in muga silkworm, *Antheraea assamensis* Helfer. *Journal of Applied Biology & Biotechnology* **2**(2): 12-15.

Somaprakash D S and Sathya Prasad K. 2008. Studies on the impact of substrata on ovipositional behaviour of eri silkworm, *Samia cynthia ricini* (Boisduval). *Indian Journal of Sericulture* **47**(2): 165-167.

Somaprakash D S and Sathya Prasad K. 2009. Effect of refrigeration on the hatchability of eri silkworm (*Samia cythia ricini* Boisduval) eggs. *Indian Journal of Sericulture* **48**(1): 56-59.

Subba Rao G and Kakoty P K. 1975. Studies on egg laying behaviour by gravid female of muga and eri. Annual Report, Central Muga and Eri Research Station, Titabar, Assam, P.69-71.

Subramanianan K, Sakthivel N and Qadri S M H. 2013. Rearing technology of eri silkworm (*Samia cynthia ricini*) under varied seasonal and host pant conditions in Tamil Nadu. *International Journal of Life Sciences Biotechnology and Pharma Research* **2**(2): 130-141.

Swamy T K N, Govindan R and Parameshwaraiah N. 1988. Effect of cocoon refrigeration in eri silkworm, *Samia cynthia ricini* Boisduval on eclosion, fecundity and hatching. *Bangladesh Journal of Zoology* **16**(1): 5-8.

Venugopal, Raghupathi M and Satyanarayana Raju Ch. 2012. 'Astra' a user-friendly compound to destroy the Pathogenic microbes causing diseases to silkworm (*Bombyx mori* L) in the rearing house and its impact on the rearing performance of new multi and bivolotine hybrids. *Journal of Experimental Zoology,* India **15**(1): 169-173.

Vijayaraghavan K, Nehru C R and Krishnaswami S. 1983. Studies on the hatching and rearing performance of *Philosamia ricini* Hutt. eggs laid on different days of oviposition. *National Seminar on Silk Research and Development,* Bangalore Pp.156.

Vishwakarma S R. 1983. Effect of refrigeration of eri silkworm, *Philosamia ricini* Hutt. eggs on the hatching (Lepidoptera: Saturniidae). *Indian Journal of Sericulture* **21**(1): 36-39.

CHAPTER 6

Rearing of Eri Silkworm on Castor

The eri silkworm, Samia cynthia ricini (Boisduval) is a polyphagous vanya silkworm and castor (Ricinus communis Linn) is the most suitable primary food plant. Castor is rich in varietal composition and many local and high yielding varieties/hybrids are widely grown throughout the country and abroad. Although, a large number of different castor genotypes exist in nature, selection and cultivation of suitable castor variety is a prerequisite for a successful ericulture for higher productivity in terms of cocoon yield. Many researchers have worked on this aspect and identified several genotypes suitable to their agro climatic zones. This chapter discusses in detail about the biochemical composition of castor leaves, their influence on consumption, digestion and utilisation indices etc., insect growth regulatory (IGR) activity of some plants on the biological performance of eri silkworm and effect of botanicals on its growth and silk production etc.

Eri silkworm is a multivoltine sericigenous insect and largely reared for its valuable silk yarn. Ericulture is an ideal subsidiary occupation for a large number of rural populations through out the world. The eri silkworms are hardy and less susceptible for diseases (Dutta, 2000). The quality of cocoons depends upon rearing technologies, seasons and silkworm breeds, quality of feeds, etc. High yielding food plant plays a significant role in development of sericulture industry (Debaraj *et al.,* 2003). With the advance of varietal improvement, a number of varieties have been developed and tested for commercial cultivation of castor in terms of seed yield in India and abroad. In India, ericulture has got great potentiality since the castor leaves are available abundantly in nature and castor cultivation is done for commercial seed production. Castor is rich in varietal composition and many local and high yielding varieties/hybrids are widely grown throughout the country. Although, a large number of different castor genotypes exist in nature, a superior castor variety has to be identified for commercial cultivation. Selection and cultivation of suitable castor variety is a prerequisite for a successful ericulture emphasizing better growth and development of eri silkworm for higher productivity in terms of cocoon yield. Systematic molecular diversity study of castor inbreeds or parental lines showed polymorphism ranged from 33.3% to 100.0%, with an average of 68.1% through ISSR primers (Gajeraa *et al.*, 2010). So, selection of a high yielding castor variety(s) is need of the hour to boost the ericulture at farmers' level. In order to evaluate suitable castor genotypes for eri silkworm rearing, several attempts have been made by different investigators (Sarkar, 1988; Chakravorty and Neog, 2006).

Studies on Rearing of Eri Silkworm on Castor in Telangana State

Identification of promising castor genotypes for rearing Eri silkworm at Regional Agricultural Research Station (RARS), Palem, Telangana state

There are many reports available on the rearing performance of eri silkworm on various food plants under North Eastern conditions of the country (Saratchandra and Joshi, 1985; Biswas and Das, 2001; Hazarika *et al.*, 2003). But in Telangana, there is a need for identification of suitable castor varieties / hybrids that meet the agro climatic needs of this region and yield better for successful rearing of eri silkworms to establish ericulture industry in the state. Hence, Lakshmi Narayanamma *et al.*, (2014) conducted an experiment to identify the promising genotypes for rearing eri silkworm at Regional Agricultural Research Station, Palem during 2011-12 *Rabi* with three brushings (rearings). Three varieties (Haritha, Kranthi and Kiran) and three hybrids (PCH-111, PCH-222 and GCH-4) along with control using the mixed leaf were selected for the study. To estimate the consumption indices, at every feeding time, the food was weighed and offered to the larvae. The standard gravimetric method (Waldbauer, 1968) was followed to measure the food consumption. Once in 24 hours, the excreta and uneaten food were carefully separated and weighed. Fifty larvae at random were collected once at the beginning of each instar (before giving the first feed) and again at the end of each instar. The consumption, digestion, gain in larval weight and consumption index were calculated by adopting the standard formulae.

Food consumption = weight of fresh food offered to larvae – weight of fresh remnants

Food digestion = weight of fresh food consumed – weight of fresh excreta

$$\text{Consumption Index CI} = \frac{F}{TA}$$

CI = Consumption Index

F = Weight of food offered

T = Duration of feeding

A = Mean weight of larva during the feeding period

$$AD\,(\%) = \frac{\text{Weight of food digested}}{\text{Weight of food ingested}} \times 100$$

AD = Approximate digestibility

The results of the experiment showed that the eri silkworms nourished with the leaves of PCH-111 genotype registered significantly higher leaf consumption during IV, V instars and during total larval period (226.8, 1526.3 and 1786.8 g/50 larvae respectively). While it was lower in control during V and total of five instars (1332.1 and 1562.9 g/50 larvae respectively) (Table 6.1). Similarly, the eri silkworms nourished with leaves of PCH-111 genotype also recorded higher food digestion (1094.2g) along with PCH-222 (1131.0g), GCH-4 (1059.5g) and Haritha (1025.5g) during the total larval period (Table 6.2). The amount of food converted to body matter was more with PCH-111 (572.2g) and GCH-4 (552.5g/50 larvae) during V instar (Table 6.3). The consumption index was more with PCH-111 (1.3g) followed by GCH-4 (0.9g), Haritha (0.9g) and PCH-222 (0.9g) in the mean of the larval period (Table 6.4). The fifth instar worms nourished with the leaves of PCH-111 (65.0g) and GCH-4 (67.0g) recorded significantly higher approximate digestibility (Table 6.5).

From the study it can be inferred that the leaves of the castor genotypes PCH-111, GCH-4 are consumed in larger quantity by the eri worms and the higher amount of dry matter converted into body matter which in turn results in higher cocoon production (Reddy and Narayanaswamy, 1998). Hence PCH-111 and GCH-4 hybrids of castor can be considered to be the best hosts with reference to their nutritional values for rearing eri silk worms.

Table 6.1: Effect of selected castor genotypes on food consumption among different larval instars of Eri silkworm, *Samia cynthia ricini* (*Rabi*, 2011-12) (Adopted from Lakshmi Narayanamma *et al.*, 2014).

Genotype	Quantity of food consumed (g/50 larvae)					
	I	II	III	IV	V	Total
Haritha	2.8[a]	4.6[d]	26.4[c]	195.3[e]	1422.3[d]	1651.4[d]
Kranthi	2.7[ab]	4.8[a]	28.0[a]	198.4[c]	1403.8[e]	1637.7[e]
Kiran	2.5[c]	4.7[c]	25.4[d]	175.8[g]	1358.5[f]	1566.9[f]
PCH-111	2.8[a]	4.7[c]	26.3[c]	226.8[a]	1526.3[a]	1786.9[a]
PCH-222	2.5[c]	4.9[a]	26.7[b]	203.8[b]	1425.5[c]	1663.4[c]
GCH-4	2.6[bc]	4.6[d]	25.5[d]	190.6[f]	1457.6[b]	1680.9[b]
Control (Mixed leaf)	2.64[c]	4.8[b]	26.7[b]	196.8[d]	1332.1[g]	1563.0[g]
F-test	Sig	Sig	Sig	Sig	Sig	Sig
CD at 5%	0.106	0.028	0.113	0.103	0.386	0.787
SEM ±	0.036	0.010	0.030	0.035	0.131	0.267

Figures followed by the same letter did not differ significantly

Table 6.2: Effect of selected castor genotypes on food digestion among different larval instars of Eri silkworm, *Samia cynthia ricini* (*Rabi*, 2011-12) (Adopted from Lakshmi Narayanamma *et al.*, 2014).

Genotype	Quantity of food digested (g/50 larvae)					
	I	II	III	IV	V	Total
Haritha	2.4	3.5[c]	18.0	149.7[b]	852.0[ab]	1025.5[ab]
Kranthi	2.3	3.8[ab]	20.4	166.0[ab]	808.5[b]	1001.1[b]
Kiran	2.3	3.9[a]	18.6	133.4[c]	838.2[b]	996.5[b]
PCH-111	2.3	4.1[a]	18.2	186.9[a]	882.7[a]	1094.2[a]
PCH-222	2.2	3.7[b]	16.1	153.5[b]	955.5[a]	1131.0[a]
GCH-4	2.2	3.6[bc]	17.4	144.7[bc]	891.6[a]	1059.5[a]
Control (Mixed leaf)	2.1	3.5[c]	16.0	147.0[b]	830.1[b]	998.6[b]
F-test	NS	Sig	NS	Sig	Sig	Sig
CD at 5%	-	0.342	-	23.84	110.23	120.42
SEM ±	0.073	0.116	1.007	8.080	37.37	40.82

Figures followed by the same letter did not differ significantly.

Table 6.3: Effect of selected castor genotypes on gain in larval weight during different larval stages of Eri silkworm, *Samia cynthia ricini* (*Rabi,* 2011-12) (Adopted from Lakshmi Narayanamma *et al.,* 2014).

Genotype	Gain in larval weight (g/50 larvae)					
	I	II	III	IV	V	Total
Haritha	0.7^a	1.5^c	16.9^a	62.2^d	414.0^c	495.4
Kranthi	0.5^c	1.6^c	13.9^b	101.2^a	464.1^b	581.4
Kiran	0.5^c	2.2^a	15.1^a	85.2^c	360.8^d	463.9
PCH-111	0.6^c	2.3^a	13.1^b	68.1^d	572.2^a	656.3
PCH-222	0.7^{ab}	2.0^{ab}	14.9^{ab}	92.1^{bc}	424.0^c	533.9
GCH-4	0.7^a	1.7^{bc}	13.9^b	88.2^c	552.5^a	656.9
Control	0.6^c	1.6^c	15.2^a	99.1^{ab}	459.2^b	575.6
F-test	Sig	Sig	Sig	Sig	Sig	NS
CD at 5%	0.084	0.430	2.939	7.769	34.542	-
SEM ±	0.026	0.133	0.907	2.398	10.661	-

Figures followed by the same letter did not differ significantly

Table 6.4: Effect of selected castor genotypes on consumption index among different larval instars of Eri silkworm, *Samia cynthia ricini* (*Rabi,* 2011-12) (Adopted from Lakshmi Narayanamma *et al.,* 2014).

Genotype	Consumption Index (g/50 larvae)					
	I	II	III	IV	V	Mean
Haritha	1.1^c	1.7^b	0.4^b	0.8^d	0.5^c	0.9^b
Kranthi	1.2^b	1.4^c	0.4^b	0.9^b	0.5^d	0.9^c
Kiran	1.2^b	1.4^c	0.3^c	0.8^c	0.5^{bc}	0.8^c
PCH-111	1.6^a	2.8^a	0.4^a	1.0^b	0.6^a	1.3^a
PCH-222	0.8^d	1.8^b	0.4^b	1.0^b	0.6^b	0.9^{bc}
GCH-4	0.8^d	1.8^b	0.4^{bc}	1.0^a	0.7^a	0.9^b
Control (Mixed leaf)	0.8^d	0.8^d	0.3^d	0.8^d	0.6^b	0.7^d
F-test	Sig	Sig	Sig	Sig	Sig	Sig
CD at 5%	0.053	0.192	0.040	0.038	0.030	0.048
SEM ±	0.018	0.065	0.013	0.013	0.010	0.016

Figures followed by the same letter did not differ significantly

Table 6.5: Effect of selected castor genotypes on approximate digestibility among different larval instars of Eri silkworm, *Samia cynthia ricini* (*Rabi,* 2011-12) (Adopted from Lakshmi Narayanamma *et al.,* 2014).

Genotype	Fresh weight (g/50 larvae)					
	I	II	III	IV	V	Mean
Haritha	84.2	74.3^d	68.3^b	76.7	59.9^{cd}	72.7
	(66.7)	(59.53)	(55.75)	(61.14)	(50.71)	(58.53)
Kranthi	85.8	79.5^{bc}	73.2^a	83.7	57.6^d	76.0
	(67.92)	(63.2)	(58.84)	(66.25)	(49.37)	(60.67)
Kiran	93.2	82. 6^b	73.1^a	75.9	54.9^e	75.9
	(71.33)	(65.35)	(58.74)	(60.63)	(47.82)	(61.49)
PCH-111	84.4	86.0^a	69.1^b	82.4	65.0^a	77.4
	(66.74)	(68.07)	(56.25)	(65.30)	(53.73)	(60.00)
PCH-222	85.8	76.6^c	60.3^c	75.4	61.2^{bc}	71.9
	(67.93)	(61.10)	(50.93)	(60.26)	(51.46)	(58.73)
GCH-4	85.7	77.7^c	68.6^b	75.9	67.0^a	75.0
	(67.78)	(61.8)	(55.94)	(60.6)	(54.97)	(59.24)
Control	82.5	72.1^d	60.1^c	74.7	62.3^b	70.3
	(65.27)	(58.10)	(50.84)	(59.83)	(52.14)	(56.99)
F-test	NS	Sig	Sig	NS	Sig	NS
CD at 5%	-	3.172	3.611	-	2.398	-
SEM ±	1.625	1.075	1.224	1.837	0.813	1.151

Figures followed by the same letter did not differ significantly

Figures in the parenthesis are angular transformed values

Effect of castor genotypes on rearing and cocoon parameters of eri silkworm at Regional Agricultural Research Station (RARS), Palem, Telangana state

Lakshmi Narayanamma and Dharma Reddy (2016) conducted an experiment in the research farm of Regional Agricultural Research Station (RARS), Palem, Telangana State during 2011-2013 (*Kharif* and *Rabi* seasons of 2011-12, 2012-13). The biomass of eri silkworms on eight castor genotypes *viz.,* Haritha, Kranthi, Kiran, DPC-9, PCH-111, PCH-222, GCH-4 and DCH-177 were studied to find out the promising castor genotypes in terms of rearing and cocoon parameters of eri silkworm. The results showed that the genotype PCH-111 recorded highest mature larval weight (6.012g), ERR (%) (92.7%) and shortest larval duration (19.33) closely followed by PCH-222 (5.982g, 92.23% and 19.54 days respectively) and GCH-4 (5.948g, 91.46% and 19.6 days respectively) (Table 6.6). The above three favourable parameters were the important criteria required for selection of a genotype. As per the rearing performance is concerned the genotypes PCH-111, PCH-222 and GCH-4 were considered as the better performers having higher ERR (91.46 to 92.70%). Highest cocoon weight (2.612g), shell weight (0.981g) and shell ratio (37.56%) was recorded in PCH-111 (Table 6.7). Eri silkworm nourished with leaves of GCH-4 recorded 2.484g cocoon weight, 0.778 g shell weight and 31.32% shell ratio being the next best treatment followed by PCH-222. The genotypes PCH-111, GCH-4 and PCH-222 ranked the highest for their rearing and cocoon parameters with no statistical variation among them. Hence PCH-111, GCH-4 and PCH-222 genotypes could be judged as the most suitable performers for eri silk culture and were recommended for cultivation and utilization in eri silkworm rearing (Lakshmi Narayanamma *et al.,* 2013).

Table 6.6: Rearing Parameters of Eri silkworm, *Samia cynthia ricini* on selected castor genotypes during 2011-13 (Adopted from Lakshmi Narayanamma and Dharma Reddy, 2016)

Genotype	Larval weight (10th day (g)	Mature larval weight (g)	Larval duration (days)	Pupal duration (days)	ERR (%)
Haritha	0.621[ab]	5.482[b]	20.21[c]	18.12	90.83[c] (72.37)
Kranthi	0.525[d]	5.252[c]	20.69[e]	18.06	88.35[e] (70.05)
Kiran	0.514[d]	5.380[bc]	20.90[f]	18.09	89.94[d] (71.51)
DPC-9	0.560[c]	5.160[d]	21.2[g]	17.98	90.54[cd] (72.09)
PCH-111	0.645[a]	6.012[a]	19.33[a]	18.21	92.70[a] (74.32)
PCH-222	0.651[a]	5.982[a]	19.54[b]	18.01	92.23[ab] (73.81)
GCH-4	0.639[a]	5.948[a]	19.6[b]	17.94	91.46[bc] (73.01)
DCH-177	0.529[cd]	5.451[b]	20.45[d]	18.08	88.5[e] (70.18)
F-test	Sig	Sig	Sig	NS	Sig
CD at 5%	0.048	0.142	0.158	-	1.081
SEM ±	0.024	0.038	0.063	0.083	0.056

Figures followed by the same letter did not differ significantly

Figures in the parenthesis are angular transformed values

Table 6.7: Cocoon Parameters of Eri silkworm, *Samia cynthia ricini* on selected castor genotypes during 2011-13 (Adopted from Lakshmi Narayanamma and Dharma Reddy, 2016)

Genotype	Cocoon weight (g)	Rate of pupation (%)	Pupal weight (g)	Shell weight (g)	Shell ratio (%)
Haritha	2.012^e	97.67^b (81.26)	1.609^c	0.403^{fg}	20.03^g
Kranthi	1.984^f	95^d (77.12)	1.526^a	0.458^e	23.08^e
Kiran	1.990^f	95^d (77.12)	1.604^c	0.386^h	19.40^h
DPC-9	2.014^e	96^c (78.52)	1.701^e	0.313^i	15.54^i
PCH-111	2.612^a	98.68^a (83.46)	1.631^c	0.981^a	37.56^a
PCH-222	2.415^c	97.14^b (81.02)	1.686^{de}	0.729^c	30.19^c
GCH-4	2.484^b	96.98^b (79.94)	1.706^e	0.778^b	31.32^b
DCH-177	2.212^d	94^e (75.24)	1.667^d	0.545^d	24.64^d
F-test	Sig	Sig	Sig	Sig	Sig
CD at 5%	0.012	0.692	0.032	0.016	0.168
SEM ±	0.009	1.984	0.016	0.014	0.048

Figures followed by the same letter did not differ significantly

Figures in the parenthesis are angular transformed values

Evaluation of castor varieties based on their performance for rearing of eri silkworm at Sarojini Naidu Vanita Maha Vidyalaya, Hyderabad

Dasari Prasanna *et al.,* (2013) have attempted to evaluate the rearing performance of eri silkworm in different seasons in relation to its primary food plant, castor genotypes. The experiment was conducted during May, 2009 at research farm of Sarojini Naidu Vanita Maha Vidyalaya, Hyderabad with four castor varieties/hybrids along with local variety. Three replications were maintained with fifty larvae for each genotype and two rearings were conducted for two years. Optimum temperature of 28 ± 2°C and 80-90% Relative humidity were maintained constantly during rearing period.

The data on comparative rearing performance of eri silkworm on castor genotypes showed that shorter larval duration (21.51 and 22.17 days) was noticed in the silkworms fed with local castor variety and DCH-519 and longer with DCS-9, 48-1, DCH-177 (24.19, 23.08 and 22.96 days) castor genotypes (Table 6.8). Cocoon weight, shell weight and shell ratio of all worms which were offered with five castor genotypes as food exhibited significant variation. The silkworms fed with DCH-519 and local variety of castor recorded higher cocoon weights (2.67 and 2.53 g/each), shell weight (0.41 and 0.39 g/each) and shell ratio (15.52 and 15.26%) (Table 6.9). In grainage parameters also, the worms offered with DCH-519 and local castor leaves recorded superior values (Table 6.10). Hence the study by Dasari prasanna *et al.,* (2013) divulge that DCH-519 and local genotype were superior in yield and rearing performance. The castor farmer can choose DCH-519 hybrid for ericulture as it provides substantial additional income and gainful employment to the poor dry land cultivators.

Studies on Rearing of Eri Silkworm on Castor in Assam State

Effect of castor varieties on performance of eri silkworm at Regional Sericultural Research Station, Titabar, Assam:

Generally, varieties may often exert variable effects on the relative survival of a herbivore insect by influencing differentially the food intake, digestion and efficiency of conversion. Many local (Green, red, red petioles and powdery) and high yielding varieties of castor (RC-8, Aruna, GCH-2, GCH-4, Co-1 and TNV-5) are widely grown in Assam and rest of India. Therefore Hazarika and Hazarika, (1996) conducted a study to demonstrate the effect of ten varieties of castor on the performance and silk production of *S. c. ricini*. The silkworms were reared according to the regular procedure by providing the food as per the age and stage of development. To account for the loss in weight of the leaves due to dessication, another set of leaves of equal weight, as given to the worms, was kept separately under the same conditions without a worm and weighed after 24 hrs. The difference in weight represented weight loss due to dessication. Moisture content of castor leaves for each of the varieties was determined by oven drying method and the carbon content of the leaves was determined by AOAC (1970), method. Samples were analysed for nitrogen with the help of Kjeltec-auto system after digesting the samples.

Table 6.8: Larval parameters of eri silkworm at different instars by feeding with five castor genotypes (Adopted from Dasari prasanna *et al.,* 2013)

Treatments	Instar I (days)	Instar II (days)	Instar III (days)	Instar IV (days)	Instar V (days)	Total larval duration (days)	Mature larval weight (g/10larvae)
Local	2.84	3.60	4.13	4.41	6.52	21.51	69.67
DCS-9	3.45	3.98	4.84	4.96	6.96	24.19	65.61
48-1	3.27	4.09	4.58	4.60	6.54	23.08	66.02
DCH-177	3.28	3.86	4.33	4.67	6.83	22.96	67.57
DCH-519	2.99	3.77	4.39	4.56	6.47	22.17	69.18
F test	Sig	Sig	Sig	Sig	Sig	Sig	Sig
SE ±	0.07	0.07	0.07	0.10	0.09	0.15	0.27

Table 6.9: Cocoon parameters of eri silkworm by feeding with five castor genotypes (Adopted from Dasari prasanna *et al.,* 2013)

Treatments	Cocoon weight (g/each)	Shell weight (g/each)	Shell ratio (%)	ERR (%)
Local	2.67	0.41	15.52	93.5
DCS-9	2.15	0.29	13.53	89.06
48-1	2.24	0.30	13.28	89.22
DCH-177	2.37	0.34	14.24	88.22
DCH-519	2.53	0.39	15.26	92.06
F test	Sig	Sig	Sig	Sig
SE ±	0.01	0.00	0.06	0.38

Table 6.10: Grainage parameters of eri silkworm by feeding with five castor genotypes

(Adopted from Dasari prasanna *et al.*, 2013)

Treatments	Pupal period (days)	Pupal weight (g/each)	Average fecundity	Hatchability (%)	Moth emergence (%)
Local	13.89	2.26	365.53	99.17	96.00
DCS-9	14.67	1.86	356.78	98.72	96.53
48-1	14.44	1.94	350.89	98.92	96.08
DCH-177	14.11	2.03	341.78	98.76	97.00
DCH-519	14.00	2.15	340.56	99.14	98.04
F test	Sig	Sig	Sig	Sig	Sig
SE ±	0.13	0.01	1.24	0.11	0.31

The results of the experiment showed that the larval period of *S. c. ricini* was completed within 21 to 23 days with varying stadia periods on different varieties. The weight of leaf consumed by each instar and total weight of leaf consumed for completing five instars in case of each of the ten varieties was given in Table 6.11. The total amount of food consumed by the eri silkworm was more in TNV-5 and least in GCH-4 variety of castor. The larvae fed with red petiole variety attained their maximum weight (7.09g) during fifth instar at its maturity stage. Some varieties produced positive effects differentially on certain instars without producing similar effects on the fifth instar larvae (Table 6.12). The larvae of *S. c. ricini* fed on GCH-2 and Aruna produced significantly lighter cocoons in comparison with the rest of the varieties. Silk content per cocoon on different varieties varied from 0.21 to 0.35 g. The red petiole variety fed larvae produced highest silk ratio of 14.06% (Table 6.13). Besides the red petiole being rich in nitrogen, may also contain some growth stimulatory substance for *S. c. ricini* which requires further investigation (Table 6.14). As a result of which even by consuming significantly least amount of red petiole leaves, larvae have produced the heaviest matured worms which subsequently have yielded superior cocoons with higher silk content in comparison with those fed on the rest of the varieties. Therefore, red petiole is the nutritionally superior variety, which can be one of the most economically important varieties. Hence red petiole variety can be recommended for commercial eri production.

Table 6.11: Weight (g) of food consumed by larvae of *S. c. ricini* during Autumn 1987 and 1988 (Adopted from Hazarika and Hazarika, 1996)

Varieties	Mean ±S.D. weight (g) of food consumed					
	Instars					
	1st	2nd	3rd	4th	5th	Total
RC-8	0.15± 007	0.38± 0.03	0.87± 0.04	1.15 ±0.23	12.96± 1.04	15.52± 1.31
GCH-2	0.18 ±0.92	0.34 ±0.02	0.86± 0.00	1.18± 0.21	13.98± 1.13	16.56 ±1.14
GCH-4	0.13± 0.04	0.38± 0.00	0.94 ±0.00	1.25± 0.16	12.73± 0.34	15.45± 0.49
TVN-5	0.13± 0.03	0.36 ±0.00	0.98 ±0.03	1.39± 0.01	15.20± 1.13	18.07± 1.09
Aruna	0.12± 0.04	0.40± 0.02	0.98 ±0.04	1.36±0.09	12.73± 0.32	15.61 ±0.42
Co-1	0.09 ±0.05	0.37± 0.02	0.87± 0.02	1.32 ±0.10	14.34 ±1.17	17.01±1.61
Green	0.14 ±0.02	0.35 ±0.01	0.81±0.01	1.11± 0.80	13.97± 0.84	16.39± 0.82
Red petiole	0.15± 0.03	0.36± 0.01	0.89± 0.11	1.14 ±0.05	10.52± 2.25	13.04 ±2.40
Red	0.19± 0.09	0.36± 0.00	0.75± 0.01	1.15± 0.10	13.61 ±2.38	16.08± 2.53
CD at 1%	NS	0.01	NS	0.06	0.02	NS

NS = Not significant at 1% level

Table 6.12: Influence of castor varieties on instar wise gain in weight of *S. c. ricini* during Autumn 1987 and 1988 (Adopted from Hazarika and Hazarika, 1996)

Varieties	Gain in weight of *S.c.ricini* (Mean ±S.D.)				
	Instars				
	1st	2nd	3rd	4th	5th
RC-8	0.003± 0.001	0.02 ±0.01	0.15± 0.06	1.13± 0.19	5.94 ±0.61
GCH-2	0.003± 0.001	0.01 ±0.01	0.15 ±0.03	1.16± 0.11	5.85 ±0.42
GCH-4	0.003 ±0.001	0.02 ±0.01	0.14± 0.02	1.34 ± 0.04	6.49 ±0.31
TVN-5	0.004 ±0.001	0.02 ±0.01	0.24± 0.06	1.34± 0.18	6.52± 0.76
Aruna	0.004± 0.001	0.03±0.02	0.20 ±0.10	1.30 ±0.03	5.68 ±0.93
Co-1	0.004± 0.002	0.03± 0.02	0.17 ±0.06	1.42 ±0.02	6.63 ±0.97
Powdery	0.003±0.002	0.02 ±0.01	0.14± 0.02	0.86± 0.00	6.31±1.17
Green	0.004 ±0.002	0.03± 0.02	0.18± 0.04	0.89± 0.08	6.16 ±0.89
Red petiole	0.004± 0.002	0.02± 0.02	0.17± 0.05	1.05± 0.10	7.09 ±0.27
Red	0.005±0.002	0.02± 0.01	0.14 ±0.08	1.17± 0.10	6.53 ±0.57
CD at 1%	0.005	NS	0.01	0.00	NS

NS= Not significant at 1% level

Table 6.13: Influence of castor varieties on Cocoon weight (CW), Silk content (SC) and Silk ratio (SR) during Autumn 1987 and 1988 (Adopted from Hazarika and Hazarika, 1996)

Varieties	Cocoon weight (Mean ± S.D)		
	CW (g)	SC (g/cocoon)	SR (%)
RC-8	2.16[a] ± 0.41	0.30 ± 0.05	13.94 ± 1.02
GCH-2	2.15[b] ± 0.13	0.29 ± 0.05	13.99 ± 2.29
GCH-4	2.24[a] ± 0.12	0.29 ± 0.04	13.31 ± 1.47
TVN-5	2.23[a] ± 0.25	0.30 ± 0.07	13.89 ± 2.03
Aruna	1.88[b] ± 0.36	0.21 ± 0.09	10.44 ± 3.65
Co-1	2.50[a] ± 0.23	0.31 ± 0.03	12.34 ± 1.67
Powdery	2.32[a] ± 0.45	0.31 ± 0.03	13.55 ± 0.91
Green	2.51[a] ± 0.49	0.31 ± 0.06	12.60 ± 1.85
Red petiole	2.79[a] ± 0.42	0.35 ± 0.04	14.06 ± 1.14
Red	2.22[a] ± 0.59	0.28 ± 0.07	12.66 ± 1.45

Table 6.14: Mean moisture and C:N content of varieties of *R. communis* during Autumn 1987 and 1988 (Adopted from Hazarika and Hazarika, 1996)

Varieties	Moisture content (%)	Carbon and Nitrogen ratio (C:N)
RC-8	85.05	3.27
GCH-2	81.97	3.29
GCH-4	83.95	3.56
TVN-5	84.64	3.49
Aruna	85.67	3.94
Co-1	85.73	4.17
Powdery	64.25	2.52
Green	71.10	2.64
Red petiole	65.80	1.74
Red	74.05	1.43

Evaluation of castor varieties through bioassay and cocoon characters of Eri silkworm at Department of Sericulture, Assam Agricultural University, Jorhat

Bordoloi *et al.,* (2005) carried out an experiment during 2000-2001 and 2001-2002 at Department of Sericulture, Assam Agricultural University, Jorhat during Spring (March-April) and Autumn (November-December) to assess the best castor variety suited for eri silkworm rearing. Larvae were reared on five castor varieties *viz.,* red petiole, DCS-9, DCH-32, DCH-177 and 48-1 by taking 50 eri silkworm larvae for each variety. Observations were recorded on average weight of mature larvae, total larval period, mortality percentage, and cocoon parameters *viz.,* average cocoon yield in terms of effective rate of rearing, shell weight, cocoon weight and silk ratio.

Results of the experiment revealed that among the castor varieties eri silkworm fed on 48-1 exhibited significantly the highest mature larval weight (5.74 and 5.29 g in summer and autumn seasons respectively) with the shortest larval duration (19.13 and 27.53 days) and comparatively better cocoon yield in both spring and autumn seasons (Table 6.15). In respect of cocoon parameters, eri silkworm larvae reared on 48-1 registered significantly the highest cocoon weight (3.454 and 3.52 g), shell weight (0.41 and 0.39 g) and silk ratio (11.90 and 12.30% in summer and autumn respectively) in both the seasons (Table 6.16). Thus it could be concluded that among the hybrid castor varieties tested, 48-1 was best suited one for rearing of eri silkworm followed by DCH-177, DCH-32 and DCS-9 particularly for North Eastern region of India.

Table 6.15: Larval growth characteristics of eri silkworm on different varieties of castor (Adopted from Bordoloi *et al.,* 2005)

Castor variety	Total larval duration (days)		Mature larval weight (g)		Mortality (%)	
	Spring	Autumn	Spring	Autumn	Spring	Autumn
Red	19.17	30.8	5.25	4.94	13.20	15.63
petiole	20.17	30.4	5.25	4.98	18.76	13.50
DCS-9	19.57	30.03	4.57	4.89	22.69	16.79
DCH-32	20.40	30.03	4.56	4.61	5.13	16.17
DCH-177	19.13	27.53	5.74	5.29	8.33	10.57
48-1	0.12	0.39	0.09	0.16	-	-
SEd ±	0.27	0.90	0.20	0.36	-	-
CD (5%)						

Evaluation of promising castor genotypes on rearing performance of eri silkworm in Central Muga Eri Research and Training Institute (CMER&TI), Assam (India):

A systematic study of castor accessions is need of the hour to find out a superior genotype for leaf biomass and silkworm rearing as well as seed oil yield. Therefore, Sarmah *et al.,* (2011) have undertaken a study to evaluate promising castor genotypes in terms of agronomical and yield attributing traits, biochemical properties of the leaves and rearing performance of eri silkworm in terms of larval weight, cocoon weight, shell weight etc.

The experiment was carried out in Central Muga Eri Research and Training Institute (CMER&TI), Assam (India), which is the native place of eri silkworm during 2007–2010. Eight promising accessions *viz.,* Ac01, Ac03, Ac04, Ac11, Ac20, Ac30, Ac36 and Ac56 (Table 6.17) were selected based on initial screening (Sarmah *et al.,* 2002). Among eight accessions, Ac01 (NBR-1) was taken as control and the experiment was conducted in three successive years i.e. 2007, 2008 and 2009. Biochemical compositions of leaves were analyzed in all the selected accessions of castor. Total soluble protein was determined by using the method of Lowry *et al.,* (1951). The total carbohydrates and total soluble sugars were determined by following Anthrone method (Sadasivam and Manickam, 2005). The crude fibre content of leaf samples was determined by the method of AOAC (1965, 1970). The method described by Price and

Butler, (1977) was employed to determine total phenol contents. Tannins in the leaf samples were estimated following the Folin–Dennis method. Sadasivam and Manickam, (2005) method was employed to determine amino acids and lipids.

Experimental results revealed that the highest larval weight (7.34±1.21g) was observed in Ac01, highest cocoon weight (3.65±0.14 g) in Ac36; highest shell weight (0.50±0.07g), largest SR% (15.52) and greatest ERR (85.67%) in Ac03 (Table 6.17). As per the silkworm rearing performance is concerned, the accession Ac03 and Ac04 were considered as the better performing accessions having higher ERR. However, there was no significant difference in larval duration when reared on different accessions of castor leaves. The results of the study indicated that Ac03 was the best accession and hence Ac03can be recommended for cultivation and utilization in eri silkworm rearing. However, the rearing performances were compared with the biochemical compositions of the different castor accessions leaves to draw definite conclusions.

Leaf biochemical composition and larval performance: Biochemical parameters like total nitrogen, crude protein, lipids, crude fibre, sugars, carbohydrates, phenols, tannins and free amino acids of all castor accessions were analyzed (Table 6.18). It was observed that the leaf biochemical compositions were also significantly different among the accessions. Although, if we compare the biochemical parameters among the accessions, Ac03 and Ac04 were closely similar to each other ($r^2 = 0.994$). Higher nitrogen, crude protein and free amino acids and a moderate range of phenol content was found in these two accessions. However, high phenol content affects feeding during larval stage. Anjani *et al.*, (2010) established a significant positive association between total phenol content and insect resistance where phenols were known to play an important role in plant defence against harmful insects which has been established in several crops (Vijaykumar *et al.*, 2009 and Rao *et al.*, 2009).

Analyses of correlation coefficients of biochemical properties and rearing performances are presented in Table 6.19. A critical look into the correlation coefficient values revealed no appreciable relationships among biochemical parameters and rearing performance parameters. The main reason could be that correlation coefficient values were worked out based on the data pertaining to all the castor accessions which include both favourable and unfavourable ones for eri silkworm rearing. It may be apt to conclude that silkworm rearing performances like hatching (hatchability), larval weight, cocoon weight, shell weight, SR and ERR were significantly different when the larvae reared on different accessions of castor plants. However, accession no Ac03 and Ac04 were found to be better performing genotypes in terms of agronomical and yield attributing traits together with rearing performances. Hence, accession no Ac03 and Ac04 of castor can be recommended for cultivation in Assam to fulfil the requirement of the sericulture industry along with castor seed production.

Further studies on Grainage performance of eri silkworm fed on different accessions of castor food plants at Central Muga Eri Research and Training Institute, Lahdoigarh, Jorhat, Assam, India

Sarkar *et al.,* (2015) continued further research on the same aspect by rearing of eri silkworm in four accessions of castor food plants *viz.,* Acc. 003, Acc. 004, Acc. 011 and Acc. 056 in two favourable seasons autumn and spring at Central Muga Eri Research and Training Institute, Lahdoigarh, Jorhat, Assam, India. The study of grainage, that is, seed production and rearing performance of eri silkmoth was conducted in three replications by keeping 100 (1:1 male female ratio) cocoons each in one replication. The data on different parameters such as moth emergence pattern, percentage of healthy and invalid moth emergence, weight of moth, wing span per treatment were recorded. The experimental study revealed that Acc.003 of castor was the most suited host plant for eri silkworm in all aspects. The percentage of un-emerged cocoon was comparatively less in case of Acc.003 of castor than that in other food plants (Table 6.20). Both potential and effective fecundity in terms of eggs, egg size, fertilization of eggs and hatching % were found higher in case of Acc.003 of castor (Table 6.21). It was observed from the data that the oviposional behaviour of eri silkworm varies with respect to feeding of different accession of castor food plants which in turn has an impact on fecundity also. In all aspects, i.e. larval weight, cocoon weight, shell weight and SR (%), Acc. 056 was observed to be less suitable than Acc. 004, Acc. 011 and Acc. 003 accessions (Table 6.22). The reason might be the food quality which reduced the actual performance of an insect below its physiological potential (Slansky and Scriber, 1984).

Hence, the study by Sarkar *et al.,* (2015) at Central Muga Eri Research and Training Institute, Lahdoigarh confirmed that Acc. 003 accession of castor found to be the best suited to the larvae of eri silkworm and provides all possible nutrition to the worms for better growth and development. The most favourable reproductive ability, oviposition potential and overall seed production of eri silkworm were found in the accession Acc.003 of castor. Hence Acc.003 of castor can be planted extensively for better growth of eri silk industry in Assam.

Table 6.16: Cocoon characters of eri silkworm on different varieties of castor (Adopted from Bordoloi *et al.*, 2005)

Castor variety	Cocoon yield of ERR %		Cocoon weight (g)		Shell weight (g)		Silk ratio (%)	
	Spring	Autumn	Spring	Autumn	Spring	Autumn	Spring	Autumn
Red petiole	86.80	84.37	2.72	3.03	0.30	0.37	11.17	12.11
DCS-9	81.24	86.50	3.44	3.07	0.41	0.34	11.88	11.22
DCH-32	77.31	83.21	2.57	3.24	0.30	0.39	11.55	11.06
DCH-177	94.87	83.83	2.77	3.41	0.31	0.33	11.10	9.95
48-1	91.67	89.43	3.45	3.52	0.41	0.39	11.90	12.30
SEd ±	0.26	0.88	0.07	0.14	0.01	0.02	NS	0.67
CD (5%)	0.59	2.02	0.17	0.32	0.02	0.04	NS	1.55

Table 6.17: Eri silkworm rearing performance on eight castor accessions (Sarmah *et al.*, 2011)

Accession no.	Larval weight (g)	Cocoon weight (g)	Shell weight (g)	SR%	ERR
Ac01	7.34 ± 1.21	3.07 ± 0.24	0.47 ± 0.05	15.30	81.66
Ac03	7.06 ± 0.81	3.22 ± 0.20	0.50 ± 0.07	15.52	85.67
Ac04	7.25 ± 0.82	3.00 ± 0.31	0.43 ± 0.08	14.33	82.67
Ac11	6.79 ± 0.75	2.84 ± 0.13	0.35 ± 0.05	12.32	79.67
Ac20	7.30 ± 0.55	3.17 ± 0.13	0.46 ± 0.08	14.51	82.00
Ac30	6.71 ± 0.36	3.54 ± 0.41	0.38 ± 0.04	10.73	80.67
Ac36	7.27 ± 0.66	3.65 ± 0.14	0.37 ± 0.07	10.13	77.33
Ac56	7.07 ± 0.63	3.19 ± 0.41	0.40 ± 0.07	12.53	79.33
Mean	7.09 ± 8.75	3.21	0.42	13.17	81.13
SD	0.24	0.27	0.05	2.05	2.51

SR=Shell ratio; ERR = Effective rate of rearing; SD= Standard deviation

Table 6.18: Biochemical analysis of foliages of eight castor accessions (Sarmah *et al.*, 2011)

Acc no.	Total N (%)	Crude protein (%)	Lipid (%)	Crude fibre (%)	Sugar (%)	Ch (%)	Phenol (%)	Tannin (%)	FAA (mg/100 g)
Ac01	1.63	10.21	11.81	5.59	3.64	18.89	18.53	1.42	0.94
	(0.05)	(1.5)	(1.8)	(0.8)	(0.4)	(2.8)	(1.6)	(0.2)	(0.02)
Ac03	1.80	11.25	9.68	4.57	4.35	21.47	19.26	1.29	2.11
	(0.05)	(1.7)	(1.5)	(0.6)	(0.7)	(2.9)	(1.6)	(0.1)	(0.1)
Ac04	1.63	10.21	10.96	5.74	4.76	20.81	21.34	1.81	1.83
	(0.06)	(1.3)	(1.9)	(0.7)	(0.4)	(2.3)	(2.0)	(0.3)	(0.2)
Ac11	1.89	11.79	12.84	5.58	3.68	26.84	16.52	1.49	2.01
	(0.03)	(1.5)	(1.5)	(0.4)	(0.3)	(2.5)	(1.3)	(0.3)	(0.3)
Ac20	1.32	8.25	20.26	7.59	5.18	27.25	15.05	1.24	1.26
	(0.04)	(1.7)	(1.9)	(0.7)	(0.5)	(3.0)	(1.6)	(0.2)	(0.1)
Ac30	1.63	10.17	7.74	6.14	3.59	19.08	17.20	1.87	0.80
	(0.06)	(1.9)	(1.0)	(0.8)	(0.3)	(1.8)	(1.7)	(0.4)	(0.03)
Ac36	1.27	7.95	22.41	9.54	4.44	22.60	40.16	1.92	0.54
	(0.06)	(1.3)	(1.4)	(1.5)	(0.4)	(2.2)	(3.8)	(0.3)	(0.01)
Ac56	1.87	11.67	23.46	5.93	5.00	25.90	22.12	1.63	0.84
	(0.08)	(1.7)	(1.9)	(1.0)	(0.7)	(2.0)	(2.1)	(0.2)	(0.01)
Mean	1.63	10.18	14.89	6.33	4.33	22.85	21.27	1.58	1.29
SD	0.23	1.45	6.17	1.54	0.63	3.39	7.99	0.26	0.61

Table 6.19: Correlation coefficients between rearing parameters and biochemical constituents of castor foliages (Sarmah *et al.*, 2011)

Biochemical parameters	Larval weight	Cocoon weight	Shell weight	SR%	ERR
Total nitrogen	-0.559	-0.560	-0.043	0.210	0.258
Crude protein	-0.552	-0.563	-0.037	0.216	0.263
Lipid	0.426	0.217	-0.222	-0.285	-0.594
Crude fibre	0.356	0.618	-0.391	-0.591	-0.682
Sugar	0.524	0.003	0.259	0.198	0.071
Carbohydrate	-0.039	-0.328	-0.259	-0.072	-0.273
Phenol	0.331	0.619	-0.338	-0.546	-0.572
Tannin	-0.201	0.528	-0.665	-0.765	-0.587
Free amino acid	-0.183	-0.700	0.297	0.560	0.669

Table 6.20: Moth emergence pattern of eri silkmoth fed on different accession of castor food plants (Adopted from Sarkar *et al.*, 2015)

Characters	Different accessions of castor food plant			
	Acc.003	Acc.004	Acc.011	Acc.056
	Nos. of seed cocoons utilized			
	300	300	300	300
1. Moth emergence (%)				
(i) Healthy male moth	45.0	42.0	38.0	44.0
(ii) Healthy female moth	44.0	39.0	37.0	32.0
(iii) Total	89.0	81.0	75.0	76.0
(iv) Invalid male moth	4.0	6.0	6.0	7.0
(v) Invalid female moth	3.0	3.0	4.0	4.0
Total	6.0	9.0	10.0	11.0
2. Un-emerged cocoon (%)	5.0	10.0	15.0	13.0
3. Single moth weight (g)				
(i) Male	1.28	1.23	1.20	1.19
(ii) Female	1.94	1.89	1.83	1.81
4. Wing span (mm) one side				
(i) Male	59-64	56-59	53-58	52-57
(ii) Female	64-69	59-63	58-60	55-58

Table 6.21: Ovipositional behaviour and egg characters of eri silkworm fed on different accession of castor food plants (Adopted from Sarkar *et al.,* 2015)

Characters	Different accessions of castor food plant			
	Acc.003	Acc.004	Acc.011	Acc.056
Coupling (%)	40.0	35.0	32.0	31.0
Effective fecundity (no.)	325	310	306	301
Av. egg retained (no.) (%)	27 (7.67)	39 (11.17)	25 (7.55)	36 (10.68)
Potential fecundity (no.)	352	349	331	337
Diameter of egg (mm)	1.62±0.02	1.53±0.03	1.54±0.02	1.51±0.03
Eggs per gram (no.)	564-599	569-605	573-604	581-619
Fertile eggs (%)	95	92	92	91
Hatching (%)	90	88	86	85

Table 6.22: Impact of different accession of castor food plants on rearing performance of eri silkworm (Adopted from Sarkar *et al.,* 2015)

Castor accessions	Larval wt. (g)	Cocoon wt. (g)	Shell wt. (g)	SR (%)
Acc.003	9.50	3.38	0.51	15.08
Acc.004	8.60	3.20	0.46	14.37
Acc.011	8.42	3.10	0.44	14.19
Acc.056	8.45	3.00	0.43	14.33

Studies on Rearing of Eri Silkworm on Castor in Karnataka State

Studies on the performance of castor genotypes on rearing and cocoon parameters of eri silkworm at Department of sericulture, UAS, GKVK, Bangalore

The selection of castor genotypes is an important factor for better growth and development of eri silkworm higher productivity in terms of cocoon yield. In this context, Chandrasekhar and Govindan, (2010) as well as Chandrasekhar *et al.,* (2012) have undertaken a study to find out a castor genotype suitable for both ericulture and seed production. The experiment was conducted during 2005-06 at Department of sericulture, UAS, GKVK, Bangalore with eight castor genotypes *viz.,* 48-1, Kranti, local green non-powdery, DCS-9, DCH-177, GCH-4, DCH-32 and DCS-85.

Eri silkworms nourished with leaves of DCS-85 genotype registered highest ERR of 92.00 % while, it was lowest with GCH-4 (83.33%) (Tables 6.23, 6.24 and 6.25). Similarly, silkworms reared on DCS-85 genotype recorded higher cocoon weight (27.78g/10 cocoons), cocoon yield (250.5g/100 worms and 75.14 kg/100 dfls), shell weight (3.415 g/10) and shell yield (9.237 kg/100 dfls) and these traits were lower in silkworms reared on GCH-4 (24.50g, 200.1g, 60.02 kg, 2.715 g and 6.651 kg respectively). The eri worms fed on leaves of DCS-85 were found

superior in respect of shell ratio (12.29%), silk productivity (4.879 cg/day) and fibroin (78.25%) and sericin (21.55%), while these parameters were inferior with GCH-4 (11.08%, 3.620 cg day, 72.10% and 27.65% respectively) (Table 6.26). Eri pupae formed by the worms nourished with leaves of DCS-9 registered higher pupal weight of 25.90 g/10 pupae and the same was lower with GCH-4 (21.76 g/10) (Table 6.27). However, fecundity of eri silk worms was more when reared with DCS-85 genotype of castor (340 eggs/laying) and least fecundity was recorded with GCH-4 (275 eggs). Further, egg hatching ranged between 95.00 to 98.00 per cent in eri silk worms reared on different castor genotypes. Hence it can be inferred that the castor genotype DCS-85 was highly suitable for eri silkworm rearing in Bangalore.

Table 6.23: Effect of feeding leaves of selected castor genotypes on larval weight of young and late-age eri silkworms (Adopted from Chandrasekhar and Govindan, 2010)

Genotype	Young age (g/10 larvae)			Late age (g/10 larvae)	
	I Instar	II Instar	III Instar	IV Instar	V Instar
DCS-9	0.060	0.162	1.461	8.251	68.19
48-1	0.059	0.160	1.425	8.190	67.77
Kranti	0.064	0.164	1.510	9.592	68.10
DCH-177	0.065	0.166	1.525	9.765	70.15
GCH-4	0.056	0.155	1.390	7.100	62.25
DCH-32	0.059	0.159	1.398	7.260	64.19
DCS-85	0.074	0.179	1.725	10.26	75.28
Local	0.070	0.172	1.600	9.821	74.00
SE ±	0.003	0.005	0.047	0.171	4.457
CD at 5%	0.008	0.014	0.140	0.513	NS

NS= Not significant at 5% level

Table 6.24: Influence of feeding leaves of selected castor genotypes on larval duration of young and late-age eri silkworms (Adopted from Chandrasekhar and Govindan, 2010)

Genotype	Young age (g/10 larvae)			Late age (g/10 larvae)		Total
	I Instar	II Instar	III Instar	IV Instar	V Instar	(days)
DCS-9	4.50	4.00	4.50	4.50	7.50	25.00
48-1	4.50	4.00	4.50	4.50	7.50	25.00
Kranti	4.00	3.50	4.00	4.25	7.00	22.75
DCH-177	4.00	3.50	4.00	4.25	7.00	22.75
GCH-4	4.50	4.00	4.50	4.50	7.50	25.00
DCH-32	4.50	4.00	4.50	4.50	7.50	25.00
DCS-85	4.00	3.50	4.00	4.25	7.00	22.75
Local	4.00	3.50	4.00	4.25	7.00	22.75
SE ±	0.191	0.144	0.144	0.144	0.206	0.775
CD at 5%	NS -	0.433	0.433	NS	NS	NS

NS= Not significant at 5% level

Leaf anatomical characters of castor genotypes and their impact on economic parameters of eri silkworm at GKVK campus of UAS, Bangalore:

Eri silkworm has feeding potential on a number of hosts with varying degree of acceptability and their utilization to convert into body matter. However, castor has been considered as the most accepted food plant of eri silkworms. In view of the manipulation in genetic composition to make them suitable to produce higher yield under varying conditions, it is also likely that the structural changes may vary due to stress factor. Further, it is possible that the anatomical characters could also vary in those developed varieties. With these changes whether the eri silkworms react differently in their acceptability to feed and utilize in converting it into better cocoon yield is the subject to be understood precisely. Keeping this in view, Sannappa and Jayaramaiah, (2001) have undertaken trials to identify the best suited genotype(s) which could go well with integration of ericulture along with castor seed production in rain fed areas of Karnataka state. The study was conducted with six castor genotypes *viz.,*

Aruna, RC-8, DCS-72, Local, SL-1 and PCS-121 at GKVK campus of UAS, Bangalore during 1996. The anatomical characters of castor leaves such as frequency, length and breadth of stomata and vascular bundle were measured as per standard procedures. The thickness of cuticular wax layer, epidermal layers, parenchymatous tissue and total leaf tissue were studied by following the procedure by Jenson, (1962). Observations on economic parameters *viz.,* feeding duration, moulting duration, larval duration, larval weight, cocoon weight, shell weight, shell ratio, fecundity and hatchability were recorded.

The results showed that the stomatal frequency was more in Aruna (16.830/mm^2) followed by DCS-72 (16.330/mm^2) genotypes (Table 6.28). The length of stomata was maximum in PCS-121 (4.467 µm), its breadth was highest in DCS-72 (2.733 µm). The length and breadth of vascular bundles were maximum in SL-1 (102.00 and 115.00 µm). The wax layer was thick in PCS-121 (2.00 µm) followed by local (1.734 µm), but the parenchymatous tissue and total leaf thickness were more in local (13.353 and 16.487 µm) and the same was on par with PCS-121 (12.00 and 15.800 µm) (Table 6.29). A significant negative relationship was observed between length & breadth of vascular bundles thickness of wax and epidermal layers of leaf with feeding duration, moulting duration, larval weight, cocoon weight, shell weight, shell ratio, fecundity and hatchability (Table 6.30). Thus, the leaf anatomical characters were different among tested castor genotypes and could influence certain economic characters in eri silkworm.

Table 6.25: Effect of feeding leaves of selected castor genotypes on rearing parameters of eri silkworms
(Adopted from Chandrasekhar and Govindan, 2010)

Genotype	Effective rate of rearing (ERR %)	Cocoon weight (g/10)	Cocoon yield		Shell weight (g/10)	Shell yield (kg/100 dfls)
			g/100 worms	Kg/100 dfls		
DCS-9	85.33 (67.55)	25.85	216.2	64.85	2.925	7.338
48-1	85.00 (67.34)	25.61	213.3	63.99	2.890	7.222
Kranti	88.00 (69.74)	26.49	228.4	68.53	3.100	8.020
DCH-177	88.33 (68.86)	26.65	230.7	69.21	3.190	8.284
GCH-4	83.33 (65.96)	24.50	200.1	60.02	2.715	6.651
DCH-32	84.00 (66.52)	24.77	203.9	61.17	2.800	6.915
DCS-85	92.00 (73.86)	27.78	250.5	75.14	3.415	9.237
Local	91.00 (72.56)	27.63	246.4	73.92	3.390	9.070
SE ±	1.733	0.476	8.956	1.32	0.064	0.069
CD at 5%	5.197	1.426	26.67	3.95	0.191	0.207

Table 6.26: Influence of feeding leaves of selected castor genotypes on cocoon traits of eri silkworms (Adopted from Chandrasekhar and Govindan, 2010)

Genotype	Shell ratio (%)	Silk productivity (cg/day)	Fibroin (%)	Sericin (%)
DCS-9	11.32 (19.66)	3.900	76.00 (60.68)	23.20 (28.79)
48-1	11.28 (19.62)	3.853	74.00 (59.35)	25.10 (30.04)
Kranti	11.70 (20.00)	4.429	76.10 (60.73)	23.70 (29.13)
DCH-177	11.96 (20.19)	4.557	77.26 (61.57)	22.45 (28.28)
GCH-4	11.08 (19.44)	3.620	72.10 (59.76)	27.65 (31.72)
DCH-32	11.30 (19.64)	3.733	73.20 (58.83)	26.60 (31.04)
DCS-85	12.29 (20.52)	4.879	78.25 (62.26)	21.55 (27.66)
Local	12.26 (20.50)	4.843	78.10 (62.11)	21.70 (27.76)
SE ±	0.080	0.156	1.231	0.337
CD at 5%	0.239	0.467	NS	1.010

NS= Not significant at 5% level

Table 6.27: Influence of feeding leaves of selected castor genotypes on grainage parameters of eri silkworms (Adopted from Chandrasekhar and Govindan, 2010)

Genotype	Rate of pupation (%)	Pupal weight (g/10)	Pupal duration (days)	Rate of moth emergence (%)	Fecundity (eggs/laying)	Egg hatching (%)
DCS-9	94.66 (77.64)	25.90	17.00	96.00 (80.49)	301	98.00 (82.05)
48-1	94.00 (75.95)	22.69	17.00	96.66 (79.55)	297	97.00 (80.12)
Kranti	95.33 (78.47)	23.37	16.00	97.33 (80.61)	312	96.00 (78.72)
DCH-177	96.00 (77.24)	23.43	16.00	97.33 (80.71)	316	95.00 (77.12)
GCH-4	91.00 (76.85)	21.76	18.00	94.33 (76.35)	275	97.00 (80.12)
DCH-32	92.00 (73.73)	21.95	18.00	94.00 (75.95)	290	96.00 (78.52)
DCS-85	98.33 (82.81)	24.34	16.00	98.00 (82.05)	340	98.00 (81.91)
Local	98.33 (82.45)	24.22	16.00	98.00 (81.95)	328	98.00 (82.05)
SE ±	2.546	0.177	0.812	1.939	11.08	1.101
CD at 5%	NS	0.530	NS	NS	33.22	3.296

NS= Not significant at 5% level

Table 6.28: Stomatal features in leaves of some castor genotypes (Adopted from Sannappa and Jayaramaiah, 2001)

Genotype	Frequency (No./mm)	Length (µm)	Breadth (µm)
Aruna	16.830	3.867	2.667
RC-8	11.830	3.733	2.467
DCS-72	16.330	3.267	2.733
Local	8.167	3.000	2.064
SL-1	8.667	3.367	2.667
PCS-121	11.170	4.467	2.667
F-test	Sig	Sig	Sig
S.Em±	0.316	0.108	0.071
CD at 5%	0.877	0.300	0.211

Table 6.29: Histological features in leaves of some castor genotypes (Adopted from Sannappa and Jayaramaiah, 2001)

Genotypes	Thickness of vascular bundle (µm)		Wax layer (µm)			Epidermal layer (µm)			Parenchymatous tissue (µm)			Total leaf thickness (µm)
	Length	Breadth	Upper	Lower	Total	Upper	Lower	Total	Upper	Lower	Total	
Aruna	64.00	78.00	0.600	0.400	1.000	0.533	0.800	1.333	5.667	5.467	11.134	13.467
RC-8	56.25	61.25	0.667	0.400	1.067	0.400	0.800	1.200	4.267	4.956	9.223	11.490
DCS-72	78.25	80.25	1.067	0.400	1.467	0.733	0.800	1.533	3.467	5.000	8.467	11.467
Local	81.25	96.25	1.267	0.467	1.734	0.600	0.800	1.400	6.200	7.153	13.353	16.487
SL-1	102.00	115.00	1.133	0.400	1.533	0.800	0.800	1.600	4.000	5.200	9.200	12.333
PCS-121	87.50	96.00	1.200	0.800	2.000	0.800	0.800	1.600	6.267	5.933	12.200	15.800
F-test	Sig	Sig	Sig	Sig	Sig	Sig	NS	NS	Sig	NS	Sig	Sig
S.Em±	2.361	1.995	0.091	0.027	0.100	0.091	-	-	0.153	-	0.678	0.749
CD at 5%	6.544	5.531	0.259	0.075	0.266	0.257	-	-	0.424	-	1.880	2.078

NS= not significant

Table 6.30: correlation co-efficients between leaf anatomical characters of castor with economic parameters of eri silkworm (Adopted from Sannappa and Jayaramaiah, 2001)

Source	Stomatal features			Vascular bundle		Leaf layers			
	Frequency	Length	Breadth	Length	Breadth	Wax	Epidermal layers	Parenchy-matous tissue	Total leaf thickness
Larval parameters									
Feeding duration	0.5203	0.2024	0.0429	-0.9126[*]	-0.8295[*]	-0.8902[*]	-0.8848[*]	-0.1897	-0.4338
Moulting duration	0.1564	0.0172	-0.2334	-0.8580[*]	-0.8244[*]	-0.8240[*]	-0.9409[*]	-0.2340	-0.4595
Larval duration	0.4906	0.1767	0.0036	-0.8150[*]	-0.8150[*]	-0.8906[*]	-0.8988[*]	-0.1641	-0.4263
Larval weight	0.4625	0.1694	0.0054	-0.8186[*]	-0.8186[*]	-0.8943[*]	-0.9000[*]	-0.1636	-0.4308
Cocoon weight									
Cocoon weight	0.3506	0.3503	0.2904	-0.8730[*]	-0.8730[*]	-0.8763[*]	-0.9140[*]	-0.1745	-0.4796
Shell weight	0.4515	0.2504	0.4692	-0.8796[*]	-0.8796[*]	-0.8864[*]	-0.9345[*]	-0.1345	-0.4703
Shell ratio	0.1697	0.6214	0.4704	-0.8615[*]	-0.8615[*]	-0.8815[*]	-0.9107[*]	-0.2625	-0.4965
Egg parameters									
Fecundity	0.4961	0.5901	0.5103	-0.9017[*]	-0.8146[*]	-0.8514[*]	-0.9615[*]	-0.1593	-0.4614
Hatchability	0.5143	0.4706	0.3805	-0.8510[*]	-0.8340[*]	-0.8716[*]	-0.9633[*]	-0.1634	-0.4107

*Significant at P = 0.05

Table 6.31: Total tannin and phenol contents of leaf in different categories of selected castor genotypes
(Adopted from Sannappa and Jayaramaiah, 1999)

Genotype	Total tannins (%)				Total phenols (%)			
	Tender	**Middle**	**Mature**	**Mean**	**Tender**	**Middle**	**Mature**	**Mean**
Aruna	0.387 (3.565)[a]	0.359 (3.433)[f]	0.322 (3.251)[c]	0.355 (3.417)[f]	0.517 (4.122)[f]	0.454 (3.865)[e]	0.408 (3.661)[e]	0.459 (3.886)[f]
RC-8	0.402 (3.637)[c]	0.369 (3.484)[c]	0.328 (3.283)[c]	0.367 (3.472)[e]	0.540 (4.213)[e]	0.465 (3.910)[a]	0.420 (3.716)[d]	0.475 (3.951)[e]
DCS-72	0.617 (4.506)[b]	0.563 (4.303)[d]	0.518 (4.127)[b]	0.566 (4.315)[d]	0.761 (5.005)[d]	0.704 (4.814)[c]	0.665 (4.679)[c]	0.710 (4.835)[d]
Local	0.624 (4.532)[b]	0.577 (4.355)[c]	0.524 (4.152)[b]	0.575 (4.349)[c]	0.786 (5.086)[b]	0.721 (4.871)[b]	0.680 (4.731)[b]	0.729 (4.898)[c]
SL-1	0.806 (5.151)[a]	0.764 (5.015)[b]	0.728 (4.894)[c]	0.766 (5.021)[b]	0.954 (5.606)[b]	0.873 (4.361)[a]	0.827 (5.219)[a]	0.885 (5.398)[b]
PCS-121	0.817 (5.187)[a]	0.780 (5.068)[a]	0.734 (4.915)[a]	0.777 (5.058)[a]	0.988 (5.705)[a]	0.886 (5.397)[a]	0.835 (5.250)[a]	0.903 (5.454)[a]
F-test	Sig	Sig	Sig	Sig	Sig	Sig	Sig	Sig
S.Em±	0.014	0.008	0.011	0.007	0.008	0.008	0.008	0.004
CD at 5%	0.063	0.033	0.048	0.031	0.036	0.036	0.037	0.018

Influence of tannins and phenols in leaves of castor genotypes on economic parameters of eri silkworm in a study at GKVK campus, Bangalore:

The quality of the leaves provided to the worms for feeding has been considered as the prime factor governing the production of good cocoon crop. In fact the nutritional status of the leaves has been implicated as a major factor in the survival of non-mulberry silkworms. In this context, Sannappa and Jayaramaiah, (1999) have undertaken investigations to determine the tannin and phenol contents in six selected castor genotypes and to understand their effect of feeding such castor leaves on the growth and development of eri silkworms. The castor genotypes *viz.,* Aruna, RC-8, DCS-72, Local, SL-1 and PCS-121 were included for assessment of tannin and phenol contents in the leaves and the work was carried out at GKVK campus, Bangalore during 1996. The leaf samples were collected after 60 days of sowing at three different heights of the plant *viz.,* top, middle and bottom in paper bags. They were shade dried for three days and then transferred to hot air oven maintained at 70°C until constant weight was obtained. The total tannin and phenol contents of the leaf were estimated separately using catachin and chlorogenic acid as standard and expressed in percentage on oven dry basis. Correlation co-efficients were worked out at P=0.05 to know the relationship between foliar constituents of castor genotypes and economic parameters.

The results showed that the tannin content varied significantly in tender, middle and mature leaves of different varieties and their means. The tannin content decreased as age of the leaves advanced. In tender, middle, mature and mean of three age groups of leaves, the highest tannin content was observed in PCS-121 genotype (0.817, 0.780, 0.734 and 0.777% respectively) and lowest in Aruna (0.387, 0.359, 0.322 and 0.555% respectively). However, SL-1 showed its proximity to PCS-121 in tender (0.806 %) and mature (0.728 %) leaves. So also, RC-8 showed parity with aruna in mature (0.328 %) leaves. Tannins are antinutritional secondary plant substances. It is understood that the larval mortality results from starvation while feeding on antinutritional foliage for non-mulberry silkworms. The presence of tannins is connected with lower biological availability of proteins, carbohydrates, amino acids, vitamins and minerals (Pandey, 1995). It was also hypothesized that tannins affect larvae of oak tasar silkworms in two ways. 1) Tannins form insoluble complexes with leaf protein in the insect gut and thus reduce the availability of plant nitrogen to the larvae. 2) Tannins can also bind the digestive enzymes and thus affect the larval ability to digest the food.

Conversely, in the results of Sannappa and Jayaramaiah, (1999) it has been observed that tannins show no adverse effect on eri silkworm.

Considerable variation was found in respect of total phenols in different castor genotypes and in leaves of different age groups. Interestingly significantly highest and significantly lowest phenol content was noticed in PCS-121 and Aruna in tender (0.988 and 0.517 %), middle (0.855 and 0.454 %) and mature (0.837 and 0.408 %) leaves. The same trend was reflected in their mean vales (0.903 and 0.459 %) also. The variation existing in respect of total phenol content might be attributable to inherent characters of the genotypes (Table 6.31). Correlation co-efficients showed that the total tannins and total phenols had significant negative relationship (-0.8867 to -0.9955 and -0.8611 to -0.9976) with all the economic parameters of eri silkworm such as larval duration, larval weigt, larval survivability, ERR, cocoon weight, pupal weight etc. However the negative impact of tannins and phenols was strongest on ERR % and least on hatchability (Table 6.32).

The above indices help to decide precisely the type of leaves to be selected for feeding to different age groups of eri silkworms in order to obtain better growth, development and higher cocoon yield.

Table 6.32: Correlation co-efficients between total tannins and phenols with economic parameters of eri silkworm (Adopted from Sannappa and Jayaramaiah, 1999)

Source	Total tannins	Total phenols
Larval duration	-0.9328[*]	-0.9613[*]
Larval weight	-0.9246[*]	-0.9549[*]
Larval survivability	-0.9427[*]	-0.9376[*]
ERR	-0.9955[*]	-0.9976[*]
Cocoon weight	-0.9174[*]	-0.9433[*]
Pupal weight	-0.9078[*]	-0.9339[*]
Shell weight	-0.9499[*]	-0.9741[*]
Shell percentage	-0.9590[*]	-0.9779[*]
Moth emergence	-0.9134[*]	-0.8928[*]
Fecundity	-0.9935[*]	-0.9867[*]
Hatchability	-0.8867[*]	-0.8611[*]

*Significant at P = 0.05

Studies on Rearing of Eri Silkworm on Castor in Gujarat State

Food consumption and utilization by Eri silkworm, *Samia cynthia ricini* Boisduval on castor cultivars at Sardar Krushinagar, Gujarat:

Patel and Patel (2010) conducted a study to know the food consumption by eri silkworm larvae on six castor genotypes *viz.,* GCH-2, GCH-4, GCH-5, GCH-6, GAUCH-1 and Assam local. The study was undertaken during November-December 2003-04. Forty healthy larvae from stock culture of eri worms were randomly selected and weighed individually on an electronic balance having sensitivity of 0.01 mg and the larvae were then transferred to a 10 cm diameter petri dish with the help of fine camel hair brush. Each larva was regarded as a replication. Moisture content of castor leaves of each variety was determined by oven drying method and the carbon content of leaves was determined by the AOAC, (1970) method. Young tender leaves of each cultivar were offered five times in a day at 6.00 a.m., 10.00 a.m., 2.00 p.m., 6.00 p.m. and 10.00 p.m. to the first three instars. Later on, the third, 4th and 5th instar larvae were transferred in to 16 cm diameter Petri dishes for convenience, and fresh food was provided four times a day at 6.00 a.m., 11.00 a.m., 5.00 p.m. and 10.00 p.m. Weight of excreta was also recorded for each worm at 24 hours interval throughout the larval stage till pupation. Observations on economic traits of eri silkworm were also recorded. Variation in leaf consumption by different instars was apparent. However, there was progressive increase in food consumption by larvae as their age advanced. The total food consumption was lower when fed on GCH-5 (15.38 g) than those fed on GCH-2 (20.86 g). Maximum increase in weight occurred in the fifth instar i.e. 48 hours prior to spinning. The larval weight of full grown larva was maximum when reared on leaves of Assam local (8.18 g), whereas, it was minimum when reared on leaves of GCH-2 (6.56 g). The cocoon weight (3.328 g), as well as shell weight (0.410 g) and shell ratio (13.11 %) were maximum when eri worms were reared on GCH-4 and GCH-6, respectively, whereas, the corresponding parameters were minimum (3.004 g, 0.37 g and 11.66 %) when they were reared on GCH-2 and GAUCH-1, respectively. Hence from the study of Patel and Patel (2010), it can be inferred that GCH-4 and GCH-6 were the most suited castor genotypes for rearing of eri silkworms in Gujarat state.

Biology of Eri Silkworm, *Samia cynthia ricini* Boisduval under ambient conditions in Sardar krushinagar, Gujarat

It is not always possible to create optimum environment for rearing eri silk worm. But it is essential to find out the feasibility for economical rearing of eri silk worm under ambient conditions. Hence, Patel and Patel (2009), studied the biology of eri silkworm, *Samia cynthia ricini* Boisduval by providing leaves of castor as larval food. The eri silk worms were reared in ordinary room condition in Entomology Laboratory of Main Castor and Mustard Research Station, S. D. Agricultural University, Sardar krushinagar by adopting complete hygienic conditions. Observations on all cocoon and grainage parameters were recorded along with growth index. The mean incubation, larval, pre-pupal and pupal period lasted for 12.79 ± 0.51, 22.88 ± 0.43, 1.08 ± 0.33 and 26.09 ± 7.35 days, respectively, with hatchability per cent of 90.68 ± 7.93. Fully fed eri worms measured 85.52 ± 4.50 mm in length and 13.48 ± 1.32 mm in breadth. The effective rearing rate was 96 per cent and the growth index was 3.62 when eri worms were reared on castor. The female cocoons were bigger and heavier than male cocoons. The mean weight of female and male cocoon was 3.53 ± 0.21 and 3.19 ± 0.27 g, respectively. The female pupae were bigger and had 'x' mark while male pupae were smaller with dot. The female moth was larger than its male counterpart. The sex ratio of laboratory reared eri moths (Female: Male) was 1:1.10, indicating preponderance of males over females. The mean pre-oviposition, oviposition and post-oviposition period were 1.16 ± 0.37, 1.68 ± 0.62 and 1.16 ± 0.37 days, respectively. The female moths laid 113.25 ± 27.81 eggs and lived longer (4.08 ± 1.07 days) than male moths (3.00 ± 0.56 days). One generation of female and male eri worms was completed in 68.44 ± 1.83 and 67.36 ± 1.64 days, respectively (Patel and Patel, 2009). These results clearly indicated the feasibility of rearing eri silk worms under ambient conditions in Gujarat revealing the possibility of extending this concept to other areas and identifying suitable seasons in different areas of India.

Consumption, Digestion and Utilization of Castor Leaves by Larvae of Eri Silkworm-Theory and Practice

Food occupies the key position in the ecology of an organism with respect to the qualitative and quantitative nutritional parameters. In food consumption and utilization studies, the qualitative requirements of food material are easy enough to workout in insects but to find out the quantitative requirement is a very difficult task. Accurate measurement of food ingestion, feaces egestion, and the resulting assimilation or utilization of food and relating it to the body substance produced (conversion efficiency) and the metabolic energy it gives not only time consuming exercises, but also very difficult to compute accurately. The dietary constituents in the form of carbohydrates, proteins, fats and nitrogen are essential nutrients required for the growth during development. Insects utilize these nutrients from their food differently, depending upon the nature of the food, mode of life and the abiotic environmental factors. Nothing is known of the effect of temperature, humidity and photoperiod. Therefore, in the present study an attempt has been made to investigate the effect of environmental conditions on the consumption and utilization of food material and their dietary nutrients.

The laboratory culture of *S. c. ricini* larvae was maintained in glass petridishes at three different experimental conditions (A, B and C) on castor leaves. Ten replicates, each consisting of ten larvae under each conditions were maintained. The experimental conditions were

Condition 'A' – Temp 24^0C, RH 70%, photoperiod - 6 hrs, light intensity 2.15×10^{-3} luman/sq.m. The photoperiod was maintained artificially.

Condition 'B' – Temp 28^0C, RH 80%, photoperiod - 12 hrs, light intensity 4.50×10^{-3} luman/sq.m.

Condition 'C' – Temp 32^0C, RH 90%, photoperiod - 18 hrs, light intensity 6.84×10^{-3} luman/sq.m.

The second instar larvae were transferred to glass petridishes with food. For mature larval instars (IV and V), it was essential to provide fresh food at two intervals to avoid the starvation. As far as possible, the leaves of same growth and area were selected as food to avoid error in the computation of mean per cent dry weight. The per cent moisture loss from the leaves during the feeding duration was determined by the method adopted by Mehrotra *et al.,* (1972). Measurements of various nutritional parameters were made on the basis of dry as well as fresh weight. The left over food and feaces were carefully separated and dried at 100^0C immediately to avoid decomposition. The dry weight of food eaten was calculated by substracting the total dry weight of the left over food from the total estimated dry weight of food introduced to the larva. The nutritional parameters of food material in the form of approximate digestibility (AD), growth rate (GR), consumption index (CI), efficiency of conversion of ingested food (ECI) and efficiency of conversion of digested food (ECD) were calculated as proposed by Waldbauer, (1968).

The co-efficient of approximate digestibility (CAD), efficiency of conversion of ingested nutrients (ECI) and efficiency of conversion of digestible (ECD) portion of carbohydrates (total and reducing sugars), proteins, fats and nitrogen were calculated as proposed by waldbauer, (1968). The quantitative estimation of these nutrients were determined by using the following techniques.

Reducing sugar: It was estimated by the method of Somogyi, (1952) at 520 g using glucose as a standard.

Total sugar: Sugar was estimated by employing the method of Seifter *et al.*, (1950) at 620 g using glucose as a standard.

Nitrogen: Nitrogen was estimated by employing the method of Mukhopadhyay and Nandi, (1975) at 430 g using ammonium sulphate as a standard.

Fat: The soluble fat in dry food and excreta was estimated by employing the Soxhlet unit method.

Nutritional ratio (NR) =

$$\frac{\text{Per cent digestible carbohydrate} + (\text{per cent digestible fat} \times 2.25)}{\text{Per cent digestible crude protein}}$$

The observation with regard to consumption index (CI), growth rate, digestibility and efficiency of conversion of ingested and digested food are given in Table 6.33. The CI and GR were lowest during first instar and gradually increased with the advancement of larval stages. These parameters were maximum under condition 'B' of fifth instar larva. The maximum AD being under experimental condition A followed by B and C conditions. As temperature and relative humidity increased, a gradual fall in AD was recorded in larval instars (Table 6.33). The maximum ECI and ECD were observed in early instars followed by later instars. These parameters fluctuate day to day, usually reaching maximum one day before the pre-moult. There existed direct relationship between AD and ECD as shown in Table 6.33. The interaction between the three experimental conditions, larval stages and the nutritional parameters showed that the differences of the mean values of the various nutritional parameters *viz.,* CI, GR, ECI and ECD were significant while not significant in case of AD. In larval stages, the CAD, ECI and ECD values were directly proportional to larval instar and maximum in fifth instar except for total sugar in which fourth instar larva having the maximum values. In experimental condition 'C' the CAD values were maximum for nitrogen, ECI for total sugar and ECD for nitrogen where as minimum values were observed in the reducing sugar. In larval instars the values of CAD, ECI and ECD were performed in the similar pattern as in condition-B. From the analysis of variance and interaction, it was observed that condition B has the maximum values for dietary constituent parameters. Whereas in interaction fat shows the non-significant results under three conditions.

There is little information available in variation of ECD from larval instar to instar in insects. The ECD was affected by the amount of energy devoted for maintenance of physiological functions of the insect. A comparison of ECD revealed that it increased in early larval instars with the increase in temperature from 24^0C to 28^0C. Further these values decreased with the increase in temperature from 28^0C to 32^0C, respectively. The ECD bears inverse relationship to AD in the early instars when the larvae digested their food better but fed only on small quantities. This results in low utilization for growth because most of the energy is utilized for maintenance. In the later instars (IV and V) of *S. c. ricini* the growth took place at a higher rate but the digestibility was lower, resulting in maximum food utilization.

Utilization of dietary constituents and their parameters

Nutrients are divided into two categories *viz.,* essential and non-essential. The major nutrients required for the growth of insects were carbohydrates, proteins, fats and nitrogen. Insects utilize nutrients from food in different ways depending upon the nature of food, mode of life and abiotic environmental factors.

Carbohydrates (Reducing and total sugars): Comparing the interactions between dietary constituents and their parameters in larval stages revealed that the values of CAD, ECI and ECD for total and reducing were maximum under experimental condition B. They increased from II to V larval instars because more energy was required for survival and growth in later larval stages. The activity of hydrolytic enzymes in the midgut increased and most of the carbohydrates were utilized to form body substances by various intermediary metabolic pathways in glycolysis and in the citric acid cycle. The utilization and digestion of a number of mono, di, tri and polysaccharides found that there was a correlation between the corresponding enzymes present in the gut region. In conformity with this in larval stages of *S. c. ricini*, the CAD, ECI and ECD amounts for reducing sugars and total sugars increased steadily during larval development and reached maximum during the last larval instar, due to increase in amylase activity and concurrent glyconeogenesis. Accumulation of this enzyme in high concentration in gut during the early instars (II and III) suggests that adequate amount of carbohydrates were not available from the ingested food for digestion. As development proceeds (IV and V), the amount of this constituent increased with intake of more food (mature leaves) by the larvae and the enzyme concentration declined after utilization.

Proteins: The value of proteins in nutrition, however, depends upon the qualitative and quantitative amino acids composition. Analysis of interaction showed that the Protein digestibility and conversion efficiency (CAD, ECI and ECD) parameters were increasing in magnitude with larval instar because of increase in the enzymes activity involved in digestion of proteins in the midgut. In *S. c. ricini* larval stages, the proteolytic enzymes consist of two main groups *viz.,* proteinases and peptidases. The former causes the degradation of large protein molecules into smaller fragments and peptidases which split the peptides and thus lead to the various free amino acids freely absorbs for various functions in *S. c. ricini* (Pant and Morris, 1969). In the fifth instar larvae of eri silkworm, the silk gland synthesized silk proteins (sericin and fibroin) which were essentially constituted by nineteen free amino acids (Poonia, 1976). The free amino acids were freely available either in midgut and haemolymph after degradation of large protein molecules by proteolytic enzymes.

Fats: Insects apparently are highly capable of utilizing fats and when necessary can synthesize these from proteins and carbohydrates. Poonia, (1985) for the first time gave detailed information about the efficiency with which the digestible portion of the fat was converte to body substances (ECI and ECD). The interaction (Tables 6.34, 6.35 and 6.36) showed that in early larval instars (II and III) less amount of fats were utilized and converted into body matter due to availability of low amount of lipase in the midgut. As larval development progressed (IV and V), the activity of lipase increased in midgut resulting maximum utilization of fats available from material. The results further infer that the amount of fats were less in quantity in the ingested food as compared to insect body.

Nitrogen: Measurements of utilization of nitrogen by insects were complicated due to presence of urine in their feaces. In the study of Poonia, (1985) three parameters of nitrogen were based on total nitrogen not corrected for uric acid in faeces. The interaction (Tables 6.34, 6.35 and 6.36) showed that the later larval instars (IV and V) utilized more nitrogen as compared to early larval instars (II and III). The conversion efficiencies (ECI and ECD) for nitrogen were higher in later larval instars because extra amount of digested nitrogen was converted to body matter for cocoon construction. This observation was on the basis of Hiratsuka, (1920) who reported that in case of mulberry silkworm, average nitrogen content of the silk in one cocoon was about 41 mg. However, it was observed that the amount of nitrogen digested and absorbed were not the same. The difference was due to the fact that some of the amino acids and guanidine derivatives passed out in the excreta of eri silkworm (Pant and Agarwal, 1963).

During the present study, there was no way to tell just how much is difference in the amount of digested nitrogen to converted for body substance without corrected the amount of uric acid present in the excreta. Larva of *S. c. ricini* was reared at three experimental conditions A, B and C. The nutritional parameters like consumption index (CI), growth rate (GR), approximate digestibility (AD), efficiency with which the ingested and digested food is converted to body substance (ECI and ECD) were computed at above mentioned conditions. Observations indicated that AD was maximum under condition-A in II instar larva where as the CI and GR were maximum under condition-B in V instar larva. The gross and net conversion efficiencies (ECI and ECD) were maximum in the early instars (II and III) under condition-B. The three parameters CAD, ECI and ECD were observed in each of the dietary constituents under each condition. The maximum value of these parameters in all dietary constituents was observed under condition-B. The value of CAD for total and reducing sugars, proteins, fat and nitrogen was directly proportional to larval life.

Table 6.33: Relationship between nutritional parameters (AD, CI, GR, ECI and ECD) with experimental conditions (A, B and C) and larval stages (II to V) and their interaction of eri silkworm (Poonia, 1985)

Source of variation	Nutrition parameter				
	AD	CI	GR	ECI	ECD
Experimental conditions					
A	54.3	1.20	0.301	33.9	56.6
B	46.7	1.94	0.479	39.7	65.3
C	44.9	1.40	0.358	36.7	59.5
SEm±	0.77	0.025	0.002	0.14	0.12
F test	Sig	Sig	Sig	Sig	Sig
CD at 5%	2.16	0.071	0.005	0.33	0.34
Larval stages					
II	63.9	0.84	0.183	47.8	66.1
III	52.9	1.20	0.266	40.3	63.3
IV	44.0	1.65	0.425	33.7	59.5
V	33.8	2.36	0.642	25.2	53.0
SEm±	0.89	0.029	0.002	0.14	0.14
F test	Sig	Sig	Sig	Sig	Sig
CD at 5%	2.50	0.082	0.005	0.38	0.39

Interaction larval stages	Experimental conditions														
	A	B	C	A	B	C	A	B	C	A	B	C	A	B	C
II	71.0	62.3	58.3	0.59	1.13	0.80	0.11	0.29	0.14	45.3	50.5	47.6	61.8	72.1	64.4
III	58.1	50.4	50.0	0.81	1.83	0.96	0.20	0.37	0.22	38.1	43.2	39.7	60.1	67.5	62.1
IV	48.8	42.0	41.3	1.40	2.10	1.44	0.35	0.51	0.43	29.5	36.6	34.9	57.1	63.3	58.0
V	39.3	31.9	30.0	2.00	2.69	2.39	0.55	0.75	0.64	22.6	28.5	24.5	47.4	58.3	53.4
SEm±		1.54			0.051			0.003			0.23			0.24	
F test		NS			Sig			Sig			Sig			Sig	
CD at 5%		NS			0.142			0.009			0.66			0.68	

SEm = Standard error, CD = Critical Difference; NS = Not significant; Sig = Significant

Table 6.34: Relationship between dietary constituent parameters (CAD, ECI and ECD) with larval stages (II to V) and their interaction of eri silkworm under experimental condition 'A' (Poonia, 1985)

Source of variation	Dietary constituent				
	Total sugar	Reducing sugar	Protein	Fat	Nitrogen
CAD	62.79	40.08	59.49	54.84	64.01
ECI	42.78	28.02	47.58	39.28	43.27
ECD	61.87	49.29	59.07	67.89	75.65
SEm±	0.13	0.09	0.14	0.22	0.14
F test	Sig	Sig	Sig	Sig	Sig
Larval stages	0.97	0.25	0.50	0.61	0.40
II	41.68	31.87	41.69	42.26	48.63
III	50.74	35.82	52.07	50.75	57.35
IV	61.83	41.63	60.54	57.51	66.53
V	69.00	50.46	67.21	65.64	71.51
SEm±	0.15	0.08	0.12	0.19	0.12
F test	Sig	Sig	Sig	Sig	Sig
CD at 5%	0.43	0.21	0.33	0.53	0.35

Interaction larval stages	Total sugar			Reducing sugar			Protein			Fat			Nitrogen		
	CAD	ECI	ECD	CAD	ECI	ECD	CAD	ECI	ECD	CAD	ECI	ECD	CAD	ECI	ECD
II	49.1	28.6	47.6	28.4	19.7	37.6	49.6	31.9	43.9	41.4	29.3	56.0	55.3	26.3	56.1
III	56.7	42.0	53.6	37.6	23.5	46.3	54.2	45.6	56.3	50.4	37.1	64.6	61.2	32.6	68.7
IV	69.5	48.3	67.8	42.5	29.6	52.8	61.9	53.5	66.1	59.2	41.3	72.1	68.8	41.5	76.0
V	76.1	52.5	78.5	51.9	39.3	61.3	72.4	59.3	70.6	68.2	49.5	79.3	71.3	50.0	92.1
SEm±		0.26			0.15			0.24			0.38			0.25	
F test		Sig			Sig			Sig			NS			Sig	
CD at 5%		0.74			0.43			0.66			1.00			0.70	

SEm = Standard error; CD = Critical Difference; NS = Not significant; Sig = Significant

Table 6.35: Relationship between dietary constituent parameters (CAD, ECI and ECD) and its larval stages (II to V) with respect to dietary constituents and their interaction of eri silkworm under experimental condition 'B' (Poonia, 1985)

Source of variation	Dietary constituent				
	Total sugar	Reducing sugar	Protein	Fat	Nitrogen
Parameters					
CAD	64.73	49.26	70.32	70.34	73.54
ECI	57.33	33.73	52.79	44.76	43.51
ECD	65.02	59.51	72.08	74.23	75.63
SEm±	0.13	0.15	0.88	0.20	0.18
F test	Sig	Sig	Sig	Sig	Sig
CD at 5%	0.47	0.42	2.13	0.55	0.49
Larval stages					
II	60.67	35.3	52.6	48.47	50.71
III	68.82	43.2	61.5	59.92	60.27
IV	76.95	51.4	69.5	67.50	69.19
V	42.99	60.0	76.8	76.53	76.73
SEm±	0.11	0.13	0.76	0.17	0.15
F test	Sig	Sig	Sig	Sig	Sig
CD at 5%	0.31	0.36	2.13	0.49	0.43

Interaction larval stages	Parameters														
	CAD	ECI	ECD	CAD	ECI	ECD	CAD	ECI	ECD	CAD	ECI	ECD	CAD	ECI	ECD
II	56.5	36.3	53.9	38.7	23.5	43.6	59.2	38.5	59.9	53.0	33.1	59.3	61.7	31.6	59.1
III	64.9	52.5	64.6	43.1	29.7	56.7	67.4	50.7	66.3	67.8	43.5	68.6	69.3	39.5	72.0
IV	73.3	59.3	74.1	53.6	34.8	65.9	74.0	57.9	76.5	74.1	48.9	79.5	76.6	47.9	83.1
V	84.3	63.7	82.5	61.2	46.9	72.0	80.7	64.0	85.6	86.3	53.6	89.5	87.4	55.1	88.3
SEm±		0.22			0.26			0.48			0.35			0.31	
F test		Sig			Sig			Sig			NS			Sig	
CD at 5%		0.61			0.73			0.98			5.31			0.86	

SEm = Standard error; CD = Critical Difference; NS = Not significant; Sig = Significant

Table 6.36: Relationship between dietary constituent parameters (CAD, ECI and ECD) and larval stages (II to V) and their interaction of eri silkworm under experimental condition 'C' (Poonia, 1985)

Source of variation	Dietary constituent													
	Total sugar			Reducing sugar			Protein			Fat			Nitrogen		
Parameters															
CAD	53.50			44.37			66.11			66.13			70.07		
ECI	59.80			30.62			50.30			41.65			40.07		
ECD	64.62			55.35			63.76			71.14			72.17		
SEm±	4.44			0.60			0.16			1.31			0.18		
F test	Sig			Sig			Sig			Sig			Sig		
CD at 5%	12.45			1.67			3.68			3.68			0.51		
Larval stages															
II	61.49			32.14			44.16			45.73			48.03		
III	54.88			39.86			57.22			56.55			56.30		
IV	64.10			46.41			64.79			64.25			65.50		
V	56.76			55.46			72.08			72.03			73.25		
SEm±	3.85			0.16			0.14			1.51			0.21		
F test	NS			Sig			Sig			Sig			Sig		
CD at 5%	10.78			0.46			0.40			4.25			0.59		
Interaction larval stages	Parameters														
	CAD	ECI	ECD	CAD	ECI	ECD	CAD	ECI	ECD	CAD	ECI	ECD	CAD	ECI	ECD
II	52.3	31.6	51.3	32.6	22.1	41.9	54.1	35.5	49.1	48.8	31.1	57.3	57.4	28.6	58.1
III	59.2	47.3	57.6	39.1	26.9	53.6	63.9	48.9	58.9	65.3	39.3	65.1	66.6	35.3	67.1
IV	70.1	53.1	69.1	48.4	32.3	58.1	699	55.5	69.0	70.6	45.1	77.1	73.9	44.1	78.3
V	79.9	58.3	80.5	57.5	41.1	67.9	76.6	76.6	78.1	78.9	51.3	85.1	82.7	52.0	84.6
SEm±		7.69			0.33			0.28			2.62			0.36	
F test		Sig			Sig			Sig			NS			Sig	
CD at 5%		21.57			0.92			0.79			7.34			1.02	

SEm = Standard error; CD = Critical Difference; NS = Not significant; Sig = Significant

Studies on Fortification of Castor Leaves for Better Nutrition of Eri Silkworm

The effect of feeding castor leaf fortified with *Mikania micrantha* leaf extracts on the cocoon characters of eri silkworm

It has been seen that many plants contain a variety of secondary metabolites which are harmful or beneficial to biotic stresses. Some compounds in plants are beneficial to insects also for better growth and development and to perpetuate the life-cycle. There are number of plants which are having insect growth regulatory (IGR) activity. When used in higher concentration such plants are detrimental to the insects but useful at lower concentrations particularly for productive insects (Mane and Patil, 2000b). The weed plants are being tried to increase the silk and egg production in mulberry silk worm *Bombyx mori* L. Shivkumar and Anantharaman, (1995) reported weed plant *Cassiatora* extracts in accelerating the maturity of *Bombyx mori*. Further, dusting of *Lantana camara* and *Clearodendron inermae* at 5% has increased silk and fecundity by eri silk worm (Santosh Kumar, 1997). *Mikania micrantha* commonly known as Japanese weed is used by sizable rearers in Karbi Anglong district of Assam for want of host plants especially at the latter larval stages when the larvae starts to eat voraciously. The rearers normally expect good quality yarn by doing so. But so far no literature is available on Japanese weed as a host plant of eri-silk worm and hence Mainu Devi, (2011) made a study to know the effect of fortification of Japanese weed extracts on castor leaves on the cocoon traits of eri silkworm *Samia c. ricini*.

The experiment consisted of twelve treatments (T1 – T10 at different concentrations of Japanese weed extract), water (T11) and absolute control (T12) and replicated thrice with 100 larvae in each replication. Japanese weed extracts of different concentrations were prepared in distilled water and sprayed on castor leaves at the rate of 4ml/sq.cm leaf area. The treated leaves were fed 4 times up to III instar and later 5 times till maturity. Observations were recorded on cocoon weight, pupal weight, shell weight, shell ratio, sericin and Fibroin content. In the study of Mainu Devi, (2011), the fortification of castor leaves with different concentrations of *Mikania micrantha* extracts did not extend or reduce the larval or pupal duration. Further, among the ten different concentrations evaluated, the treatment with 40% extracts of Japanese weed was found to be significantly superior in all the parameters studied (Pupal weight 38.17; Cocoon wt.- 44.12 gr./10; Shell wt.- 6.09 gr./10; Shell ratio 13.64%, Sericin 9.94% and

Fibroin 90.01%) as compared to untreated control (35.19 gr/10, 40.4 gr/10, 5.13 gr./10, 12.75%, 13.33% and 85.13% respectively) (Table 6.37). Further, disease incidence, total mortality, non-spinning worms, effective rate of rearing, moth emergence and hatching were found to be not significant.

The gain in cocoon characters at a particular concentration (T40) may be due to the presence of feeding stimulant factors in the aqueous extracts that increased the appetite of the larvae leading to better consumption and good growth of the eri silkworm. Further, increase in sericin and Fibroin contents of eri silk obtained from the larvae treated with 40% *Mikania micrantha* extracts may be attributed to the increased protein synthesis in larval stage leading to maximum silk output of eri silkworm *Samia c. ricini*. Similarly, dusting of 20% *Cassia sericea* on 48 hr old fifth instar eri silkworm larvae significantly increased the cocoon parameters *viz.*, cocoon weight, shell weight and shell ratio (Mane and Patil, 2000b). Mane and Patil, (2000a) also observed that castor leaves extra foliated with 10% aqueous extracts of *Amaranthus spinosus* significantly improved larval and pupal weight of eri silk worm by 6.7 and 6.4% respectively over absolute control. Hence 40% concentration of Japanese weed extract appears to be optimum level as there was no appreciable benefits either at lower or higher concentrations. Thus, Japanese weed a commonly available weed all over can be exploited for increasing the egg and silk production of eri silk worm.

Table 6.37: Cocoon characters of *Samia ricini* Donovan fed on different concentrations of *Mikania micrantha* leaf extracts extra foliated on castor leaves (Adopted from Mainu Devi, 2011)

Concentrations		Pupal Wt. gm/10	Cocoon wt.gm/10	Shell wt. gm/10	Shell ratio (%)	Sericin (%)	Fibroin (%)
10%	Mean	32.55	37.14	4.66	12.58	13.19	85.37
	SE±	1.62	1.7	0.15	0.19	0.34	0.18
20%	Mean	32.68	37.46	4.68	12.52	13.36	85.37
	SE±	1.6	1.79	0.19	0.17	0.23	0.18
30%	Mean	35.06	40.18	5.03	12.58	13.3	86.52
	SE±	1.77	1.95	0.17	0.2	0.23	0.22
40%	Mean	38.17	44.11	6.09	13.64	9.94	90.01
	SE±	1.73	1.8	0.16	0.21	0.41	0.38
50%	Mean	34.08	39.23	4.85	12.42	13.2	84.71
	SE±	1.8	2.06	0.17	0.23	0.25	0.21

Table 6.37: *Contd...*

Concentrations		Pupal Wt. gm/10	Cocoon wt.gm/10	Shell wt. gm/10	Shell ratio (%)	Sericin (%)	Fibroin (%)
60%	Mean	33.74	38.92	4.69	12.11	14.25	84.71
	SE±	1.71	2.01	0.16	0.23	0.41	0.22
70%	Mean	32.95	37.43	4.38	11.7	14.26	84.55
	SE±	1.57	1.75	0.17	0.14	0.23	0.21
80%	Mean	32.45	36.62	4.05	11.43	14.27	84.07
	SE±	1.46	1.55	0.1	0.3	0.22	0.02
90%	Mean	32.82	37.14	4.2	11.33	14.73	83.68
	SE±	1.68	1.81	0.14	0.18	0.13	0.88
100%	Mean	32.86	37.09	4.13	11.17	14.78	83.01
	SE±	1.68	1.82	0.14	0.18	0.01	0.35
Water control	Mean	35.19	40.41	5.14	12.77	13.38	85.19
	SE±	1.64	1.79	0.15	0.2	0.22	0.45
Absolute control	Mean	35.19	40.4	5.13	12.75	13.33	85.13
	SE±	1.64	1.79	0.15	0.19	0.21	0.49
Average	Mean	33.98	38.84	4.75	12.25	13.5	85.19
	SE±	0.48	0.54	0.07	0.1	0.11	0.16
CD at 5%		0.57	0.46	0.13	0.33	0.33	0.13
CD at 1%		0.7	0.59	0.17	0.42	0.42	0.24

Effect of fortification of potassium iodide and copper sulphate through castor leaf of eri silkworm, *Samia cynthia ricini* Boisduval on economic traits

Leaves of *Ricinus communis* soaked in solutions of potassium iodide and copper sulfate (50, 100 and 150 µg/ml) and fed to 5th instar larvae of the saturniid *Samia cynthia ricini* increased the weight of larvae, cocoons, pupae and cocoon shells. Copper sulfate at 100 and 150 µl/ml was found to be particularly effective (Govindan *et al.,* 1989).

Effect of fortification of castor leaves with plant extracts on the biological performance of eri silkworm

The healthy growth and ultimately the larval characters, cocoon and grainage parameters of silkworm are influenced by nutritional status of the leaves fed to them (Ito, 1978). However, due to various socio-economic or environmental factors, the farmers fail to produce quality leaves and feed to silkworm, leading to its poor growth and partial or full crop failure. Botanicals are having immense ability to influence the metabolic activities of insects. In view of this, in the recent past, many attempts have been made to fortify mulberry leaf with botanical extracts so as to improve the mulberry quality of the leaf and feed efficiency of silkworm, which in turn helped to increase larval growth, cocoon production and silk quality (Krishnaprasad *et al.,* 2001; Jeyapaul *et al.,* 2003). However, such studies are scanty on other silkworms. Hence, Philip *et al.,* (2009) made an attempt to study the effect of leaf extracts of three plants on the growth and cocoon production of eri silkworm.

Fresh leaves of *Lantana camara* L., *Parthenium hysterophorus* L. and *Stachytarpheta indica* (L.) Vahl. were collected, washed in running water and shade dried. Dried leaves were powdered in a grinder and kept in sterilized polythene bags for future use. Twenty five grams of respective leaf powder were soaked in 150 ml distilled water and kept overnight. It was filtered through double-layered muslin cloth and the filtrates were centrifuged at 3000 rpm for 15 minutes. The supernatants were maintained as stock solution (100%), from which 1:1 and 1:2 concentrations were made using distilled water. Fresh extract from stored powder was prepared once in three days. Eri silkworm rearing was done from hatching of eggs up to 2nd stage on untreated, healthy castor leaves using the variety Kranthi. From third instar onwards till spinning, leaves were dipped in respective concentrations of the above plant extracts and fed to the larvae. Each treatment had four replications of 300 larvae. Two feedings were given daily at 10 AM and 4 PM with full leaves (after removing the petiole) suitable for each stage. Data was recorded on larval duration, larval weight, ERR, cocoon weight, shell weight and silk percentage. The experiment was conducted at room temperature ($28 \pm 2^{0}C$) and replicated twice and the mean values were taken.

The results of the experiment showed that at 1:1 concentration, there was statistically significant improvement in all the parameters. Larval duration was reduced by about 16 hours compared to absolute control and 12 hours compared to water control in *S. indica* extract. In *L. camara* and *P. hysterophorus* the reduction was 15 and 13 hours respectively compared to absolute control and 11 and 9 hours respectively compared to water control (Table 6.38). At 1:2 concentrations also *S. indica* has given better performance though the improvement was not significant between plant extracts and controls. Larval weight has increased from 5.88 g in absolute control and 6.53 g in water control to 6.86g in *S. indica*. Single cocoon weight was improved by 0.42 and 0.27g respectively in *S. indica* treatment at 1:1 concentration. All the plant extracts gave statistically significant improvement at 1:1 and 1:2 concentrations in shell weight over the controls. Shell percentage at 1:1 and 1:2 concentrations was maximum in the case of *S. indica* (14.01 and 13.80% respectively). No significant improvement was recorded in ERR by weight at both concentrations of all the three extracts over water control and absolute control.

Feeding of an organism supplies the energy for growth, development, reproduction and many of its other needs (Chapman, 1998). The nutritive requirements of silkworms are very different and most of it is obtained from leaves of food plants. Although the leaves are complete diet for silkworm, it is possible that some deficiencies occur for different reasons. It appears that to boost up silk production, some supplements are essential to the silkworm in addition to their natural food (Etebari *et al.*, 2004). The supplementation of the leaves results in higher yield because the production of good quality and quantity of silk depends on larval nutrition and healthiness of the larvae, which are partially influenced by the nutritive value of leaves (Ito, 1978).

The beneficial effects of many plants on mulberry and eri silkworms have been reported by many workers. A review of literature has shown that no studies have been conducted on the effect of extracts of *S. indica* even on mulberry silkworm. Hence this seems to be the first report. Leaves of *P. hysterophorus*, *L. camara* and many other plants are reported to contain sterols such as β sitosterol, campesterol, stigmasterol and other feeding stimulants such as glucose, galactose, uralic acid, ascorbic acid and oleanolic acid (Ito *et al.*, 1964; Patil *et al.*, 1997). It can be assumed that the presence of these phagostimulant factors induced eri silkworms to feed more, utilize the leaf in a better way and to convert the utilized food efficiently into body matter.

Table 6.38: Effect of plant extracts on eri silkworm rearing (Philip *et al.,* 2009)

Treatment	Conc	Larval duration		Larval weight (g)	Single cocoon weight (g)	Single shell weight (g)	Pupal weight (g)	Shell %	ERR (%)
		Days	Hrs						
T1	1:1	21	10	6.48	2.63	0.34	2.28	13.07	20.86
T2		21	12	6.07	2.68	0.35	2.38	13.08	20.68
T3		21	9	6.86	2.94	0.41	2.52	14.01	21.45
Water		21	21	6.53	2.67	0.33	2.33	12.28	19.94
Control		21	01	5.88	2.52	0.30	2.21	12.05	18.97
CV%		2.51	2.51	7.55	10.19	13.69	10.77	6.61	5.64
F test		*	*	**	**	**	*	*	NS
CD (5%)		11.05	11.05	0.44	0.16	0.03	0.17	1.01	1.75
T1	1:2	21	20	5.91	2.62	0.34	2.29	12.78	18.59
T2		21	14	5.92	2.65	0.35	2.30	13.08	20.74
T3		21	04	6.18	3.11	0.43	2.68	13.80	22.36
Water		22	02	6.03	2.86	0.35	2.52	12.14	19.72
Control		22	12	5.70	2.41	0.29	2.12	12.05	18.14
CV%		2.79	2.79	3.53	11.33	15.42	10.98	6.42	10.22
F test		NS	NS	NS	NS	**	NS	*	NS
CD (5%)		20.26	20.26	0.29	0.44	0.07	0.38	1.17	3.19

T1 = *L. camara*; T2= *P. hysterophorus*; T3= *S. indica*

ERR = Effective rate of rearing

** Significant at 1%

* Significant at 5%

NS = Non Significant

Table 6.39: Effect of botanicals on growth and silk production of eri silkworm (Adopted from Mortale *et al.,* 2013)

Treatment	Larval weight (g)			Mortality (%)	Cocoon wt (5 nos)	Pupal wt (5 nos)	Shell wt (5 nos)	Shell %	Fecundity (5 females)	Hatching (%)
	III	IV	V							
Neem oil 1%	8.08	9.97	22.43	22.00	11.87	10.56	1.32	10.98	857.66	93.04
Neem oil 2%	7.99	9.49	22.09	24.66	11.36	10.10	1.24	10.91	849.00	92.47
Pongamia oil 1%	8.34	9.65	23.20	20.00	12.00	10.65	1.34	11.20	881.66	93.87
Pongamia oil 2%	8.19	9.49	22.86	24.00	11.63	10.36	1.26	10.93	852.00	93.15
Mahua oil 1%	8.27	10.34	23.94	19.33	11.96	10.60	1.33	11.18	862.33	94.00
Mahua oil 2%	8.24	9.96	23.44	20.00	11.63	10.30	1.30	11.26	876.00	93.41
Control	8.73	10.81	25.16	8.66	12.27	10.87	1.39	11.37	897.00	94.87
F-test	Sig	Sig	Sig	Sig	NS	NS	NS	NS	Sig	Sig
SEm±	0.13	0.12	0.25	0.98	-	-	-	-	3.21	0.23
CD (5%)	0.39	0.38	0.75	2.96	-	-	-	-	9.75	0.87

Impact of Botanicals on the Growth and Silk Production of Eri Silkworm

Studies carried out at UAS, GKVK, Bangalore

The indiscriminate and excessive use of insecticides leads to development of insecticidal resistance besides leading to environmental pollution and health hazards to non target species, thus necessitating the adoption of alternative methods to manage the insect pests, keeping the health of eri silkworm and non-target species as prime concern. In order to evaluate the growth and performance of eri silkworm, worms were fed with castor leaves treated with botanicals. The lab trial was conducted by Mortale *et al.,* (2013) at University of Agricultural Sciences, GKVK, Bangalore by maintaining three replications per treatment. Upto II moult, the eri worms were fed with untreated leaves and thereafter the worms were fed with treated leaves commencing from 24 h after treatment with the botanicals. Observations on weight of eri silkworm was recorded at the end of III, IV and V instar by taking 5 worms/treatment along with cocoon and grainage parameters. The mortality of eri silkworm was recorded at 24 hours interval and per cent mortality was calculated by using the following formula.

$$\% \text{ mortality} = \frac{\text{Number of worms dead}}{\text{Total number of worms taken}} \times 100$$

The larvae fed with the treated leaves showed significant difference in weight compared to control. The least weight at III, IV and V instars was recorded in larvae fed with 2 % neem oil (7.99, 9.49 and 22.09 g respectively) and maximum larval weight (23.94 g) was observed at V instar fed with the leaves treated with 1% mahua oil. The mortality varied from 19.33 to 24.66 % in 1 % mahua oil and 2 % neem oil treatments respectively. None of the botanicals tested had any significant influence on cocoon weight, pupal weight, shell weight and shell percentage. Observation on the effect of botanicals on the pupal weight, fecundity and hatching per cent indicated that significant difference was noticed with respect to fecundity and hatching percentage. Least fecundity and hatching percentage of 849 eggs / 5 female moths and 92.47% respectively was noticed in 2 % neem oil. Highest fecundity (881 eggs/ 5 female) was observed in pongamia 1% and highest hatching percentage (94%) was observed in mahua oil 1 % treaments. Castor leaves sprayed with neem oil, pongamia oil and mahua oil (each at 1% and 2%) and fed to eri silkworm from II moult onwards showed that leaves treated with 1 % mahua oil resulted in highest larval weight, lowest mortality of worms and the highest hatching percentage as compared to neem oil and pongamia oil. However, there was significant reduction in larval weight in 3rd, 4th and 5th instars, significant enhancement in larval mortality due to feeding of castor leaves treated with all the oils at 1% and 2% concentrations compared to untreated control. Further Fecundity of the emerged moths and hatching percentage of eggs were also adversely affected (Table 6.39). Hence these factors have to be given due consideration when there is necessity of feeding eri silk worm larvae with castor leaves treated with these oils as a part of routine plant protection measures. As the feeding was given one day after treatment, further experiments can also be planned to findout the safe waiting period after treatments with these oils.

Summary and Conclusions

➢ Food occupies the key position in the ecology of an organism with respect to the qualitative and quantitative nutritional parameters.

➢ The quality of cocoons depends upon rearing technologies, seasons, silkworm breeds, quality of feeds, etc. High yielding food plant plays a significant role in development of sericulture industry.

➢ Nutrients are divided into two categories *viz.,* essential and non-essential. The major nutrients required for the growth of insects were carbohydrates, proteins, fats and nitrogen. Insects utilize nutrients from food in different ways depending upon the nature of food, mode of life and abiotic environmental factors.

➢ Although, a large number of different castor genotypes exist in nature, a superior castor variety has to be identified for commercial cultivation.

➢ The study conducted at RARS, Palem, inferred that the leaves of the castor genotypes PCH-111, GCH-4 are consumed in larger quantity by the eri worms and the higher amount of dry matter is converted into body matter which in turn results in higher cocoon production.

➢ The genotypes PCH-111, GCH-4, PCH-222 ranked the highest for the rearing and cocoon parameters of eri silkworm with no statistical variation among the genotypes. Hence these could be judged as the most suitable performers for eri silk culture and recommended for cultivation and utilization in eri silkworm rearing.

➢ Based on the study at Sarojini Naidu Vanita Maha Vidyalya, Hyderabad, the castor farmer can choose DCH-519 hybrid for ericulture as it provides substantial additional income and gainful employment to the poor dry land cultivators.

➢ In a study conducted at Central Muga Eri Research and Training Institute, Assam, the castor accessions Ac03 and Ac04 were found to be better performing genotypes in terms of agronomical and yield attributing traits.

➢ The results of the study at CMER&TI, Lahdoigarh confirmed the accession Acc. 003 of castor was the best suited to the larvae of eri silkworm and provided all possible nutrition to the worms for better growth and development.

➢ Castor genotypes having red petioles can be considered as the nutritionally superior and economically important varieties, and can be recommended for commercial eri production at Regional Sericultural Research Station, Titabar, Assam.

➢ In North Eastern region of India, 48-1 was best suited one for rearing of eri silkworm followed by DCH-177, DCH-32 and DCS-9.

➢ The eri worms fed on leaves of castor genotype DCS-85 were found superior in respect of shell ratio (12.29%), silk productivity (4.879 cg/day), fibroin (78.25%) and sericin content in Bangalore.

➢ A significant negative relationship was observed between length & breadth of vascular bundles thickness of wax and epidermal layers of castor leaf with feeding duration, moulting duration, larval weight, cocoon weight, shell weight, shell ratio, fecundity and hatchability.

➢ The presence of tannins in castor leaf is normally connected with lower biological availability of proteins, carbohydrates, amino acids, vitamins and minerals. But in the study conducted at GKVK, Bangalore, it has been observed that tannins showed no adverse effect on eri silkworm.

➢ Correlation co-efficients showed that the total tannins and phenols were found to have negative significant relationship with economic parameters of eri silkworm such as larval duration (r=-0.9328 and -0.9613), larval weight, larval survivability, ERR (r=-0.9955 and -0.9976), cocoon weight, pupal weight etc.

➢ GCH-4 and GCH-6 were the most suited castor genotypes for the rearing of eri silkworms in Gujarat state.

➢ The highest consumption index and growth rate of the eri silkworm was recorded in the experimental condition provided with 28^{0}C temp, 80% relative humidity, 12 hrs photoperiod and 4.50×10^{-3} luman/sq.m light intensity.

➢ Efficiency of conversion of digested food increases in early larval instars with the increase in temperature from 24^{0}C to 28^{0}C, further it decreases when the temperature increases from 28^{0}C to 32^{0}C.

➢ In a study conducted on consumption, digestion and utilization of castor leaves by eri silkworm larvae showed that growth took place at a faster rate during IV and V instars, but the digestibility was lower, resulting in maximum food utilization.

➤ There are number of plants which are having insect growth regulatory (IGR) activity. When used in higher concentration such plants are detrimental to the insects but useful at lower concentrations particularly for productive insects

➤ To know the effect of Japanese weed *Mikania micrantha* leaf extracts on the cocoon characters of eri silkworm, 12 different concentrations of *M. micrantha* were sprayed on castor leaves at the rate of 4ml/sq.cm leaf area.

➤ Among the different concentrations tested, the treatment with 40% extracts of Japanese weed was found to be significantly superior in all the cocoon parameters and hence Japanese weed plant can be exploited for increasing the egg and silk production of eri silk worm.

➤ Leaves of *Ricinus communis* soaked in solutions of potassium iodide and copper sulfate (100 µg/ml) and fed to 5th instar larvae of *Samia c. ricini* increased the weight of larvae, cocoons, pupae and cocoon shells.

➤ Leaves of *P. hysterophorus*, *L. camara* and many other plants are reported to contain sterols such as β sitosterol, campesterol, stigmasterol and other feeding stimulants such as glucose, galactose, uralic acid, ascorbic acid and oleanolic acid.

➤ In an experiment conducted to test the effect of *Lantana camara* L., *Parthenium hysterophorus* L. and *Stachytarpheta indica* (L.) Vahl leaf extracts on biological performance of eri silkworm, in 1:1 and 1:2 concentrations inferred that *S. indica* has given better performance in larval weight at 1:2 concentration, cocoon weight at 1:1 concentration and maximum shell (%) in both the concentrations.

➤ Effect of botanicals on growth and silk production of eri silkworm in Bangalore showed that pongamia oil at 1% concentration recorded highest fecundity (881 eggs/ 5 female) and mahua oil at 1 % recorded the highest hatching percentage (94%).

➤ Castor leaves sprayed with neem oil, pongamia oil and mahua oil (each at 1% and 2%) and fed to eri silkworm from II moult onwards showed that leaves treated with 1 % mahua oil resulted in highest larval weight, lowest mortality of worms and the highest hatching percentage as compared to neem oil and pongamia oil.

References

Anjani K, Pallavi M and Babu S N S. 2010. Biochemical basis of resistance to leaf miner in castor (*Ricinus communis* L.). *Industrial Crops and Products* **31**: 192–196.

AOAC 1965. Official Methods of Analysis. Association of Official Analytical Chemist. 9th Ed. Washington D C.

AOAC 1970. Methods of Analysis. Association of Official Agricultural Chemists. 9th Ed. Washington D C.

Biswas N and Das P K. 2001. Effect of food plant species on rearing performance of eri silkworm *Samia cynthia ricini. Bulletin of Indian Academy of Sericulture* **5**(1): 36-39.

Bordoloi S, Dutta P P and Handique P. 2005. Evaluation of castor varieties through bioassay and cocoon characters of eri silkworm (*Samia cynthia ricini* Boisduval). *New Agriculturist* **16**(1/2): 109-111.

Chakravorty R and Neog K. 2006. Food plants of eri silkworm, *Samia ricini* (Donovan), their rearing performance and prospects for exploitation. In: Plants, Chakravorty R, Rahman S A S and Neog K. (Eds.), *Proceeding on National Workshop on Eri Food.* Central Muga Eri Research and Training Institute, Jorhat, India pp 1–7.

Chandrashekhar S and Govindan R. 2010. Studies on the performance of castor genotypes on rearing cocoon of eri silkworm. *International Journal of Plant Protection* **3**(2): 219-224.

Chandrashekhar S, Sannappa B, Manjunath K G and Govindan R. 2012. Efficacy of castor genotypes on bio-assay, cocoon and grainage traits of the domesticated vanya silkworm *Samia cynthia ricini* Boisduval. *Journal of Pharmacy Research* **5**(3): 1346-1349.

Chapman R F. 1998. The insect structure and function. *Cambridge University Press*, Cambridge.

Dasari Prasanna, Gurajala Bhargavo and Manjula. 2013. Evaluation of castor varieties based on the performance of eri silkworm *Samia cynthia ricini. International Journal of Biological & Pharmaceutical Research* **4**(12): 835-839.

Debaraj Y, Singh B K, Das P K and Suryanarayan N. 2003. Payam: An evergreen host plant of eri silkworm. *Indian Silk* **42**(1): 5–6.

Dutta L C. 2000. Effect of castor varieties on growth, nutrition and cocoon characters of eri silkworm, *Samia cynthia ricini* Boisduval. *Ph D Thesis, Assam Agricultural University*, Jorhat (India).

Etebari K, Kaliwal B B and Matindoost L. 2004. Different aspects of mulberry leaves supplementation with various nutritional compounds in sericulture. *International Journal of Industrial Entomology* **9**(1): 15-28.

Gajeraa B B, Kumara N, Singha A S, Punvara B S, Ravikirana R, Subhasha N and Jadejab G C. 2010. Assessment of genetic diversity in castor (*Ricinus communis* L.) using RAPD and ISSR markers. *Industrial Crops and Products* **32**: 491–498.

Govindan R, Magadum S B, Krishna B G and Magdum V B. 1989. Effect of fortification of potassium iodide and copper sulphate through castor leaf of eri silkworm, *Samia cynthia ricini* Boisduval on economic traits. *Mysore Journal of Agricultural Sciences* **23**(1): 62-64.

Hazarika P K and Hazarika L K 1996. Effect of castor varieties on performance of eri silkworm. *Indian Journal of Entomology* **58**(4): 284-290.

Hazarika U, Barah A, Phukan J D and Benchamin K V. 2003. Studies on the effect of different food plants and seasons on the larval development and cocoon characteristics of silkworm *Samia cynthia ricini* Boisduval. *Bulletin of Indian Academy of Sericulture* 7(1): 77-85.

Hiratsuka E. 1920. Researches on the nutrition of the silkworm. *Bulletin of the Sericultural Experiment Station, Japan* 1: 257-315.

Ito T, Kawashima K, Nakanara M, Nakanishi K and Terahara A. 1964. Effect of sterol on feeding and nutrition of the silkworm *Bombyx mori* L. *Journal of Insect Physiology* 10: 225-238.

Ito T. 1978. Silkworm Nutrition. In: *The silkworm as important laboratory tool.* Tazima Y (Eds), Kodansha Ltd., Tokyo pp. 121-157.

Jenson W A. 1962. Botanical Histochemistry: Principles and Practices. *W H Freeman and Company, San Francisco* p.408.

Jeyapaul C, Padmalatha C, Singh A J A R, Murugesan A G and Dhasarathan P. 2003. Effect of plant extracts on nutritional efficiency in mulberry silkworm, *Bombyx mori* L. *Indian Journal of Sericulture* 42(2): 128-131.

Krishnaprasad N K, Sannappa B, Reshma and Keshavaprasad P K. 2001. Effect of solanaceous leaf extracts on growth, development and cocoon traits of PM x NB_4D_2 silkworm (*Bombyx mori* L.). *Bulletin of Indian Academy of Sericulture* 5: 42-47.

Lakshmi Narayanamma V and Dharma Reddy K. 2016. Evaluation of promising castor genotypes for eri silk culture. *Journal of Insect Science* 29(1): 172-178.

Lakshmi Narayanamma V, Dharma Reddy K and Vishnuvardhan Reddy A. 2013. Castor genotypes on rearing and cocoon parameters of eri silkworm, *Samia cynthia ricini. Indian Journal of Plant Protection* 41(2): 127-131.

Lakshmi Narayanamma V, Dharma Reddy K and Vishnuvardhan Reddy A. 2014. Identification of promising castor genotypes for rearing eri silkworm, *Samia cynthia ricini. Indian Journal of Plant Protection* 42(2): 135-140.

Lowry O H, Rosenbrough N J, Farr A L and Randall R J. 1951. Protein measurements with the folin phenol reagent. *Journal of Biological Chemistry* 193.

Mainu Devi 2011. The effect of feeding castor leaf fortified with *Mikania micrantha* leaf extracts on the cocoon characters of eri silkworm, *Samia ricini* Donovan. *International Journal of Science and Advanced Technology* 1(10): 176-179.

Mane J R and Patil G M. 2000a. Effect of botanicals having phagostimulant properties on the economic traits of Eri silkworm, *Samia cynthia ricini*, Boisd. *International Journal of wild silkmoth and silk* 5: 196-199.

Mane J R and Patil G M. 2000b. Effect of botanicals having IGR activity on growth & development and economic *Eri Silk worm, Samia cynthia ricini* Boisd. *International Journal of wild silkmoth and silk* 5: 200-203.

Mehrotra K N, Rao P J and Farooqui T N A. 1972. The consumption, digestion and utilization of food by locusts. *Entomologia Experimentalis et Applicata* 15: 90-96.

Mortale H T, Narayanaswamy T K, Biradar J, Jagadish K S, Chikkalingaiah and Harish Babu S. 2013. Impact of botanicals on the growth and silk production of Eri silkworm, *Samia cynthia ricini* (Boisduval). *Current Biotica* **7**(3): 233-235.

Mukhopadhyay D and Nandi B. 1975. Changes in total and soluble nitrogen contents of Jute plants. *Current Science* **44**: 519-520.

Pandey R K. 1995. Do leaf tannins affect non-mulberry silkworms. *Indian Silk* **34**: 21-23.

Pant R and Agarwal H C. 1963. Analysis of excretory material of *Attacus ricini* in the fifth instar larval and adult stages. *Archives of Entomology and Physiology* **71**: 605-13.

Pant R and Morris D. 1969. Proteolytic and amylolytic activity in *Philosamia ricini* during development. *Indian Journal of Biochemistry* **6**: 156-157.

Patel B S and Patel G M. 2009. Biology of Eri Silkworm, *Samia cynthia ricini* Boisduval in ambient conditions. *Insect Environment* **14**(4): January-March 170.

Patel B S and Patel G M. 2010. Food consumption by eri silkworm, *Samia cynthia ricini* Boisduval on castor cultivars. *Insect Environment* **16**(1): April-June, Pp 31-32.

Patil R R, Mahadevappa M, Mahesha H M and Patil V C. 1997. Phagostimulant effects of *Parthenium* on mulberry silkworm (*Bombyx mori* L). First International Conference on *Parthenium* Management, Dharwad, Karnataka, India, Oct 6-8, 1997 Pp. 81-85.

Philip T, Somaprakash D S and Qadri S M H. 2009. Effect of fortification of castor (*Ricinus communis* L.) leaves with plant extracts on the biological performance of eri silkworm. *Indian Journal of Sericulture* **48**(2): 191-193.

Poonia F S. 1976. Free amino acids in silk gland of fifth instar larva of eri silkworm, *Philosamia ricini* (Lepidoptera: Saturniidae). *Geobios* **3**: 33-34.

Poonia F S. 1985. Consumption, digestion and utilization of castor leaves by larvae of eri silkworm, *Philosamia ricini* Hutt. (Saturniidae: Lepidoptera). *Indian Journal of Entomology* **47**(3): 255-267.

Price M and Butler L G. 1977. Rapid visual estimation and spectorphotometric determination of tannin content of sorghum grain. *Journal of Agricultural and Food Chemistry* **55**: 1268– 1273.

Rao M S, Srinivas K, Vanaja M, Rao G G S N, Venkateswarlu B and Ramakrishna Y S. 2009. Host plant (*Ricinus communis* Linn.) mediated effects of elevated CO_2 on growth performance of two insect folivores. *Current Science* **97**: 1047–1054.

Reddy D N R and Narayanaswamy K C. 1998. Quantitative consumption and utilization of castor and tapioca leaves by the eri silkworm *Samia cynthia ricini* Boisduval. *Annals of Entomology* **16**(2): 13-19.

Sadasivam S and Manickam A. 2005. Biochemical Methods. *New age International Pvt. Limited,* New Delhi.

Sannappa B and Jayaramaiah M. 1999. Influence of tannins and phenols in leaves of castor genotypes on economic parameters of eri silkworm, *Samia cynthia ricini* Boisduval (Lepidoptera: Saturniidae). *Mysore Journal of Agricultural Sciences* **33**: 297-300.

Sannappa B and Jayaramaiah M. 2001. Leaf Anatomical characters of castor genotypes and their impact on economic parameters of eri silkworm, *Samia cynthia ricini* Boisduval (Lepidoptera: Saturniidae). *Mysore Journal of Agricultural Sciences* **35**: 101-107.

Santosh Kumar G H. 1997. Large scale evaluation of insect growth regulatory (IGR) activity of *Lantana camara.* M.Sc. (Agril) Thesis, University of Agricultural Science, Dharward.

Saratchandra B and Joshi K L. 1985. A comparative study of bunch and tray feeding in Ericulture using castor (*Ricinus communis*) and Kesseru (*Heteropanax fragrans*) as food plant. *Sericologia* **25**(1): 11

Sarkar B N, Sarmah M C and Giridhar K. 2015. Grainage performance of eri silkworm *Samia ricini* (Donovan) fed on different accession of castor food Plants. *International Journal of Ecology and Ecosolution* 2(2): 17-21.

Sarkar D C. 1988. Ericulture in India. *Central Silk Board*, Bangalore, India, pp 1-49.

Sarmah M C, Chutia M, Neog K, Das R, Rajkhowa G and Gogoi S N. 2011. Evaluation of promising castor genotypes in terms of agronomical and yield attributing traits, biochemical properties and rearing performance of eri silkworm, *Samia ricini* (Donovan). *Industrial Crops and Products* **34**: 1439–1446.

Sarmah M C, Datta R N, Das P K and Benchamin K V. 2002. Evaluation of certain castor genotypes for improving ericulture. *Indian Journal of Sericulture* **41**(1): 62–63.

Seifter S, Dayton S, Novic B and Myntiyler E. 1950. The estimation of glycogen with the anthrone reagent. *Archives of Biochemistry* **25**: 191.

Shivkumar G R and Anantharaman K V. 1995. Identification of locally available plant rich in phytoecdysteroid and its extraction. *Central Sericulture Research and Training Institute, Annual Report* Mysore pp 95-97.

Slansky F Jr and Scriber J M. 1984. Food consumption and utilization. *In comprehensive Insect Physiology, Biochemistry and Pharmacology* (Eds. Kerkut G A and Gilbert L I) **4**: pp. 87-163.

Somogyi M. 1952. Notes on sugar determination. *Journal of Biological Chemistry* **195**: 19-23.

Vijaykumar L, Chakravarthy A K, Patil S U and Rajanna D. 2009. Resistance mechanism in rice to the midge *Orseolia oryzae* (Diptera: Cecidomyiidae). *Journal of Economic Entomology* **102**: 1628– 1639.

Waldbauer G P. 1968. Consumption and utilization of food by insects. *Advances in Insect Physiology* **5**: 229-288.

CHAPTER 7

Rearing Performance of Eri Silkworm on Other Host Plants

The eri silkworm, Samia c. ricini (Boisduval) is a polyphagous vanya silkworm and castor (Ricinus communis) is the most suitable primary food plant. However, seasonality of the castor is the prime limiting factor to take up the ericulture throughout the year. The castor is a warm season crop, moreover, its cultivation practices depend on rainfall conditions. In this direction, search for alternate perennial tree host plants available in different parts of India for continuous eri silkworm rearing is started. There are several secondary hosts on which eri silkworm feeds. However, all the food plants are not equally good for eri silkworm rearing and the insect shows differential behaviour when reared on different food plants. This chapter covers the influence of different host plants on rearing and cocoon parameters of eri silkworm, quantitative consumption and utilization of host leaves, rearing of eri silkworm under host scarcity condition, fortification of senescent tapioca leaves on economic traits of eri silkworm etc.

Eri silkworms are multivoltine, polyphagous and feed on a wide range of host plants. The growth, development and economic characters of silkworms are influenced to a great extent by variety of food plants and nutritive contents of foliage (Singh and Das, 2006). Though castor is the main host plant of eri silkworm, its leaves are not available throughout the year and the crop has to be sown after every six months. Moreover, its cultivation practices depend on rainfall conditions resulting in uncertainty in raising the castor crop and in turn ericulture. At the time of scarcity of castor leaf, the use of secondary food plants of eri silk worm has become a regular practice amongst the rearer. Tapioca (*Manihot esculenta* Crantz), Wild Castor (*Jatropha curcas* L), Papaya (*Carica papaya* L), Barkesseru (*Ailanthus excelsa* Roxb), Kesseru (*Heteropanax fragrans* Seem), etc. are considered as secondary hosts of eri silkworm. However, all the food plants are not equally good for eri silkworm rearing and eri silkworm shows differential behaviour, when reared on different food plants. On the other hand, eri host plants are interchangeable at 3rd instar to 5th instar especially during the scarcity of castor leaves (Dutta and Khanikor, 2005; Mukul Deka *et al.*, 2011).

Influence of host plants on rearing performance and cocoon characters of eri silkworm in Nagaland

To find the suitable secondary hosts of eri silkworm in Nagaland, Kichu *et al.,* (2014) conducted a study on influence of host plants on rearing performance and cocoon characters of eri silkworm. To study the rearing performance, eri silkworm larvae were reared on leaves of five host plants *viz.,* castor (*Ricinus communis*), kesseru (*Heteropanax fragrans*), tapioca (*Manihot utilissima*), white gulancha (*Plumeria acutifolia*) and red gulancha (*Plumeria frangipani*) under laboratory conditions for three seasons in 2013 *viz.,* midsummer (21st May to 15th July), late summer (17th August to 7th October) and autumn (12th October to 28th November). The newly hatched larvae were kept on the rearing stand along with the food plants under study and these larvae were provided with the leaves from the first to fifth instars. Observations were recorded on different growth parameters such as larval duration, larval weight, growth index, spinning period, pupal period, fecundity, hatching percentage, effective rate of rearing, cocoon weight, shell weight and pupal weight.

The larval period was significantly the shortest on castor in all the seasons (20.44, 25.60 and 16.37 days respectively) followed by white gulancha and kesseru. The longest larval duration was observed in red gulancha in midsummer (23.93 days) and late summer (37.60 days), where in autumn longest larval duration was observed in tapioca (26.0 days). The larval weight was significantly highest on castor during midsummer (7.3g) and autumn (7.5g) followed by white gulancha (6.8 g and 6.79 g in midsummer and autumn respectively) (Table 7.1). In terms of growth indices, castor was the preferred host followed by white gulancha and kesseru. The pupal weight was the highest in castor during midsummer (3.07 g) and autumn (3.32 g) (Table 7.2). The results indicated that the castor is by far the best host plant for eri silkworm rearing under Nagaland conditions also. The castor plant gave shortest larval duration, highest larval weight, fecundity, hatching percentage, pupal weight, cocoon weight, shell weight, effective rate of rearing and growth index. This can be attributed to the better leaf quality, which probably suits the larvae. In all the aspects white gulancha and kesseru were found to be equally suitable, while tapioca and red gulancha showed poor performance in Nagaland region. Among the three seasons autumn is the best season followed by midsummer and late summer for eri silk worm rearing under Nagaland conditions. Hence white gulancha and kesseru can be used as secondary hosts during the scarcity of primary host leaves i.e. castor.

Table 7.1: Influence of host plants on rearing performance and cocoon characters of eri silkworm (Adopted from Kichu *et al.*, 2014)

Host plants	Larval duration (days)			Larval weight (g)			Growth indices (GI)			Spinning period (days)			Pupal period (days)			Fecundity (No. of eggs)		
	MS	LS	A	MS	LS	A	MS	LS	A	MS	LS	A	MS	LS	A	MS	LS	A
Castor	20.44	25.60	16.37	7.30	4.83	7.50	2.35	2.05	3.61	3.80	3.37	3.07	13.67	10.47	8.47	385.22	354.61	415.06
White gulancha	20.47	25.60	17.30	6.80	4.63	6.79	2.64	1.48	3.36	3.13	3.28	3.07	13.87	10.53	8.60	301.53	300.03	318.31
Tapioca	23.40	35.00	26.00	1.87	2.13	1.88	1.54	0.57	1.92	2.93	2.28	2.27	13.73	11.53	9.73	213.34	185.32	221.27
Kesseru	20.93	27.20	19.60	5.70	5.39	5.80	1.84	2.11	2.26	3.00	2.87	2.80	13.47	11.00	8.73	256.47	232.47	264.42
Red gulancha	23.93	37.60	22.73	2.00	2.33	2.53	1.26	0.49	2.16	2.13	2.62	2.87	14.17	12.47	10.60	349.33	267.86	291.70
SEM ±	0.273	1.559	1.290	0.414	0.167	0.110	0.345	0.140	0.285	0.163	0.146	0.116	0.406	0.265	0.179	20.431	14.654	18.516
CD (5%)	0.861	4.914	4.064	1.30	0.526	0.347	NS	0.440	0.897	0.515	0.460	0.367	NS	0.835	0.564	64.380	46.176	58.344

Table 7.2: Influence of host plants on rearing performance and cocoon characters of eri silkworm
(Adopted from Kichu *et al.,* 2014)

Host plants	Hatching (%)			ERR (%)			Cocoon weight (g)			Pupal weight (g)			Shell weight (g)		
	MS	LS	A	MS	LS	A	MS	LS	A	MS	LS	A	MS	LS	A
Castor	94.47	92.85	94.22	43.81	45.39	46.53	3.80	1.51	4.20	3.07	1.31	3.32	0.73	0.20	0.88
White gulancha	91.75	91.51	92.43	47.69	37.26	45.77	3.20	1.19	3.91	2.73	1.06	3.06	0.47	0.13	0.85
Tapioca	91.49	91.55	92.49	36.68	26.49	44.23	2.30	1.19	3.22	2.00	1.06	2.49	0.47	0.13	0.73
Kesseru	93.14	91.14	91.25	38.90	46.15	40.37	3.00	1.54	3.88	2.60	1.34	3.10	0.40	0.20	0.78
Red gulancha	89.71	90.05	90.44	27.30	25.02	43.46	2.43	1.09	3.05	2.08	0.97	2.34	0.35	0.12	0.71
SEM ±	1.859	1.581	1.882	4.246	2.359	1.959	0.247	0.074	0.217	0.245	0.061	0.177	0.050	0.014	0.047
CD (5%)	NS	NS	NS	NS	7.432	NS	0.778	0.233	0.684	NS	0.192	0.558	0.158	0.045	NS

MS = Midsummer; LS = Late summer; A = Autumn

Table 7.3: Performance of *Samia cynthia ricini* on few tapioca varieties (Adopted from Patil *et al.,* 2009)

Host Plants	Total larval duration (days)	Matured larval weight (10/g)	Silk gland weight (g)	Single cocoon weight (g)	Single pupal weight (g)	Single shell weight (g)	Shell ratio (%)
Castor (check)	17.9[a]	53.86[a]	0.766[a]	2.221[a]	1.941	0.280	12.58
Co-2	22.3[c]	36.46[b]	0.664[b]	2.183[b]	1.953	0.230	10.35
ME-1	20.9[b]	35.86[b]	0.508[c]	2.157[b]	1.949	0.208	9.64
ME-120	23.4[d]	35.46[b]	0.640[b]	2.161[b]	1.969	0.192	8.88
CD at 5%	0.539	1.862	0.054	0.033	NS	NS	NS

Means followed by the same letter in a column do not differ significantly

Effect of food plant species on rearing performance of eri silkworm, *Samia c. ricini*

The larval duration and growth, effective rate of rearing (ERR), cocoon parameters and fecundity of eri silkworm fed with local red variety of castor and kesseru (*Heteropanax fragrans*) leaves was investigated by Biswas and Das (2001). When the larvae were fed with castor, the larval duration was shorter by 24 h in comparison to the silkworms fed with kesseru. The eri worms reared on castor recorded higher cocoon weight, shell weight, pupal weight and shell ratio i.e. 15.95g, 20.93g, 15.66g and 3.58% respectively. The weight of a mature larva was 62.56% higher, fecundity of moths was 17.02% higher and 6.45% higher ERR was recorded on castor fed silkworms than those reared on keserru.

Effects of host plants on biology and economic traits of eri silkworm, *Samia c. ricini* in Gujarat

Patel and Patel (2009) conducted a laboratory study to work out the effects of host plants i.e. castor and tapioca separately on biology and economic traits of eri silkworm. The results revealed that the larvae when reared on castor, the egg period lasted for 15.0 ± 0.64 days. Pre-pupal period, pupal period and adult period recorded were 1.28 ± 0.45 days, 15.56 ± 0.71 days and 3.00 ± 0.40 days respectively. Larval period was completed in 25.20 ± 0.76 days in male moths. In female moths, pre-oviposition, oviposition and post oviposition period lasted for 1.60 ± 1.58 days, 3.36 ± 0.70 days and 1.28 ± 0.54 days respectively. The duration of total life cycle (63.28 ± 1.66 days) of female was longer than male (60.04 ± 1.42 days). The average egg laying capacity was greater (187.2 ± 43.17) with higher hatching (90.44 ± 8.69%) in larvae reared on castor than those reared on tapioca. The cocoon, shell and pupa of eri worms reared on castor were bigger and heavier and the shell ratio was greater than those reared on tapioca. The rearing rate (96 %) and growth index (3.08) of *S. c. ricini* were also higher when the *Samia* worms were reared on castor than tapioca.

Performance of eri silkworm, *Samia cynthia ricini* Boisd on few food plants

During the non-availability of castor plants especially during summer, *Jatropha* can be exploited as an alternate host for eri silkworm. *Jatropha* is a multipurpose plant (medicinal, manure and diesel). Simultaneous use of *Jatropha* seeds for biodiesel and leaves for eri silkworm feeding would be a great boon to the farmers to generate additional income. Further,

Jatropha curcas exhibits wide environmental tolerance and is found in seasonally dry as well as equatorial humid regions. Hence Patil *et al.,* (2009) made a study on the performance of eri silkworm on different host plants. The disease free eri silkworm layings were reared using three tapioca varieties *viz.,* Co-2, ME-1 and ME-120 in the first set following shelf rearing with four feed frequency. Aruna variety of castor was used as check. In the second set known number of eri silkworms brushed on castor immediately after hatching and after 1st, 2nd, 3rd and 4th moult were released on the leaves of *J. curcas* and *J. gossypifolia* separately. Observations were recorded on survival, larval, pupal and cocoon characters.

Among the different tapioca varieties, ME-1 registered lowest larval duration followed by Co-2 and ME-120 (Table 7.3). Castor was significantly superior to all the three tapioca varieties for 5th instar larval, silk gland and cocoon weights followed by Co-2, ME-1 and ME-120. However, tapioca varieties failed to show any significant difference among themselves for 5th instar larval and cocoon weights. The food plants tested exercised non significant differences for pupal weight, shell weight and shell ratio. In the second set of experiment irrespective of the larval stage released on *J. gossypifolia,* larvae could only scrape the leaves, survive for 2-3 days but could not complete even a single instar. Whereas, the larvae reared on *J. curcas* immediately after brushing, after 1st, 2nd, 3rd and 4th moult, scraped the leaves, survived for 3-4 days and succumbed to death indicating their non-suitability for survival. However, 32 per cent of the 3rd instar larvae fed on *J. curcas* reached the 4th instar with prolonged larval duration of 8 days but failed to undergo 4th moult. Between the two spp., *J. curcas* resulted in prolonged death, while, *J. gossypifolia* effected sudden death probably indicating higher poisonous nature of the later species. However, eri silkworms failed to grow on *J. curcas* used from other provinces including India and could not complete the life cycle (Patil *et al.,* 2009). Jayaraj (2004) has also reported severe mortality of eri silkworms on *Jatropha glandulifera.* As per Sharma (2005), extrafoliation of aqueous extract (8 %) of *J. curcas* to castor leaf proved to be highly detrimental for growth and development of 5th instar eri silkworms causing 100 % mortality. All these reports strongly revealed poor survivability and death of different stages of eri silkworms on different species of *Jatropha.*

Hence from the study of Patil *et al.,* (2009), it can be inferred that castor is the best host plant for eri silkworm. Among the tapioca varieties, Co-2 was superior to ME-1 and ME-120 for rearing and can be used singly or in combination with castor during shortage / non-availability of castor. Eri silkworms failed to complete life cycle on *J. curcas* as well as *J. gossypifolia* and do not serve as viable alternate hosts for eri silkworm.

Effect of host plants on the consumption rate, leaf-cocoon and leaf egg ratio of Eri silkworm

The polyphagous nature of the eri silkworm adverts the estimates of food required during different instars, which is inevitable to be decided the amount of rearing to be taken up and selection of the best host for egg production. Reddy and Narayanaswamy (1999) have taken up an experiment to estimate the amount of food consumed during the larval stages, leaf-cocoon and leaf-egg ratios and also to evaluate its dietetics. The eri silkworm was reared separately on four hosts *viz.,* three varieties of castor (Bangalore local, Aruna and RC-8) and Tapioca (Mangalore local variety) till cocoon formation. Food consumption was measured by the standard gravimetric method. The consumption, leaf-cocoon and leaf-egg ratio were calculated by adopting the standard formulae.

Weight of food consumed = Weight of food offered – Weight of food remained

Leaf provided to cocoon ratio: It is the ratio of gross quantity of leaves supplied to produce one unit of cocoon.

Leaf consumed to cocoon ratio: It is the ratio of quantity of leaves consumed to produce one unit of cocoon and were calculated adopting the formulae.

$$\textbf{Leaf offered to cocoon ratio} = \frac{\text{Average weight of food offered to a larva}}{\text{Average weight of acocoon}}$$

$$\textbf{Leaf consumed to cocoon ratio} = \frac{\text{Average weight of food consumed}}{\text{Average weight of a cocoon}}$$

Leaf offered toEgg ratio: It is the number of eggs produced for the unit of leaves offered

Leaf consumed to Egg ratio: It is the number of eggs produced for the unit of leaves consumed

$$\textbf{Leaf offered to egg ratio} = \frac{\text{Average fecundity of a moth}}{\text{Average weight of food offered to a larva}}$$

$$\textbf{Leaf consumed to egg ratio} = \frac{\text{Average fecundity of amoth}}{\text{Average weight of food consumed by a larva}}$$

The results showed that the average quantity of food consumed by a larva was 15.11 g on Bangalore local, 19.87 g on Aruna and 19.28 g on RC-8 varieties of castor and 17.88 g on mangalore local of tapioca (Table 7.4). Leaf provided to cocoon ratios were higher in tapioca (19.17:1) and lower in Bangalore local castor (12.17:1). The larvae provided with the leaves of Aruna and RC-8 varieties of castor did not differ significantly (Table 7.5). The data on leaf egg ratio indicated that a significantly higher number of eggs was yielded per gram of leaf offered and consumed on Bangalore local variety of castor (5.08: 1 and 3.40: 1) (Table 7.6). It could be inferred that the Bangalore local variety of castor is a suitable host for rearing eri silkworm, because of the lowest leaf-cocoon ratio and highest leaf-egg ratio. Aruna and RC-8 varieties of castor and tapioca stood in serial sequence with regard to their suitability.

Table 7.4: Instar wise consumption of food (g) by *S. c. ricini* larva (Adopted from Reddy and Narayanaswamy, 1999)

Host	Instar					Total
	I	II	III	IV	V	
Castor						
Bangalore Local	5.258	9.393	50.069	189.311	1257.595	1511.626
Aruna	5.824	8.491	52.280	231.335	1689.189	1987.119
RC-8	5.605	11.594	53.747	236.995	1620.631	1928.572
Tapioca						
Mangalore Local	4.523	22.469	61.575	213.569	1486.834	1788.970
SEM	0.320	0.739	2.810	5.988	28.710	29.592
CD (5%)	0.686	1.569	5.964	12.698	60.870	62.736

Table 7.5: Leaf-cocoon ratios of *S. c. ricini* fed on different host leaves (Adopted from Reddy and Narayanaswamy, 1999)

Host	Food offered	Food consumed
Castor		
Bangalore Local	12.17 : 1	8.17:1
Aruna	15.02 : 1	9.38:1
RC-8	14.12 : 1	8.57:1
Tapioca		
Mangalore Local	19.17 : 1	10.15:1
SEM	0.36	0.83
CD (5%)	0.77	0.75

Table 7.6: Leaf-egg ratios of *S. c. ricini* fed on different host leaves (Adopted from Reddy and Narayanaswamy, 1999)

Host	Food offered
Castor	
Bangalore Local	5.08 : 1
Aruna	4.37 : 1
RC-8	4.46 : 1
Tapioca	
Mangalore Local	3.40 : 1
SEM	0.23
CD (5%)	0.48

Evaluation of proximate composition of eri silkworm pupae fed on *Ailanthus grandis* and *A. excelsa* leaves

Generally, after getting the silk from cocoon, the pupae are thrown away as waste. But eri silkworm pupae are not considered as waste as they are consumed by some people as delicacy (Sarkar, 1980; Chowdhury, 1982). Pupal protein contains all essential amino acids including high concentration of leucine, which is important for human consumption (Rao *et al.,* 2005). Since, eri pupae are rich in protein and minerals, its consumption is very effective against protein malnutrition of the children. Besides, it can be used as an ingredient in poultry and in other livestock feed. Hence Deori *et al.,* (2016) conducted a study to evaluate the proximate composition of pupae obtained by rearing eri silkworm on Borpat and Barkesseru leaves.

Eri silkworm larvae were reared on three food plants *viz., Ailanthus grandis* (Borpat), *Ailanthus excelsa* (Barkesseru) and castor (*Ricinus communis*). Castor leaves were fed to the larvae for comparison with those of Borpat and Barkesseru fed individuals. The pupae were taken out by cutting the cocoon shells with the help of scissors and then dried in hot air oven at a temperature ranging between 80-90^{0}C for about 2 hours per day until they were dried properly. The dried pupae were then ground in an electric grinding machine to make it powder. Chemical analyses of eri pupae were done for the moisture, crude protein, crude fat, crude fibre, total mineral and total carbohydrates.

The results of the experiment showed that the percentage of moisture, crude fat, crude fibre, crude protein, total mineral and total carbohydrate content of eri pupae obtained by feeding on Barkesseru leaves were 16.77%, 19.85%, 4.62%, 53.18%, 5.75% and 5.34% respectively. In case of larvae fed on Borpat leaves the percentages were 15.14%, 21.02%, 5.21%, 56.73%, 6.18% and 5.49% respectively and in castor leaves the percentages were 14.98%, 22.43%, 5.47%, 66.85%, 6.41% and 5.72% respectively (Table 7.7).

If we compare the results of Deori *et al.,* (2016) with those of previous workers, the percentage of moisture content of eri pupae was found to be much higher than the mulberry pupae (Bose and Mazumder, 1990) and muga silkworm pupae (Choudhury, 2000). However, the percentage of crude fat content of these eri pupae in the study of Deori *et al.,* (2016) was found to be little higher than the previously reported value in the same insect (Rao *et al.,* 2005). Hence it can be inferred that the pupae obtained by feeding eri silkworm on Borpat and barkesseru could be suitably utilized as a source of food as good as pupae obtained from castor fed silkworms.

Table 7.7: Nutrient contents of eri silkworm pupae obtained by feeding on different host plants (Adopted from Deori *et al.,* 2016)

Host plant	Nutrient content					
	Moisture content (%)	Crude fat (%)	Crude fibre (%)	Crude protein (%)	Total mineral (%)	Total carbohydrate (%)
Barkesseru	16.77 (14.88-19.00)	19.85[b] (18.37-21.16)	4.62 (4.23-4.84)	53.18[b] (49.25-56.81)	5.75 (5.13-5.72)	5.34 (4.00-6.11)
Borpat	15.14 (14.58-15.63)	21.02[ab] (20.92-21.09)	5.21 (4.52-5.77)	56.73[b] (53.30-61.75)	6.18 (6.00-6.32)	5.49 (4.45-6.32)
Castor	14.98 (14.52-15.33)	22.43[a] (22.02-22.91)	5.47 (4.98-5.82)	66.85[a] (65.71-68.01)	6.41 (4.75-8.71)	5.72 (4.42-6.74)
S. Ed (±)	NS	0.69	NS	2.81	NS	NS
CD (P=0.05)	-	1.70	-	6.87	-	-
CD (P=0.01)	-	2.58	-	10.40	-	-

NB: The values presented in parentheses indicate the range

Quantitative consumption and utilization of castor and tapioca leaves by eri silkworm

The preference of food depends upon the nutritional value of leaf and perhaps also due to the presence of non-nutritional stimulants. The amount of food digested varied with reference to different hosts. It was observed that digestion of the food does not follow any trend as that was found with regard to consumption. The digestion is also influenced by factors like the physiological status of the insect, its nutritional requirement and nutritional value of the host. The chemical compound that stimulates the feeding may not be readily digested, perhaps because of its chemical complexity. Hence, the digestion indicates the amount of digestible nutrients available in the host.

Knowledge of the eco-physiological studies on the consumption and utilization of food by the eri worm seems to be demanding in the light of growing awareness and interest. The literature on the subject was so far confined and limited to the castor leaf consumption only. Hence, Reddy and Narayanaswamy, (1998) dealt with the consumption and utilization of three varieties of castor and tapioca, which would serve as alternate food during different instars of eri silkworm.

The eri larvae were reared separately on four hosts *viz.,* three varieties of castor (local, Aruna and RC-8) and of a tapioca (Mangalore local variety) till cocoon formation. Ten larvae at random were collected once at the beginning of each instar (before giving the first feed) and again at the end of each instar, the same were frozen at $0°C$ to death and oven dried at $100±5°C$ till a constant weight was obtained. Direct estimation of consumption (C), egestion (F), assimilation (A) and conversion was done by using the formulae.

$$C = P + R + F$$

Where, R = Quantity metabolized

P = Production

$$A = C - F$$

The food consumed was calculated by subtracting the food remained (FR) at the end of feeding from the food given (FG) at the beginning, using the formula

$$C = FG - FR$$

The results showed that the total food consumed during the entire larval period was significantly higher in Aruna (19.87 and 7.86 g on fresh and dry weight basis respectively) and lower on local variety of castor (15.11 g) on fresh weight basis, and on RC-8 variety (6.34 g) on dry weight basis (Table 7.8). The total amount of food digested by the whole larval stage was higher with reference to aruna variety of castor (11.92 g) followed by local variety (10.97 g) (Table 7.9). Though the amount of food consumed was less on local variety of castor, its digestibility was more than the other three hosts.

Utilization of food varied from host to host. The amount metabolized both on the basis of fresh and dry matter of the food was significantly less in the larvae fed on RC-8 variety of castor, but the amount of food converted to body matter was more with RC-8 variety than with the other hosts. On the other hand, tapioca fed larvae utilized more food for metabolism and thus fewer amounts were felt for body matter. This may be due to its poorer nutritional value. However, the net higher production of body matter was with Aruna variety of castor (6.70g) and the least with tapioca (4.45 g) on fresh weight basis. But, on dry weight basis, it was higher with RC-8 variety of castor (1.01 g) and the least with local variety (0.76 g) (Tables 7.10 and 7.11). The amount of food consumed to produce one gram of body substance was less with RC-8 (2.86 g of fresh food, 6.34 g of dry matter) and higher with tapioca (4.03 g of fresh food; 10.4 g of dry matter). Similarly, the food assimilated was less with RC-8 (1.55 g of fresh food; 3.32 g dry matter) and higher with tapioca (2.43 g of fresh food; 6.01 g of dry matter).

The local variety of castor though showed higher digestibility had the most part of it metabolized. Whereas, the RC-8 variety, which has the less dry matter content is consumed in larger quantity and the higher amount of dry matter of which was converted into body matter. Also, lesser amount of food is required to produce a gram body matter with RC-8 variety of castor, followed by Aruna variety. Hence, RC-8 and Aruna varieties of castor are considered to be the best hosts with reference to their nutritional values.

In an another experiment conducted by Basaiah *et al.,* (1990) to know the consumption and utilization of food by *Samia cynthia ricini* on 3 castor varieties (Local, Aruna and RC-8) and one tapioca variety (Mangalore Local), the rate of ingestion did not differ between plants on fresh weight basis, but cassava had the highest value (600.4 g/g body weight per day) on the basis of dry weight. Castor local had the highest approximate digestibility (80.80%) and growth rate. Conversion efficiency of ingested and digested food into body matter were highest in larvae reared on Aruna. Hence local variety of castor and Aruna were also highly suitable for eri silkworm rearing.

Table 7.8: Amount of food ingested (g) by *S. c. ricini* during different larval stages on different hosts (Adopted from Reddy and Narayanaswamy, 1998)

Host	Fresh weight Basis						Dry weight Basis					
	I	II	III	IV	V	Total	I	II	III	IV	V	Total
Castor												
Local	5.25	9.39	50.06	189.31	1257.59	1511.62	1.61	2.76	17.48	63.54	600.27	685.68
	(0.35)	(0.62)	(3.31)	(12.52)	(83.19)	(100.00)	(0.23)	(0.40)	(2.55)	(9.27)	(87.54)	(100.00)
Aruna	5.82	8.49	52.28	231.33	1689.18	1987.11	1.21	2.65	17.00	80.88	684.37	786.12
	(0.29)	(0.43)	(2.31)	(11.64)	(85.01)	(100.00)	(0.15)	(0.34)	(2.61)	(10.29)	(87.06)	(100.00)
RC-8	5.60	11.59	53.74	236.99	1620.63	1928.57	1.31	3.49	18.20	97.63	513.68	634.33
	(0.29)	(0.60)	(2.79)	(12.29)	(84.03)	(100.00)	(0.21)	(0.55)	(2.87)	(15.39)	(80.98)	(100.00)
Tapioca												
Mangalore	4.52	22.46	61.57	213.56	1486.83	1788.97	1.20	9.28	19.80	76.22	656.54	763.06
Local	(0.25)	(1.26)	(3.44)	(11.94)	(83.11)	(100.00)	(0.16)	(1.22)	(2.60)	(9.90)	(86.04)	(100.00)
SE	0.32	0.73	2.81	5.98	28.71	29.59	-	0.28	0.82	2.23	18.66	18.99
CD at 5%	0.686	1.569	5.964	12.698	60.870	62.736	-	0.595	1.739	4.731	39.56	39.889

NB: The values in parentheses are percentages

Table 7.9: Amount of food digested (g) by *S. c. ricini* during different larval stages on different hosts (Adopted from Reddy and Narayanaswamy, 1998)

Host	Fresh weight Basis						Dry weight Basis					
	I	II	III	IV	V	Total	I	II	III	IV	V	Total
Castor												
Local	4.79	8.03	40.51	142.57	901.64	1097.56	1.22	1.51	9.45	30.19	412.03	454.41
Aruna	5.21	7.22	41.36	166.20	972.05	1192.06	0.85	1.50	8.15	40.26	408.98	459.76
RC-8	4.46	9.82	42.19	170.64	821.44	1048.56	1.04	1.85	8.55	85.09	255.78	352.36
Tapioca												
Mangalore Local	3.57	20.22	47.93	138.94	870.49	1081.17	0.82	1.39	8.33	35.55	504.73	457.83
SE	0.31	0.72	2.59	4.75	20.12	21.01	0.05	0.18	-	1.68	10.52	11.47
CD at 5%	0.660	1.545	5.503	10.076	42.667	44.558	0.118	0.396	-	3.568	22.31	24.330

Table 7.10: Amount of food metabolised (g) by *S. c. ricini* during different larval stages on different hosts (Adopted from Reddy and Narayanaswamy, 1998)

Host	Fresh weight Basis						Dry weight Basis					
	I	II	III	IV	V	Total	I	II	III	IV	V	Total
Castor												
Local	4.07	6.45	23.59	80.24	487.62	574.31	1.13	1.21	6.83	22.52	348.64	380.35
Aruna	4.67	5.64	28.85	64.99	419.56	523.25	0.73	1.16	6.12	27.34	327.36	362.82
RC-8	3.91	7.61	27.08	85.42	248.28	362.36	0.98	1.47	6.34	46.12	166.72	214.86
Tapioca												
Mangalore Local	2.95	17.88	34.86	70.84	509.64	636.17	0.75	7.11	6.80	27.22	339.62	381.33
SE	0.30	0.72	2.55	4.62	20.41	28.38	0.05	0.19	0.68	1.98	10.32	11.27
CD at 5%	0.976	2.362	8.281	14.988	66.128	91.977	0.188	0.616	2.212	6.441	33.441	36.52

Table 7.11: Gain in larval weight (g) by *S. c. ricini* during different larval stages on different hosts (Adopted from Reddy and Narayanaswamy, 1998)

Host	Fresh weight Basis						Dry weight Basis					
	I	II	III	IV	V	Total	I	II	III	IV	V	Total
Castor												
Local	0.71	1.51	16.92	62.22	414.03	495.42	0.08	0.28	2.62	7.67	63.55	74.22
Aruna	0.54	1.57	13.96	101.21	552.49	669.80	0.12	0.33	1.93	12.92	81.61	96.93
RC-8	0.54	2.21	15.11	85.21	572.16	675.25	0.06	0.41	2.21	8.95	89.06	100.71
Tapioca												
Mangalore Local	0.62	2.33	13.07	68.10	360.84	444.98	0.05	0.28	1.53	8.33	66.09	76.30
SE	0.02	0.13	0.90	2.39	10.66	-	0.01	0.03	0.17	0.79	2.25	-
CD at 5%	0.084	0.430	2.939	7.769	34.542	-	0.036	0.120	0.579	2.577	7.30	-

Consumption of food by the larvae of eri silkworm, *S. cynthia ricini* under different dietary regimes

Eri silkworm has five larval stages. These instars are only the feeding stages and therefore, whatever the larva feeds during the feeding period, will be reflected in the form of cocoon quality/quantity, fecundity etc. To understand the food consumption instar wise and to know their importance, Joshi (1991) undertaken the present investigation. Larvae of eri silkworm were reared on four dietary regimes *viz.,* TT = First to fifth instar on tapioca; TC= first to third instar on tapioca and fourth to fifth instar on castor, CT= first to third instar on castor and fourth to fifth instar on tapioca and CC = First to fifth instar on castor. Experimental rearings were kept at 26±2^{0}C temperature and 90±5 per cent relative humidity up to fourth instar and 60±5 per cent RH for rest of the stages. Photoperiod consisting of 12 h photophase and 12 h scotophase was maintained.

The experiment was started with ten replicates each consisting of ten larvae. In all the replicates, the equally weighed amount of food was offered and a parallel sample of food (according to the specific dietary regime) was kept separately to know the amount of food ingested. During the experiment left-over food and faeces were quantitatively collected and dried in oven at 100^{0}C to a constant weight. The amount of food consumed was computed from the aliquot. Results showed that the consumption of food is gradually increasing (from 0.16 to 83.80% in TT, 0.12 to 84.06% in TC, 0.17 to 80.82% in CT and 0.15 to 81.60% in CC) from first instar to fifth. The larvae of fourth and fifth instars consume 96.89% of total food consumption in TT, 97.60% in TC, 95.88% in CT and 96.37% in CC (Table 7.12). These observations showed that in the larval life, the fourth and the fifth instars are the most important since they consume most of the food out of total consumption. Legay (1958) pointed out that among of all the stages of the larval life of *Bombyx mori*, the fifth instar is the most important stage for either qualitative or quantitative food consumption. Because this factor is reflected consistently through large variations in the development of silk gland and secretory intensity of the silk.

Rearing technology of eri silkworm under varied host plant conditions in Tamil Nadu

The leaves of *kesseru* were next to the castor leaves in terms of cocoon harvest and other economic parameters in South India (Kumar *et al.,* 1993). The match wood tree species, *Ailanthus excelsa* is common in the plains and hills are also an alternative host for eri silkworm. Sachan and

Bajpai (1973) observed superior larval growth, development and higher cocoon production when eri silkworms were fed with leaves of Rosy castor variety followed by S- 30, EB-31 and EB-16 varieties. Papaya plantation is also coming up on a large scale in many parts of South India and the leaves can be used productively. The progressive growth of eri silkworm was superior when fed on castor. Further, the larvae receiving the castor leaves during fifth instar had better growth irrespective of the diet used earlier, i.e., whether tapioca/castor (Joshi, 1987). So Subramanianan *et al.,* (2013) studied stage wise larval duration of eri silkworm on 14 different host plants, which include both castor and tapioca.

The larval duration of Eri silkworm on different host plants revealed that, it took 22-23 days in CNBR, CPDB and CGDB genotypes; 23-24 days were required in castor – tapioca combinations to complete the larval development and a maximum of 25 days on the tapioca host plant (Table 7.13). Depending upon the host plant the larval duration was prolonged. The moulting period for all larval stages ranged from 2.5 to 3.0 days on different host plants and combinations. It was short in favourable plants. The lowest feeding duration of 19.5 days was observed in CNBR variety followed by 20 days in CPDB genotype. The eri silkworms recorded their highest feeding duration of 24 days in TV-2 tapioca variety. Previously Dookia (1980) reported that eri silkworm reared on Rosy castor variety recorded maximum mature larval weight (7.90 g), cocoon weight (3.68 g), pupal weight (3.25 g) and shell weight (0.43 g), while shell ratio was more on EB-31 variety (13.31%). Hence it can be inferred from the studies of Subramanianan *et al.,* (2013) that the genotypes CNBR and CPDB can be recommended for Tamil Nadu region, because the eri silkworm larvae registered shortest larval duration.

Table 7.12: Per cent food intake by each instar of eri silkworm larvae, when reared on four dietary regimes (Joshi, 1991)

Dietary regimens	Per cent food consumption by instars				
	I	II	III	IV	V
TT	0.16	0.41	2.52	13.09	83.80
TC	0.12	0.31	1.95	13.54	84.06
CT	0.17	0.57	3.37	15.06	80.82
CC	0.15	0.51	2.97	14.76	81.60

Table 7.13: Stage-wise larval duration of Eri silkworm on different host plants (Adopted from Subramanianan *et al.,* (2013).

Host plant	I Instar (days)	II Instar (days)	III Instar (days)	IV Instar (days)	V Instar (days)	Moulting duration (days)	Feeding duration (days)	Larval duration (days)
CPDB	3.5	4.0	3.5	5.0	6.5	2.5	20.0	22.5
CGDB	3.5	4.0	3.5	5.0	7.0	2.5	20.5	23.0
TV1	4.0	4.5	4.0	5.0	7.5	3.0	22.0	25.0
TV2	4.0	4.5	4.0	5.0	7.5	3.0	24.0	25.0
CPDB-TV1	3.5	4.0	4.0	5.0	7.0	3.0	22.0	23.5
CPDB-TV2	3.5	4.0	4.0	5.0	7.0	3.0	22.0	23.5
CGDB-TV1	3.5	4.0	4.0	5.0	7.0	3.0	22.0	23.5
CGDB-TV2	3.5	4.0	4.0	5.0	7.0	3.0	22.0	23.5
TV1-CPDB	4.0	4.5	3.5	5.0	7.0	3.0	22.0	24.0
TV1-CGDB	4.0	4.5	3.5	5.0	7.0	3.0	22.0	24.0
TV2-CPDB	4.0	4.5	3.5	5.0	7.0	3.0	22.0	24.0
TV2-CGDB	4.0	4.5	3.5	5.0	7.0	3.0	22.0	24.0
CNBR	3.5	4.0	3.5	4.5	6.5	2.5	19.5	22.0
CNBG	3.5	4.0	4.0	5.0	6.5	2.5	22.5	23.0

Note: First mentioned variety was used up to II instar and second variety for later stages

C = Castor; P = Pink; DB = Double bloom; G = Green; TV = Tindivanam variety (Tapioca); NB = No bloom; R = Red

Table 7.14: Cocoon characteristics of *Samia ricini* reared on *Ailanthus* species and castor (Adopted from Saikia and Dutta, 2013)

Host plants	Cocoon weight (g)	Shell weight (g)	Shell ratio (%)	Pupal weight (g)	Cocoon size (cm) Length- Breadth	Degumming period (min)	Degumming loss (%)
Barkesseru	2.65	0.36	13.59	2.29	5.02- 2.66	50.40	20.00
Borpat	2.72	0.34	12.50	2.38	5.35- 2.76	53.00	19.40
Castor	3.22	0.44	13.24	2.79	6.20- 2.92	48.40	18.40
CD at 5%	0.08	0.01	0.58	0.08	0.53- 0.16	0.83	0.83

Table 7.15: Physical characteristics of spun yarn produced from *Ailanthus* species and castor fed eri cocoons (Adopted from Saikia and Dutta, 2013)

Host plants	Yarn yield (%)	Yarn size (Count 'S')	Tensile strength (kg)	Tenacity (g/tex)	Elongation (%)	Imperfection (%)
Barkesseru	72.40	13.85	0.25	5.83	23.58	4.89
Borpat	74.20	13.90	0.22	5.18	23.10	4.67
Castor	75.20	14.06	0.17	4.03	21.82	4.61
CD at 5%	0.83	NS	0.02	0.47	1.42	0.06

Evaluation of *Ailanthus* species in relation to cocoon and yarn characters of eri silkworm in North Eastern region of India

Ericulture is carried out traditionally from stray Castor or Kesseru plantations supplemented occasionally by other secondary host plants like tapioca, payam, korha etc. in case of shortage of leaves. *Ailanthus* is a small genus of lofty trees, belonging to the family Simaroubiaceae, comprising about four species in India. The eri silkworms are frequently found in natural state feeding on *Ailanthus* leaves in North Eastern region. It was reported that though castor is the most preferred food plant for commercial eri silkworm rearing and silk production, *Ailanthus* leaves can also be best utilized for rearing of the silkworm during scarcity of castor leaves (Chowdhury, 1982; Khanikor *et al.,* 1997 and Shaw, 1998). Rearing of eri silkworm on *Ailanthus* sps is advantageous because they are perennial, ever green, quick growing tall trees and are not browsed by cattle.

The literature on the effect of different *Ailanthus* species on commercial and physical characters of cocoon and yarn is meagre. In this regard, Saikia and Dutta, (2013) planned an investigation to evaluate two different *Ailanthus* species in relation to cocoon and yarn characteristics of eri silkworm. Larvae of eri silkworm were reared with leaves of two different *Ailanthus* species *viz.,* Borpat (*A. grandis*) and Borkesseru (*A. excelsa)* and castor (*Ricinus communis*) separately under laboratory conditions of Assam Agricultural University, Jorhat by maintaining temperature and humidity at 27 ± 2^{0}C and $84 \pm 4\%$ RH from first to fifth instar.

Results of the experiment revealed that castor fed eri silkworm produces cocoons with the highest cocoon weight (3.22 g), shell weight (0.44 g), pupal weight (2.79 g) and size of cocoons and differed significantly from both Borpat and Barkesseru fed cocoons except for shell ratio which was recorded highest on Barkesseru (13.59%) followed by castor (13.24%) and lowest in Borpat (12.50%) (Table 7.14). With significantly lower degumming loss and shorter degumming period though yarn yield was recorded highest on castor fed cocoons, the physical yarn characters *viz.,* tensile strength, tenacity, elongation and imperfection of the yarn produced from the Barkesseru and Borpat fed cocoons are significantly better than castor fed cocoons with no significant difference in respect of size of the yarn i.e. count (Table 7.15). These variations in cocoon and yarn characters of the silkworm might be due to the nutritional quality of the food plants and also the required balance of the nutrients within the species owing to many factors including the synthetic ability and metabolic activities of the silkworm involving specific interrelations between certain nutrients (House, 1974).

Thus, it is imperative to conclude that though castor is considered to be the best for rearing and cocoon production of eri silkworm, the *Ailanthus* species *viz.*, Borpat and Barkesseru can also be suitably utilized for commercial rearing of eri silkworm considering the economic yarn characters to enhance the production and productivity of eri silk in North Eastern region.

Rearing performance of eri silkworm on *Ailanthus excelsa* and *A. grandis* at Department of Sericulture, Assam Agricultural University, Jorhat

Deori and Khanikor (2015) conducted an experiment to study the larval duration, larval weight, effective rate of rearing, cocoon weight, shell weight, pupal weight, shell ratio, pupal duration and fecundity of eri silkworm on *A. excelsa* and *A. grandis* to find out the feasibility of substituting castor leaves for more production of eri silk. The experiment was conducted in the Department of Sericulture, Assam Agricultural University, Jorhat. Freshly hatched eri silkworm larvae were reared on two *Ailanthus* species. Castor being the primary host was also taken for the comparison of results.

Castor fed larvae had shown the shortest larval period (22.83 days), highest larval weight (7.30 g), effective rate of rearing (86.67%), cocoon weight (3.38g), shell weight (0.438g) and pupal weight (2.68g) followed by *A. grandis* and *A. excelsa*. Slightly longer larval periods were observed in *A. excelsa* (28.50 days) and *A. grandis* (29.00 days) in comparison to castor (Table 7.16). In case of shell ratio and fecundity, *Ailanthus* fed individuals were statistically at par with castor. From the present study, it can be concluded that the leaves of both *A. grandis* and *A. excelsa* can be utilized as a substitute of castor leaves during the scarcity throughout the year for more production of eri silk.

Table 7.16: Effect of host plants on larval, pupal and cocoon parameters of eri silkworm (Adopted from Deori and Khanikor, 2015)

Host plants	Larval duration (days)	Larval weight (g)	ERR (%)	Cocoon wt (g)	Shell wt (g)	Shell ratio (%)	Pupal wt (g)	Pupal duration (days)	Fecundity (No. of eggs/ female)
Barkesseru	28.50[a]	6.02[b]	63.33[b]	2.79[b]	0.323[b]	11.59	2.17[b]	18.66[a]	350.00
Borpat	29.00[a]	6.11[b]	72.67[b]	2.81[b]	0.353[b]	12.61	2.23[b]	18.33[a]	310.00
Castor	22.83[b]	7.30[a]	86.67[a]	3.38[a]	0.438[a]	12.96	2.68[a]	15.66[b]	313.33
S.Ed ±	0.49	0.44	5.33	0.12	0.02	NS	0.18	0.231	NS
CD at 5%	1.20	1.06	13.05	0.28	0.04	-	0.43	0.57	-

Figures followed by the same letter did not differ significantly at P=0.05   NS = Not Significant

Evaluation of new host plant species for ericulture at University of Agricultural Sciences, Dharwad

Eri silkworm is a polyphagous insect and feeds on a wide range of host plants. But only on few host plants developmental biology and other related aspects have been studied. Patil and Savanurmath (1994) reported that there are a number of host plants till left unexploited due to lack of technology. Accordingly, Naik and Murthy, (2013) conducted a study to know the growth, development and economic cocoon parameters of eri silkworm, *Samia cynthia ricini* on 23 new hosts at DBT Ericulture Laboratory, University of Agricultural Sciences, Dharwad, Karnataka during November-December and January-February. The leaves of new host plant species were fed to the eri silkworm at its different instars. Feeding response like no feeding, slight feeding, moderately feeding and good feeding were recorded. Among 23 plant species tried for ericulture, five plant species have been accepted by the eri silkworm as food plants. Out of five plant species, eri larvae had good feeding response and survivability on fountain tree, banyan tree and Indian almond, moderate response on carrot leaves and slight feeding response and survivability on jack fruit leaves. It may be due to stimulant action (gustatory stimulants) and nutritional content of leaf. Non- acceptance of host may be due to antifeedant (feeding deterrent), poor phagostimulant and the presence of alkaloids in the leaf. For no feeding, it may be due to the odour (olfactory stimulants) which repelled the worms from feeding site and morphological feature of leaves. Among five plant species only four plant species showed the good feeding and survivability of larvae (100%) on banyan tree, fountain tree, carrot leaves and Indian almond. Whereas, poor survivability of larvae (32.0%) was noticed on jack fruit leaves during first instar (Table 7.17).

Weight of the eri silkworm recorded during different instars on different host plants is presented in Table 7.18. Highest mature larval weight of 4.55 and 4.45g was recorded in castor and fountain tree followed by banyan tree (4.01g) and Indian almond (3.87g). November-December is the most suitable season compared to January-February for the rearing of eri silkworm.

Rearing performance of eri silkworm on potential new hosts under Western Maharashtra conditions

Even though the eri silkworm has several secondary host plants, all the food plants are not equally good for eri silkworm rearing and the silkworms show differential behaviour, when reared on different food plants. In India, Maharashtra is a non-traditional state for eri silkworm and it has suitable climate to rear the *Samia cynthia ricini* silkworm for raw silk production. The information on rearing performance of eri silkworm on secondary host plants is scanty under Western Maharashtra conditions. Hence, Kavane (2014a; 2014b and 2015) made an attempt to rear the insect from 1st instar to adult formation on *Terminalia arjuna* (Arjun), *Carica papaya* (Papaya) and *Terminalia catappa* (Badam). The data were recorded for the study of biology, which includes larval duration, adult longevity and fecundity, effective rate of rearing (ERR), cocoon characters, colour, shape, single cocoon weight, shell weight etc. The results showed that the growth parameters and cocoon characters of the eri silkworm reared on arjun, papaya and badam leaves were normal. The eri silkworm successfully completed its life cycle ranging from 52 to 55 days on arjun, 47 to 53 days on papaya and badam leaves (Table 7.19). The effective rate of rearing, weight of pre-spinning larvae, cocoon, shell and pupae in *T. arjuna* ranged from 40 to 45 per cent, 6 to 8 g, 2.80 g, 0.40 g, 2.35 g, respectively, while in papaya these values are 45 to 55 per cent, 6 to 7 g, 2.49 g, 0.30 g, 2.10 g, respectively. When eri silkworm was reared on badam leaves, the ERR ranged from 90-95%. Parameters such as length of cocoon shell, width of shell, shell thickness, etc. were recorded. The fecundity ranged from 400 to 425 eggs in arjuna and badam, while in papaya the number of eggs laid by the female was 350 to 415 (Table 7.20). The findings of badam, papaya and arjun as potential new hosts for eri silkworm have opened new vistas in promoting vanya silk industry in Maharashtra.

Table 7.17: Survivability of eri silkworm on new host plants (Adopted from Naik and Murthy, 2013)

S.No.	Host plant	Survivability percentage						Feeding response
		I Instar	II Instar	III Instar	IV Instar	V Instar	Mean	
1.	*Artocarpus heterophyllus* Lam.	32.00 (34.44)	36.00 (36.86)	40.00 (39.22)	32.00 (34.44)	28.00 (31.94)	33.60 (35.41)	A (SF)
2.	*Acacia ferruginea* DC.	0.00	0.00	0.00	0.00	0.00	0.00	NA
3.	*Bauhinia purpurea* L.	0.00	0.00	0.00	0.00	0.00	0.00	NA
4.	*Bambusa arundinaceae* Retz.	0.00	0.00	0.00	0.00	0.00	0.00	NF
5.	*Bambusa vulgaris* L.	0.00	0.00	0.00	0.00	0.00	0.00	DF
6.	*Ficus bengalensis* L.	100.0 (89.96)	100.0 (89.96)	100.0 (89.96)	100.0 (89.96)	96.0 (78.43)	99.2 (84.83)	A (GF)
7.	*Sapindus emarginatus*	0.00	0.00	0.00	0.00	0.00	0.00	NA
8.	*Peltophorum ferrugineum* Decne	0.00	0.00	0.00	0.00	0.00	0.00	DF
9.	*Pongamia pinnata* Linn.	0.00	0.00	0.00	0.00	0.00	0.00	DF
10.	*Samanea saman* Jacq.	0.00	0.00	0.00	0.00	0.00	0.00	NA
11.	*Spathodea companulata* Pal.	100.0 (89.96)	100.0 (89.96)	100.0 (89.96)	96.0 (78.43)	96.0 (78.43)	98.4 (82.70)	A (GF)
12.	*Swietenia mahogoni* Marophylla	0.00	0.00	0.00	0.00	0.00	0.00	DF
13.	*Tabubia argentina*	0.00	0.00	0.00	0.00	0.00	0.00	DF
14.	*Terminalia catappa* Linn.	100.0 (89.96)	100.0 (89.96)	100.0 (89.96)	92.00 (73.54)	88.0 (70.06)	96.00 (78.43)	A (GF)
15.	*Cassia auriculata* Linn.	0.00	0.00	0.00	0.00	0.00	0.00	DF
16.	*Ficus religiosa* L.	0.00	0.00	0.00	0.00	0.00	0.00	DF
17.	*Hardwickia binata* Roxb.	0.00	0.00	0.00	0.00	0.00	0.00	DF
18.	*Manilkara hexandra* Roxb.	0.00	0.00	0.00	0.00	0.00	0.00	DF
19.	*Simarouba glauca* DC.	0.00	0.00	0.00	0.00	0.00	0.00	DF
20.	*Daucus carota* Linn.	100.0 (89.96)	100.0 (89.96)	96.00 (78.43)	72.00 (58.03)	68.00 (55.53)	87.20 (69.01)	A (MF)
21.	*Acalypha gracilens* A.	0.00	0.00	0.00	0.00	0.00	0.00	DF
22.	*Lantana camara* L.	0.00	0.00	0.00	0.00	0.00	0.00	DF
23.	*Cassia fistula* Linn.	0.00	0.00	0.00	0.00	0.00	0.00	DF
24.	Castor (control)	100.0	100.0	100.0	100.0	100.0	100.0	A (GF)

A = Acceptance; SF – Slight Feeding, *Artocarpus heterophyllus* Lam 0-50%; MF = Moderately feeding, *Daucus carota* Linn. 50-90%;
GF = Good feeding, *Ficus bengalensis* L., *Spathodea companulata* Pal., *Terminalia catappa* Linn. – 90-100%

NA = Non-acceptance, DF = No feeding

Table 7.18: Weight of Eri silkworm at different instars on different host plants during (Nov.–Dec. and Jan.-Feb.) (Adopted from Naik and Murthy, 2013)

Treatment	1st instar wt (g)	2nd instar wt (g)	3rd instar wt (g)	4th instar wt (g)	5th instar wt (g)	Mature larval wt (g)
Host						
Castor (GCH-4) (H-1)	0.018[a]	0.477[a]	1.609[a]	2.651[a]	6.270[a]	4.55[a]
Fountain tree (H-2)	0.015[ab]	0.459[abc]	1.515[a]	2.616[b]	6.200[a]	4.45[a]
Banyan tree (H-3)	0.014[ab]	0.448[bc]	1.354[b]	2.394[c]	6.004[a]	4.01[b]
Indian almond (H-4)	0.013[b]	0.441[c]	1.347[b]	2.339[d]	5.994[a]	3.87[b]
Carrot (H-5)	0.017[ab]	0.469[ab]	1.553[a]	2.098[e]	5.512[b]	3.60[c]
S.Em ±	0.002	0.002	0.007	0.0007	0.0278	0.02
CD at 1%	0.008	0.007	0.028	0.0026	0.1082	0.06
Season						
Nov-Dec (S1)	0.017	0.466	1.493	2.477	6.036	4.35
Jan-Feb (S2)	0.014	0.451	1.390	2.360	5.956	3.84
S.Em ±	0.0002	0.001	0.005	0.0004	0.0176	0.01
CD at 1%	0.0008	0.004	0.0018	0.0016	0.0684	0.04
Interactions						
H1 x S1	0.019[a]	0.484[a]	1.658[a]	2.723[a]	6.285[a]	4.78[a]
H1 x S2	0.017[abc]	0.470[b]	1.560[b]	2.579[c]	6.254[ab]	4.32[b]
H2 x S1	0.017[abc]	0.472[ab]	1.642[a]	2.685[b]	6.248[ab]	4.69[a]
H2 x S2	0.014[cd]	0.446[d]	1.388[d]	2.540[d]	6.153[abc]	4.22[bc]
H3 x S1	0.015[bcd]	0.453[cd]	1.408[cd]	2.440[e]	6.020[abc]	4.28[bc]
H3 x S1	0.012[d]	0.443[dc]	1.301[e]	2.348[g]	5.988[bc]	3.74[d]
H4 x S2	0.014[cd]	0.450[d]	1.401[cd]	2.379[f]	6.012[bc]	4.18[c]
H4 x S1	0.012[d]	0.432[e]	1.293[e]	2.298[h]	5.975[c]	3.56[e]
H5 x S1	0.018[ab]	0.473[ab]	1.650[a]	2.160[i]	5.645[d]	3.84[d]
H5 x S2	0.017[abc]	0.465[bc]	1.457[c]	2.036[j]	5.379[e]	3.37[f]
S.Em ±	0.0003	0.002	0.0010	0.0009	0.0393	0.02
CD at 1%	0.0011	0.009	0.0038	0.0037	0.1529	0.09

Figures followed by same letter did not differ significantly

Table 7.19: Cocoon characters of *S. c. ricini* silkworm on *Terminalia arjuna* (Arjun), *Carica papaya* (Papaya) and *Terminalia catappa* (Badam) (Adopted from Kavane 2014a; 2014b and 2015)

Host	Fecundity No/Worm	Hatching %	Larval wt (g)	ERR (%)	Cocoon wt (g)	Shell wt (g)	SR (%)	Larval period (days)	Pupal wt (g)
Badam	400-425	100	7 to 9	90-95%	3.02	0.45	14.90	21.5	2.60
Papaya	350-415	100	6 to 7	45-55%	2.49	0.30	12.04	24-25	2.10
Arjuna	400-425	98	6 to 8	40-45%	2.80	0.40	10.95	26.0	2.35

Table 7.20: Rearing performance of *S. c. ricini* silkworm on *Terminalia arjuna* (Arjun), *Carica papaya* (Papaya) and *Terminalia catappa* (Badam) (Adopted from Kavane 2014a; 2014b and 2015)

Life stages	Duration (days)			Duration of shedding cuticle (hrs)		
	Arjuna	Badam	Papaya	Arjuna	Badam	Papaya
Eggs	10	8-9	10	-	-	-
1st instar	3.9	3	3.5	12	10	12
2nd instar	2.9	2.1	2	18	15	18
3rd instar	4.2	3.5	4.5	16	15	18
4th instar	6.4	5.5	6	24	20	24
5th instar	8.6	7	8	-	-	-
Pupal	15	15	15	-	-	-
Adult longevity	3-4	3-4	3-4	-	-	-

Rearing of Eri Silkworm Under Host Scarcity Conditions

Impact of independent and sequential feeding of different host plants on economic traits of eri silkworm at Department of Sericulture, Bangalore

Eri silkworm is known to feed on the leaves of more than 50 host plant species, among which, castor is considered as the primary food plant of eri silkworm. However, during the scarcity of castor leaves especially at fourth and fifth instars when the larvae consume more quantity of leaves, secondary food plants like tapioca, *Ailanthus* and jatropha can be used for successful rearing of eri silkworm. Eri host plants are interchangeable especially at 3rd instar to 5th instar stages (Dutta and Khanikor, 2005; Mukul Deka *et al.*, 2011). However, this new practice of utilizing sequential use of food plants for commercial rearing has not been worked out. In this regard, Venu and Munirajappa (2013) conducted a study on the rearing performance of eri silkworm by feeding the leaves of castor, tapioca, *Ailanthus* and jatropha, independently as well as sequentially at Department of Sericulture, Bangalore University, Bangalore with the following treatments.

T1 = Castor (1st to 5th instar); T2 = Tapioca (1st to 5th instar); T3 = *Ailanthus* (1st to 5th instar); T4 = Jatropha (1st to 5th instar); T5 = Castor (1st to 3rd instar) + Tapioca (4th to 5th instar); T6 = Castor (1st to 3rd instar) + *Ailanthus* (4th to 5th instar); T7 = Castor (1st to 3rd instar) + Jatropha (4th to 5th instar). During rearing observations on larval duration, larval weight, effective rate of rearing (ERR), cocoon weight, pupal weight, shell weight and shell ratio percentage were recorded.

Eri silkworms fed on castor, tapioca, *Ailanthus* and jatropha leaves showed the mean larval duration of 19 days, 20 days, 23 days and 24.33 days respectively (Table 7.21). However, eri larvae which were fed with castor leaves during 1st to 3rd instar and feeding with tapioca, *Ailanthus* and jatropha leaves from 4th instar to spinning recorded a mean larval duration of 20.67days, 23.33days and 24.67days respectively (Table 7.22). Eri silkworms fed on castor, tapioca, *Ailanthus* and jatropha leaves recorded the mean cocoon weight of 3.58g, 3.18g, 3.05g, and 2.81g respectively. Further, eri larvae which were fed with castor leaves from 1st to 3rd instar periods and with tapioca, *Ailanthus* and jatropha leaves during 4th and 5th instar periods recorded the mean cocoon weight of 3.31g, 3.16g and 2.67g respectively. It has been observed that the castor leaves serve as chief feed for rearing of eri silkworms, the combination of castor and tapioca leaves could also be beneficially used for commercial rearing of eri silkworm. When the worms were reared by feeding the leaves of castor up to 3rd instar and interchanging with tapioca leaves during 4th and 5th instar resulted in better eri silk recovery. The study of Venu and Munirajappa (2013) clearly proved that, the rearing of eri larvae by feeding the castor leaves from 1st to 3rd instar and interchanging with tapioca leaves during 4th and 5th instars improved the commercial characteristic features such as larval weight, larval duration, ERR, cocoon weight and shell weight.

Table 7.21: Effect of independent treatment of host plants on growth and development of eri silkworm (Adopted from Venu and Munirajappa, 2013)

Parameters	Food Plants (M±SE)			
	T1	T2	T3	T4
Larval duration (days)	19.00± 0.58	20.00± 0.58	23.00 ± 0.58	24.33± 0.33
Larval weight (g)	7.53 ±0.15	6.76 ±0.08	6.08 ± 0.01	6.06 ± 0.01
ERR (%)	88.67 ±0.88	83.00 ±1.53	75.67 ±1.45	73.00± 1.53
Cocoon weight (g)	3.58 ±0.04	3.18 ± 0.04	3.05 ± 0.04	2.81 ±0.09
Pupal weight (g)	3.04 ±0.03	2.68 ± 0.03	2.58 ± 0.03	2.37 ±0.08
Shell weight (g)	0.54 ±0.02	0.50 ± 0.01	0.48 ±0.01	0.44 ± 0.01
Shell ratio (%)	15.08 ± 0.32	15.72 ± 0.34	15.73 ± 0.30	15.65 ± 0.31

Table 7.22: Effect of sequential treatment of host plants on growth and development of eri silkworm (Adopted from Venu and Munirajappa, 2013)

Parameters	Food Plants (M±SE)		
	T5	T6	T7
Larval duration (days)	20.67± 0.88	23.33 ±0.67	24.67 ±0.33
Larval weight (g)	6.48 ±0.23	6.00± 0.06	5.78 ±0.09
ERR (%)	80.00± 0.58	76.00± 0.58	71.33 ±0.88
Cocoon weight (g)	3.31±0.03	3.16 ±0.01	2.67± 0.02
Pupal weight (g)	2.86 ±0.03	2.69 ±0.01	2.25± 0.02
Shell weight (g)	0.46 ±0.01	0.46± 0.01	0.41± 0.01
Shell ratio (%)	13.86± 0.21	14.55± 0.27	15.35± 0.24

Utilization of minerals by larvae of the eri silkmoth, *Samia c. ricini*

Larvae of *Samia c. ricini* were reared on 4 experimental diets *viz.,* 1st to 5th instar on tapioca, 1st to 3rd instar on tapioca and 4th to 5th instar on *Ricinus communis*, 1st to 3rd instar on *R. communis* and 4th to 5th instar on tapioca, and 1st to 5th instar on *R. communis*. From 1st to 4th instar larval development of the experiment, the temperature and relative humidity maintained were 26°C and 90% RH, and for the rest of the experimental period the RH was 60%. The eri silkworm larvae that were fed with castor leaves ingested maximum amounts of phosphorus and potassium (64.02 and 122.4 mg, respectively), followed by the larvae fed with tapioca during early instars and castor in later instars (60.42 and 116.69 mg respectively), castor in early instars followed by tapioca in older instars (55.42 and 89.37 mg respectively) and the larvae fed with tapioca throughout their larval period (45.78 and 73.38 mg). The order of magnitude of the coefficient of apparent digestibility of potassium was castor only (69.38%), tapioca followed by castor (62.97%), castor followed by tapioca (62.97%) and tapioca only (62.92%). The corresponding values for phosphorus were 41.58%, 37.33%, 44.22% and 42.74% in all the four treatments (Joshi, 1986).

Effect of host plants on some biochemical parameters of eri silkworm during its development

Host preference of forb chewers does not correspond precisely with host nutritional suitability but can be influenced by ecological factors such as natural enemies, host phenology and microclimate, nutritional and allelochemical composition. Host plant directly influences growth and development of larvae and quality and quantity of silk produced by them. As the bio-components of leaves vary with age and variety, the enzyme activity also varies, and exerts a direct influence on quality and quantity of silk produced. Amino acid composition of leaves may also influence the synthesis of silk. In view of this, the enzymes alanine aminotransferase (AAT), aspartate aminotransferase (AsAT), and glutamine synthetase (GS) have been studied during developmental stages of eri silkworm larvae fed on castor and kesseru leaves (Gogoi and Yadav, 1995). The worms were reared in the laboratory on castor and kesseru leaves and were allowed to develop at 25±2°C and 80±2% RH with 12:12 hr L:D cycle. The larvae of desired stage were chilled, dried and weighed. Homogenates were diluted and assayed for enzyme activities. The water content was estimated by measuring the fresh and dry weight of the larvae.

Activity of all the three enzymes predominated in larvae fed on kesseru leaves than those fed on castor leaves, except aspartate aminotransferase, that recorded highest activity during third instar larvae fed on castor leaves. All the three enzymes showed intermittent fluctuations of similar pattern in their activity throughout the development of both types of larvae except in glutamine synthetase which showed highest level during third instar stage of development in case of castor fed larvae and during fifth instar stage of kesseru fed leaves. In contrast, the amount of protein, DNA, RNA, free amino acids and body wet weight were recorded higher in castor fed larvae as compared to those fed on kesseru. The water content was higher in kesseru fed larvae.

Castor leaves have low moisture and crude fat and higher amount of crude protein, phosphorus, calcium and carbohydrate than the kesseru leaves. Due to the high amount of carbohydrate the activities of all the three enzymes studied were lower in castor fed larvae than the kesseru fed larvae. As per the findings of Gogoi and Yadav, (1995) the protein content was more in castor fed larva when compared to kesseru fed larvae. Low amount of protein in kesseru fed larvae was probably because of increased proteolysis which may also explain the high levels of these enzymes to maintain amino acids pool for protein synthesis and keto acid levels for energy metabolism by providing links between protein and carbohydrate metabolism. Gradual increase in DNA of larvae fed on castor leaves indicate steady increase in cell number due to cell multiplication. The amount of DNA in larvae fed on kesseru leaves showed maximum level in second instar and thereafter it decreased. It may be due to in proportionate growth of kesseru fed larvae, or alternatively due to high amount of water in the kesseru fed larvae as compared to castor fed larvae (Table 7.23). The amount of RNA in kesseru fed larvae showed lower value when compared to the larvae fed on castor leaves throughout the development except in second and fifth instar stage. This is difficult to explain at present, however the high RNA level during these stages may be necessary for protein synthesis. This assumption was further supported by the finding that in second and fifth intsar larvae the amount of FAA was also high as compared to other developmental stages of the larvae (Hoque and Khan, 1987). With the increase of food intake during successive developmental stages the body weight increased gradually in both types of larvae. The larval duration was longer in kesseru fed larvae whereas cocoon weight was higher in castor fed larvae. From the present result it can be confirmed that castor is the principal host plant than kesseru.

Table 7.23: Effect of food plant on various biochemical parameters of *S. c. ricini* (Adopted from Gogoi and Yadav, 1995)

(values are mean ± SD of 4-5 larvae in each group)

Parameter	Food plant	Instar				
		1st	2nd	3rd	4th	5th
Protein	A	113.00±3.00	80.00± 1.00	58.00 ±1.00	62.00 ±1.00	73.00± 3.00
(mg/ larvae)	B	61.00±2.00	78.00± 1.00	73.00 ±2.00	49.00± 2.00	51.00 ±1.00
DNA	A	27.88±0.04	23.62± 0.05	56.30± 0.03	58.76 ±0.02	63.58± 0.03
(mg/ larvae)	B	33.88± 0.38	63.13 ±0.52	33.50 ±0.31	30.78± 0.36	39.50± 0.38
RNA	A	246.79± 3.48	252.79 ±1.41	556.42± 0.98	786.61± 0.61	371.61± 0.69
(mg/ larvae)	B	189.26 ±3.43	297.99 ±3.58	261.67± 3.81	525.78± 2.06	403.90± 1.63
FAA	A	14.41 ±0.11	11.51 ±0.33	27.88± 0.23	7.33 ±0.03	8.88 ±0.01
(mg/ larvae)	B	7.38 ±0.02	8.08± 0.05	3.70± 0.10	4.70 ±0.87	5.15 ±0.08
Larval wt.	A	2.21± 0.01	59.90± 0.90	411.50 ±1.11	815.00± 1.90	3050.00±3.00
(mg/larva)	B	2.26±0.01	63.70± 0.03	175.90± 1.00	594.00 ±1.40	2550.00 ±3.40
Water content	A	86.00± 2.00	85.00± 1.00	87.00 ±1.00	81.00± 0.99	86.00 ±1.00
(% larva)	B	92.00± 1.00	94.00± 2.00	89.00 ±0.99	92.00± 2.00	94.00 ±2.00

A = Larvae fed on castor leaves

B = Larvae fed on kesseru leaves

FAA = Total free amino acids

Table 7.24: Effect of fortification of senescent tapioca leaves with different dosages of jaggery and soya flour on economic parameters of eri silkworm (Adopted from Sakthivel and Qadri, 2013)

Treatment	5th instar larval period (Hrs)	Matured larval weight (g)	ERR (%)	Cocoon yield (kg/100dfls)	Shell yield (kg/100dfls)	SCW (g)	SSW (g)	Silk ratio (%)	Fecundity (No)	Fecundity (%)
T_1	188	4.92	54.27	27.96	2.63	1.79	0.17	9.39	188.66	59.17
T_2	188	5.15	55.16	28.89	27.32	1.82	0.17	9.48	209.00	63.51
T_3	188	5.00	52.01	26.68	2.53	1.78	0.17	9.48	195.33	64.84
T_4	188	5.12	53.48	30.16	2.96	1.96	0.19	9.80	221.00	68.39
T_5	182	5.15	64.70	36.71	4.02	1.97	0.22	10.96	256.66	76.00
T_6	182	5.23	68.34	40.11	4.63	2.04	0.24	11.53	255.33	74.45
T_7	182	5.63	70.37	43.33	4.64	2.14	0.23	10.71	261.33	74.92
T_8	182	5.79	72.94	44.89	5.65	2.14	0.27	12.60	275.00	82.14
T_9	202	3.96	46.24	21.21	1.82	1.59	0.14	8.60	161.00	47.84
T_{10}	156	6.56	96.77	74.11	11.29	2.66	0.41	15.23	364.33	96.14
CD (5%)	4.56	0.185	1.647	1.057	0.131	0.030	0.007	0.302	6.669	3.007

SCW = Single Cocoon weight

SSW = Single shell weight

Table 7.25: Changes caused due to infection of cassava mosaic virus in tapioca leaf (Adopted from Philip, 2010)

Parameters	Type of leaf		% increase / decrease	T test
	Healthy	Infected		
Physiological changes				
Stomatal diffusive resistance (S/cm)	1.63	3.31	+103.1	**
Transpiration rate ($\mu g/cm^2/S$)	31.21	14.53	-53.44	**
Leaf temperature (^{o}C)	33.30	33.50	+0.60	NS
Relative water content (%)	82.78	77.66	-6.18	**
Leaf pigment value (SPAD value)	37.30	26.10	-30.02	**
Biochemical changes				
Moisture %	77.51	73.46	-5.23	**
Moisture retention capacity (%)	76.61	73.56	-3.98	*
Chlorophyll a (mg/g fresh wt.)	0.47	0.24	-48.94	**
Chlorophyll b (mg/g fresh wt.)	0.74	0.35	-52.70	**
Total chlorophyll (mg/g fresh wt.)	1.20	0.59	-50.83	**
Carotenoid (mg/g fresh wt.)	1.08	0.60	-44.44	**
Total protein (mg/g fresh wt.)	122.97	105.86	-13.91	**
Total sugar (mg/g fresh wt.)	82.03	92.55	+12.82	*

*and ** indicate of significance of values at P=0.05 and 0.01 respectively,
NS= Non-significant

Effect of fortification of senescent tapioca leaves on economic traits of eri silkworm:

In tapioca plantation, a huge amount of senescent foliage is available at the time of tuber harvest and is wasted by the farmers (Sakthivel *et al.,* 2010). These senescent leaves generally have poor nutritive values which affect the growth and development of eri silkworm and cocoon yield. In order to utilise those leaves, Sakthivel and Qadri, (2013) have undertaken an experiment to know the effect of feeding senescent tapioca leaves on economic traits of eri silkworm and the feasibility of supplementing sugarcane jaggery and soya powder as the source of carbohydrate and protein to enhance eri cocoon production and additional income to the tapioca farmers through ericulture. The senescent leaves from 10 months old tapioca plants of variety Mullavadi (MVD1) was selected for the fortification study. The larvae were mass reared up to 4th instar using normal tapioca leaves following the normal package of practices. After resumption from the fourth moult uniform size of worms were selected and fed with the fortified senescent leaves till maturity. The treatments included T_1= senescent leaves (SL) with 5% jaggery solution, T_2 = SL with 10% jaggery solution, T_3 = SL with soya powder @ 5g/100g leaves, T_4 = SL with soya powder @ 10g/100g leaves, T_5 = SL with 5% jaggery solution + soya powder @ 5g/100g leaves, T_6 = SL with 5% jaggery solution + soya powder @ 10g/100g leaves, T_7 = SL with 10% jaggery solution + soya powder @ 5g/100g leaves, T_8 = SL with 10% jaggery solution + soya powder @ 10g/100g leaves, T_9 = Senescent leaves without any treatment (control) and T_{10}= Normal matured leaves (standard check). The economic traits such as larval period (h), weight of matured larva (g), ERR (%), cocoon yield/100dfls (kg), shell yield/100 dfls (kg), shell ratio (%), fecundity (no) and hatchability (%) of eggs were recorded (Table 7.24).

The results of Sakthivel and Qadri, (2013) showed that the important economic parameters such as ERR (%), cocoon yield, shell yield and SR (%) of eri silkworm reared on senescent tapioca leaves were drastically reduced (46.24, 21.21, 1.82 and 8.60) compared to the normal matured leaf (96.77, 74.11, 11.29 and 15.23). Fortification of the senescent leaf with jaggery solution and soya powder exhibited significant improvement in all the treatments. However, maximum enhancement (72.94, 44.85,

5.65 and 12.60) was observed when the leaves were supplemented with 10% jaggery solution + soya powder @ 10g/100g leaves followed by 10% jaggery solution + soya powder @ 5g/100g leaves (70.37, 43.33, 4.64 and 10.71) (Table 7.24). In the study of Sakthivel and Qadri, (2013), the fifth instar larvae of eri silkworm fed on normal matured tapioca leaves recorded higher values in economic traits while on senescent leaves exhibited drastic reduction in all parameters. This may be due to the inadequate nutritional value in senescent leaves. The larvae fed on the senescent leaves supplemented with jaggery or soya protein alone elicited comparatively poor results among the treatments. But fortification of senescent tapioca leaves with the combination of jaggery and soya flour as a source of carbohydrate and protein which are essential for growth and development of silkworm and silk production, exhibited marked improvement in all the economic traits over the senescent leaves. Rani *et al.*, (2011) observed significant enhancement in larval weight, cocoon weight, shell weight, shell ratio, filament length, denier and fibroin content of mulberry silkworm fed on the leaves enriched with Amway protein.

It was estimated that the sugarcane jaggery contains 65-68% sucrose, 10-15% reducing sugars, 0.4% protein, 0.1% fat, 0.6-1.0% minerals and trace of vitamins (Rao *et al.*, 2007) and soya flour contains 40.73% protein, 26.30% fat and 17.76% carbohydrate contents and trace minerals (Famurewa *et al.*, 2006) and assimilation of these essential nutrients adequately by eri silkworm when supplemented with senescent tapioca leaves could be attributed to increased larval weight and silk yield when compared to the larvae fed on senescent leaves without fortification. Horie and Watanbe, (1983) had confirmed that the supplementation of soya bean protein increased the protein and amino acid content in the haemolymph of silkworm larvae which resulted marked improvement in silk yield. The study thus suggests that the huge quantity of senescent tapioca leaves wasted at the time of tuber harvest could be successfully utilized by supplementing with 10% jaggery solution and soya powder @ 10g/100g of leaves for rearing of eri silkworm which ensured considerable improvement in economic traits of the worms and silk yield (Sakthivel and Qadri, 2013).

Effect of feeding CMV infected Tapioca leaves on the cocoon and grainage parameters of eri silkworm at Central Sericultural Research and Training Institute, Mysore

Economic characters and reproductive performance of eri silkworm were also earlier found adversely affected when fed with diseased castor leaves (Nagaveni *et al.,* 2002; Shree and Nagaveni, 2002). Reduction in total chlorophyll, chlorophyll a, chlorophyll b, amino acids, total proteins, phenols, sugar and mineral elements are also reported in castor leaves infected with various diseases. Studies conducted with mulberry silkworm also showed that diseased mulberry leaves are not fit for silkworm rearing (Sullia and Padma, 1987; Siddaramaiah and Hegde, 1988). Nutritional quality of leaves plays a vital role in the robust growth of silkworm larvae, cocoon production and reproductive performance. All the nutrients in balanced proportion are necessary for the healthy growth of silkworm and for producing quality cocoons.

Tapioca (*Manihot utilissima* Pohl) leaf is one of the major foods of the eri silkworm. It is one of the staple foods of the low income group and also serves as a raw material for starch based industries. However, tapioca is very badly affected by Cassava Mosaic Virus (CMV) disease making the leaf chlorotic, distorted and curled with reduced leaf area. The infection is widely prevalent in all tapioca cultivating areas. In tapioca, the CMV virus was found to cause significant reduction in almost all the physiological and biochemical parameters of tapioca leaf. Total leaf protein content was reduced by about 13.9%, while there was an increase of total sugar content by 12.8% (Table 7.25) (Philip, 2010).

Studies have shown that various diseases cause reduction in the nutritional components of food pants of silkworm and affect the economic characters of cocoons and reproductive performance. However, no reports are available on the effect of feeding CMV infected tapioca leaves on eri silkworm. In this regard, Philip *et al.,* (2008) have taken up the study to find out the effect of feeding of eri silkworm with CMV infected leaves on the cocoon and grainage parameters. The experiment was conducted during 2006-2007. Cassava Mosaic Virus infected leaves were collected from tapioca variety H-165 from Tamil Nadu and were grown at Central Sericultural Research and Training Institute, Mysore. Eri worms were reared up to II moult on disease free tapioca leaves. From III instar through initiation of spinning, they were fed separately with equal quantities of virus infected and healthy leaves (control). Two feedings were given daily at 10 AM and 4 PM by providing whole leaves (after removing the petiole) of suitable maturity for each stage. Data on larval duration, larval weight, ERR, cocoon weight, shell weight and shell percentage were recorded. After cocooning (pupation) an equal number of cocoons were selected randomly from each replication, and were cut opened to find out the number of cocoons with live pupae and their sex. The experiment was repeated thrice and the mean values were taken. All the experiments were conducted at room temperature ($30 \pm 2^{\circ}$C).

The results of the experiment showed that in case of larvae fed on diseased leaves, larval duration was prolonged in excess of 2 days with concomitant decline in cocoon weight, shell weight, silk percentage and ERR. The average weight of fully matured larva was 5.24g in healthy leaf fed batch, whereas it was 3.69g in diseased leaf fed lot. The larvae provided with diseased tapioca leaves took more time to mature. There was a significant difference in all the cocoon characters between healthy and diseased leaf fed batches (Table 7.26). Further, the reproductive performance of the silkworm was also adversely affected when the larvae were fed with CMV infected leaves. There was a significant difference in fecundity as well as hatching between the treated and control larvae. Fecundity was 396 eggs and 308 eggs, respectively in healthy and diseased leaf fed batches. There was a slight increase in the weight of eggs in favour of treated batches. However, significant difference was observed in hatching which was 90.75 and 84.75% respectively in healthy leaf fed and diseased leaf fed lots (Table 7.27). Hence the study suggests that the CMV infected tapioca leaves are inferior in nutritional value and unsuitable for eri silkworm rearing.

Increase in free amino acids and amides, total phosphorous and potassium and reduction in phenols, flavonoids, ascorbic acid, calcium and total lipids are reported in CMV infected tapioca leaves (Ninan *et al.*, 1976; Daniel *et al.*, 1977). The infected leaves are also reported to contain less chlorophyll and carotenoid contents and have lower photosynthetic rate. Obviously, the nutritional quality of CMV infected leaves is inferior to healthy leaves and is not suitable for eri silkworm rearing. This could be the reason for reduced larval growth, larval weight, longer larval duration and inferior cocoons. For quality silkworm cocoon production, protein level in the range of 20-25% is required (Table 7.26 and 7.27).

Table 7.26: Effect of feeding CMV infected tapioca leaves on larval development and cocoon parameters of eri silkworm (Adopted from Philip *et al.*, 2008)

Treatment	Initial wt of III instar larvae (g)	V instar larval weight (g)	Larval duration Day:Hours	ERR by weight (g)	Single cocoon weight (g)	Single shell weight (g)	Pupal weight (g)	Shell percentage
Feeding with Healthy leaf (control)	1.37	5.24	29 : 7	19.90	2.56	0.33	2.23	12.90
SD	0.03	0.06	11.36	0.15	0.02	0.01	0.01	0.24
Feeding with Diseased leaf	1.37	3.69	31 : 21	13.16	1.99	0.23	1.77	11.29
SD	0.03	0.06	10.39	0.19	0.02	0.01	0.02	0.65
t=	NS	36.7**	7.99**	56.9**	42.21**	13.75**	36.08**	4.67**

** Significant at 1% (P<0.01);   NS – Non-significant

Table 7.27: Effect of feeding CMV infected tapioca leaves on grainage parameters of eri silkworm (Adopted from Philip *et al.*, 2008)

Treatment	Pupation period Days:Hours	Pupal period Days:Hours	Moth emergence (%)	Incubation period of eggs Days:Hours	Fecundity	Hatching (%)	No. of Eggs/g
Healthy leaf feeding (control)	8 : 00	20 : 03	87.75	10 : 00	396	90.75	584
SD	0.03	3.46	2.50	0.06	10.13	1.50	6.78
Diseased leaf feeding	9.12	24.21	70.75	11 : 06	308	84.75	591
SD	13.86	6.00	2.50	12.00	7.07	0.97	8.60
t=	5.19**	32.91**	9.62**	5.00**	14.24**	6.74**	NS

**Significant at 1% (P<0.01);   NS – Non-significant

Use of Japanese weed (*Mikania micrantha*) as food plant of eri silkworm (*Samia c. ricini*)

Devi and Kalita (2009) conducted a study to know the impact of feeding of Japanese weed leaves (*Mikania micrantha*) on the larval duration, larval growth, cocoon parameter and fecundity of eri silk worm. The rearing was conducted at laboratory condition. The larvae were reared on three experimental treatments (Ex tr I - host plant. *Mikania micrantha* from 1st instar till maturity, Ex tr II - host plant-mixed-*Ricinus communis* from 1st to 4th instar larva, the 5th instar larva fed with *Mikania micrantha* and *Ricinus communis* in equal proportion till maturity, Ex tr III - the host plant *Ricinus communis*- the control). In spite of heavy mortality (90%) of 1st instar larva shown by Ex tr-I, the survivors (10%) were highly adapted to the new environment of food plant *Mikania micrantha*. The larval and cocoon parameters studied in the first set of experiment were superior over the other two sets of experiment. The larval, pupal, cocoon, shell weight and fecundity were recorded to be highest in Ex tr-I (88.54 g/10, 24.57 g/10, 20.50 g/10, 3.63 g/10 and 250 eggs) respectively followed by Ex tr - II and III (77.39 g/10, 24.52 g/10, 27.53 g/10, 2.71 g/10 and 248 eggs; 77.58 g/10, 23.89 g/10, 27.68 g/10, 3.48 g/10 and 248 eggs respectively). Further, the highest shell ratio (12.63%) was recorded in the larva treated with Japanese weed from 1st instar till maturity. This experiment clearly revealed that eri silkworm, (*Samia c. ricini*) has very high potential to adapt to suitable new hosts like Japanese weed (*Mikania micrantha*). Further such adapted insects can become more potent strains with higher degree economic traits. Further this experiment also sheds light for high scope to develop new strains of eri silkworm with better economic traits by continuous rearing on other host plants like Japanese weed and utilizing the same host as food plant in countries like Japan. Alternatively Japanese weed can become cultivated host plant in other countries like India.

Impact of feeding disease affected castor leaves on food consumption and utilization in eri silkworm, (*Samia c. ricini*)

The effect of feeding castor leaves infected with *Cercospora* leaf spot, *Alternaria* leaf blight and by both the diseases on the food consumption and utilization of *Samia cynthia ricini* larvae were investigated by Shree *et al.*, (2001). The amount of food ingested, digested, and oxidized did not vary significantly whether the larvae were fed with diseased or healthy leaves. The approximate digestibility, was not significantly affected in larvae fed with leaves infected by leaf spot (*Cercospora ricinella*). The approximate digestibility increased in larvae fed with leaves infected by blight (*Alternaria ricini*) and leaves with mixed infection. Growth rate was significantly low only in larvae supplied with leaves affected by mixed infection. The coefficient of metabolism, efficiency of conversion of ingested and digested food, and larval weight were lower in larvae reared on infected leaves. Larvae fed with diseased leaves consumed more food, had longer larval period (2-3 days longer), and had unequal sizes compared with the control.

Summary and Conclusions

➢ Though castor is the main host plant of eri silkworm, its leaves are not available throughout the year and the crop has to be sown after every six months.

➢ At the time of scarcity of castor leaf, the use of secondary food plants of eri silk worm has become a regular practice amongst the rearer.

➢ In a study conducted to find the suitable secondary hosts of eri silkworm in Nagaland, white gulancha and kesseru were found to be equally suitable like castor, while tapioca and red gulancha showed poor performance.

➢ When eri silkworm larvae were fed with castor and kesseru, castor fed larvae recorded higher cocoon weight, shell weight, pupal weight and shell ratio compared to kesseru.

➢ In a study conducted to know the effects of host plants on biology and economic traits of eri silkworm, *Samia c. ricini* in Gujarat, the cocoon, shell and pupae of eri worms reared on castor were bigger and heavier and the shell ratio was greater than those reared on tapioca.

➢ Among the tapioca varieties, Co-2 was superior to ME-1 and ME-120 for rearing of eri silkworm and can be used singly or in combination with castor during shortage / non-availability of castor.

➢ The food preference in eri silkworm is mainly governed by its physio-chemical properties, which are responsible for variations in leaf consumption.

➢ Bangalore local variety of castor is a suitable host for rearing eri silkworm, because of the lowest leaf-cocoon ratio and highest leaf-egg ratio.

➢ The pupae obtained by feeding eri silkworm on Borpat and barkesseru could be suitably utilized as a source of food as good as pupae obtained from castor fed silkworms.

➢ Among the different castor and tapioca varieties tested for their quantitative consumption and utilization, RC-8 and Aruna varieties of castor are considered to be the best hosts with reference to their nutritional values.

➢ Among the 14 host plants tested for their suitability in Tamil Nadu, the eri silkworm larvae that were reared on CNBR and CPDB genotypes registered shortest larval duration. Hence these genotypes can be recommended for Tamil Nadu region.

➢ In North Eastern region of India, *Ailanthus* species *viz.*, Borpat and Barkesseru can be suitably utilized for commercial rearing of eri silkworm considering the economic yarn characters to enhance the production and productivity of eri silk.

➢ Eri silkworm has several secondary host plants. Out of these two members of *Ailanthus* species namely *Ailanthus grandis* (Borpat) and *Ailanthus excelsa* (Barkesseru) are advantageous because they are perennial, ever green, quick growing tall trees and are not browsed by cattle.

➢ In an experiment conducted to evaluate new host plant species for ericulture in Dharwad, among the 23 host plants tested, good feeding and 100% survivability of larvae was recorded on banyan tree, fountain tree, carrot leaves and Indian almond.

➢ The findings of badam, papaya and arjun as potential new hosts for eri silkworm have opened new vistas in promoting vanya silk industry in Maharashtra.

➢ Eri silkworm has more than 50 host plant species and these are interchangeable especially at 3rd instar to 5th instar stages.

➢ The rearing of eri larvae by feeding the castor leaves from 1st to 3rd instar and interchanging with tapioca leaves during 4th and 5th instars improved the commercial characteristic features such as larval weight, larval duration, ERR, cocoon weight and shell weight.

➢ The eri silkworm larvae that were fed with castor leaves ingested maximum amounts of phosphorus and potassium (64.02 and 122.4 mg, respectively), followed by the larvae fed with tapioca during early instars and castor in later instars (60.42 and 116.69 mg respectively)

➢ Host plant directly influences growth and development of larvae and quality and quantity of silk produced by them. As the bio-components of leaves vary with age and variety, the enzyme activity also varies, and exerts a direct influence on quality and quantity of silk produced.

➤ The activity of enzymes alanine aminotransferase (AAT), aspartate aminotransferase (AsAT), and glutamine synthetase (GS) studied in castor and kesseru fed larvae showed that the activity of all the three enzymes predominated in larvae fed on kesseru leaves than those fed on castor leaves, except aspartate aminotransferase, that recorded highest activity during third instar larvae fed on castor leaves.

➤ The huge quantity of senescent tapioca leaves wasted at the time of tuber harvest could be successfully utilized by fortifying with 10% jaggery solution and soya powder @ 10g/100g of leaves for rearing of eri silkworm. This process ensured considerable improvement in economic traits of the worms and silk yield.

➤ In eri silkworm larvae when fed on CMV infected diseased leaves, the larval duration was prolonged in excess of 2 days with concomitant decline in cocoon weight, shell weight, silk percentage and ERR. This clearly demonstrated that because the nutritional quality of CMV infected leaves is inferior to healthy leaves and hence those are not suitable for eri silkworm rearing.

➤ Eri silk worm larvae when fed with Japanese weed leaves (*Mikania micrantha*) from 1st instar to maturity, 90% of the 1st instar larvae were died, but the 10% survivors were highly adapted to the new environment of food plant *Mikania micrantha* revealing the scope for development of better strains on other hosts.

➤ In an experiment conducted to know the impact of feeding disease affected castor leaves on food consumption and utilization in eri silkworm showed that the amount of food ingested, digested, and oxidized by the eri silkworm larvae did not vary significantly whether the larvae were fed with diseased or healthy castor leaves.

➤ The coefficient of metabolism, efficiency of conversion of ingested and digested food, and larval weight were lower in larvae reared on infected leaves.

References

Basaiah J M M, Reddy D N and Krishnamurthy R V. 1990. Consumption and utilization of food by eri silkworm *Samia cynthia ricini* Boisduval. *Entomon* **15**(1-2): 95-98.

Biswas N and Das P K. 2001. Effect of food plant species on rearing performance of eri silkworm *Samia cynthia ricini*. *Bulletin of Indian Academy of Sericulture* **5**(1): 36-39.

Bose P C and Mazumder S K. 1990. Biochemical composition of pupae waste and its utilization. *Indian Silk* **27**(2): 45-46.

Choudhury K C. 2000. Study on the waste muga pupae for its possible utilization. *Ph D. Thesis*, Guwahati University, Assam.

Chowdhury S N. 1982. Eri silk industry. Directorate of Sericulture and Weaving, Government of Assam, Guwahati.

Daniel R S, Thankappan M and Hrishi N. 1977. Metabolic changes in cassava (*Manihot utilissima* Crantz.) due to cassava mosaic virus infection. *Journal of Root Crops* **3**(1): 39-42.

Deori G and Khanikor D P. 2015. Rearing performance of Eri silkworm (*Samia ricini* Boisd.) on *Ailanthus excelsa* and *Ailanthus grandis*. *Journal of Experimental Zoology* **18**(2): 839-841.

Deori G, Khanikor D P and Sarkar C R. 2016. Evaluation of Proximate composition of eri silkworm pupae on *Ailanthus grandis* and *Ailanthus excelsa* leaves. *Journal of Experimental Zoology, India* **19**(1): 515-516.

Devi M and Kalita J. 2009. Use of Japanese weed (*Mikania micrantha*) as food plant of eri silkworm (*Samia ricini* Donovan). *Environment and Ecology* **27**(4): 1635-1637.

Dookia B R. 1980. Varied silk ratio in cocoons of eri silkworm (*Philosamia ricini* Hutt.) reared on different castor varieties in Rajasthan. *Indian Journal of Sericulture* **19**: 38-40.

Dutta P P and Khanikor D P. 2005. Rearing performance of eri silkworm (*Samia ricini*) with interchange of food plants in different seasons. *Bulletin of Indian Academy of Sericulture* **9**(2): 31-35.

Famurewa J A V, Oluwalana I B and Oludahunsi O F. 2006. Effect of drying methods on the physic-chemical properties of soyflour. *Journal of Food Technology* **4**(4): 283-286.

Gogoi R and Yadav R N S. 1995. Effect of host plants on some biochemical parameters of eri silkworm, *Philosamia ricini* during its development. *Indian Journal of Experimental Biology* **33**: 372-374.

Hoque A A M and Khan A R. 1987. *Acta Entomologica Bohemoslovaca* **84**: 397.

Horie Y and Watanbe K. 1983. Daily utilization and consumption of dry matter in food by the silkworm *B. mori* L (Lepidoptera : Bombycoidae). *Applied Entomology and Zoology* **18**: 70-80.

House H L. 1974. Insect Nutrition *In: The physiology of Insecta, Vol. V (2nd Edn.),* Rock stein (Ed.), Academic Press, New York and London, pp.1-53.

Jayaraj S. 2004, Ericulture in Tamil Nadu, Kerala and Pondicherry states: Some experiences. In: *Proce. Workshop Prospects Dev. Ericulture,* Karnataka, UAS, Dharwad, pp: 10-29.

Joshi K L. 1986. Utilization of minerals by larvae of the eri silkmoth, *Philosamia ricini* Hutt (Lepidoptera: Saturniidae). *Indian Journal of Sericulture* **25**(2): 54-57.

Joshi K L. 1987. Progression Factor for Growth in Eri Silk Moth in Relation to Diet. *Indian Journal of Sericulture* 26: 98-99.

Joshi K L. 1991. Consumption of food by the larvae of eri silkmoth *Philosamia ricini* Hutt. (Lepidoptera: Saturniidae). *Indian Journal of Entomology* **53**(4): 628-629.

Kavane R P. 2014a. *Carica Papaya* – A Potential host of *Philoamia ricini* eri silkworm under Western Maharashtra condition. *Indian Journal of Applied Research* **4**(11): 472-474.

Kavane R P. 2014b. *Terminalia catappa* – A Potential new host of *Philoamia ricini* eri silkworm under Western Maharashtra condition. *International Journal of Advanced Research* **2**(11): 433-438.

Kavane R P. 2015. *Terminalia arjuna* – A new host of *Philosamia ricini* eri silkworm under Western Maharashtra condition. *Indian Journal of Pharma and Bio Sciences* **6**(1): 787-792.

Khanikor D P, Dutta S K and Khound J. 1997. Exploitation of Barkesseru, *A. excelsa* leaves as a substitute to castor leaves for rearing eri silkworm. *New Agriculturist* **8**(1): 7-12.

Kichu N, Imtinaro L and Pankaj Neog 2014. Influence of host plants on rearing performance and cocoon characters of eri silkworm, *Samia cynthia ricini* Boisduval. *Indian Journal of Entomology* 77 (1): 88-91.

Kumar R, Gargi, Prasad D N and Saha L M. 1993. A Note on Secondary Food Plants of Eri Silkworm. *Indian Silk* **31**(9): 20-21.

Legay J M. 1958. Recent advances in silkworm nutrition. *Annual Review of Entomology* **3**: 75-86.

Mukul Deka, Saranga Dutta and Dipali Devi. 2011. Impact of feeding of *Samia cynthia ricini* Boisduval (red variety) (Lepidoptera: Saturniidae) in respect of larval growth and spinning. *International Journal of Pure and Applied Sciences and Technology* **5**(2): 131-140.

Nagaveni V, Shree M P and Ravi Kumar K. 2002. Effect of feeding eri silkworms with diseased castor leaves on the economic parameters of cocoon. *Indian Journal of Sericulture* **41**(2): 155-156.

Naik M and Murthy C. 2013. Evaluation of new host plant species for ericulture. *International Journal of Plant protection* **6**(2): 444-448.

Ninan C A, Abraham S, Sathyavathi Bai L, Chadrasekharan Nair P N and Kuriachen P. 1976. Lipid metabolism in mosaic infected cassava. In: *Proceedings of the 4th Symposium of International Society for Tropical Root*

Crops. Coch J, Mac Inlyre R and Graham M (Eds.) IDRC-080c CIAT, Pp. 173-1740.

Patel B S and Patel G M. 2009. Effects of host plants on biology and economic traits of Eri silkworm, *Samia cynthia ricini* Boisduval. *Insect Environment* **14**(4): January-March 172.

Patil G M and Savanurmath C J. 1994. Eri silkworm – The poorman's friend. *Indian Silk* **33** (4): 41-45.

Patil R R, Kusugal S and Ankad G. 2009. Performance of eri silkworm, *Samia cynthia ricini* Boisd on few food plants. *Karnataka Journal of Agricultural Sciences* **22**(1): 220-221.

Philip T, Qadri S M H, Somaprakash D S, Shekhar M A and Kamble C K. 2008. Effect of feeding 'CMV' infected tapioca leaves on the cocoon and grainage parameters of eri silkworm, *Samia cynthia ricini* Boisduval. *Indian Journal of Sericulture* **47**(2): 161-164.

Philip T. 2010. Cassava mosaic virus induced physio-biochemical changes in the leaves of tapioca (*Manihot utilissima* Pohl.). *International Journal of Plant Protection* **3**(1): 31-33.

Rani A G, Padmalatha C, Raj R S and Ranjith Singh A J A. 2011. Impact of supplementation of Amway protein on the economic characters and energy budget of silkworm *Bombyx mori* L. *Asian Journal of Animal Science* **10**: 1-6.

Rao M, Prasad R N and Suryanarayan N. 2005. Ericulture-An additional income for tapioca (cassava) growers and good nourishment to the malnutritional tribal populace. *The 20th Congress of the International Sericulture Commission*, Bangalore 2, 94-98.

Rao P K V J, Das M and Das S K. 2007. Jaggery – A traditional Indian Sweetener. *Indian Journal of Traditional Knowledge* **6**(1): 95-102.

Reddy D N R and Narayanaswamy K C. 1998. Quantitative consumption and utilization of castor and tapioca leaves by the eri silkworm *Samia cynthia ricini* Boisduval. *Annals of Entomology* **16**(2): 13-19.

Reddy D N R and Narayanaswamy K C. 1999. Effect of host on the consumption rate, leaf-cocoon and leaf egg ratio of Eri silkworm, *Samia cynthia ricini* Boisduval. *Entomon* **24**(1): 67-70.

Sachan J N and Bajpai S P. 1973. Response of castor (*Ricinus communis* L.) varieties on growth and silk production of Eri silkworm *Philosamia ricini* Hutt. (Lepidoptera: Saturniidae). *Annals of Arid Zone* **11**: 112-115.

Saikia P and Dutta L C. 2013. Evaluation of Ailanthus species in relation to cocoon and yarn characters of eri silkworm, *Samia ricini* Boisd. *Journal of Applied Zoological Researches* **24**(2): 173-175.

Sakthivel N and Qadri S M H, Anbazhagan R, Krishnamoorthi T S and Jayaraj S. 2010. Cropping system of eri food plants for economically viable ericulture in Tamil Nadu. *Asian Textile Journal* **19**(4): 70-76.

Sakthivel N and Qadri S M H. 2013. Effect of fortification of senescent tapioca leaves on economic traits of Eri silkworm. *Indian Journal of Sericulture* **52**(1): 44-47.

Sarkar D C. 1980. Ericulture in India. *Central Silk Board*, Bangalore P 4-23.

Sharma H P. 2005. Effect of plant extract and fertilization on biologicl parameter of *Bombyx mori* Linn. And *Samia cynthia ricini* Boisd. *M. Sc. (Agril.) Thesis*, G. B. Pant University of Agriculture and Technology, Pantnagar, India.

Shaw C. 1998. Evaluation of *Ailanthus* species in relation to nutrition, growth and cocoon characters of eri silkworm, *Philosamia ricini* Hutt. M.*Sc. (Agri) Thesis Assam Agricultural University, Jorhat, Assam,* pp121.

Shree M P and Nagaveni V. 2002. Development and reproductive performance of eri silkworms (*Samia cynthia ricini* Boisduval) fed with diseased castor leaves (*Ricinus communis* L.). *Sericologia* **42**(2): 209-215.

Shree M P, Nagaveni V, Chandra M and Kumar K R. 2001. Impact of feeding disease affected castor leaves on food consumption and utilization in eri silkworm, (*Samia ricini* Donovan). *Bulletin of Indian Academy of Sericulture* **5**(2): 67-71.

Siddaramaiah A L and Hegde R K. 1988. Effect of feeding *Cercospora* infected leaves of mulberry on silkworm development. *Mysore Journal of Agricultural Sciences* **22**(4): 498-504.

Singh B K and Das P K. 2006. Prospects and problems for development of ericulture in non- traditional states. In: Proceedings of Regional Seminar on *"Prospects and problems of Sericulture – An economic enterprise in North West India"* Dehradun, 2006 pp. 312-315.

Subramanianan K, Sakthivel N and Qadri S M H. 2013. Rearing technology of eri silkworm (*Samia cynthia ricini*) under varied seasonal and host plant conditions in Tamil Nadu. *International Journal of Life Sciences Biotechnology and Pharma Research* **2**(2): 130-141.

Sullia S B and Padma S D. 1987. Acceptance of mildew affected mulberry leaves by silkworm (*Bombyx mori* L.) and its effect on cocoon characters. *Sericologia* **27**(4): 693-696.

Venu N and Munirajappa. 2013. Impact of Independent and sequential feeding of different host pants on economic traits of Eri silkworm, *Philosamia ricini* Hutt. *International Journal of Science and Nature* **4**(1): 51-56.

CHAPTER 8

Biology and Rearing Performance of Other Silk Worms

The silkworms are very important economic insects which contribute substantially to the national economy and Gross Domestic Production (GDP) of many countries. The food plant and its nutritional status in sericulture play a pivotal role for the successful larval rearing resulting to higher cocoon number of better quality. This success further leads to either silk or seed production for the commercial sustenance and thus, the industry has to aim for optimal utilization of seri biodiversity as flora and fauna. Though, there is no dearth of food plants, the accessibility for the utilization makes the silkworm rearers to search for alternatives. India is the only country that produces all the four types of silks. This chapter discusses in detail about the rearing and grainage performances of Indian Tropical tasar silkworm, muga silkworm and wild eri silkworm, effect of coupling duration on fecundity and fertility of muga silkworm, nutritional efficiency of mulberry and tasar silkworms etc.

India is the only country that produces all the types of silks namely, Mulberry, Tropical Tasar, Oak Tasar, Eri and Muga. Silk obtained from sources other than mulberry are generally termed as non-mulberry or "vanya" silks. The requirement of nutrition quantity and quality are highly specific in sericigenous insects for optimal physiological status and sustainable productivity. The availability of essential nutrients in food plant is vital for successful life cycle, cocoon quality, metamorphosis to moth stage and their reproductive activity. The larval development and their feeding status have impact on genital development, fecundity and egg fertility. These insects being polyphagous face a challenge of differences in their life cycle pace and pattern depending on the host plant they feed. Studies correlating larval nutrition with pupal and adult physiology have shown that the oocyte production depends on nutrient status of host plant, as the nitrogen and carbon components of eggs are obtained through feed.

Among the vanya silks, the tropical tasar is an important variety produced by a wild silkworm of *Antheraea mylitta* Drury, which is polyphagous and feeds primarily on *Terminalia tomentosa* (asan), *Terminalia arjuna* (arjun) and *Shorea robusta* (sal) and secondarily on *Lagerstroemia parviflora, Zizyphus mauritiana, Anogeissus latifolia, Syzigium cumini, Careya arborea* and *Hardwickia binata*, besides several other food plants of minor importance. The *L. parviflora* is widely available in the tropical tasar rearing areas of Jharkhand and Chhattisgarh states of India, which is considered to be the best secondary food plant for tasar silkworm, *A. mylitta*. The leaf nutrition of tasar food plant can enhance the effective rate of rearing (ERR), health and growth of larvae and better crop yields as the feed quality has direct correlation with cocoon and shell weights, silk ratio and silk filament. Even if, the harvested cocoons are less in number, their quality can compensate the crop economics either with higher silk or egg productivity or quality.

Muga silkworm, *Anthereae assamensis* Helfer is an endogenous sericigenous species of North-East India distributed mainly in sub-Himalayan region of Assam and Meghalaya. The silkworm is semi domesticated, polyphagous and multivoltine in nature having five to six generations in a year. It feeds on a wide range of host plants among which Som, *Persea bombycina* Kost. and Soalu, *Litsea polyantha* Juss are the primary food plants of this silkworm. Most of the traditional rearers prefer to rear on these two food plants for production of muga silk cocoons.

Samia canningi Hutton, a wild eri silkworm is bivoltine in nature and undergoes eight months pupal hibernation in Nagaland climatic conditions. It was found to be completely free from silk worm diseases like pebrine, muscardine but minor infestation of Flacherie was observed in the month of August-September. Infestation of uzi fly and enemies like ant, wasp and preying bird were also recorded.

Rearing and grainage performance of Indian Tropical Tasar Silkworm, *Antheraea mylitta* Drury at Department of Sericulture, Government of Jharkhand, India

Cocoon yield and seed quality of tropical tasar silkworm, *Antheraea mylitta* Drury depends on the food plant variety and its nutritional status, but, farmers prefer to conduct rearings on alternative tasar food plants based on availability and accessibility. This practice helps them in managing the time-bound agricultural works for economic advantage. However, the leaf production rate, quantity and season, gestation period of the plant needs consideration in comparison with primary host plants for the commercial feasibility. In this regard, Manohar Reddy *et al.,* (2010) conducted an experiment to test the commercial feasibility for rearing and grainage parameters of Indian Tropical Tasar Silkworm, *Antheraea mylitta* in tasar rearing areas of Jharkhand and Chattisgarh states of India.

Manohar Reddy and his co-workers (2010) have chosen to analyze the impact of two different tasar food plants like *T. tomentosa* and *L. parviflora* on rearing, cocoon, economic traits and grainage performance of Daba ecorace of *A. mylitta.* The study has been carried out in West Singhbhum district, Department of sericulture, Government of Jharkhand, India during July to September, 2010. The observations were recorded on different rearing parameters like larval weight, larval duration, effective rate of rearing (ERR), cocoon yield and melt cocoon percentage. The observations on cocoon commercial traits like single cocoon weight, single shell weight, single pupal weight, silk ratio percentage and silk yield per dfl, were also recorded. Further, grainage parameters like moth emergence percentage, moth emergence duration, coupling percentage, fecundity and egg fertility have also been recorded.

The silkworm rearing and grainage performance being important for sustenance of tropical tasarculture, the cocoons generated on *Lagerstroemia parviflora* food plant were analyzed for economic viability in comparison with *Terminalia tomentosa* plant. Though, the overall performance of tasar rearings and grainage behaviour was found better in *T. tomentosa* food plant, the important commercial traits like cocoon weight, shell weight, silk ratio and egg fertility are positive in *L. parviflora* fed cocoons (Tables 8.1 and 8.2). This indicates the better availability of nutrients required for silk production and egg fertility in *L. parviflora* food plant than *T. tomentosa*. Hence, the leaf of *L. parviflora* can be fed during final larval stages to attain optimal silk and seed productivity in *A. mylitta* (Tables 8.3 and 8.4). It was concluded that the silkworm rearing and grainage performance of the cocoons generated on *L. parviflora* food plant has shown lesser feasibility in general than *T. tomentosa* plant. But, the traits like cocoon weight, shell weight, silk ratio and egg fertility are positive in *L. parviflora* fed cocoons over the *T. tomentosa* fed. Hence *L. parviflora* leaf can be fed to final instar larva to attain the commercial advantage of silk productivity and reproductive efficiency. However, the leaf production rate, quantity and season, gestation period of this plant needs to be explored in comparison with other primary host plants for assessing the comprehensive commercial feasibility of *L. parviflora* as viable tasar food plant.

Table 8.1: Comparative performance on rearing parameters of *A. mylitta* fed on *T. tomentosa* and *L. parviflora* food plants (Adopted from Manohar Reddy *et al.*, 2010)

Parameter	*T. tomentosa* (Asan)		*L. parviflora* (Sidha)		t-test
	Mean	**S.D.**	**Mean**	**S.D.**	
Larval weight (g)	33.800	1.924	35.000	1.581	1.078[NS]
Larval span (days)	34.600	1.140	45.800	1.304	14.459[**]
Effective rate of rearing (%)	61.000	9.454	29.000	4.183	6.921[**]
Cocoon yield (nos/dfl)	48.800	7.563	23.200	3.347	6.921[**]
Melt cocoon (%)	9.400	2.074	16.000	2.646	4.390[**]

Table 8.2: Comparative performance on cocoon economic traits of *A. mylitta* fed on *T. tomentosa* and *L. parviflora* food plants (Adopted from Manohar Reddy *et al.*, 2010)

Parameter	*T. tomentosa* (Asan)		*L. parviflora* (Sidha)		t-test
	Mean	**S.D.**	**Mean**	**S.D.**	
Single cocoon weight (g)	10.916	0.161	11.008	0.325	0.568[NS]
Single shell weight (g)	1.192	0.029	1.334	0.047	5.749[**]
Single pupa weight (g)	9.630	0.130	9.600	0.292	0.210[NS]
Silk ratio (%)	10.918	0.186	12.120	0.406	6.0142[**]
Absolute silk yield (g)	58.964	8.113	30.904	3.736	7.025[**]

Table 8.3: Comparative performance on grainage parameters of *A. mylitta* fed on *T. tomentosa* and *L. parviflora* food plants (Adopted from Manohar Reddy *et al.*, 2010)

Parameter	*T. tomentosa* (Asan)		*L. parviflora* (Sidha)		t-test
	Mean	**S.D.**	**Mean**	**S.D.**	
Moth emergence span (days)	17.800	1.304	20.400	2.408	2.123[NS]
Moth emergence (%)	87.600	3.715	79.000	2.739	4.167[**]
Moth coupling (%)	70.400	2.408	60.800	3.633	4.925[*]
Fecundity (nos)	244.200	20.633	195.400	15.947	4.185[*]
Egg fertility (%)	83.100	2.231	89.500	3.102	3.934[*]

Table 8.4: Comparison for commercial impact in rearing performance, cocoon traits and grainage behaviour of *A. mylitta* fed on *T. tomentosa* and *L. parviflora* food plants (Adopted from Manohar Reddy *et al.*, 2010)

Parameters	*T. tomentosa* (Control)	*L. parviflora* (Treatment)	% change over control	Commercial impact
Larval weight (g)	33.80	35.00	+03.55	Positive
Larval span (days)	34.60	45.80	+32.37	Negative
Effective rate of rearing (%)	61.43	29.69	-51.67	Negative
Cocoon yield (nos/dfl)	49	23	-53.07	Negative
Melt cocoon (%)	9.40	16.00	+70.21	Negative
Single cocoon weight (g)	10.92	11.01	+0.824	Positive
Single shell weight (g)	1.19	1.33	+11.76	Positive
Single pupa weight (g)	9.63	9.60	+0.312	-
Silk ratio (%)	10.90	12.12	+11.19	Positive
Absolute silk yield (g)	58.96	30.90	-47.59	Negative
Moth emergence span (days)	17.80	20.40	+14.61	Negative
Moth emergence (%)	87.60	79.00	-09.82	Negative
Moth coupling (%)	70.40	60.80	-13.64	Negative
Fecundity (nos)	244	195	-20.08	Negative
Egg fertility (%)	83.10	89.50	+07.70	Positive

Rearing and grainage parameters of Muga silkworm, *Anthereae assamensis* Helfer

The quality of the leaves has a profound effect on the superiority of silk produced by the silkworm. Leaves of superior quality enhance the chances of good cocoon crop (Ravikumar, 1988). The quality of feed plays a remarkable role for growth and development of the silkworm and ultimately on the economic traits of cocoons (Hazarika *et al.,* 2003, 2005; Gangwar, 2010; Kumar and Vadamalai, 2010). In muga silkworm, *Anthereae assamensis* Helfer also nutrition plays an important role in improving the growth and development of the silkworm.

Singh and Goswami (2012) made a comparative evaluation of the two popular varieties of muga food plants *viz.,* Som, *Persea bombycina* Kost. and Soalu, *Litsea polyantha* Juss through bioassay with regard to the rearing parameters and grainage performance particularly valid moth percentage and fecundity during five different rearing seasons. The rearing was conducted in outdoor on the foliages of som and Soalu during five seasons *viz.,* Aherua (June-July), Bhadia (August-September), Early kotia (September-October), Agherua (November-December) and late Jarua (January-February) under nylon net cover by providing sufficient space between food plants and nets for better aeration. Nylon nets were used as shade during day time and to protect from pests and predators during night. Chawki rearing was conducted on the tender leaves sprouted from the timely pruned plants while the late stage rearing was conducted on semi mature leaves. The rearing plots were well disinfected before rearing. The bioassay consisted of four replications in each food plant variety with two hundred worms per replication. For evaluation of rearing performance, four yield contributing parameters *viz.,* effective rate of rearing (ERR %), single cocoon weight, cocoon shell weight, cocoon shell ratio (SR %) were considered. To examine the varietal effect of the food plants on grainage performance of the harvested cocoons, valid moth percentage of the emerged moths and fecundity were recorded. The ranking of the different species and breeds were worked out by following the multiple trait evaluation index method of Mano *et al.,* (1992, 1993) as follows

$$\text{Evaluation Index (EI)} = \frac{A-B}{C} \times 10 + 50$$

where,

A= individual value of the genotype

B = average value of the particular trait of the genotypes

C = standard deviation of the particular trait

10 = standard unit,

50 = fixed value.

Evaluation index values for individual characters for each genotype were calculated and average cumulative index value of the five characters under study was obtained. The genotype which recorded average index values of >50 was considered for selection and the genotypes which recorded average index value of <50 were considered as inferior.

Results of Singh and Goswami (2012) revealed that none of the food plants were superior in all the characters during the five rearing seasons. Each plant exhibited superiority in certain characters during particular season only (Tables 8.5 and 8.6). Similar observation has been reported in mulberry silkworm (Maqbool *et al.*, 2005). The rearing on soalu plant showed shorter larval duration and higher cocoon weight than that of som while in silk quality cocoons from som exhibited superior quality (Saikia *et al.,* 2004). Chakravorty *et al.,* (2004) reported higher weight of first and second instar larvae fed on soalu than that of the larvae fed on som and Digloti (*Litsea salicifolia*) while higher moulting percentage was observed in first and second instar larvae fed on som. Singh and Goswami (2012) observed higher cocoon weight in soalu than that of som while shell ratio was significantly higher in som than that of soalu in most of the seasons. In grainage parameters, soalu exhibited significant higher values than som indicating its superiority as seed crop. It can be deduced that commercial rearing of muga silkworm *Anthereae assamensis* Helfer can be taken up on both the food plants. But for seed crop, rearing should be conducted on soalu plants irrespective of seasons, because the rearing on soalu has shown better performance.

Table 8.5: Rearing and grainage parameters (Mean ± standard deviation) of *A. assamensis* Helfer during different rearing seasons on two different food plants (Adopted from Singh and Goswami, 2012)

Season	Type of food plant	ERR (%)	Cocoon weight (g)	Shell weight (g)	SR (%)	Valid moth (%)	Fecundity
Aherua	Som	47.70±2.2	5.25±0.24	0.4±0.03	7.65±0.22	55.33±3.05	127±4
	Soalu	52±2	6.10±0.2	0.46±0.02	6.64±0.2	67±3.25	140±3
	't' value	2.47	2.50	1.25	6.24	11.50	3.46
Bhodia	Som	37.90±3.3	8.00±0.5	0.52±0.03	6.25±0.27	60±2	135±5
	Soalu	39.50±3.5	8.5±0.2	0.51±0.02	5.92±0.3	80.00±4	142±3
	't' value	1.41	1.32	1.25	1.55	9.5	6.75
Early kotia	Som	62.00±2	7.70±0.03	0.52±0.03	6.48±0.22	64.00±3	142±4
	Soalu	60.00±3	8.60±0.2	0.57±0.02	6.90±0.3	83±4	150±6
	't' value	1.16	1.90	0.69	3.2	11.42	3.17
Agherua	Som	60.00±2	5.20±0.24	0.42±0.03	8.08±0.22	72±5	142±5
	Soalu	63.30±3.5	5.50±0.32	0.43±0.04	7.8±0.20	75±3	150±3
	't' value	2.49	2.86	0.5	12.5	7.68	3.98
Late Jarua	Som	36.60±3.5	6.30±0.02	0.44±0.02	6.98±0.24	61.50±4.2	140±3
	Soalu	41±2	6.50±0.25	0.5±0.03	7.69±0.3	80.50±3	145±5
	't' value	8.5	1.19	1.5	1.43	7.36	8.53

Table 8.6: Evaluation indices for six traits of muga silkworm on two different food plants during five seasons (Adopted from Singh and Goswami, 2012)

Season	Type of food plant	ERR (%)	Cocoon weight (g)	Shell weight (g)	SR (%)	Valid moth (%)	Fecundity	Cumulative E.I
Aherua	Som	51.71	42.33	52.08	66.16	43.00	46.95	50.28
	Soalu	55.15	48.72	54.17	52.33	54.90	66.12	55.23
Bhodia	Som	42.08	63.00	66.67	46.98	47.77	58.75	54.20
	Soalu	43.56	66.76	64.58	42.46	68.15	69.07	59.09
Early kotia	Som	64.40	60.75	66.67	50.14	51.84	69.07	60.48
	Soalu	62.57	67.52	77.08	55.89	71.21	80.87	69.19
Agherua	Som	62.57	41.95	45.83	72.05	60.00	69.07	58.58
	Soalu	65.63	44.21	47.92	68.49	63.05	80.87	61.69
Late Jarua	Som	40.88	50.23	50.00	56.98	42.67	46.95	47.95
	Soalu	44.95	51.73	62.50	66.71	50.83	73.49	58.37

Effect of coupling duration on fecundity and fertility of Muga silkworm moth *Anthereae assamensis*

The grainage period of muga silkworm is remarkably long and experiences an unbalanced sex ratio during emergence. As in other lepidopteran insects, the males mature early and their population is proportionately high in the early part of grainage and just reverse in the later part. As a result a considerable number of female moths are usually wasted due to lack of their male counterparts. However, such situations were successfully overcome in *Bombyx mori*, *S. cynthia ricini* and *Antheraea mylitta* by neutralization of male moths for multiple coupling.

Barah and Sahu, (2003) reported that 4 hrs of mating duration is necessary in muga silkworm to get maximum fertility in both the single use and multiple use male moths. However, low hatchability was encountered during the grainage operation in the multiple uses of male moths during summer grainage and low hatching during winter grainage in the 4 hrs mating duration. In muga silkworm, seasonal influence of coupling duration on fecundity and hatching (%) has not been properly studied. The optimum coupling duration to get maximum fertility is both temperature and season specific. Hence, it is felt that the optimum coupling duration in muga culture needs to be re-evaluated during winter and summer separately and also to explore the possibility of re-neutralization of male moths to meet the shortage of male moths during grainage. Therefore, Goswami and Singh (2012), have undertaken a study to find out the optimum coupling duration to get maximum hatching at different climatic conditions.

The seed cocoons of *A. assama* were consigned in the grainage room at the ambient temperature of 29-32^{0}C and RH 85-90% during summer (May to September, 2008-2009) and temperature 15-25^{0}C and RH 72-88% during winter (November-February, 2008-2010) till emergence of moths. The emerged moths were allowed for natural coupling. The coupled moths were tied to straw made sticks, locally called kharika by binding the female moth with the help of cotton thread and after de-coupling the individual female moths were allowed to lay eggs on kharika. For neutralization of male moths, a resting period of 2-4 hrs was given before the second mating. The decoupled moths were kept for egg laying for 72 hours. Disease free layings were collected after individual mother moth examination, washed, disinfected, dried and kept for hatching. The fecundity and hatching were recorded for each individual moth and treatment.

The impact of mating duration on fecundity and fertility during the winter and summer seasons of the muga silkworm is presented in

Table 8.7. Significant variations in fecundity and hatching due to coupling durations were observed. It was observed that during summer 180 min coupling duration was adequate to obtain optimum hatching while during winter 300 min was required. During summer season hatching ranged from 59.70 to 85.36% among the different coupling durations. Average fecundity during summer crop ranges from 159 to 179 eggs. The effect of re-using male moths in second mating with virgin female moths in different coupling durations (30 min to 480 min) on fecundity and hatching percentage during different seasons revealed that minimum coupling duration of 180 mins is necessary to get optimum fecundity. Male moths can be reused for second mating during winter after resting for 2-4 hrs of first decoupling, without affecting the fertility while, multiple utilization of male moths during summer cannot be recommended. For multiple uses of male moths during winter season, 300 minutes coupling duration was adequate to obtain maximum hatching. The mating duration, activeness of the male moth, temperature and relative humidity in silkworm essentially affects the silkworm seed quality. In mulberry silkworms, it has been accepted that coupling duration of 3-5 hrs is necessary for optimum fecundity and fertility.

The study of Goswami and Singh (2012) indicated that there were significant variations in fecundity of the different coupling durations in muga silkworm *A. assamensis*, while the hatching was also significantly affected by coupling duration. The optimum period of coupling duration to get maximum hatching in muga silkworm varies in different seasons. They observed that during summer 180 min is adequate to obtain optimum hatching while during winter 300 min is enough to get maximum hatching. For obtaining hatching of > 50%, coupling duration of 30 min is sufficient. In muga silkworm the first ejaculation occurring within 30 min of coupling might be enough to fertilize at least 50% of the eggs. During winter, ejaculation time might be delayed, for which longer duration of 300 min coupling is essential.

With this study, we can infer that the male moths can be utilized two times in winter without affecting the fecundity and fertility. However, the use of male moths for re-mating in muga silkworm during summer season is not recommended since the hatching is <50%, which might be due to decline in the potential of spermatozoa caused by the high temperature (Table 8.8). Many workers have reported that high temperature of >32^{0}C causes male sterility in silkworms (Sugai and Kiguchi, 1967; Katsuno, 1977).

Table 8.7: Effect of coupling duration on fecundity and hatching of muga silkworm, *Antheraea assama* (Adopted from Goswami and Singh, 2012)

Treatment and coupling duration (Min)	Winter season		Summer season	
	Fecundity (nos)	Hatching (%)	Fecundity (nos)	Hatching (%)
T1-30	110	61.50	160	59.70
T2-60	109	61.22	163	60.84
T3-90	117	64.82	173	75.22
T4-120	114	62.68	172	75.02
T5-150	118	69.12	169	76.48
T6-180	128	76.62	176	82.40
T7-210	128	76.46	178	84.64
T8-240	137	75.07	179	85.36
T9-270	128	77.94	172	83.92
T10-300	128	83.48	168	81.42
T11-330	132	83.61	159	81.46
T12-360	130	83.51	159	81.41
T13-390	131	84.02	173	81.01
T14-420	133	83.49	168	80.21
T15-450	130	84.09	168	80.21
T-16-480	130	82.51	174	81.72
CD at 1%	15.86	12.24	12.32	11.89
CD at 5%	9.99	7.43	7.91	6.98

Table 8.8: Effect of reused male on fecundity and hatching of muga silkworm, *Antheraea assama* (Adopted from Goswami and Singh, 2012)

Treatment and coupling duration (Min)	Winter season		Summer season	
	Fecundity (nos)	Hatching (%)	Fecundity (nos)	Hatching (%)
T1-30	114	48.94	114	31.08
T2-60	125	55.38	108	31.0
T3-90	118	54.54	114	33.0
T4-120	118	59.84	114	32.7
T5-150	118	62.04	116	34.26
T6-180	129	60.28	120	34.82
T7-210	126	61.46	122	36.48
T8-240	135	63.94	121	37.38
T9-270	131	64.08	119	36.24
T10-300	124	72.16	122	38.32
T11-330	128	72.52	122	39.08
T12-360	127	71.90	124	38.64
T13-390	129	74.45	124	41.24
T14-420	130	73.20	124	41.04
T15-450	128	73.06	123	39.54
T-16-480	127	72.82	123	39.54
CD at 5%	6.73	9.46	7.09	2.14
CD at 1%	10.47	14.73	11.03	3.97

Biology and rearing performance of *Samia canningi* Hutton - a wild eri silk moth in Nagaland

Samia canningi worms feed on nine host plants. Kakati and Chutia (2010), conducted an experiment to study the biology and rearing performance of *Samia canningi* on three host plants *viz., Heteropanax fragrans, Evodia fraxinifolia* and *Litsea citrata* in two different seasons. The total larval duration was recorded as 24.6±0.97 to 27.2±0.81 days in *H. fragrans*, while in *E. fraxinifolia* and *L. citrata* it was recorded from 26.6±0.94 to 28.6±0.70 and 29.6±0.68 to 32.1±0.87 days respectively. The cocoon exhibited colour morphism and varied in length (3.38±0.15-3.87±0.07 cm. in male to 4.06±0.06-4.49±0.04 cm. in female), breadth (1.13±0.06-1.31±0.07 cm. in male to 1.32±0.03-1.66±0.05 cm. in female) and weight (1.46±0.09-1.92±0.07 gm. in male to 2.36±0.15-2.84±0.23 in female). The highest fecundity was recorded in *E. fraxinifolia* (315.90±19.33) whereas hatching percentage, effective rate of rearing and cocoon:dfl ratio was recorded maximum in *H. fragrans* as 80.90±2.25, 59.83±3.73 and 150±14.59 respectively during first season. During second season, highest shell weight was recorded in *L. citrata* (0.26±0.01 gm in male and 0.36±0.02 gm in female) while shell ratio was maximum in *E. fraxinifolia* (3.71±0.10% in male and 15.17±0.68% in female).

Nutritional Efficiency of Mulberry and Tasar Silkworms - A Comparative Analysis and Need for Such Analysis in Case of Eri Silk Worm

Tasar silkworm is the sericigenous wild variety insects of tropical India and *Bombyx mori* (Mulberry silkworms) is completely domesticated species. Mulberry silkworm larvae obtain nutrients from mulberry leaves to build up body, sustain life, spin cocoons and egg production. Such nutritional requirements in food consumption have direct impact on the overall physical and genetical traits such as larval and cocoon weight, quantity of silk production, pupation and reproductive traits. Silkworm nutrition refers to the substances required by silkworm for its growth and metabolic functions. Some of these substances are obtained from ingested food, while the remaining other nutritional components are synthesized by silk worm itself. The synthesis is done through various biochemical pathways including proteinaceous silk fiber of commercial interest. The nutritional adequacy of the food materials can be judged only by its ability to support growth in successive instars. In silkworm rearing food is a factor of paramount importance which regulates its growth, development and silk yield in sericulture. Silkworm breeds also differ in their nutritional requirements depending on the variety, rearing environment, season and quantum of nutrition. The efficiency of conversion of ingested food will vary with both the digestibility of food and the proportional amount of the digestible portion of that food converted into body substance and metabolized for energy to maintain larval life.

It was stated that the assimilation efficiency did not vary significantly as a function of reduced food consumption. The insect growth rate is related to the capacity of food intake and nutrients absorbed by the body from different host plants. Hiratsuka (1920) gathered detailed information on the ingestion, digestion and utilization of mulberry leaves as well as consumption of nutrients for both the growth and maintenance of life of *Bombyx mori*. The nutritional studies in mulberry silkworm with respect to food utilization have been experimented in relation to growth, body weight, food digested and ingested by Ueda and Suzuki (1967) as well as Horie and Watanabe (1983). Thus, the food consumption, assimilation and growth of silkworms have been elaborately studied in *Bombyx mori*. However, practically no information is available on the nutritional biology of tasar silkworms due to its wild nature. Therefore, Nitu Kumari and Roy, (2011) conducted a study with the objective of adding some information on the comparative nutritional efficiency of the tasar and mulberry silkworms under laboratory conditions.

The study was undertaken with larval growth and their efficiency with which ingested food is converted to biomass which is calculated by dividing the dry weight of food ingested into the dry weight gained by the larvae. For experiment 10 healthy larvae of *B. mori* (Nistari race) and *A. mylitta* (Daba ecorace) were taken in plastic trays separately. The weighed fresh mulberry and arjun leaves were provided for *B. mori* and *A. mylitta* respectively. The moisture content of leaves was maintained with utmost care by covering with wax paper. A parallel batch of larvae of each species was also maintained to replenish the missing larvae and also to determine the dry weight and subsequent increment in larval weight separately. The weight was taken and healthy silkworm larvae were counted daily and the missed larvae were replaced from reserved batch left over. The missing larvae were collected on subsequent day on daily basis and kept in a paper cover after separating manually. The left-over leaves, excreta and reserve batch larvae were dried in hot air oven daily at about 60°C to assess the dry weight or more scientifically the weight at constant moisture level. For further observations the dry weight of left over leaves, excreta and larvae were recorded. Then, the scheme for determination of nutritional efficiency was done by following the modified IBP formula (Petrusewicz and Macfadyen, 1970). The most applicable scheme has been adopted in the study of Nitu Kumari and Roy, (2011) was mentioned below:

$$C = P + R + F$$

Where, C= Food consumed; P = Growth; R = Metabolism (during wt loss); F = Faecal matter (undigested food).

I The food consumed (C) was calculated by total intake of dry weight of leaves by silkworm larvae during each instar.

Consumption (ingesta) = Dry weight of leaf – Dry wt of leaf left over

II Assimilation (A) was estimated by subtracting feed (excreta) from consumption (ingesta).

Assimilation = Consumption – Faecal matter (Digesta) = Ingesta – Excreta

$$A = C - F$$

III Larval growth (P) was estimated by subtracting the initial dry weight of larvae in each instar from the final dry weight of respective instar.

$$P = \text{Final dry wt} - \text{Initial dry wt}$$

IV Metabolism is measured as total biochemical reactions involving both catabolic as well as anabolic reactions of an organism. Metabolism (R) was estimated as difference between larval growth (P) and assimilation (A).

$$\text{Metabolism} = \text{Larval growth} - \text{Assimilation}$$

$$R = P - A$$

The rates of consumption (Cr), assimilation (Ar), growth (Pr) and metabolic rate (Mr) were calculated by dividing the respective amount of weight by the product of mid body wt or mean wt (g) of the larvae and durations (days) required for the completion of respective instars (Waldbauer, 1968).

For estimation of various rates the following formulae have been considered

$$\text{Consumption rate (Cr)} = \frac{C}{\text{Mid body wt.} \times \text{days}} \ \text{g/g/day}$$

$$\text{Assimilation rate (Ar)} = \frac{A}{\text{Mid body wt.} \times \text{days}} \ \text{g/g/day}$$

$$\text{Growth rate (Pr)} = \frac{P}{\text{Mid body wt.} \times \text{days}} \ \text{g/g/day}$$

$$\text{Metabolic rate (Mr)} = \frac{R}{\text{Mid body wt.} \times \text{days}} \ \text{g/g/day}$$

All consumption and growth parameters were measured on dry weight basis. The indices used were relative consumption rate (Cr), growth rate (Gr), approximate digestibility (AD), efficiency of conversion of ingested food (ECI) and efficiency of conversion of digested food (ECD) to biomass according to Waldbauer (1968).

The following indices are assessed as follows:

1. Consumption index (CI) =C\T*M

2. Relative growth rate (Gr) =P\T*M

3. Apporoximate Digestibility (AD) = {(C-F\C)}*100

4. Efficiency of conversion of ingested (ECI) = (Gr\CI)\100

    food to body substance.

5. Efficiency of conversion of digested (ECD) = {P\(C-F)}*100

    food to body substance.

Where,

   M = mean weight of larvae during feeding

   F = weight of faeces produced

   C = weight of food consumed

   P = weight gain by larvae

   T = duration of feeding (days)

The results of Nitu Kumari and Roy, (2011) indicated that the tasar silkworms have highest activities in terms of their relative rate of consumption (5.333g), assimilation (3.183g), growth (0.220g) and metabolism (2.963g) recorded in 4th larval stage (Table 8.9). But their highest consumption rate (0.4907g/g/day), assimilation rate (0.3493g/g/day), growth rate (0.01730g/g/day) and metabolic rate (0.3319g/g/day) were observed in 2nd larval stage (Table 8.11).

Interestingly in mulberry silkworm the maximum consumption, assimilation, growth and metabolism were observed in 3rd instar larval stage. The maximum values of consumption, assimilation, growth and metabolism were 0.740g, 0.693g, 0.065g and 0.628g respectively (Table 8.10). But the rates of consumption (0.1612g/g/day), assimilation (0.1520g/g/day), growth (0.010g/g/day) and metabolism (0.1405g/g/day) were found maximum in 1st instar larval stage (Table 8.12). In mulberry silkworms the cocoon weight (1.480-1.731g), reelability (70%) and filament length (389m) were lesser than the tasar silkworm cocoon weight (7.61g-10.25g), reelability (80%) and filament length (683m) (Table 8.13). Thus, on the comparative basis the tasar silkworms have more metabolic rate and have more silk production capacity than the mulberry silkworms. Food consumption and utilization were lowest in young instars and increased gradually as the growth progressed in both mulberry and tasar silkworms. In tasar silkworms the highest rate of consumption (5.333g), assimilation (3.183g), growth (2.220g) and metabolism (2.963g) were observed in 4th instar larvae. In 5th instar the value of these parameters was lowest due to arrested feeding behaviour. The lowest values of consumption (0.250g), assimilation (0.233g), growth (0.120g) and metabolism (0.113g) were observed in 5th instar of tasar silkworm larvae.

Food utilization parameters have been studied in many insects. The nutritive value of mulberry leaves depends on various agro–climatic factors and any deficiency in food and nutrients affect silk synthesis in the silkworms. Nutritional management directly influences the quality and quantity of silk production. This finding clearly indicates that the varieties with highest conversion efficiencies may reduce the larval span and consequently less quantity of the food is needed to support optimal growth.

Nutritional factors influencing silk production in *B. mori*

Nutrition of leaves plays an important role in the silkworm growth and overall silk production (Adolkar *et al.,* 2007). Any effort to improve the yield requires considerations of cumulative effect of the major traits which influences the silk yield. Vyjayanthi and Subramanyam, (2002) stated that in the silkworm, *B. mori* feeding behaviour depends on the niche, amount of food offered, quality of food, age and health of larvae. As most phytophagous Lepidopterans are voracious feeders any imbalance in the inputs from various factors affect food intake and result in poor larval development. Gangwar *et al.,* (1993) reported that *B. mori* has been found to ingest more mulberry leaves during night hours as compared to day hours. Basu *et al.,* (1995) evaluated the food quality relevant to all the aspects of insect performances including growth, development and reproductive potential of mulberry silkworms. In the study conducted by Raman *et al.,* (1994) approximate digestibility (AD) and efficiency of conversion of ingested food to body substance (ECI) were inversely correlated with the larval age. Koilpillai (1995) observed that the increase in feeding duration in silk worm *B. mori* from 3 to 24 hr a day resulted in a decrease in consumption rate, assimilation rate and metabolic rate but an increase in conversion rate and gross and net conversion efficiencies. Restriction of feeding duration affected the egg production in terms of number and biomass and weight of first instar larvae emerging from those eggs considerably. Production efficiency increased with increase in feeding duration. Correlation studies were carried out on the economic parameters *viz.,* larval weight, cocoon weight, shell weight and nutritional parameters *viz.,* ingesta, digesta and approximate digestibility (AD) of dry matters by Raman *et al.,* (1995). The patterns of correlation were different in bivoltine and multivoltine hybrids, which clearly indicated that the contributory factors for the productivity were very different. The study conducted by Singh and Ninangi (1995) indicated that ingestion, digestion and utilization of food were mostly dependent on the feeding level and genotype of the hybrid. The quiescence in the absence of food in mulberry silk worm larvae also suggests that long term diet deprivation might induce a reduced metabolic state in the larva that would minimize energy loss or nutrient depletion (Nagata and Nagasawa, 2006). Srivastava *et al.,* (1982) studied the effects of food deprivation on larval duration, weight of cocoon and fecundity of silkworms. They found that larvae took longer time to pupate when they were deprived of food than those feeding 24 hr/day Srivastava *et al.,* (1982) also observed a decrease in weight of cocoon (shell) and a marked loss of fecundity due to food deprivation during feeding period.

In *B. mori* it was reported that the growth is heterogenic and varied according to silkworm race, quality and quantity of food intake and climatic conditions (Krishnaswami *et al.,* 1973). Sharma *et al.,* (1986) have undertaken to investigate various parameters such as comparative weight gain, total larval period, pupation %, weight of the cocoon formed, amount of food utilized, consumption index, growth rate, approximate digestibility, efficiency of conversion of ingested and digested food etc. in the case of mulberry silkworms fed on different varieties of mulberry.

Thus, the results of Nitu Kumari and Roy, (2011) revealed that on the comparative basis the Saturniidae species (tasar silkworms) have more active metabolism and gave more silk production than the mulberry silkworms. Hence, tasar silkworms will be more beneficial and may be exploited for their commercialization as cottage industry for their sustainable development as energy resources. It is desirable to conduct such comparative studies with eri silk worm as well as other vanya silk worms along with *B.mori*. However, the fact remains that apart from highly domesticable nature, the quality of silk produced by *B. mori* is far superior to the silk produced by other silk worms including eri silk worm. This fact attracted the attention of scientists to conduct more studies on *B. mori* than any other silk worms.

Table 8.9: Measurement of food consumption, assimilation, growth and metabolism of Tasar (*Antheraea mylitta*) silkworm (Adopted from Nitu Kumari and Roy, 2011)

Stage	Consumption (C–g)	Assimilation (A–g)	Growth (P–g)	Metabolism (R–g)
1st Instar	1.133	0.783	0.070	0.713
2nd Instar	2.550	1.816	0.090	1.726
3rd Instar	3.616	2.334	0.140	2.194
4th Instar	5.333	3.183	0.220	2.963
5th Instar	0.250	0.233	0.120	0.113

Table 8.10: Measurement of food consumption, assimilation, growth and metabolism of Mulberry (*Bombyx mori*) silkworm (Adopted from Nitu Kumari and Roy, 2011)

Stage	Consumption (C–g)	Assimilation (A–g)	Growth (P–g)	Metabolism (R–g)
1st Instar	0.500	0.466	0.030	0.406
2nd Instar	0.625	0.582	0.040	0.543
3rd Instar	0.740	0.693	0.065	0.628
4th Instar	0.625	0.576	0.060	0.516
5th Instar	0.100	0.008	0.050	0.042

Table 8.11: Measurement of food consumption rate, assimilation rate, growth rate and metabolism rate of Tasar (*Antheraea mylitta*) silkworm (Adopted from Nitu Kumari and Roy, 2011)

Stage	Consumption rate (C–g)	Assimilation rate (A–g)	Growth rate (P–g)	Metabolism rate (R–g)
1st Instar	0.4053	0.2797	0.0143	0.2546
2nd Instar	0.4907	0.3493	0.0173	0.3319
3rd Instar	0.3690	0.2381	0.0143	0.2239
4th Instar	0.2693	0.1608	0.0111	0.1496
5th Instar	0.0080	0.0074	0.0038	0.0036

Table 8.12: Measurement of food consumption rate, assimilation rate, growth rate and metabolism rate of Mulberry (*Bombyx mori*) silkworm (Adopted from Nitu Kumari and Roy, 2011)

Stage	Consumption rate (C–g)	Assimilation rate (A–g)	Growth rate (P–g)	Metabolism rate (R–g)
1st Instar	0.1612	0.1520	0.0100	0.1405
2nd Instar	0.1011	0.0943	0.0065	0.0879
3rd Instar	0.0839	0.0785	0.0075	0.0712
4th Instar	0.0534	0.0493	0.0050	0.0441
5th Instar	0.0057	0.0005	0.0025	0.0024

Table 8.13: Comparative values of silk production (Adopted from Nitu Kumari and Roy, 2011)

Species	Temperature (°C)		Relative humidity (%)		Cocoon wt (g)		Shell wt (q)		Reelability (%)	Filament Length (m)
	Min.	Max.	Min.	Max.	Min.	Max.	Min.	Max.		
Mulberry	20	20	60	70	1.480	1.731	0.255	0.327	70	389
Tasar	28	30	75	85	7.61	10.25	0.85	1.46	80	683

Summary and Conclusions

➤ India is the only country that produces all the types of silks namely, Mulberry, Tropical Tasar, Oak Tasar, Eri and Muga.

➤ The availability of essential nutrients in food plant is vital for successful life cycle, cocoon quality, metamorphosis to moth stage and their reproductive activity.

➤ Among the vanya silks, the tropical tasar is an important variety produced by a wild silkworm of *Antheraea mylitta* Drury. It is polyphagous and feeds on several food plants.

➤ Even though the tasar silkworm produces less cocoons, their quality can compensate the crop economics either with higher silk or egg productivity or quality.

➤ Muga silkworm, *Anthereae assamensis* Helfer is an endogenous sericigenous species. It is semi domesticated, polyphagous and multivoltine in nature having five to six generations in a year.

➤ *Samia canningi* Hutton, a wild eri silk moth is bivoltine in nature and undergoes eight months pupal hibernation in Nagaland climatic conditions.

➤ Cocoon yield and seed quality of tropical tasar silkworm, *Antheraea mylitta* Drury depends on the food plant variety and its nutritional status. The leaf of *L. parviflora* can be fed during final larval stages to attain optimal silk and seed productivity in *A. mylitta.*

➤ Commercial rearing of muga silkworm *Anthereae assamensis* Helfer can be taken up on the foliages of som, *Persea bombycina* Kost. and Soalu, *Litsea polyantha* Juss. The rearing on soalu plant showed shorter larval duration and higher cocoon weight than that of som while in silk quality cocoons from som exhibited superior quality.

➤ The commercial rearing of muga silkworm can be taken up on both the food plants even though rearing on soalu has shown better performance while for seed crop, rearing should be conducted on soalu plants irrespective of seasons.

➤ The grainage period of muga silkworm is remarkably long and experiences an unbalanced sex ratio during emergence.

➤ The males mature early and their population is proportionately high in the early part of grainage and just reverse in the later part. As a result a considerable number of female moths are usually wasted due to lack of their male counterparts.

➤ In Muga silk worm, coupling duration of 180 min during summer and 300 min during winter is adequate to obtain maximum hatching.

➤ The male moths of muga silkworm can be utilized two times in winter without affecting the fecundity and fertility. However, the use of male moths for re-mating in muga silkworm during summer season is not recommended since the hatching is <50%.

➢ In an experiment conducted to know the biology and rearing performance of *Samia canningi* Hutton, in two different seasons on three host plants *viz., Heteropanax fragrans, Evodia fraxinifolia* and *Litsea citrata* revealed that highest fecundity was recorded in *E. fraxinifolia* whereas hatching percentage, effective rate of rearing and cocoon:dfl ratio was recorded maximum in *H. fragrans* during first season.

➢ In second season, highest shell weight was recorded in *L. citrata* (0.26±0.01 gm in male and 0.36±0.02 gm in female) while shell ratio was maximum in *E. fraxinifolia* (3.71±0.10% in male and 15.17±0.68% in female).

➢ Silkworm nutrition refers to the substances required by silkworm for its growth and metabolic functions. Some of these substances are obtained from ingested food, while the remaining other nutritional components are synthesized by silk worm itself.

➢ In an experiment conducted to know the information on the comparative nutritional efficiency of the tasar and mulberry silkworms using Nistari race and Daba ecoraces respectively revealed that the tasar silkworms have highest activities in terms of relative rate of consumption (5.333g), assimilation (3.183g), growth (0.220g) and metabolism (2.963g) in 4th instar larval stage.

References

Adolkar V V, Raina S K and Kimbu D M. 2007. Evaluation of various mulberry *Morus* Spp. (Moraceae) cultivars for the rearing of the bivoltine hybrid race Shaanshi BV-333 of the silkworm *B.mori. International Journal of Tropical Insect Science* **27**: 6-14.

Barah A and Sahu A K. 2003. Utilisation of male moths of muga silkworm *Antheraea assama* Ww (Lepidoptera: Saturniidae) for multiple coupling. *Bulletin of Indian Academy of Science* **7**(1): 94-98.

Basu R, Roychoudhury N, Shamsuddin M, Sen S K and Sinha S S. 1995. Effect of leaf quality on rearing and reproductive potential of *Bombyx mori* L. *Indian Silk* **95**: 21-22.

Chakravorty R, Neog K, Suryanarayana N and Hazarika L K. 2004. Feeding and moulting behaviour of muga silkworm (*Anthereae assama* Ww) on different food plants. *Sericologia* **44**(2): 145-152.

Gangwar S K, Samasundaram P and Thangavelu K. 1993. Feeding behaviour of silkworm *Bombyx mori* L. *Journal of Advanced Zoology* **14**(2): 115-118.

Gangwar S K. 2010. Impact of varietal feeding of eight Mulberry varieties on *Bombyx mori* L. *Agricultural Biological Journal* **1**(3): 350-354.

Goswami D and Singh N I. 2012. Effect of coupling duration on fecundity and fertility of muga silk moth *Antheraea assama* Ww. *Indian Journal of Entomology* **74**(2): 132-135.

Hazarika U, Barah A and Chakravorty R. 2005. Physiological and biochemical response of castor (*Ricinus communis* Linn.) to application of NPK and their correlation with economic parameters of eri silkworm (*Samia ricini* Donovan). In: *Proceeding of 20ᵗʰ Congress of the International sericulture commission volume II, sec, 3 Non-mulberry silkworms*, 2005, Bangalore India, 94-98.

Hazarika U, Barah a, Phukan J D and Benchamin K V. 2003. Studies on the effect of different food plants and seasons on the larval development and cocoon characteristics of silkworm *Samia cynthia ricini* Boisduval. *Bulletin of Indian Academy of Sericulture* **7**(1): 77-85.

Hiratsuka E. 1920. Researches on the nutrition of the silkworm. *Bulletin of the Sericultural Experiment Station, Japan* **1**: 257-315.

Horie Y and Watanabe K. 1983. Daily utilization and consumption of dry matter in food by the silkworm *B. mori* L (Lepidoptera : Bombycoidae). *Applied Entomology and Zoology* **18**: 70-80.

Kakati L N and Chutia B C. 2010. Biology and rearing performance of *Samia canningi* Hutton - a wild eri silk moth in Nagaland. *Indian Journal of Entomology* **72**(4): 343-351.

Katsuno S. 1977. Studies on eupyrene and apyrene spermatozoa in the silkworm, *Bombyx mori* L. I. The intra testicular behaviour of the spermatozoa at various stages from the pupa to adult. *Applied Entomology and Zoology* **12**: 142-153.

Koilpillai 1995. Impact of nutritional modulation on survival, growth bioenergetics and reproductive potential in silkworm, *Bombyx mori* L. *Journal of Entomological Research* **19**: 223-227.

Krishnaswami S, Narasimhanna M N, Suryanarayan S K and Kumararaj S. 1973. Manual of Sericulture, Vol. II (Silkworm Rearing) Published by FAO, USA, Rome, Agricultural Service Bulletin AGS, ASB/15 pp. 64.

Kumar R and Vadamalai E. 2010. Rearing Performance of Eri Silkworm *Philosamia ricini* in Monsoon Season of Uttar Pradesh. *Asian Journal of Experimental Biological Sciences* **1**(2): 303-310.

Mano Y, Nirmal Kumar S, Basavaraja H K, Mal Reddy N and Datta R K. 1992. An index for multiple trait selection for silk yield improvement in *Bombyx mori* L. In: *National Conference on Mulberry Sericultural Research*. Central Sericultural Research and Training Institute, Mysore, India, Dec, 1992, p.116.

Mano Y, Nirmal Kumar S, Basavaraja H K, Mal Reddy N and Datta R K. 1993. A new method to select promising silkworm breeds/combinations. *Indian Silk* **31**: 53.

Manohar Reddy R, Charan R, Prasad B C, Siva Reddy C, Manjula A and Sivaprasad V. 2010. Rearing and grainage performance of Indian Tropical Tasar Silkworm, *Antheraea mylitta* Drury fed on *Terminalia tomentosa* (W&A) and *Lagerstroemia parviflora* (Roxb.) food plants. *Academic Journal of Entomology* **3**(3): 69-74.

Maqbool A, Malik G N, Dar H U, Afifa S, Kamili and Gul Zaffar. 2005. Evaluation of some bivoltine silkworm genotypes under different seasons. *Indian Journal of Sericulture* **44**(2): 147-155.

Nagata S and Nagasawa H. 2006. Effect of diet-deprivation and physical stimulation on the feeding behaviour of the larvae of the silkworm *Bombyx mori*. *Journal of Insect Physiology* **52**: 807-815.

Nitu kumari and Roy S P. 2011. Some aspects of the identification of nutritionally efficient silkworms (Insecta: Lepidoptera: Bombycoidea), their metabolic rate and sustainable development as energy resources. *The Bioscan* **6**(3): 475-481.

Petrusewicz K and Macfadyen A. 1970. Productivity of terrestrial animals Principles and method. IBP Hand book, No 13 *Blackwell Scientific Publication*, Oxford and Edinburgh 190 pages.

Raman A K V, Magdum S B, Shivakumar G R, Giridhar K and Datta R K. 1995. Correlation studies on different economic and nutritional parameters in *Bombyx mori* L. hybrids. *Indian Journal of Sericulture* **34**: 118-121.

Raman K V A, Mgdum S B and Datta R K. 1994. Feed efficiency of the silkworm *Bombyx mori* L. hybrid (NB4DC x KA). *Insect Science and Application* **15**(2): 111-116.

Ravikumar C. 1988. Western ghat as a bivoltine region prospects, challenges and strategies for its development. *Indian Silk* **26**(9): 39-54.

Saikia S, Handique R, Pathak A and Das K. 2004. Rearing performance of muga on the primary and secondary food plants with an attempt for the survival of now extinct Mejankari silk heritage of Assam. *Sericologia* **44**(3): 373-376.

Sharma B, Badan P and Tara J S. 1986. Comparative consumption, population and silk production in silkworm, *Bombyx mori* fed on various varieties of mulberry existing in Jammu division of Kashmir State. *Sericologia* **26**(26): 419-429.

Singh G B and Ninangi O. 1995. Comparative studies on food utilization efficiency in some silkworm strains under different feeding level. *Sericologia* **35**(4): 667-675.

Singh N I and Goswami D. 2012. Food plant varietal effect on the rearing and grainage of Muga silkworm, *Antheraea assamensis* Helfer, 1837. *Munis Entomology and Zoology* **7**(2): 1023-1027.

Srivastava A D, Mishra S D and Poonia F S. 1982. Effect of food deprivation on larval duration, cocoon (shell) weight and fecundity of eri silkmoth, *Philosamia ricini* Hutt (Lepidoptera:Saturnidae). *Indian Journal of Sericulture* **21-22**: 11-15.

Sugai E and Kiguchi K. 1967. Male sterility of the silkworm, *Bombyx mori* L. induced by high temperature during pupal period. *Journal of Sericulture Science, Japan* **36**: 491-496.

Ueda S and Suzuki K. 1967. Studies on the growth of silkworm, *Bombyx mori* L. I. Chromo logical changes of the amount of food ingested and digested, body weight and water content of the body and their mutual relationships. *Bulletin of the Sericultural Experiment Station, Japan* **22**(1): 65-67.

Vyjayanthi N and Subramanyam M V V. 2002. Effect of fenvalerate- 20EC on sericigenous insects: I. food utilization in the late-age larva of the silkworm, *Bombyx mori* L. *Ecotoxic Environment and Safety* **53**: 206-211.

Waldbauer G P. 1968. Consumption and utilization of food by insects. *Advances in Insect Physiology* **5**: 229-288.

CHAPTER 9

Exploitation of Genetic Diversity in Eri Silkworm for Rearing Performance and Amenability

Higher productivity in ericulture can be achieved through a combination of ideal castor varieties and eri silkworm breeds associated with proper climatic conditions including suitable agronomic practices. Eri silkworm Samia cynthia ricini (Boisduval) has 26 eco races and 12 strains based on the heritable morphological characters. The eri silkworms are cultured in different parts of India but to get the best yield, people should have some knowledge that which breed is more beneficial than others. To identify the best strain for rearing in commercial purpose and identify suitable strains/eco races to that particular locality, several workers have researched. Accordingly they identified different eco races. This chapter discusses in detail about the rearing performance of different eco races suitable to different states, studies on combining ability analysis of inbred lines, interaction of castor genotypes and breeds, leaf anatomical characters and their impact on economic parameters of breeds, rearing performance of different strains on different castor genotypes, effect of bio pesticides on different strains and genetic variability of other related silkworms etc.

As discussed in chapter 2. Nomenclature and Biodiversity, eri silkworm, *Samia cynthia ricini* (Boisduval) has 26 eco races and 12 strains based on the heritable morphological characters. Among them, six pure-line strains were isolated from Borduar and Titabar eco-races on the basis of larval markings and colour. These are yellow plain (YP), yellow spotted (YS), yellow zebra (YZ), green blue plain (GBP), green blue spotted (GBS) and green blue zebra (GBZ). The remaining six strains *viz.,* white plain (WP), green plain (GP), white spotted (WS), green spotted (GS), white semi zebra (WSZ) and yellow semi zebra (YSZ) were isolated from wild type, *P. ricini* through selection and breeding over 10 generations by rearing under standard laboratory condition.

Comparative Study of Eri Silkworm Strains Based on Morphological Traits in Different Parts of India

Identification of eri silkworm strains suitable for Assam

The silkworm are cultured in different parts of Assam but to get the best yield, people should have some knowledge that which breed is more beneficial than others. To identify the best strain for rearing in commercial purpose, Sharma and Kalita, (2013) conducted a comparative study on six strains of eri silkworm based on morphological traits. The studies help to compare different strains on the basis of their larval weight, silk gland weight and cocoon characters and will help them to select best breed for rearing as well as to choose parents for hybridization. Data on growth, rearing performance and cocoon characters of eri silkworm *viz.*, larval weight (g), silk gland weight, SGTSI, effective rate of rearing (ERR %), cocoon weight (g), shell weight (g) and shell ratio (SR%) of different strains of eri silkworm were recorded and presented in Tables 9.1 and 9.2. The experimental findings revealed highest ERR in greenish blue spotted worm (94.7%) followed by Greenish blue zebra (91.3%) and performance of both the strains were significantly higher than the other strains. The larval growth rate is found to be highest in all the strains in first instar but decreased gradually with advancement of larval stage and recorded least in fifth instar. The growth rate is highest in YZ followed by GBP. The strain YZ (14.09) recorded highest shell ratio followed by YS (13.92). The single cocoon weight (3.83g), shell weight (0.54g) and shell ratio (SR%) were found to be higher in Yellow zebra followed by Yellow spotted (13.92%) and Greenish blue spotted (13.60%) strains. Correlation is observed between larval weight and silk gland weight, cocoon weight and shell weight. The study reported that the strain with higher larval weight has higher silk gland weight and also shows high shell ratio. YZ was better in terms of larval and silk gland weight and also has higher shell ratio and therefore expected to produce more fibers than other strains. This study revealed YZ as the best strain in terms of growth and cocoon characters. GBS and YS were also recorded as better strains for rearing and may be the better strains for hybridization with YZ.

Table 9.1: Larval growth rate at all instars of different strains of eri silkworm in Assam (Sharma and Kalita, 2013)

Strains	Larval growth rate in five instars				
	I	II	III	IV	V
GBP	0.733	0.588	0.430	0.342	0.317
GBS	0.720	0.548	0.501	0.415	0.312
GBZ	0.703	0.535	0.482	0.320	0.310
YP	0.680	0.525	0.365	0.263	0.230
YS	0.693	0.512	0.405	0.388	0.295
YZ	0.760	0.586	0.455	0.398	0.321

Table 9.2: The ERR, shell weight, shell ratio, cocoon weight and fecundity of different strains of eri silkworm in Assam (Sharma and Kalita, 2013)

Strains	ERR (%)	Fecundity (nos)	Single cocoon weight (g)	Single shell weight (g)	Shell ratio (%)
GBP	88.9	357.54	3.44	0.45	13.08
GBS	94.7	458.34	3.45	0.47	13.60
GBZ	91.3	435.60	3.56	0.44	12.36
YP	90.4	456.70	3.34	0.45	13.47
YS	86.7	445.20	3.52	0.49	13.92
YZ	88.9	430.23	3.83	0.54	14.09

Evaluation of some promising strains of Eri silkworm in Vidarbha region of Maharashtra

Although, several workers attempted evaluation of best eri silkworm strains under different climatic conditions in India, no works had been done on this line under climatic condition of Vidarbha region of Maharashtra until 2014. In this context, Wankhade *et al.,* (2014) have made an attempt to evaluate the best strain of eri silkworm under climatic conditions of Vidarbha region for future establishment of ericulture.

They conducted experiment with 4 strains of eri silkworm *viz.,* Greenish Blue Zebra (GBZ), Yellow Zebra (YZ), Greenish Blue Plain (GBP), Yellow Plain (YP) during October-November 2013 at Wardha District of Vidarbha region of Maharashtra. The first instar larvae were reared with tender castor leaves twice a day, and then thrice a day up to II and III instars. The IV and V instar larvae were given equal feeding with mature castor leaves four times a day. The strains of eri silkworm were evaluated with different parameters of complete life cycle covering from hatching of eggs, larval rearing, cocoon spinning, emergence of moth from cocoon, mating of moths to egg laying. The temperature, relative humidity and other parameters *viz.,* larval weight, larval duration, ERR%, cocoon weight, pupal weight, shell weight, shell ratio, mortality, moth weight and size, fecundity and hatching were recorded. The results showed that the Greenish Blue Zebra (GBZ) and Yellow Zebra (YZ) exhibited better performance in all economic characters like fecundity and better adaptability than Yellow Plain (YP) and Greenish Blue Plain (GBP) strains in the climatic conditions of Vidarbha region of Maharashtra (Table 9.3). It is thus confirmed from the study of Wankhade *et al.,* (2014) that GBZ and YZ strains are better performers in all important economic characters under the climatic conditions of Vidarbha region of Maharashtra. So, these strains can be established for commercial exploitation for the interest of castor cultivating farmers of the region.

Table 9.3: Rearing performance of 4 strains of eri silkworm, *Samia ricini* during rearing in early winter (October-November, 2013) in Vidarbha, Maharashtra (Wankhade *et al.,* 2014)

Sr. No	Parameters		GBZ (Greenish Blue Zebra)	YZ (Yellow Zebra)	YP (Yellow Plain)	GBP (Greenish Blue Plain)
1.	Larval weight (g)		7.34±0.07	7.25±0.07	6.68±0.04	6.67±0.12
2.	Larval duration (days.hrs)		18.02	18.03	18.06	18.06
3.	Cocoon weight (g)		2.80±0.03	2.73±0.01	2.40±0.02	2.36±0.03
4.	Shell weight (g)		0.38±0.00	0.38±0.00	0.32±0.00	0.32±0.00
5.	Shell ratio (%)		13.55±0.23	13.77±0.26	13.32±0.26	13.41±0.34
6.	Pupal weight (g)		2.42±0.04	2.35±0.02	2.08±0.03	2.04±0.03
7.	Pupal duration (days.hrs)		11.08	11.10	11.16	11.18
8.	Larval mortality (%)		1.5%	1.0%	3.0%	2.5%
9.	ERR (%)		98.50%	99.00%	97.00%	97.5%
10.	Moth weight	♂	0.75±0.04	0.68±0.04	0.57±0.01	0.57±0.02
		♀	1.63±0.06	1.52±0.02	1.33±0.02	1.28±0.02
11.	Moth wing expansion	♂	11.2±0.05	11.1±0.03	10.9±0.04	10.9±0.08
		♀	10.6±0.05	10.5±0.06	10.4±0.04	10.4±0.05
12.	Moth body length	♂	2.6±0.024	2.6±0.023	2.5±0.013	2.5±0.016
		♀	3.1±0.013	3.1±0.013	3.0±0.013	3.0±0.015
13.	No. egg laying		369±3.03	384±3.50	362±2.11	311±4.43
14.	Hatching (%)		91.4±0.86	92.2±0.75	92.8±0.56	89.90±0.61
15.	Life cycle duration (days.hrs)		40.04	40.06	40.16	40.18

± Standard error

Rearing performance of different eco-races of eri silkworm in Uttar Pradesh

The rearing performance of different eco-races mainly depends on the combined action of hereditary potential and environment of which they are exposed. It is well known that the dynamic environment conditions prevail at different seasons bring about profound changes in the physical as well as biotic factors influencing the growth, development and the expression of economic characters in different silkworm races. The selection of silkworm races suitable for particular location is very important for higher cocoon production (Periaswamy and Radhakrishnan 1984; Jaiswal *et al.,* 2003).

The state of Uttar Pradesh possesses different geographical regions and the main season of eri silkworm rearing in UP is autumn, which is characterised by low temperature and humidity. In autumn rearing some of the eco races have low cocoon weight which can be attributed due to cold shock. Benchamin (2000) reported that fecundity was high in winter. Low temperature adversely affects the hatching percentage of eri silkworm. It was also found that in autumn season some eco races of the eri silkworm had prolonged larval duration because of low temperature, reduced rate of metabolism resulting in slow growth (Somaprakash and Sathya Prakash, 2009). Eri silkworm is delicate insect and is prone to many infectious diseases caused by various pathogenic micro-organisms. Therefore, selection of most promising eco-race for autumn season may minimise the disease susceptibility in eri silkworms in UP. Hence Rajesh Kumar and Elangovan (2013) conducted a study to identify the best performing and disease resistant eco-race of eri silkworm for autumn season of UP in the Department of Applied Animal Sciences, Babasaheb Bhimrao Ambedkar University, Lucknow, Uttar Pradesh during autumn season (October-November) of 2006 and 2007 with four eco-races of eri silkworm *viz.,* Borduar, Titabar, Dhanubhanga and Mendipathar. Three replicates of each 400 larvae were maintained separately for each eco-race since II moult. Locally available castor leaves were offered to the silkworms. Tender leaves were fed four times a day until III instar and semi tender and mature leaves were fed four times a day during IV and V instar stage. The observations were recorded on rearing, cocoon and grainage parameters including the leaf silk conversion rate (%) (LSCR %).

$$LSCR\% = \frac{\text{Single shell weight}}{V - \text{instars larvae ingesta} / 400} \times 100$$

The values of various rearing parameters such as hatching %, yield of larvae (by number and weight), larval weight, larval duration, cocoon weight, shell weight, shell ratio, cocoon shape variability, pupal period, pupation rate (%) and leaf silk conversion rate (%) of different eco-races during autumn rearing are given in the Tables 9.4 and 9.5. The highest hatching (%) of 94.73%, minimum larval duration of 20.66 days, highest larval weight of 7.39g and highest yield of cocoon by number (365) was recorded in the Borduar eco race followed by the other races. The highest cocoon weight, shell weight and shell ratio (%) were observed in Borduar eco race both during 2006 and 2007 autumn rearing. Highest cocoon shape variability of 270.49% was recorded in the Borduar eco race. The rearing parameters differed significantly among the different eco-races of eri silkworm. Hence in the present study of Rajesh Kumar and Elangovan (2013) based on the recorded parameters, Borduar eco race of *S. cynthia ricini* showed better rearing performance than Titabar, Dhanubhanga and Mendipathar eco-races. Hence it can be concluded that Borduar eco-race of eri silkworm is superior and suitable for rearing during autumn season of Uttar Pradesh.

In another study Rajesh Kumar and Elangovan (2012) conducted an experiment to identify superior eco-races of eri silkworm which is suitable for commercial rearing in summer season in Uttar Pradesh during 2007 and 2008. The experimental procedure, ecoraces that were studied for the suitability, site of experimentation and observations recorded were same as the previous experiment. The Titabar eco-race showed better rearing performance compared to other eco-races in all the rearing parameters. The highest hatching (%) of 94.65%, minimum larval duration of 18.75 days, highest larval weight of 6.75g, highest yield of cocoon by number (345) and highest yield of cocoon by weight of 1.159 kg was recorded in the Titabar eco race followed by the other races. The highest cocoon weight, shell weight and shell ratio (%) were observed in Titabar eco race both during 2007 and 2008 summer rearing. Highest cocoon shape variability of 250.44% was recorded in the Titabar eco race with minimum pupal period of 8.75 days (Tables 9.6 and 9.7). The rearing parameters such as weight of full grown larvae, yield by number of larvae, yield by weight of larvae, single cocoon weight, shell weight, shell ratio, cocoon shape variability, pupation rate differed significantly among different eco races of eri silkworm. Hence the results of the study by Rajesh Kumar and Elangovan (2012) showed the superiority and suitability of Titabar eco-race for summer rearing in Uttar Pradesh.

Table 9.4: Rearing performance of different eco-races of eri silkworm during autumn season, 2006 (Rajesh Kumar and Elangovan, 2013)

Eco-race	Hatching (%)	Larval duration (Days)	Larval wt (g)	Yield/400 larvae		Single cocoon wt (g)	Shell wt (g)	Shell ratio (%)	Cocoon shape variability	Pupal period (day)	Pupation rate (%)	LSCR (%)
				By No.	By Wt. (Kg)							
Borduar	94.73	20.66 ±0.040	7.39 ±0.026	365 ±1.73	1.260± .0050	3.46± .010	0.56 ±.010	16.20± .020	270.49 ±5.56	11.33± .040	85.81± .764	3.06 ±.010
Titabar	94.26	21.50 ±0.060	7.37± 0.026	361 ±1.73	1.210 ±.0040	3.40 ±.017	0.52± .010	15.36 ±.010	261.68 ±6.21	11.62± .050	84.95± .500	2.94 ±.010
Dhanubhanga	93.63	21.75 ±0.492	6.82 ±0.062	349± 4.36	1.183 ±.0030	3.39± .036	0.51 ±.020	15.15± .017	250.75± 5.87	12.20 ±.036	83.47 ±.281	2.88 ±.026
Mendipathar	94.66	21.55 ±0.056	7.36± 0.043	362 ±3.61	1.241 ±.0075	3.43± .040	0.55± .011	16.10± .065	268.98 ±5.89	11.75± .474	85.78± .197	2.90 ±.036
F-Value	-	14.69	127.83	15.66	127.98	3.64	8.77	649.47	6.99	11.52	15.22	35.45
P-Value	-	0.001	0.00	0.001	0.00	0.060	0.007	0.00	0.013	0.003	0.001	0.00
CD (5%)	-	0.508	0.099	7.225	0.012	0.060	0.031	0.073	0.646	0.481	0.926	0.054

Table 9.5: Rearing performance of different eco-races of eri silkworm during autumn season, 2007
(Rajesh Kumar and Elangovan, 2013)

Eco-race	Hatching (%)	Larval duration (Days)	Larval wt (g)	Yield/400 larvae		Single cocoon wt (g)	Shell wt (g)	Shell ratio (%)	Cocoon shape variability	Pupal period (day)	Pupation rate (%)	LSCR (%)
				By No.	By Wt. (Kg)							
Borduar	94.82	20.40 ±0.055	7.45 ±0.040	368 ±1.73	1.295± .043	3.54± .026	0.60 ±.017	17.00± .111	266.08 ±5.51	10.75± .497	87.50± .045	3.11 ±.036
Titabar	94.76	20.62 ±0.030	7.43± 0.060	363 ±2.65	1.275 ±.017	3.52 ±.017	0.59± .010	16.83 ±.010	265.00 ±6.21	11.50± .050	86.76± .500	2.99 ±.043
Dhanubhanga	93.70	21.75 ±0.045	6.86 ±0.036	352± 1.73	1.220 ±.026	3.45± .017	0.52 ±.017	15.17± .045	246.48± 5.88	11.75±.492	83.38 ±.080	2.95 ±.036
Mendipathar	94.69	20.62 ±0.030	7.43± 0.020	365 ±1.73	1.270 ±.095	3.48± .026	0.57± .026	16.49± .055	258.28 ±5.82	11.75± .491	86.32± .052	2.93 ±.026
F-Value	-	470.46	142.67	36.50	119.29	9.29	10.86	450.90	7.32	3.18	14.71	15.00
P-Value	-	0.00	0.00	0.00	0.486	0.006	0.003	0.00	0.011	0.085	0.00	0.001
CD (5%)	-	0.084	0.093	1.018	0.116	0.047	0.043	0.113	0.257	0.568	0.118	0.081

Table 9.6: Rearing performance of different eco-races of eri silkworm during summer season, 2007
(Rajesh Kumar and Elangovan, 2012)

Eco-race	Hatching (%)	Larval duration (Days)	Larval wt (g)	Yield/400 larvae		Single cocoon wt (g)	Shell wt (g)	Shell ratio (%)	Cocoon shape variability	Pupal period (day)	Pupation rate (%)	LSCR (%)
				By No.	By Wt. (Kg)							
Borduar	93.50	19.33 ±0.092	6.52±0.030	342 ±2.64	1.043± .012	3.05± .043	0.45 ±.036	14.82± .036	240.28±5.78	9.62± .030	85.15± .082	2.88±.095
Titabar	94.65	18.75±0.474	6.75± 0.043	345 ±1.73	1.159 ±.009	3.36 ±.036	0.50± .036	14.96 ±.036	250.44 ±5.67	9.50± .040	86.98± .169	2.89±.030
Dhanubhanga	92.00	19.40 ±0.050	6.23 ±0.046	336± 1.73	0.870±.008	2.59± .040	0.37±.010	14.28± .036	233.33± 5.51	9.40 ±.060	82.50 ±.197	2.82±.065
Mendipathar	92.45	19.20 ±0.050	6.47± 0.010	339 ±2.65	0.935 ±.005	2.76± .065	0.41± .017	14.93± .045	244.48 ±5.93	9.40± .050	84.65± .197	2.88 ±.075
F-Value	-	1.39	108.70	9.00	555.22	149.38	12.37	201.52	4.74	0.78	362.44	0.61
P-Value	-	0.313	0.01	0.006	0.01	0.01	0.002	0.01	0.035	0.538	0.01	0.624
CD (5%)	-	0.509	0.079	5.226	0.011	0.104	0.060	0.075	0.404	0.090	0.248	0.157

Table 9.7: Rearing performance of different eco-races of eri silkworm during summer season, 2008 (Rajesh Kumar and Elangovan, 2012)

Eco-race	Hatching (%)	Larval duration (Days)	Larval wt (g)	Yield/400 larvae		Single cocoon wt (g)	Shell wt (g)	Shell ratio (%)	Cocoon shape variability	Pupal period (day)	Pupation rate (%)	LSCR (%)
				By No.	By Wt. (Kg)							
Borduar	93.59	19.33±0.040	6.60±0.026	346 ±1.15	1.100± .014	3.15± .017	0.48 ±.020	15.27± .026	241.18±5.71	9.33±0 .040	86.13± 1.04	2.91 ±.036
Titabar	94.82	19.08 ±0.020	6.82± 0.032	350±4.36	1.190 ±.006	3.41 ±.036	0.52± .010	15.31 ±.030	251.49 ±5.66	8.75± 0.474	87.95± 0.96	2.93±.026
Dhanubhanga	91.15	19.50±0.040	6.36 ±0.017	338± 1.73	0.900 ±.013	2.61± .030	0.38 ±.026	14.61± .070	243.05± 5.68	9.65±0.040	83.36±0.56	2.87 ±.020
Mendipathar	92.75	19.40 ±0.055	6.54± 0.020	342±4.58	1.020±.004	2.89± .036	0.43± .017	14.87± .080	246.56±5.75	9.24± 0.040	85.16± 1.04	2.90 ±.060
F-Value	-	3.34	182.85	7.49	414.35	372.53	29.53	104.53	1.90	2.03	12.89	1.23
P-Value	-	0.076	0.01	0.010	0.01	0.01	0.01	0.01	0.209	0.188	0.002	0.361
CD (5%)	-	0.082	0.047	7.621	0.023	0.071	0.045	0.130	0.158	0.463	0.087	0.076

Comparative studies on rearing performance of some eco races of eri silkworm under Odisha conditions

Compared to the expansion of the sericulture industry in the post-independence era, Ericulture (eri farming) in India has not been as popular among the farmers (Kar, 2004). With 18,000 hectares of food plants, and ample manpower, ericulture holds a huge potential in Odisha. In their natural habitats in North East India the silkworms are geographically isolated (ecoraces) and have morphological differences in the states of Assam and Meghalaya (Singh *et al.,* 2003). The ecoraces (types) of *S. cynthia ricini* are commercially exploited in these places because of their high silk yield potential. However, the eri silkworm dfls supplied by the earlier authors to the farmers of Odisha was of mixed (unknown) origin and was collected from a place called Mirza in Assam. It has a very low yield rate, making the Eri farming not so lucrative. One of the reasons for the low yield may be attributed to the climatic conditions of Odisha. Hota *et al.,*(2005) have reported under-performance by the Eri silkworm in Odisha with regard to the larval characters like silk yield, pupation rate, moth emergence and fecundity. Hence Ray *et al.,* (2010) conducted experiments with three pure line races of *S. cynthia ricini* namely Borduar, Titabar and Mendipathar in different seasons in the School of Life Sciences, Sambalpur University, Burla during 2009-2010. Rearing and grainage experiments were conducted with five replicates of 30 larvae each in separate bamboo trays to maintain purity of the eco races. Cocoon harvesting was done after fifth and sixth day of spinning. Commercial traits like cocoon weight (g), shell weight (g), shell ratio (%), pupal weight (g) and ERR (%) were recorded.

The results of the investigation showed that the pure line ecoraces of *S. c. ricini* showed significant increase over the mixed type in terms of morphological and quantitative characters such as absolute silk content, larval weight, cocoon weight, pupal weight, cocoon shell weight and shell ratio (Ray *et al.,* 2010). It was observed that commercial productivity and reproductive parameters were influenced by the seasonal variations of the rearing environment. Values of productivity parameters like cocoon weight, shell weight, shell ratio percentage and total silk production were found to be highest in winter followed by spring, autumn and rainy seasons while the lowest values were recorded in the summer crop. The two- way analysis of variance showed significant difference in rearing parameters among the different eco races of Eri silkworm in different seasons. Among the varieties, the rearing efficiency and silk ratio of Borduar eco race was found to be the highest in comparison to others in Odisha.

Rearing of eri silkworm strain on different plants under Punjab conditions

Ericulture is extended up to the altitude of 1500 m from the lowest level of plains and is being practiced in different states of India. Mavi and Singh (1989) have determined the feasibility of ericulture under Punjab conditions. Except castor, the other known hosts of eri silkworm are not prevalent in Punjab. In this direction, Rabinder Kaur and Virk (2012) evaluated different plants *viz.*, papaya, cauliflower, brinjal and potato in addition to castor, the primary food plant of eri silkworm. The experiment was conducted with the Borduar strain of eri silkworm at Entomology research farm, Punjab Agricultural University, Ludhiana. These larvae were given fresh leaves of respective plants daily and were observed for survival and development. The data for larval duration of first and third instar, per cent mortality and cocoon formation were recorded. The results show that there was 100 per cent mortality of first instar larvae on papaya and cauliflower leaves in 7.55±1.99 days and 6.10±0.31 days, respectively. Only 20 and 10 per cent third instar larvae were able to form the cocoons in 14.0±2.67 and 11.67±0.58 days on brinjal and potato leaves, respectively. No third instar larva completed the larval period on brinjal and potato leaves and died after 9.55±4.19 and 5.50±1.28 days, respectively. All the larvae successfully completed the larval period in 25.4±1.80 days on castor leaves with 100 per cent cocoon formation (Table 9.8).

Table 9.8: Per cent survival and mortality of eri silkworm strain on different hosts (Rabinder Kaur and Virk, 2012)

Host	Larval mortality (%)		Cocoon formation (%)	
	First instar	**Third instar**	**First instar**	**Third instar**
Papaya	100 (7.55 ±1.99)	80 (10.33±3.96)	0	20 (14.0±2.67)
Cauliflower	100 (6.10± 0.31)	90 (10.93±3.39)	0	10 (11.67±0.58)
Potato	-	100 (5.50±1.28)	-	-
Brinjal	-	100 (9.55±4.19)	0	0
Castor	0	0	100 (25.4±1.80)	100 (7.55±1.99)

So, it can be concluded that the castor is the only host plant for successful commercial rearing of eri silkworm in Punjab. The castor plants are available throughout the year in Punjab and therefore the rearing of eri silkworm on castor leaves is possible during the whole year.

Genetic Studies on Eri Silkworm for Enhanced Silk Production

Genetic analysis of larval and cocoon traits in eri silkworm by diallel cross

Generation of information on the nature of gene action and the quantity of genes influencing the inheritance of a metric trait is an essential pre-requisite for planning an effective breeding programme. As the metric traits are under the control of poly genes, their inheritance is complex. This necessitates the study by employing biometrical techniques amongst which diallel analysis is one. This technique has been used in case of *B. mori* L. But the information on genetic analysis of polygenically inherited characters in eri silkworm is lacking. In this direction Govindan *et al.,* (1997) studied the genetic analysis of larval and cocoon traits in eri silkworm, *Samia cynthia ricini* by diallel cross in the laboratory. Five pure breeds *viz.,* WP, WZ, GP, GS and GZ were crossed in 5 x 5 diallel fashion. The component analysis revealed the additive gene action for rate of progression to fourth instar, total larval duration, effective rate of rearing, maximum larval weight, cocoon yield by number and weight, cocoon weight, shell weight and silk productivity. Hence, the most appropriate breeding programme that can be followed for the improvement of these traits is selection. These traits are under the control of one to two groups of genes.

Studies on combining ability analysis of inbred lines of eri silkworm

Combining ability analysis is extensively employed in many plants and animals for the selection of promising parents and hybrids as well as to design and formulate efficient breeding plans. The documented literature on analysis of combining abilities and exploitation of hybrid vigour in eri silkworm to increase the productivity is scanty. Nagaraja and Govindan (1994) studied the general and specific combining abilities of five strains of eri silkworm and pointed out the prospects of exploiting heterosis for the productivity related traits like effective rate of rearing, cocoon yield by number and weight. However, further attempts have not been made to exploit the hybrid vigour and evolution of high yielding breeds in ericulture in spite of having high potential for production. Under these circumstances, Singh *et al.,* (2012a) conducted an experiment to study the combining ability analysis of 6 inbred lines of eri silkworm *viz.,* yellow plain (YP), yellow spotted (YS), yellow zebra (YZ), Green blue plain

(GBP), green blue spotted (GBS) and Green blue zebra (GBZ). The inbred lines were crossed in a 6 x 6 diallel fashion and 30 crosses were raised. These crosses along with the parents were reared indoor on castor leaves in a randomised block design with three replications each. After III moult, 300 larvae were retained in each replication. Data were recorded for fecundity, hatching, larval weight, effective rate of rearing, cocoon weight, cocoon shell weight, cocoon shell percentage and absolute silk yield in both parents and hybrids. Eri cocoons being unreelable, whole cocoon shells are utilized in spinning for silk threads after taking out the pupa and as such weight of whole green cocoon is not considered as an important parameter in respect of silk yield, instead absolute silk yield (ASY) is considered as important parameter and is calculated as follows,
ASY (g) = Single cocoon shell weight (g) X ERR

The mean data pertaining to the ten quantitative traits of six parental breeds of eri silkworm are presented in Table 9.8. Each breed showed superiority in certain traits only. The highest mean of fecundity, male cocoon shell weight, male cocoon shell percentage and absolute silk yield were exhibited by the breed YZ. YP showed highest hatching percentage, larval weight and male cocoon weight while highest mean female cocoon weight and female cocoon shell weight were shown by GBP. Highest cocoon shell percentage was found in the breed YS.

The analysis of variance for combining ability of nine yield contributing traits manifested significant GCA variances in all the traits except absolute silk yield, while significant SCA variances were observed in seven parameters. Significant reciprocal variances were observed in all the nine parameters which were the indication of maternal effects in these traits in eri silkworm. Among the six breeds, yellow zebra was found to be the best general combiner which exhibited significant effects in seven traits followed by yellow plain for six traits and greenish blue zebra in five traits. Among the hybrids, YP x GBZ and YZ x GBS showed significant SCA effects for six traits. Observations on reciprocal crosses revealed significant effects by GBZ x GBP in seven traits. The hybrids YP x GBZ and GBZ x GBP have been identified for further trial rearing for higher production of eri silk. The results obtained from the Nagaraja and Govindan (1994) study indicated that the eri silkworm breeds like yellow plain, yellow zebra and greenish blue zebra possessing significant GCA effects for several characters may be utilized in future breeding programme for the development of promising eri silkworm breeds. The promising hybrids *viz.,* YP x GBZ and GBZ x GBP exhibiting high and significant SCA and reciprocal effects for several economic traits may be exploited on commercial scale for the production of eri silk.

Estimation of combining ability in eri silkworm *Samia c. ricini* for pupal and allied traits

The ultimate goal of the most breeding programmes is to increase the yielding ability. Hybridization is one of the means of obtaining increased cocoon yield and the choice of breeds in such a programme is difficult owing to the differential behaviour of the inbred lines in selection for combining ability. In this regard, Nagaraja *et al.,* (1996) studied the combining ability of five parental breeds of *Samia cynthia ricini*, i.e. WP, WZ, GP, GS and GZ differed in their general combining ability for rate of pupation, pupal weight, pupal duration, moth emergence, sex ratio, fecundity and hatching. The results of the experiment show that the WP breed of eri silkworm was a good general combiner. Additive gene action was predominant for all the characters except for moth emergence as the estimates of GCA variance were greater than SCA variance.

Screening for eri silkworm ecoraces using morphological characters, growth, yields and ISSR markers

The selection of eri silkworm ecoraces with high yield and distinct morphological characters is necessary for variety improvement. In this regard, Wongsorn *et al.,* (2015) screened five ecoraces *viz.,* SaKKU1, SaKKU2, SaKKU3, SaKKU4 and SaKKU5 using ISSR markers. These ecoraces were derived from international academic cooperation, Songklanakarin. They were cultured using castor leaves of TCO 101 cultivar as food plant at 25±2°C, 80±5% R.H. Based on morphological characters, they are similar, except the body of the 5th instar larva of SaKKU1 is clearly covered with more creamy white powder and the mature larva has a shiny dominant yellow colour. The duration of the life cycle among ecoraces was also similar; 46-53 days (SaKKU1), 42-53 days (SaKKU2), 42-52 days (SaKKU3), 40-56 days (SaKKU4) and 41-52 days (SaKKU5). SaKKU1 had the highest survival rate at larval stage (1st-5th instar) (100.00%) and larva (1st-5th instar) - adult (88.89%), including the predominant heaviest average larva weight of all instars, 0.0317 g (2nd instar), 0.2206 g (3rd instar), 1.0788 g (4th instar), 4.0102 g (5th instar) and

8.9940 g (5 days of 5th instar), which was significantly different to other ecoraces. Moreover, this ecorace recorded the highest fresh cocoon weight (3.8016 g), pupa weight (3.2532 g), shell weight (0.5287 g), shell ratio (14.01%), fresh cocoon weight/10,000 larvae (38.01 kg), eggs/moth (531.13 eggs), total eggs (6,375.27 eggs) and total hatching eggs (6,006.13 eggs). Among the different observations recorded, especially highest survival rate and yields favoured the ecorace, SaKKU1 for further varietal improvement programme. In parallel, genetic relationship analysis of eri silkworm ecoraces using inter-simple sequence repeat (ISSR) technique was also carried out. The result revealed from dendrogram analysis that SaKKU1 was at the farthest distance than other ecoraces, especially against SaKKU3. Based on all above results, the SaKKU1 ecorace was considered to be the most suitable for heat tolerant variety improvement.

Correlation of larval parameters with economic traits of eri silkworm, *Samia c. ricini*

Sannappa *et al.*, (1999) studied the correlation of larval parameters with economic traits of eri silkworm, *Samia cynthia ricini* by rearing three breeds of eri silkworm *viz.*, White-Plain, Blue-Plain and White-Zebra on 6 castor genotypes, Aruna, RC-8, DCS-72, Local, SL-1 and PCS-121. Results of the experiment showed that the larval duration and mature larval weight were correlated with larval survival, effective rate of rearing, cocoon weight, pupal weight, shell weight, shell ratio, adult emergence, fecundity and hatchability. Hence it is concluded that increased larval duration and mature larval weight have an influence on rearing performance and cocoon and grainage parameters.

Rearing of Eri Silkworm Races on Castor Varieties

Interaction of Castor Genotypes and Breeds of Eri Silkworm in Relation to Cocoon and Grainage Parameters

The healthy growth of the eri silkworm and ultimately the economic traits such as, cocoon yield, shell weight and silk percentage are influenced greatly by the nutritional level of the leaves fed to the worms (Krishnaswami *et al.*, 1971). Pandey (1995) also opined that the nutritional value of the leaf produced is influenced by several factors such as variety cultivated, agronomic practices adopted, harvesting and preservation of leaf, abiotic and biotic stress factors. It has also been stated that the silk ratio varied with the type of host and breed used in eri silkworm rearing. In this context, Sannappa and Jayaramaiah (1999) have undertaken a study to identify castor genotypes and breeds of eri silkworm suitable for maximizing the cocoon and egg production.

Three breeds of eri silkworm *viz.,* White plain, Blue plain and white zebra were reared on six castor genotypes *viz.,* Aruna, RC-8, Local, DCS-72, SL-1 and PCS-121 grown during 1996 at GKVK farm of UAS, Bangalore. The observations such as larval duration, effective rate of rearing, cocoon weight, pupal weight, shell weight, shell ratio, sex ratio, moth emergence, normal moth formation, fecundity and hatchability were recorded. The results showed that the feeding of leaves of RC-8 genotype to worms resulted in longer larval duration (22.26 days), closely followed by Aruna (22.21 days) (Table 9.9). Among the breeds, white zebra recorded longer larval duration (20.84 days). In interaction of genotypes and breeds, white zebra fed on RC-8 recorded longer larval duration (22.77 days). Feeding leaves of RC-8 genotype to worms resulted in higher cocoon weight (2.234g) and pupal weight (1.873g). However, the shell weight (0.339g) and shell ratio (15.89%) were higher in cocoons formed from the worms fed on Aruna genotype. The white plain breed of eri silkworm recorded higher cocoon weight (1.967g), shell weight (0.277g) and shell ratio (13.97%) irrespective of the genotype fed to the worms. In interaction of breeds and genotypes, the cocoons of white plain breed formed from the worms fed on RC-8 (2.272g) and Aruna were heavier with shell weight of 0.353g. The percentage of moth emergence (99.90) and normal moths' formation (99.43) were higher in case of cocoons formed from the worms fed on Aruna genotype. The fecundity was also higher in the moths which resulted from the cocoons spun by the worms fed on Aruna (361.89). The above parameters were on par when the worms fed on RC-8 genotype (99.90%, 99.11 and 353.22 respectively). The hatchability was maximum in the eggs laid by the moths emerged from the cocoons spun by worms which had Aruna leaves (98.94%) as food. Among the breeds, the white plain showed higher normal moths formation (98.66%) and also hatchability of eggs (97.68%). The breed blue-plain (98.03 and 96.97%) was on par with the former. From the study of Sannappa and Jayaramaiah (1999) it may be inferred that castor genotypes Aruna and RC-8 could be fed to white plain eri silkworm breed to maximise the cocoon yield and eri eggs production.

Table 9.9: Mean performance of Parental breeds of eri silkworm utilized in diallel crosses (Singh *et al.,* 2012a).

Parental Breed	Fecundity (No.)	Hatching (%)	Larval weight	Cocoon weight (g)		Shell weight (g)		Shell (%)		ASY
				Male	Female	Male	Female	Male	Female	
YP	473	95.07	6.30	3.16	3.87	0.51	0.53	16.25	13.71	210.59
GBP	372	88.76	5.92	3.06	4.00	0.48	0.57	16.24	13.49	161.57
YZ	501	93.73	6.19	3.11	3.85	0.51	0.55	16.29	14.21	225.63
GBZ	433	91.39	5.92	2.81	3.55	0.43	0.48	15.19	13.59	143.53
YS	444	88.85	5.30	2.73	3.46	0.43	0.49	15.62	14.25	159.51
GBS	384	92.93	5.48	2.59	2.84	0.41	0.44	15.19	13.85	135.73
Mean	435	91.79	5.85	2.91	3.60	0.46	0.51	15.80	13.85	172.76
SD	99.54	4.98	0.42	0.236	0.457	0.048	0.052	0.63	0.57	47.53
CV(%)	22.89	5.43	7.08	8.09	12.70	10.41	10.15	3.97	4.14	27.51

Table 9.10: Effect of Castor genotypes on rearing, cocoon and grainage parameters in different eri silkworm breeds (Sannappa and Jayaramaiah, 1999)

Treatments	Larval duration (days)	ERR (%)	Cocoon weight (g)	Pupal weight (g)	Shell weight (g)	Shell ratio (%)	Sex ratio (M:F)	Moth emergence (%)	Normal moths (%)	Fecundity	Hatchability (%)
Genotypes											
Aruna	22.21	90.83(72.37)	2.132	1.793	0.339	15.89(23.48)	1.107	99.90(87.44)	99.43(85.74)	361.89	98.94(84.88)
RC-8	22.26	89.90(71.49)	2.234	1.873	0.338	15.12(22.87)	1.044	99.90(87.44)	99.11(84.89)	352.22	99.17(86.16)
DCS-72	20.03	84.72(66.95)	1.849	1.604	0.245	13.23(21.32)	1.089	98.74(83.80)	99.06(84.99)	303.89	98.09(82.32)
Local	20.00	84.70(66.93)	1.877	1.629	0.248	13.22(21.31)	1.162	98.47(83.19)	98.21(82.52)	312.78	97.48(81.22)
SL-1	19.59	80.88(64.05)	1.814	1.586	0.227	12.52(20.71)	1.062	94.39(79.88)	97.49(81.07)	263.78	95.27(77.38)
PCS-121	19.62	79.88(63.31)	1.784	1.566	0.218	12.22(20.45)	1.200	94.39(76.36)	95.13(77.19)	252.11	92.00(73.61)
CD at 5%	0.146	-	0.007	0.026	0.004	0.120	-	2.232	2.485	43.04	2.093
Breeds											
White-Plain	20.34	86.72(68.84)	1.967	1.679	0.277	13.97(21.91)	1.194	98.66(84.14)	98.66(84.14)	314.56	97.68(82.37)
Blue-Plain	20.67	85.59(67.85)	1.946	1.678	0.269	13.70(21.69)	1.122	98.03(82.65)	98.03(82.65)	307.56	96.97(81.02)
White-Zebra	20.84	83.14(65.86)	1.932	1.670	0.262	13.44(21.48)	1.116	97.52(81.52)	97.52(81.52)	301.22	95.83(79.39)
CD at 5%	0.104	-	0.005	-	0.003	0.085	-	1.757	1.757	-	1.480

Leaf anatomical characters of castor genotypes and their impact on economic parameters of White-Plain breed of eri silkworm

Sannappa and Jayaramaiah, (2001) reared the White-Plain breed of eri silkworm (*S. cynthia ricini*) on leaves of 6 castor bean cultivars *viz.,* Aruna, RC-8, DCS-72, Local, SL-1 and PCS-121 to determine the most suitable castor cultivar for eri silkworm rearing. Alongside this, the leaf anatomy of the castor cultivars was determined. Aruna, PCS-121 and DCS-72 recorded the highest stomatal frequency (16.83/mm), length (4.47 µM) and breadth (2.73 µM), respectively. SL-1 recorded the highest length (102.00 µM) and breadth (115.00 µM) of vascular bundle, whereas Local recorded the highest total wax layer (1.73 µM), parenchymatous tissue (13.35 µM) and leaf thickness (16.487 µM). Negative significant relationship was established between the anatomical features of castor leaves such as vascular bundle length and breadth, and wax and epidermal layers and economic parameters of eri silkworm such as feeding and moulting duration; larval, cocoon and shell weight; shell ratio; fecundity and hatchability.

Influence of castor varieties on rearing and grainage performance of different breeds of eri silkworm, *Samia cynthia ricini*

Rearing and grainage performance of different breeds of eri silkworm, may differ on different varieties of host plant, castor. In this context, Ramakrishna Naika *et al.,* (2003) conducted studies to determine the suitable castor cultivar (between Aruna and Local cultivars) for feeding worms of four eri silkworm breeds (White-Plain, Blue-Plain, White Zebra and White Semi-Zebra) based on their rearing and grainage parameters. Eri silkworm breed White semi-Zebra fed with leaves of Local (with non-bloomy green petioles) castor cultivar registered the highest, mature larval weight (72.88 g/10 larvae). White-Plain breed reared on Local variety recorded the least larval duration (923.2 days), and the highest effective rate of rearing (93.2%), cocoon weight (3.02 g), shell weight (0.326 g), pupal weight (2.69 g) and fecundity (361.7 eggs per laying). Shell ratio was highest with the combination of Aruna × Blue-Plain (11.20%), while the moth emergence (i.e. egg hatching) percentage was higher with the Local × White-Plain (99.31%) and Local × White Semi-Zebra breeds (99.31%).

Rearing performance of different strains of eri silkworm, *Samia cynthia ricini* on different castor genotypes

Lakshmi Narayanamma and Dharma Reddy (2013) conducted a study to evaluate the rearing performance of brick red (Kokrajhar) (Plate 39) and plain white (Ambagoan) (Plate 40) strains of Eri silkworm during *Kharif* and *Rabi* seasons of 2012-13 on nine castor genotypes *viz.,* Haritha, Kranthi, Kiran, DPC-9, PCS-262, PCH-111, PCH-222, GCH-4, DCH-177 along with control (using the mixed leaf) in the Ericulture laboratory of Regional Agricultural Research Station, Palem. The rearing performance of Eri silkworms was analyzed based on their larval weight, larval duration, pupal duration, effective rate of rearing (ERR) (%), cocoon weight, rate of pupation, pupal weight, shell weight, shell ratio, moth emergence, fecundity and hatchability (%). Brick red strains showed better performance of rearing, cocoon and grainage parameters compared to plain white strains except for fecundity (Tables 9.13, 9.14 and 9.15). Brick red strains of Eri silkworms nourished with leaves of PCH-111 (92.7%) and PCH-222 (92.23%) registered significantly higher effective rate of rearing (ERR), while it was lower with kranthi (88.35%), DCH-177 (88.5%) and control (88.29%). Both the strains of Eri worms fed on leaves of PCH-111 were found superior in respect of cocoon weight (2.61 and 2.24g respectively), shell weight (0.981 and 0.586 g respectively) and shell ratio (37.56 and 26.15% respectively) closely followed by GCH-4 and PCH-222. Brick red strains of eri silkworms nourished with leaves of PCH-111 (99.9%), PCH-222 (98.74%) and GCH-4 (99.9%) recorded significantly higher moth emergence, while it was lower with PCS-262 (92%), whereas plain white strains fed with leaves of PCH-111 (97.6%) and GCH-4 (98%) showed higher moth emergence. Though the fecundity of the plain white strains fed with leaves of Haritha (305.3), Kiran (361.89), DPC-9 (352.22), PCS-262 (303.89) and DCH-177 (318.48) was higher, but the egg hatchability percentage was more with brick red strains. The results of Lakshmi Narayanamma and Dharma Reddy (2013) study inferred that the eri silkworm strain particularly Kokrajhar race reared on the foliage of PCH-111 castor genotype showed better larval weight, effective rate of rearing, cocoon weight, shell weight and higher fecundity with shorter larval duration. Hence rearing of Kokrajhar strain on PCH-111 genotype is suitable to reap higher returns in Telangana region.

Plate 39 Brick red cocoons

Plate 40 Plain white cocoons

Table 9.11: Rearing parameters of two Eri silkworm, *Samia cynthia ricini* brick red (Kokrajhar) and plain white (Ambagoan) strains on selected castor genotypes during *Kharif* and *Rabi* 2012-13 (Lakshmi Narayanmma and Dharma Reddy, 2013)

Genotype	Larval weight (10th day (g))		Mature larval weight (g)		Larval duration (days)		Pupal duration (days)		ERR (%)	
	Brick Red	Plain White	Brick Red	Plain White	Brick Red	Plain White	Brick Red	Plain White	Brick Red	Plain White
Haritha	0.621[ab]	0.584[a]	5.482[b]	4.682[c]	20.21[c]	21.64[c]	18.12	18.02	91.46[bc] (73.01)	84.72[b] (66.95)
Kranthi	0.525[d]	0.490[bc]	5.252[c]	4.752[c]	20.69[e]	22.18[e]	18.06	18.00	88.35[e] (70.05)	84.7[b] (66.93)
Kiran	0.514[d]	0.475[c]	5.380[bc]	4.421[d]	20.90[f]	22.26[ef]	18.09	17.95	89.94[d] (71.51)	80.12[d] (64.01)
DPC-9	0.560[c]	0.458[cd]	5.160[d]	4.364[de]	21.2[g]	22.38[f]	17.98	18.09	90.54[cd] (72.09)	85.21[b] (65.07)
PCS-262	0.575[bc]	0.518[b]	4.891[e]	4.482[d]	21.18[g]	21.98[d]	18.18	18.06	90.83[c] (72.37)	90.83[a] (72.37)
PCH-111	0.645[a]	0.598[a]	6.012[a]	5.162[a]	19.33[a]	19.84[a]	18.21	18.09	92.70[a] (74.32)	89.9[a] (71.49)
PCH-222	0.651[a]	0.601[a]	5.982[a]	4.962[b]	19.54[b]	20.98[b]	18.01	18.01	92.23[ab] (73.81)	80.88[d] (64.05)
GCH-4	0.639[a]	0.610[a]	5.948[a]	5.011[b]	19.6[b]	20.94[b]	17.94	17.98	90.83[c] (72.37)	82.24[c] (65.21)
DCH-177	0.529[cd]	0.445[d]	5.451[b]	4.681[c]	20.45[d]	21.04[b]	18.08	18.07	88.5[e] (70.18)	80.12[d] (64.01)
Control	0.518[d]	0.421[d]	5.252[cd]	4.325[e]	21.28[g]	21.89[d]	18.12	18.00	88.29[e] (70.01)	79.88[d] (63.31)
CD at 5%	0.048	0.036	0.142	0.120	0.158	0.149	-	-	1.081	1.035

Figures followed by the same letter did not differ significantly; Figures in the parenthesis are angular transformed values

Table 9.12: Cocoon parameters of two Eri silkworm, *Samia cynthia ricini* brick red (Kokrajhar) and plain white (Ambagoan)strains on selected castor genotypes during *Kharif* and *Rabi* 2012-13 (Lakshmi Narayanmma and Dharma Reddy, 2013)

Genotype	Cocoon weight (g)		Rate of pupation (%)		Pupal weight (g)		Shell weight (g)		Shell ratio (%)	
	Brick Red	Plain White	Brick Red	Plain White	Brick Red	Plain White	Brick Red	Plain White	Brick Red	Plain White
Haritha	2.012[e]	1.894[f]	97.67[b] (81.26)	96[b] (78.48)	1.609[c]	1.636[b]	0.403[fg]	0.258[e]	20.03[g]	13.62[e]
Kranthi	1.984[f]	1.794[i]	95[d] (77.12)	96[b] (78.48)	1.526[a]	1.598[a]	0.458[e]	0.196[f]	23.08[e]	10.93[f]
Kiran	1.990[f]	1.828[h]	95[d] (77.12)	95[c] (77.12)	1.604[c]	1.634[b]	0.386[h]	0.194[f]	19.40[h]	10.61[g]
DPC-9	2.014[e]	1.921[e]	96[c] (78.52)	95[c] (77.12)	1.701[e]	1.764[d]	0.313[i]	0.157[h]	15.54[i]	8.17[i]
PCS-262	2.018[e]	2.001[d]	96[c] (78.52)	96[b] (78.48)	1.628[c]	1.652[b]	0.390[gh]	0.349[d]	19.33[h]	17.44[d]
PCH-111	2.612[a]	2.241[a]	98.68[a] (83.46)	97.67[a] (81.26)	1.631[c]	1.655[b]	0.981[a]	0.586[a]	37.56[a]	26.15[a]
PCH-222	2.415[c]	2.189[c]	97.14[b] (81.02)	96.18[b] (78.50)	1.686[de]	1.698[c]	0.729[c]	0.491[b]	30.19[c]	22.43[b]
GCH-4	2.484[b]	2.208[b]	96.98[b] (79.94)	96.21[b] (78.50)	1.706[e]	1.805[d]	0.778[b]	0.403[c]	31.32[b]	18.25[c]
DCH-177	2.212[d]	2.010[d]	94[e] (75.24)	94[d] (75.24)	1.667[d]	1.838[e]	0.545[d]	0.172[g]	24.64[d]	8.56[h]
Control	1.985[f]	1.871[g]	95[d] (77.12)	94[d] (75.24)	1.567[b]	1.614[a]	0.418[f]	0.257[e]	21.06[f]	13.74[e]
CD at 5%	0.012	0.014	0.692	0.698	0.032	0.029	0.016	0.018	0.168	0.159

Figures followed by the same letter did not differ significantly; Figures in the parentheses are angular transformed values

Table 9.13: Grainage parameters of two Eri silkworm, *Samia cynthia ricini* brick red (Kokrajhar) and plain white (Ambagoan) strains on selected castor genotypes during *Kharif* and *Rabi* 2012-13 (Lakshmi Narayanmma and Dharma Reddy, 2013)

Genotype	Sex ratio (M:F)		Moth emergence (%)		Normal moths (%)		Fecundity		Hatchability (%)	
	Brick Red	Plain White	Brick Red	Plain White	Brick Red	Plain White	Brick Red	Plain White	Brick Red	Plain White
Haritha	1:108	1:104	94.39^c (76.36)	93.50^{bc} (74.41)	98.21 (82.52)	96.00 (78.52)	300^b	305.3^b	98.6 (83.21)	96.9 (79.86)
Kranthi	1:069	1:085	93.25^{cd} (74.32)	94.00^b (75.24)	95.13 (77.19)	94.18 (75.25)	325.3^a	312.89^a	98.6 (83.21)	98.3 (82.51)
Kiran	1:101	1:121	94.00^c (75.24)	92.00^c (73.21)	96.42 (80.12)	95.28 (77.21)	325.6^a	361.89^a	98.1 (82.08)	97.48 (81.22)
DPC-9	1:112	1:108	94.00^c (75.24)	93.50^{bc} (74.41)	95.00 (77.12)	94.00 (75.24)	305.3^b	352.22^a	98.5 (82.97)	92.0 (73.61)
PCS-262	1:106	1:106	92.00^d (73.21)	92.00^c (73.21)	94.39 (79.88)	92.00 (73.21)	290.0^b	303.89^{bc}	98.0 (81.87)	92.0 (73.61)
PCH-111	1:160	1:145	99.90^a (87.44)	97.69^a (81.26)	99.43 (85.74)	96.58 (80.20)	341.0^a	312.78^{ab}	99.17 (86.16)	96.9 (79.86)
PCH-222	1:125	1:121	98.74^{ab} (83.80)	95.20^b (77.14)	99.06 (84.99)	95.14 (77.19)	320.0^a	263.78^{cd}	98.9 (83.98)	95.27 (77.38)
GCH-4	1:131	1:062	99.90^a (87.44)	98.00^a (81.87)	99.11 (84.89)	96.00 (78.48)	302.0^b	252.11^d	98.0 (81.87)	96.9 (79.86)
DCH-177	1:124	1:164	96.48^b (80.14)	95.28^b (77.19)	95.13 (77.19)	94.39 (75.16)	315.0^b	318.48^a	98.3 (82.51)	95.83 (78.22)
Control	1:101	1:121	94.39^c (79.88)	92.80^c (73.20)	96.72 (80.21)	93.50 (74.41)	327.67^a	325.33^a	96.9 (79.86)	94.28 (75.14)
CD at 5%	-	-	2.261	2.169	-	-	35.01	42.18	-	-

Figures followed by the same letter did not differ significantly Figures in the parenthesis are angular transformed values

Feed utilization efficiency of white plain eri-silkworm breed on eight castor genotypes in Ethiopia

In Ethiopia, rearing of eri and mulberry silkworm is practiced sparsely. However, eri silkworm rearing for eri silk production is better practiced compared to the mulberry one. In general, the quality of feed given to insects generally affects the economic products of insects. Therefore, the suitability of feed can be determined through estimation of rate of ingestion, digestibility, conversion efficiency of food and growth rate of the insect. So, the quantitative nutritional approach consists of measuring the amount of food consumed, digested and assimilated, excreted, metabolized, and converted into biomass is of high importance together with qualitative approaches. In addition, identifying the factors that contribute more to how organisms respond to different foods and food components is highly essential. However, very little information is available in Ethiopia. Hence, Kedir Shifa *et al.,* (2014) made an attempt to assess the influence of castor genotypes on feeding efficiency of eri silkworm, *S. c. ricini* and to find out promising castor genotype (s) for better eri silk productivity at Melkassa Agricultural Research Center (MARC), Ethiopia. A white plain eri-silkworm breed was used for this experiment and it was reared on the eight castor genotypes namely Abaro, Acc 106584, Acc 203241, Acc 208624, Ar sel, Bako, Hiruy and local check. This breed was reared following cellular rearing techniques starting from brushing till cocoon spinning on the eight genotypes. As used by Waldbauer (1968), to find out feeding efficiency of the breed on the castor genotypes, the food offered were weighed daily for all replicated treatments (50 worms per feeding tray). However, a parallel sample of food (according to specific dietary routine) was kept separately to know the amount of food ingested by splitting leaves along the midrib. That is, one half weighed and offered to the larvae and the other half weighed and used to determine the percent dry matter in the leaf material. According to Reddy and Swamy (1999), after twenty four hours of feeding, the uneaten food and pellets/faecal matter in all the replicates were collected separately and dried at 75°C to a constant weight. The amount of food ingested was determined by deducting the weight of uneaten food from the weight of parallel sample. The dry weight attained by the larvae was also determined by drying a single larvae per day, one from each replicate and the larvae that is picked up from the replicates was again replaced (of the same age and weight) from the parallel culture. From the data obtained *viz.,* ingesta, digesta, excreta, Approximate Digestibility (AD), Reference Ratio (RR), Relative Consumption Indices (CI), Relative Growth Rate

(RGR), Efficiency of Conversion of Ingesta (ECI) and Digesta (ECD) for larva and cocoon were calculated as per the gravimetric method.

The castor genotypes showed wide variation in feeding efficiency of eri silkworms when leaves of these genotypes were used as a feed source. Analysis of feeding efficiency of eri silkworms fed on castor genotypes was carried out by using the 5th instar larva. This is because 97 % of the total food intake in silkworms is made during the last two instars. Among castor genotypes, Abaro and Acc 208624 expressed better performance in all evaluated variables simultaneously. These genotypes yielded 6.672 and 6.904 g/larva of ingesta, 3.379 and 3.574 g/larva of digesta, 50.637 and 51.766 % approximate digestibility, 2.026 and 2.073 reference ratio, 0.814 and 0.862 relative consumption rate, 33.164 and 31.406 % efficiency to convert ingested leaves to larval biomass, 19.811 and 19.236 % efficiency to convert ingested food to cocoon as well as 39.126 and 37.173 % efficiency to convert digested food to cocoon, respectively (Tables 9.11 and 9.12). On the other hand, local genotype showed lower conversion efficiency to larval biomass and cocoon weight though it was good enough in other parameters. These differences were because of the quality of leaf fed has a great influence on the amount ingested food due to the physico-chemical characteristics in relation to the age of the leaf and the characteristics of the genotypes (Legay, 1958 and Friend, 1958).

The experiment results of Kedir Shifa *et al.,* (2014) indicated that selection and/or evaluation of castor genotypes based up on feed utilization efficiency of eri silkworms is very essential to achieve better eri silk productivity. Final analysis of feed utilization efficiency of eri silkworms from different castor genotypes revealed that genotypes Abaro and Acc 208624 as superior genotypes compared to others. As a result, these genotypes will be recommended for eri silkworm research and development works in the future in Ethiopia.

Table 9.14: Effect of castor genotypes on the amount of eri silkworm ingestion, digestion and excretion (Kedir Shifa *et al.,* 2014)

Castor genotypes	Ingesta	Digesta	Excreta
Abaro	6.6717[abc]	3.3783[abc]	3.29333[c]
Acc 106584	6.6760[abc]	3.5893[ab]	3.08667[d]
Acc 203241	6.6170[bc]	3.1770[bc]	3.44000[b]
Acc 208624	6.9043[ab]	3.5743[ab]	3.33000[bc]
Ar sel	6.3707[cd]	3.0640[c]	3.30667[c]
Bako	5.9250[d]	2.5983[d]	3.32667[bc]
Hiruy	7.1423[a]	3.4290[abc]	3.71333[a]
Local check	7.0873[ab]	3.6607[a]	3.42667[b]
SE	0.1711	0.1532	0.0395
CV	4.4411	8.0188	2.0344

Table 9.15: Influence of castor genotypes on efficiency of conversion of ingesta and digesta to larval biomass and cocoon by eri silkworm (Kedir Shifa *et al.,* 2014)

Castor genotypes	ECI to larva	ECD to larva	ECI to cocoon	ECD to cocoon
Abaro	33.164[abc]	63.189[abc]	19.8110[a]	39.126[ab]
Acc 106584	28.950[d]	58.194[bc]	17.1543[bc]	31.989[de]
Acc 203241	33.898[ab]	67.837[abc]	16.7040[bc]	35.109[bcde]
Acc 208624	31.406[bdc]	64.649[abc]	19.2360[a]	37.173[bc]
Ar sel	34.728[a]	75.953[a]	17.5507[b]	36.493[bcd]
Bako	35.191[a]	72.213[ab]	18.8970[a]	43.136[a]
Hiruy	31.273[bcd]	65.898[abc]	15.8333[c]	33.032[cde]
Local check	30.386[cd]	55.638[c]	16.0640[c]	31.30[e]
SE	0.9941	4.8026	0.4423	1.5229
CV	5.3182	12.7102	4.339	7.3434

Studies on Adaptability of Eri Silkworm Strains on Different Host Plants

Impact of feeding of *Samia cynthia ricini* (red variety) in respect of larval growth and spinning

The N.E. region of India is one of the ten bio-geographic regions of India harbouring varied flora and fauna including silk producing insects. A separate eco-race of eri silkworm is confined to a small geographical pocket of Assam i.e. Kokrajhar which produced red colour cocoon. The colour sustained in successive generations at Kokrajhar area which has been segregated when shifted from that particular region. Although many workers reported feeding habit of eri silkworm, study on the red coloured eri silkworm is very scanty. Therefore these investigations were carried out by Deka *et al.,* (2011) to know the rearing performance of eri silkworm (red variety) on different types of food plants *viz.,* castor, kesseru and tapioca during 2008 to 2010 for 12 generations with different treatments at Government Eri Seed Grainage, Adabari, Kokrajhar, Assam for selection of food plants for better production and productivity.

The experiment was conducted with nine treatments

T1 : Castor (first to fifth instar)

T2 : Kesseru (first to fifth instar)

T3 : Tapioca (first to fifth instar)

T4 : Castor (first to third instar) + Kesseru (fourth to fifth instar)

T5 : Castor (first to third instar) + Tapioca (fourth to fifth instar)

T6 : Kesseru (first to third instar) + Castor (fourth to fifth instar)

T7 : Kesseru (first to third instar) + Tapioca (fourth to fifth instar)

T8 : Tapioca (first to third instar) + Castor (fourth to fifth instar)

T9: Tapioca (first to third instar) + Kesseru (fourth to fifth instar)

Observations were recorded on larval duration, weight of full grown larva, weight of mature larva, ERR (%), Fecundity and cocoon parameters like colour of the cocoon, single cocoon weight, single shell weight, silk ratio (%). The evaluation of data revealed that castor (red variety) has shown supremacy over the other food plants i.e. kesseru and tapioca for larval growth as well as spinning characters.

Performance of C_2 breed of eri silkworm, *Samia ricini* (Donovan) in different host plants

Very recently, Central Muga Eri Research Training Institute, Lahdoigarh, Jorhat, Assam has developed a high productive eri silkworm breed through hybridization programme named as C_2 with higher shell weight and fecundity at its subordinate unit Regional Eri Research Station, Mendipathar. The performance of C_2 breed observed during hybrid authorization programme is presented in the Table 9.16. Rearing performance of high yielding eri silkworm breed 'C_2' has been studied on kesseru *Heteropanax fragrans*, Borkesseru *Ailanthus excelsa* and Borpat (*Ailanthus grandis*) leaves. High yielding castor variety, Acc 003 later named as NBR-2 was taken as control. Young instar worms were fed with castor and late instars worms during 4th and 5th stage fed with perennial food plants. The control was maintained by feeding with castor throughout the rearing period. Rearing characters *viz.,* Larval wt. (g), Larval duration (days), Cocoon wt (g), Shell wt (g), Shell ratio (SR %) and Effective rate of rearing (ERR %) were recorded during the rearing period. The results revealed that the breed showed its best performance in castor followed by kesseru. It is already reported that castor is the primary food plant of eri silkworm and Kesseru is used as the best alternative food plant. Further, leaf from perennial species can be best utilized for a successful rearing operation. It then becomes economical with a steady source of income to a family. To overcome the seasonality issue of the castor, Kesseru and *Ailanthus* may be the better option as a perennial host plants. In the study both *Ailanthus* species showed impressive performance in eri rearing except longer larval duration and little lower economic characters. During the study all food plants showed positive economic character in rearing performance (Table 9.17) (Sarmah *et al.,* 2015).

Table 9.16: Performance of Eri silkworm C_2 breed at Regional Eri Research Station, Mendipathar (Sarmah *et al.*, 2015)

Sl. No.	Particulars	Norms fixed by Hybrid Authorization Committee of Central Silk Board	Local eco-races	C_2 breed
1	Fecundity (No)	350	322	356
2	Hatching (%)	85	79.65	83.91
3	Cocoon yield by number/dfl	250	203	247
4	Cocoon yield by weight/dfl (kg)	0.750	0.590	0.900
5	ERR (%)	85	79.38	84.02
6	Single Cocoon weight (g)	3.0	2.89	3.67
7	Single Shell weight (g)	0.45	0.38	0.54
8	Cocoon shell ratio (%)	14.00	13.09	14.80

Barkesseru (*Ailanthus excelsa* Roxb.), an ideal substitute for castor to rear eri silkworm

Narayanaswamy *et al.*, (2006) undertaken an experiment, to study barkesseru as an ideal substitute for castor to rear eri silkworm during fifth instar which is evaluated based on amylase activity and its relationship with consumption indices and economic parameters. The experiment was conducted with brick red cocoon breed and plain white cocoon breed. The bulk population of these two breeds were raised on castor leaves till fourth instar. Immediately after the fourth moult the larvae of both the breeds were provided with leaves of castor, tapioca and barkesseru. Three replications of 100 larvae each were maintained for every host plant. The digestive juice and hemolymph were collected separately, breed wise and host plant wise from day one till the onset of spinning. The consumption indices were computed gravimetrically and observations such as mature larval weight, cocoon weight, pupal weight and shell ratio were recorded.

Among the host plants, the activity of digestive amylase and hemolymph amylase was maximum when the larvae were fed with leaves of castor (21.83±2.48 to 143.09±5.64 mg/ml per hour) followed by barkesseru (21.82 to 140.05 mg/ml per hour), while it was minimum in tapioca (17.57 to 135.15 mg/ml per hour). It is already known that secretion of digestive enzymes in some insects is influenced by the quantity of food consumed and also is in response to the specific substrate in the food (Ishaaya *et al.*, 1971). These variations might be due to nutritional status of the host plants.

Table 9.17: Performance of eri silkworm C2 breed on different host plants at Regional Eri Research Station, Mendipathar (Sarnah *et al.,* 2015)

Food plants	No. of dfls reared	Fecundity (No.)	Hatching (%)	Weight of mature larvae (g)	Larval duration (days)	Pupation (%)	Cocoon yield / dfl (by No.)	Total cocoon harvested (No.)	ERR (%)	Single cocoon weight (g)	Single shell weight (g)	Shell ratio (%)
Kesseru, *H. fragrans*	20	355	85	8.45	22	92	235	4700	77.87	4.05	0.51	12.59
Borpat, *Ailanthus Grandis*	20	355	85	8.50	28	90	226	4520	74.89	4.06	0.56	13.79
Borkesseru, *Ailanthus excels*	20	355	85	8.50	28	90	220	4400	72.90	3.93	0.50	12.72
Castor, *Ricinus communis* (control)	20	355	85	9.50	20	95	265	5300	87.82	4.12	0.61	14.82
Mean	20	355	85	8.74	24.50	91.75	236.50	4730	78.37	4.04	0.55	13.48
C. Level (95.0%)	0.00	0.00	0.00	0.81	6.56	3.76	31.78	635.69	10.54	0.13	0.08	1.66

Among the eri silkworm breeds, the brick red cocoon breed had significantly higher consumption (791.62 g/50 larvae), digestion (502.70 g/50 larvae), approximate digestibility (67.19%) and reference ratio (3.45), but the consumption index (0.95) and efficiency of conversion of digested food (32.38%) were significantly maximum in plain white cocoon breed. However, the growth rate and efficiency of conversion of ingested food were non-significant among breeds (Table 9.18). Among the different hosts, worms reared on castor registered significantly greater values for digestion (505.57 g/50 larvae), growth rate (0.19), approximate digestibility (74.65%), efficiency of conversion of ingested food (24.63%) and efficiency of conversion of digested food (33.55%). With tapioca, though the consumption was maximum (934.89 g/50 larvae) all other indices were significantly lower. In the interaction between breeds and hosts, all the consumption indices were found to be significantly different. The consumption was higher when the larvae of brick red cocoon breed were reared on tapioca (970.38 g/50 larvae), but it was minimum in plain white cocoon breed fed with castor (647.31 g/50 larvae).

As the age advances the activity of amylase in both digestive juice and hemolymph increased steeply from first to fourth day of fifth instar followed by a decrease towards the onset of spinning. Similar trend was observed in respect of consumption, digestion and approximate digestibility. Thus a positive correlation was observed between the activity of digestive and hemolymph amylase with consumption (r=0.6869 and 0.8474 respectively), digestion (r=0.8631 and 0.9308 respectively) and approximate digestibility (r=0.7553 and 0.6928 respectively) (Table 9.19). However, the activity of both the amylases had inverse relationship with the consumption index, growth rate, efficiency of conversion of ingested food and efficiency of conversion of digested food, when the larvae were reared on castor.

In growth and economic parameters, brick red cocoon breed registered maximum mature larval weight (4.54 g), cocoon weight (2.090 g), pupal weight (1.792 g) and shell ratio (12.01%) compared to plain white cocoon breed. In host plants, the eri silkworm breeds reared on castor and barkesseru exhibited maximum mature larval weight, cocoon and pupal weight and shell ratio. Among the interactions, the larvae of brick red cocoon breed reared on castor showed better performance (Table 9.20). However, most of the economic parameters were at par with each other when the larvae of *S. c. ricini* were reared on either castor or barkesseru during fifth instar.

Table 9.18: Consumption indices during fifth instar among eri silkworm breeds as influenced by host plants (Narayanaswamy *et al.,* 2006)

Breeds/Host plants/Interactions	Consumption (g/50 larvae)	Digestion (g/50 larvae)	Consumption index	Growth rate	AD (%)	ECI (%)	ECD (%)	RR
Eri silkworm breeds								
Brick red cocoon breed (BR)	791.62	502.70	0.85	0.16	67.19	20.01	30.75	3.45
Plain white cocoon breed	731.79	434.16	0.95	0.17	58.93	19.29	32.38	3.72
(PW)	6.13	4.50	0.02	0.04	0.69	0.02	0.40	0.06
SE±	18.91	13.88	0.06	-	2.14	-	1.25	0.18
CD at 5%								
Host plants								
Castor (C)	669.20	505.57	0.75	0.19	74.65	24.63	33.55	3.77
Tapioca (T)	934.89	428.99	0.94	0.14	45.82	15.07	32.87	1.82
Barkesseru (B)	681.02	470.85	0.99	0.19	68.70	19.26	28.25	5.16
SE±	7.51	5.51	0.02	0.004	0.85	0.37	0.49	0.07
CD at 5%	23.16	17.00	0.08	0.012	2.62	1.16	1.53	0.22
Interactions								
BR X C	690.80	541.70	0.75	0.17	76.82	23.17	31.16	3.92
BR X T	970.38	453.20	0.83	0.13	46.65	16.11	34.53	1.87
BR X B	713.67	513.20	0.95	0.18	78.10	20.76	26.56	4.56
PW X C	647.61	469.44	0.75	0.20	72.48	26.08	35.99	3.63
PW X T	899.40	404.59	1.06	0.15	45.00	14.03	31.20	1.78
PW X B	648.37	428.47	1.03	0.19	69.30	17.16	29.94	5.75
SE±	7.67	7.80	0.03	0.005	1.20	0.53	0.70	0.10
CD at 5%	23.01	24.04	0.11	0.018	3.70	1.65	2.16	0.32

AD: Approximate digestibility, ECI: Efficiency of conversion of ingested food,

ECD : Efficiency of conversion of digested food, RR: Reference ratio

Table 9.19: Correlation between the activity of digestive and hemolymph amylases with consumption indices in eri silkworm as identified by host plants (Narayanaswamy *et al.*, 2006)

Host plants	Amylase activity	Consumption	Consumption indices						
			Digestion	CI	GR	AD	ECI	ECD	RR
Castor	Digestive juice	0.6869[*]	0.8631[*]	-0.5310[NS]	-0.4341[NS]	0.7553[*]	-0.1704[NS]	-0.4172[NS]	0.8876[*]
	Hemolymph	0.8474[*]	0.9308[*]	-0.7080[*]	-0.6016[*]	0.6928[*]	-0.02791[NS]	-0.3256[NS]	0.7898[*]
Tapioca	Digestive juice	0.4419[NS]	0.8145[*]	-0.5205[*]	-0.4789[NS]	-0.2395[NS]	0.1614[NS]	0.3303[NS]	-0.3763[NS]
	Hemolymph	0.6009[*]	0.9107[*]	-0.6989[*]	-0.6470[*]	-0.3986[NS]	0.1654[NS]	0.4327[NS]	-0.5581[*]
Barkesseru	Digestive juice	0.6235[*]	0.8929[*]	-0.5088[NS]	-0.4768[NS]	0.9135[*]	-0.0520[NS]	-0.3838[NS]	-0.4670[NS]
	Hemolymph	0.8255[*]	0.9791[*]	-0.6821[*]	-0.6384[*]	0.8621[*]	-0.1396[NS]	-0.4564[NS]	0.8201[*]

Table 9.20: Growth and economic parameters of eri silkworm breeds as influenced by host plants (Narayanaswamy *et al.,* 2006)

Breeds/Host plants/Interactions	Mature larval weight (g)	Cocoon weight (g)	Pupal weight (g)	Shell ratio (%)
Eri silkworm breeds				
Brick red cocoon breed (BR)	4.54	2.090	1.792	12.01
Plain white cocoon breed (PW)	4.29	2.038	1.744	11.72
SE±	0.05	0.01	0.016	0.14
CD at 5%	0.16	-	-	-
Host plants				
Castor (C)	5.46	2.339	1.976	12.60
Tapioca (T)	3.38	1.779	1.541	11.06
Barkesseru (B)	4.42	2.074	1.787	11.94
SE±	0.06	0.023	0.029	0.17
CD at 5%	0.20	0.07	0.06	0.54
Interactions				
BR X C	5.77	2.354	1.998	12.64
BR X T	3.42	1.812	1.564	11.39
BR X B	4.44	2.104	1.815	12.01
PW X C	5.15	2.323	1.955	12.55
PW X T	3.33	1.746	1.519	10.73
PW X B	4.40	2.044	1.758	11.88
SE±	0.09	0.03	0.03	0.24
CD at 5%	0.28	-	-	-

Table 9.21: Correlation between amylase activity and some economic parameters of eri silkworm as identified by host plants (Narayanaswamy *et al.,* 2006)

Amylase activity	Larval weight	Cocoon weight	Pupal weight	Shell ratio
Digestive amylase	0.9703[*]	0.9631[*]	0.9585[*]	0.9583[*]
Hemolymph amylase	0.8785[*]	0.9028[*]	0.9147[*]	0.9522[*]

The activity of digestive amylase and hemolymph amylases were significantly and positively correlated with larval weight, cocoon weight, pupal weight and shell ratio (Table 9.21). Amarnatha and Narayanaswamy (2004) reported that the digestive amylase had significant positive relationship with larval, cocoon weight, shell weight and shell ratio in *Bombyx mori*. Further, the activity of digestive and hemolymph amylases was positively correlated with consumption, digestion and approximate digestibility, which might have influenced enhanced larval, cocoon and shell weights in *S. c. ricini*. Hence it is summarised that the activity of both digestive and hemolymph amylases, consumption indices and economic parameters were higher/ superior when the larvae of *S. c. ricini* were provided with castor or barkesseru leaves during fifth instar, indicating barkesseru as a good substitute for castor to rear eri silkworm during fifth instar.

Ecological Adaptability of Eri Silkworm Races to Different Seasons and Regions

Screening of superior ecoraces of eri silkworm *Samia cynthia ricini* for better economic traits in different seasons:

To screen the superior ecoraces of eri silkworm for better economic traits, Sarkar *et al.,* (2012) conducted experiment with four ecoraces *viz.,* Diphu, Borduar, Barpeta and Kokrajhar for eight generations during two consecutive years. The rearing and grainage characters of these ecoraces of eri silkworm were observed. Different grainage characters were found significantly higher such as moth emergence (97-98%), coupling realization (37-38%), fecundity (345-440 nos.) and hatching (80-89%) in respect of Diphu, Borduar and Barpeta ecoraces. Similarly, in respect of rearing characters also significantly higher hatching (89%), larval weight (8.30-9.40 g) and cocoon weight (3.00-4.74 g) were recorded in Diphu, Borduar and Barpeta ecoraces. Pupal weight (2.68-4.18 g), shell weight (0.39-0.56 g) and ERR (70-77%) was also found significantly higher in Diphu, Borduar, Barpeta and Kokrajhar ecoraces irrespective of different seasons. After the analysis of the growth and economic traits of different eri silkworm it was observed that Diphu, Borduar and Barpeta ecoraces are the most promising eri silkworm ecoraces. These screened out ecoraces may be recommended for commercial exploitation.

Studies on cocoon biometry of Bangalore local eri silkworm race reared in Assam condition

The commercial characteristics of eri cocoons *viz.,* cocoon weight, shell weight, shell thickness, silk ratio directly influences the quality of cocoon production. All the characteristics have direct relationship with pricing, marketing structure and fabric quality of eri silk. Since commercial characteristics of eri cocoons vary according to the dwelling micro climate environment, biometrical study of eri cocoon is of greater importance under different ecological conditions. Nayak *et al.,* 1998 and Nayak and

Guru, 1998 studied biometric test for tasar cocoon, but for eri cocoon the information is lacking. Keeping these points in view research on biometry of cocoon of Bangalore eri silkworm race under Assam condition was undertaken by Bordoloi *et al.,* (2004) to study the economic characters of eri silk cocoons. The experiment was conducted in the department of Sericulture, Assam Agricultural University during July-August, 2002. The larvae were reared exclusively on the castor leaves, the produced cocoons of two different colours. Average temperature (maximum and minimum) and relative humidity (morning and evening) during rearing period ranged as $32.38 \pm 1.30^{\circ}C$ and $23.29 \pm 2.25^{\circ}C$ and $89.9 \pm 0.94\%$ and $70.8 \pm 3.57\%$ respectively. Cocoons were separated into two colour groups *viz.,* red and white. Then cut open and again grouped into sex wise. The treatment for the study consisted of cocoon colour group i.e. red and white and sex group i.e. male and female. Biometry of cocoons was taken in three different aspects *viz.,* linear, gravimetric and volumetric measurements. Biometry of whole cocoons, shell, pupa and exoskeleton were taken. Observations were recorded according to the treatments and data was recorded.

The results of the experiment revealed that occurrence of red and white cocoons of Bangalore eco race was 64.28% and 35.72% respectively, which showed covariance of 5.57% and 3.10% in both the colour group cocoons. However, variation in the sexual percentage in the Bangalore eco race was found significantly different in white coloured cocoons resulting 60.01% and 39.99% of male-female population, which showed 6.48% and 9.72% co-efficient of variance respectively. While the red coloured cocoons showed no significant affect in respect of male-female percentage. Sexual percentage in the breed resulted significantly higher sex ratio in white cocoons over red cocoons (Table 9.22).

Table 9.22: Occurrence of red and white cocoons and sex ratio in Bangalore local eri silkworm race under Assam condition (Bordoloi *et al.,* 2004)

Eco race	Occurrence of cocoons (%)		Sex ratio		Sexual percentage in the ecorace (%)			
					Red		White	
	Red	White	Red	White	Male	Female	Male	Female
Bangalore	35.72	64.28	1.04	1.52	50.58	48.15	60.01	39.99
CV (5%)	3.10	5.57	16.36	12.94	NS	NS	6.48	9.72

Table 9.23: Biometry of cocoon of Bangalore eri eco race produced at Assam condition (Bordoloi *et al.,* 2004)

Variables		Cocoon				Shell			Pupa			Exoskeleton weight (g)
		Length (cm)	Diameter (cm)	Volume (cc)	Weight (g)	Thickness (mm)	Weight (g)	Silk ratio (%)	Length (cm)	Diameter (cm)	Weight (g)	
Male	White	3.60	1.47	10.71	2.52	0.05	0.31	12.55	3.07	1.13	2.21	0.013
	Red	3.55	1.31	11.11	2.37	0.05	0.28	11.91	2.99	0.93	2.09	0.015
	Irrespective of colour	3.60	1.39	10.91	2.45	0.05	0.30	12.23	3.03	1.03	2.15	0.014
Female	White	4.41	1.55	11.18	2.90	0.06	0.33	11.36	3.11	1.18	2.56	0.019
	Red	4.86	1.74	12.24	3.80	0.07	0.48	12.35	3.08	1.24	3.32	0.021
	Irrespective of colour	4.64	1.64	11.71	3.35	0.07	0.41	11.86	3.10	1.21	2.94	0.020
Irrespective of sex	White	4.04	1.51	10.94	2.71	0.06	0.32	11.96	3.09	1.16	2.38	0.016
	Red	4.21	1.52	11.68	3.09	0.07	0.38	12.13	3.03	1.09	2.71	0.018
	Irrespective of colour	4.12	1.52	11.31	2.90	0.06	0.36	12.05	3.07	1.13	2.55	0.017
CD (5%)	Sex	0.01	0.01	0.02	0.01	0.01	0.01	0.03	0.02	0.01	0.01	NS
	Colour	0.01	0.01	0.02	0.01	0.01	0.01	0.03	0.02	0.01	0.01	NS
	Interaction	0.01	0.01	0.02	0.02	0.01	0.01	0.04	0.02	0.01	0.02	NS

Results on biometry of cocoon parameters showed that irrespective of sex, red coloured cocoon had significantly higher cocoon, shell and pupal parameters over white cocoons. However, irrespective of colour, female cocoon were found superior than male cocoons in respect of cocoon length, diameter, volume and weight. Similar response was observed in respect of shell thickness and weight, pupal length, diameter and weight. Exoskeleton weight showed no significant differences in respect of colour and sex. Thus it can be concluded that the red and female cocoons produced by the Bangalore eri eco race in Assam condition, were superior over white and male cocoons (Table 9.23).

Influence of Botanicals and Biopesticides on Economic Traits of Eri Silkworm

Effect of botanical insecticides on the growth and silk production of Ambagoan and Kokrajhar strains of eri silkworm

To study the effect of botanical insecticides on the growth and silk production of eri silkworm, Lakshmi Narayanamma *et al.,* (2013) conducted a lab trial at Ericulture laboratory of RARS, Palem during *Kharif* and *Rabi* seasons of 2011-12. The experiment was conducted with six treatments *viz.,* neem oil @ 3 ml, 5 ml and 10 ml/lit and karanj oil @ 2 ml and 5 ml/lit along with untreated control on two eri silkworm strains viz., plain white cocoon type (Ambagoan) and brick red cocoon type (Kokrajhar) using the most popular PCH-111 hybrid leaves. The worms were fed with untreated castor leaves by maintaining 50 larvae per treatment up to second moult. From second instar onwards, the eri worms were fed with treated leaves commencing from 24 hours after treatment with the botanicals. The leaves were dipped in the botanical solution, drained out the solution, shade dried the leaves and fed to the eri silkworms.

The results showed that during fourth instar least larval weight of 0.38g was recorded in the plain white strain of eri silkworm larvae that were fed with karanj oil (5 ml/lit) treated leaves followed by the larvae fed with 2 ml/lit karanj oil (0.46g) and 10 ml/lit neem oil (0.5g). In brick red strain the weight of the larva was significantly higher compared to the plain white strains and this strain has shown some resistance to karanj oil treatment. Effective rate of rearing varied from 76.84 to 89.9% in plain white strains, whereas in brick red strains it ranged from 79.88% to 91.46%. Comparatively shorter larval duration was recorded in brick red strains than plain white strains. In plain white strains, significantly higher weight was recorded in the cocoons formed out of worms fed with 3 ml/lit neem oil (1.89g) and 2 ml/lit karanj oil (1.82g) treated leaves. In brick red strains, higher shell weight of 0.64g was recorded with control treatment closely followed by 2 ml/lit karanj oil (0.50g) treatment showing its loss of efficacy (Tables 9.24 and 9.25). The treatmental differences were significant with respect to fecundity and no significant differences were noticed with hatching per cent.

This is a clear indication that, though neem oil and karanj oil were significantly superior in reducing the pest population, they are having adverse effect on the growth and rearing performance of eri silkworm (Mortale, 2004). Further, determining the safety period for botanicals, particularly with reference to ericulture needs immediate attention in order to ensure good performance from the eri silkworm, in spite of adapting eco friendly pest management measures.

Effect of Biopesticides on diverse breeds of eri silkworm, *Samia cynthia ricini* (Boisduval)

A study was undertaken by Lakshmi Narayanamma (2013) to evaluate the effect of biopesticides *viz.,* neem oil and karanj oil in different concentrations, on the rearing and cocoon parameters of six diverse breeds of eri silkworm viz., Blue green plain (BGP), Yellow plain (YP), Blue green spotted (BGS), Yellow spotted (YS), Blue green zebra (BGZ) and Yellow zebra (YZ) using the most popular Haritha variety leaves during *Kharif* and *Rabi* seasons of 2012-13 at Ericulture Laboratory, Regional Agricultural Research Station, Palem because the dietary nutritional management influences the rearing and commercial parameters of the silkworm. The experiment was conducted in completely randomized design by taking 50 larvae per treatment using five treatments of botanicals along with an untreated control. The observations were recorded on the rearing parameters i.e. larval duration and effective rate of rearing (ERR) and cocoon parameters *viz.,* cocoon weight, pupal weight, shell weight, shell ratio percentage by taking 10 cocoon samples from each replication. No doubt, out of all treatments, untreated control proved its superiority in rearing and cocoon parameters. The BGP breed nourished with 2 mL/lit karanj oil treated leaves was the next best treatment to control (89.9%) with 84.72% effective rate of rearing and recorded lowest larval duration of 19.59 days. In BGP breed significantly high weight of cocoon was recorded in worms fed on untreated control (2.11g), 2 mL/lit karanj oil (2.12g) and 5 mL/lit karanj oil (2.27g) with no statistical variation among them. This breed has shown some tolerance to karanj oil treatment (Tables 9.26, 9.27 and 9.28). In YP breed 3 mL/lit and 5 mL/lit neem oil treatments recorded highest effective rate of rearing (84.29 and 83.21% respectively), while least larval duration of 20.21 days was recorded with 3 mL/lit neem oil. BGP breed worms fed with leaves from untreated control recorded significantly higher shell ratio (25.6%) followed by 10 mL/lit neem oil (16.6%), while in YP breed 3 mL/lit neem oil treatment recorded the highest shell ratio of 14.4%. From the study it is inferred that the strains BGP and YP are the better performers in spite of feeding on the treated foliage and can be exploited for commercial rearing

Feasibility of Exploiting Genetic Variability in Other Related Silk Worms

Breeding perspective for silk yield and quality in Indian tropical tasar silkworm, *Antheraea mylitta* Drury

India enjoys the availability and practice of mulberry, tasar (tropical or temperate), eri and muga silks. The utilization of tropical tasar seri-biodiversity, however, requires appropriate breeding methods so as to exploit the global demand for vanya silk, besides bringing reforms to tribal communities and weaker sections, and improving the socioeconomic status of landless rural populace. Among the 44 existing ecoraces of *Antheraea mylitta*, only Daba and Sukinda are commercially used for cocoon production and in situ conservation, and ex situ stabilization of additional ecoraces, particularly *Shorea robusta* (sal), as amenable parental base for hybridization and silk production are needed. The breeding for disease resistance, correlating different traits of commercial value in tasar silkworm are important, since the tasar silkworm is an eco-insect reared outdoor and exploring the biotechnological tools for transgenic application appears pragmatic and worthy of testing. The management of genotype and environment interaction through multilocational breeding stations, irrespective of rearing seasons, can be applied for compatible ecorace breeds or lines. This can involve the benefits for indigenous knowledge along with trained breeders, and which can be considered as the indispensable strategy to achieve tasar raw silk production and quality (Reddy *et al.,* 2010).

Table 9.24: Effect of Botanicals on growth and silk production of plain white strain of eri silkworm, *Samia cynthia ricini* during *Kharif* and *Rabi* 2011-12 (Lakshmi Narayanamma *et al.,* 2013)

Treatment	Larval weight (g)			ERR (%)	Larval duration (days)	Cocoon wt (g)	Pupal wt (g)	Shell wt (g)	Shell ratio (%)	Fecundity	Hatching (%)
	III	IV	V								
Neem oil (3 ml/lit)	0.49	0.98^b	2.68^{ab}	82.24^c (65.21)	21.64^b	1.89^a	1.60	0.29^b	15.34^b	228^b	92.0 (73.61)
Neem oil (5 ml/lit)	0.40	0.78^b	2.15^b	80.12^d (64.01)	22.18^d	1.75^b	1.52	0.23^d	13.14^d	200^{bc}	92.0 (73.61)
Neem oil (10 ml/lit)	0.25	0.50^c	1.95^{bc}	79.88^e (63.31)	22.26^e	1.50^b	1.35	0.15^f	10.0^e	180^c	85.0 (70.42)
Karanj oil (2 ml/lit)	0.38	0.46^c	1.75^c	84.72^b (66.95)	21.98^c	1.82^{ab}	1.58	0.24^c	13.19^d	205^b	94.28 (75.14)
Karanj oil (5ml/lit)	0.26	0.38^d	1.50^c	76.84^f (60.32)	22.38^f	1.48^b	1.26	0.22^e	14.86^c	192^c	90.0 (72.86)
Untreated Control	0.64	1.50^a	3.42^a	89.9^b (71.49)	21.04^a	2.21^a	1.64	0.37^a	16.74^a	275^a	94.28 (75.14)
CD at 5%	-	0.38	0.75	1.035	0.146	0.42	-	0.014	0.16	28.01	-

Figures in the parenthesis are angular transformed values; Figures followed by the same letter did not differ significantly

Table 9.25: Effect of Botanicals on growth and silk production of brick red strain of eri silkworm, *Samia cynthia ricini* during *Kharif* and *Rabi* 2011-12 (Lakshmi Narayanamma *et al.*, 2013)

Treatment	Larval weight (g)			ERR (%)	Larval duration (days)	Cocoon wt (g)	Pupal wt (g)	Shell wt (g)	Shell ratio (%)	Fecundity	Hatching (%)
	III	IV	V								
Neem oil (3 ml/lit)	0.38	1.02^b	3.32^b	88.5^b (70.18)	20.21^a	2.01^d	1.63^b	0.38^d	18.9^f	305^b	94.28 (75.14)
Neem oil (5 ml/lit)	0.35	0.88^b	3.01^c	84.72^c (66.95)	20.69^c	1.98^e	1.52^d	0.36^e	20.6^c	300^b	92.0 (73.61)
Neem oil (10 ml/lit)	0.28	0.65^c	2.54^e	80.12^d (64.01)	20.9^d	1.74^f	1.32^f	0.32^f	18.4^e	290^b	92.0 (73.61)
Karanj oil (2 ml/lit)	0.32	0.94^{bc}	2.82^d	91.46^a (73.01)	21.18^e	2.21^b	1.76^b	0.50^b	22.9^b	320^{ab}	95.83 (78.22)
Karanj oil (5ml/lit)	0.25	0.80^b	2.85^d	79.88^e (63.31)	21.2^e	2.18^c	1.38^c	0.45^c	20.4^d	302^b	92.0 (73.61)
Untreated Control	0.62	1.72^a	3.95^a	91.46^a (73.01)	20.45^b	2.61^a	1.58^a	0.6^a	24.5^a	341^a	96.9 (79.86)
CD at 5%	-	0.35	0.12	1.081	0.152	0.014	0.029	0.018	0.15	35.01	-

Figures followed by the same letter did not differ significantly; Figures in the parenthesis are angular transformed values

Table 9.26: Effective rate of rearing and larval duration of different breeds of eri silkworm, *Samia cynthia ricini* as influenced by biopesticides during 2012-13 (Lakshmi Narayanamma, 2013)

Treatments	Effective Rate of Rearing (%)						Larval Duration (days)					
	BGP	YP	BGS	YS	BGZ	YZ	BGP	YP	BGS	YS	BGZ	YZ
Neem oil (3 ml/lit)	82.24[c]	84.29[b]	80.5[c]	81.33[c]	76.33[e]	80.00[c]	20.03[b]	20.21[a]	21.64[a]	21.68[b]	19.87[a]	19.46
Neem oil (5 ml/lit)	80.14[d]	83.21[b]	82.72[b]	78.0[e]	74.83[f]	81.67[b]	20.00[b]	20.69[c]	22.18[bc]	21.73[b]	20.67[d]	19.65
Neem oil (10 ml/lit)	79.88[e]	77.67[d]	78.12[d]	76.32[f]	78.33[d]	74.00[e]	22.21[d]	20.9[d]	22.26[c]	22.25[c]	20.84[e]	19.61
Karanj oil (2 ml/lit)	84.72[b]	81.69[c]	81.46[bc]	83.77[b]	82.63[b]	79.24[d]	19.59[a]	21.18[e]	21.98[ab]	20.01[a]	20.12[b]	19.51
Karanj oil (5ml/lit)	76.84[f]	72.46[e]	79.88[c]	78.63[d]	81.00[c]	70.97[f]	21.26[c]	21.2[e]	22.38[c]	22.69[d]	20.34[c]	19.75
Untreated Control	89.9[a]	92.23[a]	88.46[a]	85.67[a]	86.00[a]	88.5[a]	19.62[a]	20.45[b]	21.04[a]	20.10[a]	20.00[ab]	19.65
CD at 5%	1.235	1.165	1.681	0.021	1.024	1.438	0.254	0.152	0.246	0.192	0.168	-

Table 9.27: Cocoon weight and shell weight of different breeds of eri silkworm, *Samia cynthia ricini* as influenced by biopesticides during 2012-13 (Lakshmi Narayanamma, 2013)

Treatments	Cocoon weight (g)						Shell weight (g)					
	BGP	YP	BGS	YS	BGZ	YZ	BGP	YP	BGS	YS	BGZ	YZ
Neem oil (3 ml/lit)	1.96[bc]	2.01[d]	1.89[a]	2.16[a]	1.81	2.13	0.24	0.29[b]	0.38[d]	0.35[b]	0.34[ab]	0.24[c]
Neem oil (5 ml/lit)	1.94[c]	1.98[e]	1.75[b]	2.11[a]	1.79	2.23	0.28	0.23[c]	0.36[e]	0.33[cd]	0.33[b]	0.24[c]
Neem oil (10 ml/lit)	1.93[e]	1.74[f]	1.50[b]	1.74[c]	1.78	1.84	0.32	0.15[d]	0.32[f]	0.32[d]	0.24[d]	0.35[b]
Karanj oil (2 ml/lit)	2.12[a]	2.21[b]	1.82[a]	1.95[b]	1.77	1.87	0.27	0.24[c]	0.50[b]	0.35[b]	0.25[cd]	0.36[b]
Karanj oil (5ml/lit)	2.27[a]	2.18[c]	1.48[b]	1.82[bc]	2.2	1.81	0.31	0.22[c]	0.45[c]	0.34[bc]	0.24[d]	0.24[c]
Untreated Control	2.11[ab]	2.61[a]	2.21[a]	1.99[b]	2.22	2.20	0.54	0.37[a]	0.60[a]	0.51[a]	0.35[a]	0.55[a]
CD at 5%	0.16	0.014	0.42	0.24	-	-	-	0.014	0.018	0.016	0.015	0.016

Table 9.28: Shell ratio of different breeds of eri silkworm, *Samia cynthia ricini* as influenced by biopesticides during 2012-13 (Lakshmi Narayanamma, 2013)

Treatment	Shell Ratio (%)					
	BGP	**YP**	**BGS**	**YS**	**BGZ**	**YZ**
Neem oil (3 ml/lit)	12.2[f]	14.4[b]	20.1[f]	16.2[e]	18.8[a]	11.3[e]
Neem oil (5 ml/lit)	14.4[c]	11.6[c]	20.6[e]	15.6[f]	18.4[b]	10.8[f]
Neem oil (10 ml/lit)	16.6[b]	8.6[f]	21.3[d]	18.4[c]	13.5[e]	19.0[c]
Karanj oil (2 ml/lit)	12.7[e]	10.9[d]	27.5[b]	17.9[d]	14.1[d]	19.3[b]
Karanj oil (5ml/lit)	13.7[d]	10.1[e]	30.4[a]	18.7[b]	10.9[f]	13.3[d]
Untreated Control	25.6[a]	14.2[a]	27.1[c]	25.6[a]	15.8[c]	25.0[a]
CD at 5%	0.15	0.16	0.12	0.14	0.08	0.14

Genetic variability and heritability of a few quantitative characters in *Antheraea assamensis* Helfer

Singh *et al.,* (2012b) made an attempt to estimate genetic variability, heritability, genetic advance and correlation of a few traits in the available germplasm of muga silkworm, *Antheraea assamensis*- Aa-SD, Aa-Blue, Aa-MM, Aa-Ga, Aa-TM, Aa-IM, Aa-KA and Aa-SM. Important yield-contributing characters, i.e. fecundity, hatching percentage, number of cocoons per dfl, effective rate of rearing (ERR%), average cocoon weight, single cocoon weight, single shell weight and single cocoon filament length were recorded. For improvement of the muga silkworm, simple phenotypic selection based on fecundity, female cocoon shell weight, male cocoon shell weight and filament length will be highly effective. Progeny testing and recurrent selection will be an effective means to improve cocoon weight and ERR%.

Bionomics of Indian oak Tasar silkmoth, *Antheraea roylei* Moore and its potential for breeding in North East India

During the survey of oak fed silkmoth fauna in the forest of North-Eastern part of India, the Indian oak fed silkmoth, *Antheraea roylei* Moore was recorded by Singh and Debaraj, (2011) from oak growing areas of this region. It is distributed in the Indian Sub-Himalayan range and South East Asian countries. It feeds on about 12 varieties of oak tasar food plants. The bionomics and potential of the silkmoth was studied in the laboratory condition. *A. roylei* behaves as bivoltine in nature. The female moth lays 185-230 eggs. The average hatching per cent, larval period, cocoon yield, effective rate of rearing (ERR %), cocoon weight, cocoon shell weight, cocoon shell ratio % and average single filament length in the laboratory condition were 61.40%, 38 days, 30 cocoons/disease free laying (dfl), 20.36%, 7.94 g, 0.85 g, 10.71% and 250 m respectively. *A. roylei* shows high genetic compatibility with *A. pernyi* and *A. proylei* in interspecific hybridization which shows the genetic potential of this species. The F_1 hybrids of *A. roylei* with *A. pernyi* and *A. proylei* showed high vigour in all the yield contributing characters. Their study can be very helpful for future breeding programs in oak tasar culture.

Summary and Conclusions

➢ Eri silkworm *Samia cynthia ricini* (Boisduval) has 26 eco races and 12 strains based on the heritable morphological characters.

➢ Rearing performance of several eco-races of eri silkworm in Vidarbha region of Maharashtra revealed that Greenish Blue Zebra (GBZ) and Yellow Zebra (YZ) as the best strains. Molecular study by using RAPD and SDS-PAGE showed differences in DNA and Protein Profiles among the eri strains.

➢ Comparative study on six strains of eri silkworm based on morphological traits in Assam revealed that the YZ as the best strain in terms of growth and cocoon characters. Greenish blue Spotted (GBS) and Yellow spotted (YS) were also recorded as better strains for rearing and may be the better strains for hybridization with YZ.

➢ Similar study in Uttar Pradesh to identify superior eco-races of eri silkworm suitable for commercial rearing revealed the superiority and suitability of Titabar eco-race for summer and Borduar eco race for autumn rearing.

➢ Studies were carried out on combining ability analysis of six inbred lines of eri silkworm *viz.*, yellow plain (YP), yellow spotted (YS), yellow zebra (YZ), green blue plain (GBP), green blue spotted (GBS) and green blue zebra (GBZ) in a 6 x 6 diallel fashion. The promising hybrids YP x GBZ and GBZ x GBP were exhibiting high and significant SCA and reciprocal effects for several economic traits. These can be exploited on commercial scale.

➢ Combining ability of five parental breeds of eri silkworm i.e. WP, WZ, GP, GS and GZ differed in their general combining ability for rate of pupation, pupal weight, pupal duration, moth emergence, sex ratio, fecundity and hatching showed that the WP breed of eri silkworm was a good general combiner.

- ➤ Correlation of larval parameters with economic traits of eri silkworm showed that larval duration and mature larval weight were correlated with larval survival, effective rate of rearing, cocoon weight, pupal weight, shell weight, shell ratio, adult emergence, fecundity and hatchability.

- ➤ Comparative studies on rearing performance of some eco races of eri silkworm under Odisha conditions revealed that pure line eco races exhibited significant increase over the mixed type in terms of morphological and quantitative characters. Among seasons, highest productivity was recorded in winter followed by spring, autumn and rainy seasons while the lowest values were observed in the summer crop.

- ➤ Genetic analysis of larval and cocoon traits in eri silkworm by diallel cross revealed the importance of additive gene action for rate of progression to fourth instar, total larval duration, effective rate of rearing, maximum larval weight, cocoon yield by number and weight, cocoon weight, shell weight and silk productivity.

- ➤ The most appropriate breeding programme that can be followed for the improvement of cocoon traits is selection.

- ➤ Studies on combining ability analysis of inbred lines of eri silkworm showed that YZ strain is superior in terms of fecundity, male cocoon shell weight, male cocoon shell percentage and absolute silk yield. YP strain showed highest hatching percentage, larval weight and male cocoon weight while highest mean female cocoon weight and female cocoon shell weight were shown by GBP.

- ➤ Interaction of castor genotypes and breeds of eri silkworm in relation to cocoon and grainage parameters revealed that white zebra fed on RC-8 recorded longer larval duration, higher cocoon weight and pupal weight while feeding on leaves of Aruna resulted in higher shell weight and shell ratio.

- ➤ Diphu, Borduar and Barpeta are the most promising eco races of eri silkworm with better economic traits irrespective of the seasons.

- ➤ Influence of leaf anatomical characters of castor genotypes on economic parameters of eri silkworm showed negative and significant relationship between vascular bundle length and breadth, and wax and epidermal layers on economic parameters of eri silkworm such as feeding and moulting duration; larval, cocoon and shell weight; shell ratio; fecundity and hatchability.

- Screening for eri silkworm eco races using morphological characters, growth, yields, and ISSR markers revealed that SaKKU1 was at the farthest distance than other eco races, especially SaKKU3. Based on overall results, the SaKKU1 eco race was considered to be the most suitable for heat tolerant variety improvement.

- Castor is the only host plant for successful commercial rearing of Borduar strain of eri silkworm in Punjab conditions.

- Feed utilization efficiency of white plain eri-silkworm breed on eight castor genotypes showed the suitability of Abaro and Acc 208624 genotypes in Ethiopia. Hence these genotypes will be recommended for eri silkworm research and development works in the future in Ethiopia.

- Kokrajhar ecorace of eri silkworm survived well on red variety of castor compared to kesseru and tapioca for larval growth as well as spinning characters.

- Kokrajhar race of eri silkworm reared on the foliage of PCH-111 castor genotype showed better larval weight, effective rate of rearing, cocoon weight, shell weight and higher fecundity with shorter larval duration in Telangana region.

- C2 breed of eri silkworm developed by Central Muga Eri Research Training Institute, Lahdoigarh, Jorhat, Assam showed its best performance in castor followed by kesseru among different food plants.

- Studies on cocoon biometry of Bangalore local eri silkworm race reared in Assam condition showed that the red cocoons were found superior to the white cocoons in respect of all the commercial characters and on average cocoon produced by Bangalore local eri eco race was grade marked having cocoon weight of 2.90g.

- Effect of botanical insecticides on the growth and silk production of Ambagoan and Kokrajhar strains of eri silkworm indicated that, though neem oil and karanj oil were significantly superior in reducing the pest population, they are having adverse effect on the growth and rearing performance of eri silkworm.

- Effect of Biopesticides on diverse breeds of eri silkworm shows that the strains BGP and YP are the better performers in spite of feeding on the treated foliage and can be exploited for commercial rearing.

- Regarding breeding perspectives of other related silkworms, in Indian tropical tasar silkworm *Antheraea mylitta*, among the 44 existing eco races only Daba and Sukinda are commercially used for cocoon production and in situ conservation, and ex situ stabilization of additional eco races, particularly *Shorea robusta* (sal), as amenable parental base for hybridization and silk production.

➤ For improvement of the muga silkworm, *Antheraea assamensis*, simple phenotypic selection based on fecundity, female cocoon shell weight, male cocoon shell weight and filament length will be highly effective.

➤ Indian oak fed Tasar silkmoth, *Antheraea roylei* Moore shows high genetic compatibility with *A. pernyi* and *A. proylei* in interspecific hybridization which shows the genetic potential of this species. The F_1 hybrids of *A. roylei* with *A. pernyi* and *A. proylei* showed high vigour in all the yield contributing characters.

References

Amarnatha N and Narayanaswamy K C. 2004. Relationship between the amylase activity and some metric traits in some elite bivoltine breeds of *Bombyx mori* L. *National workshop on Role of Sericulture in Cocoon crop improvement.* Central Sericulture and Germplasm Resource centre, Hosur, India.

Benchamin K V. 2000. Muga and ericulture: Subsistence to substantial farming. *Indian Silk* Jan-Feb 51-54.

Bordoloi S, Dutta P P and Handique P. 2004. Studies on cocoon biometry of Bangalore local eri silkworm race (*Samia cynthia ricini* Boisduval), (Lepidoptera: Saturniidae) reared in Assam condition. *New Agriculturist* **15**(1,2): 5-7.

Deka M, Saranga Dutta and Dipali Devi. 2011. Impact of feeding of *Samia cynthia ricini* Boisduval (red variety) (Lepidoptera: Saturniidae) in respect of larval growth and spinning. *International Journal of Pure and Applied Sciences and Technology* **5**(2): 131-140.

Govindan R, Nagaraja M, Swamy T K N and Devaiah M C. 1997. Genetic analysis of larval and cocoon traits in eri silkworm, *Samia cynthia ricini* Boisduval by diallel cross. *Mysore Journal of Agricultural Sciences* **31**(2): 180-184.

Hota M M, Pati P K, Pradhan K C and Sharma M. 2005. Ericulture in Orissa Potential Exists. *Indian Silk* **44**: 12-13.

Ishaaya I, Moore I and Joseph B. 1971. Protease and amylase activity in the larvae of Egyptian cotton worm, *Spodoptera lituralis*. *Journal of Insect Physiology* **17**: 945-953.

Jaiswal K, Dhami S S, Gangwar S K and Jain A. 2003. Performance of some bivoltine silkworm (*Bombyx mori* L.) hybrids for cocoon shape uniformity in summer season. *Proceedings of recent researches of Indian sericulture industry* 1-2 February, Lucknow, 122-125.

Kar J. 2004. Survey on Eri Resources of Orissa. *Final Report of Directorate of Textiles*, Orissa, 45-48.

Kedir Shifa, Waktole Sori and Emana Getu. 2014. Feed Utilization Efficiency of Eri-Silkworm (*Samia cynthia ricini* Boisduval) (Lepidoptera: Saturniidae) on Eight Castor (*Ricinus communis* L.) genotypes. *International Journal of Innovative and Applied Research* 2(4): 26- 33.

Krishnaswami S, Kumararaj S, Vijayaraghavan N and Kasiviswanathan K. 1971. Silkworm feeding trials for evaluating the quality of mulberry leaves as influenced by variety, spacing and nitrogen fertilization. *Indian Journal of Sericulture* 10: 78-89.

Lakshmi Narayanamma V, Dharma Reddy K and Vishnuvardhan Reddy A. 2013. Effect of botanical insecticides on the growth and silk production of Ambagoan and Kokrajhar strains of eri silkworm, *Samia cynthia ricini* (Boisduval). *Journal of Biopesticides* 6(1): 41-45.

Lakshmi Narayanamma V. 2013. Effect of Biopesticides on diverse breeds of eri silkworm, *Samia cynthia ricini* (Boisduval). Paper presented in the International Conference on Insect Science, held at UAS, Bangalore during February 14-17, 2013.

Lakshmi Narayanmma V and Dharma Reddy K. 2013. Rearing performance of different strains of eri silkworm, *Samia cynthia ricini* (Boisduval) (Lepidoptera: Saturniidae) on different castor genotypes. Paper presented in the *IV Biopesticide International Conference*, held from 28th – 30th November, 2013 at Palayamkottai, Tamil Nadu.

Mavi G S and Singh H. 1989. Scope of Ericulture under Punjab conditions. *Journal of Research, Punjab Agricultural University* 26: 437-444.

Mortale H T. 2004. Studies on the foliage pests of castor, *Ricinus communis* L. host plant of eri silkworm, *Samia cynthia ricini* (Boisduval). In: *M. Sc Thesis*, submitted to Department of Sericulture, UAS, Bangalore 62 pp.

Nagaraja M and Govindan R. 1994. Combining ability estimates in the eri silkworm, *Samia cynthia ricini* Boisduval for larval and cocoon traits. *Sericologia* 34(3): 455-460.

Nagaraja M, Govindan R and Narayanaswamy T K. 1996. Estimation of combining ability in eri silkworm *Samia cynthia ricini* Boisduval for pupal and allied traits. *Mysore Journal of Agricultural Sciences* 30(1): 48-51.

Narayanaswamy K C, Saritha Kumari K, Manjunath Gowda and Amarnatha N. 2006. Barkesseru (Ailanthus excelsa Roxb.), an ideal substitute for castor to rear eri silkworm, *Samia cynthia ricini* Boisduval. *Environment and Ecology* 24(4): 720-725.

Nayak B K and Guru B C. 1998. Studies on cocoon biometry of the bivoltine Daba ecorace of the tasar silk moth *Antheraea mylitta* D. (Lepidoptera : Saturniidae) reared at Bangiriposi. *Bulletin of Indian Academy of Sericulture* 2(1): 31-34.

Nayak B K, Das S K and Guru B C. 1998. Studies on cocoon biometry of the bivoltine Daba ecorace of the tasar silk moth *Antheraea mylitta* D. (Lepidoptera : Saturniidae) reared at Khuntgaon area of Sundargarh district, Orissa, India. *Bulletin of Indian Academy of Sericulture* **2**(2): 50-53.

Pandey R K. 1995. Do leaf tannins affect non-mulberry silkworm? *Indian Silk* **34**: 21-23.

Periaswamy K and Radhakrishnan S. 1984. Performance of exotic Chinese polyvoltine silkworm *B. mori* L. and its bivoltine hybrids in Tamil Nadu state. *Sericologia* **24**: 383-388.

Rabinder Kaur and Virk J S. 2012. Evaluation of Different plants for rearing of Eri silkworm, *Samia cynthia ricini* under Punjab conditions. *Journal of Insect Science* **25**(3): 270-271.

Rajesh Kumar and Elangovan V. 2012. Rearing performance of different ecoraces of eri silkworm (*Philosamia ricini*) Donovan during Summer season of Uttar Pradesh. *Journal of Experimental Zoology, India* **15**(1): 163-168.

Rajesh Kumar and Elangovan V. 2013. Rearing performance of eri silkworm (*Philosamia ricini*) during Autumn season of Uttar Pradesh. *Indian Journal of Sericulture* **52**(2): 131-138.

Ramakrishna Naika, Sannappa B and Govindan R. 2003. Influence of castor varieties on rearing and grainage performance of different breeds of eri silkworm, *Samia cynthia ricini*. *Journal of Ecobiology* **15**(4): 279-285.

Ray P P, Rao T V and Dash P. 2010. Comparative studies on rearing performance of some ecoraces of eri silkworm (*Philosamia ricini* H.) in different seasons. *The Bioscan* **1**: 181-186.

Reddy D N and Swamy K C. 1999. Effect of host on the consumption rate, leaf – cocoon and leaf egg ratio of eri silkworm, *Samia cynthia ricini* Boisduval (Lepidoptera:Saturniidae). *Entomon* **24**: 67-70

Reddy R M, Sinha M K and Prasad B C. 2010. Breeding perspective for silk yield and quality in Indian tropical tasar silkworm, *Antheraea mylitta* Drury (Lepidoptera: Saturniidae). *Journal of Applied Sciences* **10**(17): 1902-1909.

Sannappa B and Jayaramaiah M. 1999. Interaction of castor genotypes and breeds of eri silkworm in relation to cocoon and grainage parameters. *Mysore Journal of Agricultural Sciences* **33**: 214-223.

Sannappa B and Jayaramaiah M. 2001. Leaf anatomical characters of castor genotypes and their impact on economic parameters of eri silkworm, *Samia cynthia ricini* Boisduval (Lepidoptera: Saturniidae). *Mysore Journal of Agricultural Sciences* **35**(2): 101-107.

Sannappa B, Chinnaswamy K P, Raj S S and Chavan S S. 1999. Correlation of larval parameters with economic traits of eri silkworm, *Samia cynthia ricini* Boisduval. *Insect Environment* **5**(1): 7-8.

Sarkar B N, Sarmah M C and Dutta K. 2012. Screening of superior ecoraces of eri silkworm *Samia ricini* for better economic traits in different seasons. *Journal of Applied Bioscience* **38**(1): 73-77.

Sarmah M C, Sarkar B N, Ahmed S A and Giridhar K. 2015. Performance of C_2 breed of eri silkworm, *Samia ricini* (Donovan) in different food plants. *Entomology and Applied Science Letters* **2**(1): 47-49.

Sharma P and Kalita J C. 2013. A comparative study on six strains of eri silkworm (*Samia ricini* Donovan) based on morphological traits. *Global Journal of Bi-Science and Biotechnology* **2**(4): 506-511.

Singh B K, Debaraj Y, Sarmah M.C, Das P K and Suryanarayan N. 2003. Eco-races of eri silkworm. *Indian Silk* **5**: 7-10.

Singh L S, Debaraj Y, Singh N I, Ray B C and Singh R. 2012a. Studies on the combining ability analysis of six inbred lines of eri silkworm, *Samia ricini* Donovan. *Indian Journal of Sericulture* **51**(2): 167-172.

Singh N I and Debaraj Y. 2011. Bionomics of Indian oak tasar silkmoth, *Antheraea roylei* Moore and its potential for breeding in North East India. *Munis Entomology & Zoology* **6**(2): 987-994.

Singh N I, Goswami D, Khan Z M S and Rajan R K. 2012b. Genetic variability and heritability of a few quantitative characters in *Antheraea assamensis* Helfer - an indigenous Indian silkworm. *Indian Journal of Genetics and Plant Breeding* **72**(1): 97-99.

Somaprakash D S and Sathya Prakash K. 2009. Effect of refrigeration on the hatchability of eri silkworm (*Samia cynthia ricini* Boisduval) eggs. *Indian Journal of sericulture* **48**(1): 56-59.

Waldbauer G P. 1968. Consumption and utilization of food by insects. *Advances in Insect Physiology* **5**: 229-288.

Wankhade L N, Barman H D, Rai M M and Rathod M K. 2014. Evaluation of some promising strains of Eri Silkworm, *Samia ricini* in climatic condition of Vidarbha region of Maharashtra. *Indian Journal of Pure and Applied Biology* **29**(2): 247-253.

Wongsorn D, Saksirirat W, Sirimungkararat S and Urairong H. 2015. Screening for eri silkworm (*Samia ricini* Donovan) ecoraces using morphological characters, growth, yields, and ISSR marker. *Songklanakarin Journal of Science and Technology* **37**(5): 499-505.

CHAPTER 10

Role of Environmental Factors on Biology of Eri Silkworm

The growth of silkworm larvae mostly depends on environmental factors and rearing conditions. A striking feature of the insect life cycle is its physiological and behavioural adaptations to the seasonally changing environmental conditions such as temperature, relative humidity, rainfall, photo period and drought. Temperature influences everything that an organism does, relative humidity affects embryonic development, and rainfall affects both. It is well known that the dynamic environmental conditions prevailing in different seasons of the year bring profound changes in growth and development and the expression of economic characters in different silkworm races. Fluctuation in temperature, humidity and other seasonal factors prevents insects from attaining their physiological potential life history performance. The variations in climate, nutrient status, feeding duration and larval crowd along with environmental stimuli influence the insect body size and its biological success. Environment causes effective restrain on biological characters of silkworm like, effective rate of rearing (ERR), single cocoon weight, single shell weight, and shell ratio and filament length. This chapter discusses in detail about the role of environmental factors on rearing, cocoon and grainage parameters of eri silkworm and its eco-races.

Temperature, relative humidity, rain fall and photo period are the most important environmental factors that influence the biology of insects. Insects are cold-blooded (Poikilothermic) organisms, so their body temperature is more or less same as that of their surrounding microenvironment. Temperature influences everything that an organism does, Relative humidity affects embryonic development, and rainfall affects both. Further, photoperiod also can have profound influence on larval duration and larval weight in eri silkworm apart from its well established effect on induction and termination of diapause in many insect species.

Environmental temperature, being the major abiotic factor, regulates the body temperature of caterpillar that determines the rates of feeding, fecundity and mortality. Temperature is probably the single most important environmental factor that influences behaviour, development, survival, reproduction and geographical distribution of insects. All the insect species have their own requirement of temperatures for usual growth. Very high temperature slows down the growth that may lead to developmental malfunction in larval ecdysis and adult emergence. Temperature regimes and different levels of relative humidity are known to play an important role in the life cycle of insect and its adaptability to the local climate. Some researchers believe that the effect of temperature on insects largely overwhelms the effects of other environmental factors and that is why, most of the insect's forewarning and forecasting models are based upon accumulated degree-days from a base temperature and bio-fix point. Yamamura and Kiritani, (1998) estimated that with a 2°C rise in temperature, insects might experience one to five additional life cycles per season. Most researchers seem to agree that warmer temperatures result in more types and higher populations of insects. It has been reported that the temperature has highly significant effects on all developmental parameters of insect than humidity and rainfall. Developmental time of an insect is inversely related to temperature. Among the various environmental factors that influence the silkworm cocoon crops, the most important are temperature followed by humidity. Temperature has a direct effect on the growth, development and physiological activity, nutrient absorption, digestion, blood circulation, respiration etc. Temperature plays an important role in egg hatching, larval growth and quality of the cocoons produced, adult fertility etc. With the increase in temperature, the larval growth and development is accelerated resulting reduction in larval duration, cocoons of lower weight and quality, while at low temperature, growth and development is slow leading to prolonged larval period, abnormal growth and sensitivity against several diseases.

The summer crops of eri silkworms are most unpredictable as they are reared under formidable weather conditions (Sahu *et al.,* 2006). Silkworm rearing may be seriously affected by unfavourable temperature that affects every stage in the life cycle of eri silkworm. The physiology of silkworm, assimilation, metabolic rate, enzyme activity and nutrients conversion are influenced by ambient temperature. Silkworms are more sensitive to temperature during 4th and 5th instar larval stages. The late age worms particularly the 5th stage larvae cannot tolerate high temperature and high relative humidity. Eri silkworms are capable of growing in temperature range from 15°C to 40°C. However, the ideal temperature for the physiological activity ranges from 20°C to 30°C and maximum productivity is observed while the temperature ranged from 23°C to 28°C (Sharma *et al.,* 2003).

Second abiotic factor that has significant impact on the performance of insects in terrestrial environments is humidity. Humidity interacts with free water availability and with the water content of the food plants. Humidity shows mostly indirect effect on growth and development. Under too dry conditions, the leaves wither very fast and become unsuitable for insect feed, resulting in retarded growth of larvae, which makes them weak and easily susceptible to diseases and other adverse conditions. During egg incubation, it is important that humidity should be maintained at 80% on an average for normal growth of embryo. If humidity falls below 70% during incubation, it affects hatching. The relative humidity also plays a vital role in silkworm rearing. The humidity influences the physiological activity of silkworm such as the amount of ingestion and digestion. Further, humidity affects the rate of withering / drying of leaf in rearing beds; there of affecting its suitability as feed and consumption etc. Host preference of forb chewers does not correspond precisely with host nutritional suitability but can be influenced by ecological factors such as natural enemies, host phenology and microclimate (Gogoi and Yadav, 1995). Studies have also indicated the combined effects of temperature and relative humidity. Temperature range of 23 to 31°C and relative humidity of 70 to 80% is required for grainage operations in Assam (Samson and Barah, 1989). According to Engelmann, (1984) relative humidity ranged from 72 to 78%, was also congenial for development of different stages of eri silkworm during grainage.

Rainfall alters the functioning of microhabitat, which along with soil and other environmental factors affects foliage and water levels, which consequently affect on performance of insect. Chemically the variation from succulent leaves to mature is so large that it may halve the growth rate of lepidopteran larvae. Fluctuation in temperature, humidity and other seasonal factors prevent insects from attaining their physiological potential life history performance. However, forest insects display a remarkable range of adaptations to changing environments and maintain their internal temperature (thermoregulation) and water content within tolerable limits, despite wide fluctuations in their surroundings, because impact of temperature is modified by habitat and other physical conditions.

Temperature, humidity, air circulation, gases and photoperiod etc. show a significant interaction in their effect on the physiology of silkworm depending upon the combination of factors and developmental stage affecting growth, development, productivity and quality of silk. Seasonal and environmental interaction have great role to play on the expression of commercial characters of Eri silkworm.

Season indicates the inter-annual variation in temperature, humidity, sunshine, rainfall, etc. of a particular place, which is governed by different geographical parameters. Studies have shown the importance of seasonal variations in biology and development of insect. The potential phenotypic expression of a genotype desires suitable environment and the environment being exogenic factor influences the expressivity of gene by generating different phenotypes under different environments. The variations in climate, nutrient status, feeding duration and larval crowd along with environmental stimuli influence the insect body size and its biological success. Natural population is often heterogeneous and undergoes continuous selection by the local environmental fluctuations and by geographic transition. Green *et al.*, (1996) mentioned that the ecological range of most organisms especially the poikolothermic animals is determined by the available food, ambient temperature, humidity etc. According to Krishnaswamy *et al.*, (1973), the silkworm rearing seasons are broadly classified as spring, summer, and rainy crops, based on the temperature, humidity, and rainfall.

Identification of best rearing season is a prerequisite for the success of eri-business in a given area. Sengupta, (1988) opined that the success of ericulture in Japan could be attributed to the identification of the suitable rearing seasons. Jayswal *et al.*, (1990) reviewed the interaction between genotype and environment for economic traits in hybrid combinations of silkworm and asserted that environment causes effective restrain on biological characters of silkworm like, effective rate of rearing (ERR), single cocoon weight, single shell weight, shell ratio and filament length. Shiva Kumar *et al.*, (1997) also concluded that the quality of food and rearing temperature, largely influence silkworm productivity.

There are several workers who reported the effect of the season on the biological parameters of different silkworms. Gupta *et al.*, (1992) reported on egg yielding capacity of different races of silkworm, *B. mori*. There is extensive information available on effect of season, micronutrients and different food plants on rearing performance of eri silkworm. Ray *et al.*,

(2010b) studied on the effect of food plant and seasons on the larval development and cocoon characters of eri silkworm. Jaiswal *et al.,* (2006) made comparative analysis of rearing performance of different ecoraces of eri silkworm in monsoon season of Uttar Pradesh. Gangwar (2011) studied the effect of seasonal variation on the survival percentage and cocoon yield in some breeds of eri silkworm. The seasons of the different region influence the larval growth, larval duration, ERR%, pupation%, cocoon weight, pupal weight, shell weight and shell ratio (%). The effect of seasons on the growth and development in the insects has been reported by Reddy *et al.,* 1998 and Muthukrishnan and Pandian (1987).

Shree and Nagaveni (2002) reported minimum pupal period (13 days) during summer season compared to other rearing seasons. Sahu *et al.,* (2006) reported average hatching percentage of 75.67 % during summer season compared to spring and autumn seasons. Kichu *et al.,* (2014) reported that autumn is the best season followed by midsummer and late summer for eri silk worm rearing under Nagaland conditions. Hussain *et at.,* (2011) observed highest larval weight (5.20 g), single cocoon weight (3.26 g), shell weight (0.50 g), shell ratio (15.34 %) and ERR (94.80%) during summer season. Hazarika *et al.,* (2003) noticed that the summer season favours minimum larval period, maximum larval weight, ERR %, single cocoon weight, shell weight and hatching percentage. Phukan *et al.,* (2006) observed shortest larval period during summer season compared to autumn season. Singh and Das, (2006) found highest hatching, single cocoon weight, shell weight, shell ratio and highest ERR during summer season. Debaraj *et al.,* (2003) noticed highest larval weight, cocoon weight, shell weight and shell ratio during summer season (May-June). These economic characters of eri silkworms are not only controlled by genes, but also known to be influenced by different climatic factors such as temperature, relative humidity, photoperiodic cycle etc (Jaiswal *et al.,* 2006). Mahobia *et al.,* (2005) reported highest mature larval weight (6.12g), highest single cocoon weight, shell weight, shell ratio percentage, hatching percentage during summer (June) rearing, which was less in winter and rainy seasons. Kar *et al.,* (2004) recorded commercial traits of cocoons of eri silkworm in summer season and observed highest hatching percentage, single cocoon weight, shell weight, and shell ratio. Further, they also found the survival rate of silkworm was lower during rainy and winter seasons compared to summer season and observed minimum pupal period and highest larval weight during summer season.

Eri silkworm is a delicate, domesticated insect which cannot tolerate diurnal and seasonal fluctuations in the environmental condition. The life cycle of silkworm is influenced by bio-ecological stress during larval period. Deviations in humidity and temperature levels below and above certain critical limits affect larval growth and development of eri silkworm. Day to day change in weather during larval rearing poses great threat to the cocoon crop. The impact of temperature and relative humidity on commercial traits of cocoon as well as biological aspects has been investigated extensively in various countries. Sugai and Takashashi (1981) and Sahu *et al.,* (2006) reported that the tolerable range of temperature for the growth of silkworm *Bombyx mori* and eri silkworm is from 20 – 40°C. However, increase in temperature beyond 35°C causes less spinning, mortality of larvae and pupae, poor moth emergence and sterility at adult stage. Gohain and Borua, (1982) reported that shortest development period, highest larval weight, oviposition and silk material of eri silkworm were found in 28±2°C and 26±2°C at 70% RH. More pupal weight and higher weight of silk material were observed in 26±2°C at 70% RH. Ullal and Narasimhanna, (1987) reported that the high temperature followed by strong fluctuations results in poor quality cocoons of eri silkworm. Anonymous, (1987) reported that at low humidity larvae lose water from their body in considerable amount which results in inefficient metabolism. Further low humidity also leads to less conservation of energy for pupae and adult stages. This may adversely affect egg laying ability of the female moths. Pandey Priyanka *et al.,* (2006) observed that the variation in temperature and relative humidity has significantly influenced the larval duration of *S. c. ricini.* Pandey and Tripathi (2008) indicated that decrease in the humidity from 80% to 75% caused increase in the larval weight and survival rate of larvae but further decrease in the humidity caused decline in the survival rate, hatching percentage and larval weight of silkworm *B. mori.* Mubashar Hussain *et al.,* (2011) observed that rearing of silkworm larvae at lower level of RH resulted in lower fecundity, lower hatching, lower pupation and higher larval mortality.

Debaraj *et al.,* (2002) found highest fecundity in Titabar ecorace and minimum in Dhanubhanga ecorace. Ravishankar *et al.,* (1998) and Sakthivel *et al.,* (2004) reported that there was less fecundity during low temperature as compared to high temperature. Kar *et al.,* (2004) reported highest fecundity during winter season and during spring rearing. Brahma *et al.,* (2006) observed a maximum fecundity during winter season. Singh and Das (2006) studied the effect of temperature on fecundity and observed the highest fecundity during winter and lowest during spring season. Chakravorty and Neog (2006) observed highest fecundity in Borduar eco-race and minimum in Dhanubhanga eco-race. Many researchers have observed highest larval duration and maximum larval weight (7.64g), green cocoon weight (3.74g), shell weight (0.49g), shell ratio (13.16%) during low temperature. Benchamin, (2000) reported that fecundity was higher in winter and low in summer season. Somaprakash and Sathyaprasad, (2009) reported that low temperature adversely affected the hatching percentage of eri silkworm. It was also found that during autumn season some eco-races of the eri silkworm had prolonged larval duration because of low temperature, reduced rate of metabolism resulting in the slow growth. Singh *et al.,* (2003) and Chakravorty, (2004) observed maximum shell weight (0.53g) in Borduar ecorace and highest shell ratio percentage (14.64) in Kokrajhar eco-race.

In silkworm, which passes through a non feeding adult stage, growth attained by larvae at termination of feeding and prior to undertaking pupation represents the total growth.

The growth and development of silkworm larva depends on quality of leaves, rearing condition, environmental factors, diseases and pest attack. A striking feature of the insect life cycle is its physiological and behavioural adaptations to the seasonally changing environmental conditions such as photoperiod, temperature and drought (Yamashita and Hasegawa, 1985). The major impact of light on performance of an insect, results from the daily rhythm of day and night and the seasonal change in day length (Thomas and Philogene, 1979; Mishra, 2011) or Photoperiod. An insect serves as a clock indicating the seasonal changes influencing its life cycle, distribution and abundance. Photoperiodism is measured in terms of day length, in the strict sense, is measured as the hours and minutes elapsing between sunrise and sunset.

Influence of Environmental Factors on Biology and Economic Traits of Eri Silkworm in Different Regions of India

Effect of Temperature and Relative Humidity

Madhya Pradesh

Effect of temperature and relative humidity on rearing performance of eri silkworm at the department of Zoology, PGMVM College, Ujjain, Madhya Pradesh

To study the effect of temperature and humidity on larval parameters (larval length, larval weight, larval duration, ERR and pupation), cocoon parameters (cocoon weight, shell weight, pupal weight and shell %) and grainage parameters (fecundity and hatching) of eri silkworm, Vaidya *et al.*, (2014) conducted an experiment at four different temperatures viz., 20±3°C, 24±3°C, 28±3°C and 32±3°C, each temperature with three humidity levels (50%, 70% and 90%). The experiment was conducted in the department of Zoology, PGMVM College, Ujjain, Madhya Pradesh.

The results of Vaidya *et al.*, (2014) indicated that the larval weight, ERR%, pupation %, cocoon weight, pupal weight, shell weight, shell %, fecundity and hatching % were observed maximum in 70% relative humidity and 28±3°C temperature. The shortest larval duration of 19 days, highest cocoon and pupal weights of 3.39 g and 2.84 g respectively were recorded in 70% relative humidity at 28±3°C temperature (Tables 10.1, 10.2, 10.3 and 10.4). From the study of Vaidya *et al.*, (2014), it can be concluded that the temperature of 28±3°C and 70% relative humidity are the best for rearing of eri silkworm for sustainable development of the rearers of Ujjain district.

Table 10.1: The rearing performance (larval, cocoon and grainage parameters) of Eri silkworm in (20±3°C) at different relative humidity conditions (Vaidya *et al.*, 2014)

Larval parameter	Humidity		
	50%	70%	90%
Larval length	4.1 cm	5.3 cm	5.1 cm
Larval weight	6.5 g	8.5 g	7.8 g
Larval duration	30 days	26 days	32 days
ERR%	75%	80%	82%
Pupation	80%	85%	76%
Cocoon parameter			
Cocoon weight	2.54 g	2.90 g	2.70 g
Pupal weight	2.22 g	2.55 g	2.54 g
Shell weight	0.34 g	0.38g	0.35 g
Shell %	13.3%	13.9%	12.9%
Grainage parameter			
Fecundity	371	385	350
Hatching	60%	82%	75%

Table 10.2: The rearing performance (larval, cocoon and grainage parameters) of Eri silkworm in (24±3°C) at different relative humidity conditions (Vaidya *et al.,* 2014)

Larval parameter	Humidity		
	50%	**70%**	**90%**
Larval length	5.5 cm	7.3 cm	6.1 cm
Larval weight	8.3 g	9.5 g	8.4 g
Larval duration	28 days	22 days	24 days
ERR%	78%	96%	87%
Pupation	81%	92%	82%
Cocoon parameter			
Cocoon weight	2.92 g	3.15 g	3.08 g
Pupal weight	2.47 g	2.66 g	2.61 g
Shell weight	0.45 g	0.50 g	0.46 g
Shell %	15.4%	15.6%	14.9%
Grainage parameter			
Fecundity	376	378	355
Hatching	83%	95%	75%

Table 10.3: The rearing performance (larval, cocoon and grainage parameters) of Eri silkworm in (28±3°C) at different relative humidity conditions (Vaidya *et al.,* 2014)

Larval parameter	Humidity		
	50%	**70%**	**90%**
Larval length	5.2 cm	6.5 cm	5.5 cm
Larval weight	8.3 g	10.6 g	7.5 g
Larval duration	21 days	19 days	22 days
ERR%	80%	96%	85 %
Pupation	82%	93%	85 %
Cocoon parameter			
Cocoon weight	3.01 g	3.39 g	3.22 g
Pupal weight	2.55 g	2.84 g	2.74 g
Shell weight	0.42 g	0.55 g	0.48 g
Shell %	13.01%	16.2%	14.9%
Grainage parameter			
Fecundity	380	450	386
Hatching	82%	96%	84%

Table 10.4: The rearing performance (larval, cocoon and grainage parameters) of Eri silkworm in (32±3°C) at different relative humidity conditions (Vaidya *et al.,* 2014)

Larval parameter	Humidity		
	50%	**70%**	**90%**
Larval length	4.3 cm	4.5 cm	4.5 cm
Larval weight	6.2 g	6.4 g	6.0 g
Larval duration	24 days	25 days	25 days
ERR%	73%	82%	79 %
Pupation	60%	71%	52 %

Table 10.4: *Contd…*

Cocoon parameter			
Cocoon weight	2.98 g	3.04 g	2.93 g
Pupal weight	2.62 g	2.66 g	2.58 g
Shell weight	0.36 g	0.38 g	0.35 g
Shell %	12.07%	12.40%	11.09%
Grainage parameter			
Fecundity	310	280	277
Hatching	62%	70%	55%

Uttar Pradesh

Effect of temperature and relative humidity on moth emergence and fecundity of eri silkworm at Department of Applied Animal Sciences, Babasaheb Bhimrao Ambedkar University, Lucknow, Uttar Pradesh

The eri silkworm and its grainage techniques are quite different from that of other non mulberry silkworms namely, Tasar and Muga. The ideal temperature for the physiological activity ranges from 20°C to 30°C. However, increase in temperature beyond 35°C causes less spinning, mortality of larvae and pupae, poor moth emergence and sterility at adult stage (Sugai and Takashashi, 1981; Sahu *et al.*, 2006). Though there is plenty of information on the effect of host plants on the growth and development of eri silkworm, but the information on the effect of temperature and relative humidity on grainage parameters of eri silkworm is scanty. In this regard, Rajesh Kumar and Elangovan (2014) investigated the effect of temperature and relative humidity on moth emergence and fecundity of different eco-races of eri silkworm in the Department of Applied Animal Sciences, Babasaheb Bhimrao Ambedkar University, Lucknow, Uttar Pradesh during winter, 2009 and spring, 2010. The eco-races of eri silkworm such as Borduar, Mendipathar, Titabar and Dhanubhanga were selected for the experiment to evaluate their rearing performance. Experiments were conducted from second moult to harvesting of cocoons. For each experiment, sum of 400 larvae of each eco-race were maintained separately in wooden trays in three replicates. Ambient temperature and relative humidity of the experimental chamber were recorded regularly using thermometer and hygrometer respectively. These eco-races were subjected to different temperature and relative humidity treatments i.e. 15±1 to 21±1°C and 56±5 to 64±5% during winter season, 2009 and 22±1 to 28±1°C, 65±5 to 72±5% during spring season, 2010. Observations on moth emergence and fecundity, larval duration, larval weight, single cocoon weight, single shell weight and shell ratio were recorded.

There were no significant differences among the eco-races of eri silkworm on moth emergence during winter 2009 and spring 2010 rearing seasons. But there was a significant difference on moth emergence between winter and spring seasons when all the races were considered. The highest moth emergence was recorded in Borduar ecorace both during winter season 2009, (92.50%) and spring, 2010 (98.00%) rearing seasons (Table 10.5). The larval weight of different ecoraces ranged between 5.20 g to 6.10 g during winter rearing, while it was ranged from 6.15 g to 7.65 g during spring 2010 (Table 10.6). However, the highest larval weight was observed in Borduar ecorace in both the seasons. The highest shell ratio of 14.10% was observed in Borduar ecorace during winter 2009 and highest shell ratio of 16.91% in Mendipathar ecorace during spring 2010 rearing seasons. The results of the Rajesh Kumar and Elangovan (2014) clearly indicated that the temperature and relative humidity has profound effect on moth emergence and fecundity of different eco-races of eri silkworm. Further, lower larval duration and enhanced larval weight and higher shell ratio were observed during spring 2010 among all the ecoraces compared to winter 2009 clearly revealing the effect higher temperature prevailing during spring. However, the influence of higher temperature during spring was not considerable and consistent regarding cocoon weight and shell weight among different ecoraces of eri silkworm.

Table 10.5: Effect of temperature and relative humidity on moth emergence and fecundity of different eco-races of eri silkworm during winter (2009) and spring (2010) rearing (Rajesh Kumar and Elangovan, 2014)

Eco-race	Winter, 2009 (A) (15±1 to 21±1°C and 56±5 to 64±5% RH)		Spring, 2010 (B) (22±1 to 28±1°C, 65±5 to 72±5% RH)	
	Moth emergence (%) (Mean ± SD)	Fecundity (No.) (Mean ± SD)	Moth emergence (%) (Mean ± SD)	Fecundity (No.) (Mean ± SD)
Borduar	93.50 ± 1.80	345 ± 4.58	96.50 ± 1.80	375 ± 11.53
Mendipathar	92.00 ± 2.78	332 ± 12.16	95.50 ± 2.59	356 ± 9.53
Titabar	90.50 ± 3.12	298 ± 6.55	96.00 ± 2.69	345 ± 10.53
Dhanubhanga	91.00 ± 2.17	317 ± 10.14	95.00 ± 3.27	332 ± 6.92
F-Value	0.82	15.56	0.18	10.38
P-Value	P > 0.05	P < 0.01	P > 0.05	P < 0.01

(A) = Mean temperature 18 ± 1°C and humidity 65 ± 5 % during winter

(B) = Mean temperature 25 ± 1°C and humidity 68.5 ± 5 % during spring

Table 10.6: Effect of temperature and relative humidity on cocoon parameters of different eco-races of eri silkworm during winter (2009) and spring (2010) seasons (Rajesh Kumar and Elangovan, 2014)

Eco-race	Larval duration (days)		Larval weight (g)		Cocoon weight (g)		Shell weight (g)		Shell ratio (%)	
	Mean ± SD (2009) (A)	Mean ± SD (2010) (B)	Mean ± SD (2009) (A)	Mean ± SD (2010) (B)	Mean ± SD (2009) (A)	Mean ± SD (2010) (B)	Mean ± SD (2009) (A)	Mean ± SD (2010) (B)	Mean ± SD (2009) (A)	Mean ± SD (2010) (B)
Borduar	27.75±0.50	18.00±0.50	6.15±0.26	7.48±0.15	3.19±0.16	3.52±0.20	0.45±0.04	0.56±0.05	14.10±0.11	15.90±0.07
Mendipathar	29.00±0.50	19.25±0.75	6.00±0.24	7.25±0.14	3.45±0.12	3.15±0.28	0.48±0.08	0.51±0.08	13.92±0.11	16.19±0.13
Titabar	28.50±0.25	18.75±0.25	6.10±0.13	7.65±0.10	3.00±0.09	3.36±0.07	0.41±0.06	0.52±0.07	13.66±0.12	15.48±0.09
Dhanubhanga	29.00±0.50	19.50±0.25	5.20±0.13	6.15±0.15	2.69±0.16	2.68±0.10	0.36±0.05	0.42±0.07	13.38±0.05	15.67±0.09
F-Value	5.15	5.60	14.48	69.33	15.72	11.70	2.02	1.90	27.94	26.26
P-Value	$P < 0.05$	$P < 0.05$	$P < 0.01$	$P < 0.01$	$P < 0.01$	$P < 0.01$	$P > 0.05$	$P > 0.05$	$P < 0.01$	$P < 0.01$

(A) = Mean temperature $18 \pm 1°C$ and humidity 65 ± 5 % during winter

(B) = Mean temperature $25 \pm 1°C$ and humidity 68.5 ± 5 % during spring

Influence of photoperiod

Assam

Effect of photoperiod on larval growth of eri silkworm in two different seasons in the Department of Sericulture, Assam Agricultural University, Jorhat

Several workers have demonstrated the influence of photoperiod on behavioural pattern in mulberry silkworm (*Bombyx mori* L.) such as hatching, larval growth etc, but in eri silkworm the information is lacking. Hence the present study was carried out by Saikia *et al.,* (2014) to find out the effect of photoperiod on larval growth and development of Eri silkworm. The experiment was conducted in the Department of Sericulture, Assam Agricultural University, Jorhat during 2012-13. The experiment was conducted in three different rearing rooms *viz.,* normal room with natural dark-light condition, artificially lighted room and totally dark room. Two wooden cages (56 x 54 x 105 cm^3) in the lighted room and two other wooden cages in the dark room having three chambers each were placed. The control one was placed at normal laboratory condition. For rearing 1500 newly hatched healthy larvae were randomly selected. The larvae were distributed into fifteen rearing trays (36 x 26 x 7 cm^3) containing 100 numbers in each and they were placed inside the rearing cages and fed on Castor leaves according to their treatment only. The temperature and humidity ranged at 23^{0}C ± 3^{0}C and 80 ± 5% in spring (March-April) and 23^{0}C ± 2^{0}C and 82 ± 5% in autumn (October-November) season with the light intensity being 40 lux. The experimental larvae were exposed to five different photoperiodic regimes (daily cycles). They were:

1. T_1 : Normal laboratory condition (Control)
2. T_2: 24 hrs light.
3. T_3 : 24 hrs dark.
4. T_4 : 8 hrs light followed by 16 hrs dark
5. T_5: 16 hrs light followed by 8 hrs dark.

Different observations made were instar wise larval duration (days), total larval duration, weight (g) of full grown larva (5th instar), weight of matured larva (5th instar) and effective rate of rearing (ERR%).

Table 10.7: Larval duration of *Samia c. ricini* during spring and autumn season under different photoperiods (Saikia *et al.,* 2014)

Treatments	Instars (days)										Total larval duration (days)	Total larval duration (days)
	1st (Spring)	1st (Autumn)	2nd (Spring)	2nd (Autumn)	3rd (Spring)	3rd (Autumn)	4th (Spring)	4th (Autumn)	5th (Spring)	5th (Autumn)	(Spring)	(Autumn)
T1	2.67	2.57	3.73	3.63	3.87	5.60	6.00	6.53	8.40	9.20	24.67	27.53
T2	2.80	2.67	3.47	4.50	4.27	6.37	6.40	7.00	9.20	9.13	26.13	29.67
T3	2.07	2.77	3.13	3.60	3.13	5.14	5.73	5.47	8.53	8.20	22.59	25.17
T4	2.53	2.73	3.47	4.03	3.73	5.10	5.13	6.50	9.27	8.30	24.13	26.67
T5	2.13	2.50	2.93	3.37	3.67	4.50	4.97	5.00	7.40	8.10	21.10	23.47
SED±	0.20	NS	0.12	0.16	0.15	0.12	0.15	0.17	0.12	0.17	0.23	0.29
CD (5%)	0.45	NS	0.28	0.37	0.34	0.28	0.35	0.39	0.28	0.40	0.53	0.68

Table 10.8: Fifth Instar larval weight and ERR (%) during spring and autumn season under different photoperiods (Saikia *et al.,* 2014)

Treatments	Weight (g)				ERR (%)	
	Full grown (Spring)	Full grown (autumn)	Matured (spring)	Matured (autumn)	Spring	Autumn
T1	7.31	5.47	6.09	4.32	82.48	78.30
T2	6.20	5.29	5.10	4.16	76.93	76.45
T3	7.40	6.23	6.23	4.85	87.15	83.25
T4	6.82	6.10	5.40	4.82	83.32	80.97
T5	7.90	6.27	6.50	4.12	85.25	84.01
S.ED±	0.23	0.20	0.15	0.18	1.04	0.90
CD (5%)	0.53	0.46	0.35	0.42	2.39	2.07

The study by Saikia *et al.,* (2014) revealed that photoperiod has significant effect on the instar wise larval duration as well as total larval duration.   In both the seasons, total larval duration was found to be significantly the lowest in the fifth treatment which was shortened by 16.92% in spring and 17.30% in autumn compared to control i.e. normal laboratory conditions. Treatment three showed shorter larval duration than control and was next to T_5 (Table 10.7). However, the longest larval duration was recorded in the II treatment among the treatments including control. In case of instar wise larval duration also, in almost all the instars treatment five showed comparatively shorter larval duration than all other treatments and control. However, in autumn season the larval duration was slightly longer than that of spring season. Larval duration in *Bombyx mori* was delayed by few days when reared at continuous darkness compared to continuous light (Rajanna, 1987). The highest full grown and matured larval weights were recorded in T_5 in both the seasons (Table 10.8). Similar findings were also reported by Nuzhat and Delvi, (1998) in *Bombyx mori* L. and Rao, (1988) in *Samia c. ricini* Boisduval. The ERR% was found to be the highest in T_3 (87.15%) in spring season and T_5 (84.01%) in autumn season which were at par with T_5 (85.25%) during spring and T3 (83.25%) during autumn season and differed significantly from all other treatments including control.

The study by Saikia *et al.,* (2014), establishes that light has significant effect on growth and development of eri silkworm and among the treatments evaluated, LD 16:8 hrs emerged out to be the better photoperiodic range for rearing of eri silkworm pertaining to larval duration and matured larval weight. However further experiment would be necessary to utilize beneficially the photoperiodic effect for commercial eri silk production.

Effect of photoperiod during egg incubation on certain life parameters of eri silkworm in the Department of Sericulture, Assam Agricultural University, Jorhat

The growth of silkworm larvae mostly depends on environmental factors and rearing conditions. Photoperiod is the environmental rhythm of recurring alternation of light illumination (Photo phase) and darkness (Scoto phase) and is extremely important for biological and physiological process of many insect species. Kogure, (1930) elucidated the photoperiodicity in silkworm *Bombyx mori* and established the relationship between illumination and voltinism. It was reported that photoperiodic cycle plays an important role in determination of egg hatching and emergence of silk moth and all functions of silkworm are affected by the photoperiod from emergence to cocoon yield. Effect of photoperiod during rearing of eri silkworm has been studied by Rao, (1988) and Saikia, (2001). Bordoloi *et al.,* (2014) investigated the effect of photoperiod during egg incubation on larval growth, cocoon and reproductive parameters of eri silkworm.

The eggs of eri silkworm were collected from the culture maintained in the Department of Sericulture, Assam Agricultural University, Jorhat. Eggs were disinfected with 2% aqueous formaldehyde solution followed by washing in plain water. The disinfected eggs were distributed separately in petri dishes containing 100 eggs per each replication of light & dark hour (LD) treatments and were incubated at 25 ± 2^0C with $75 \pm 2\%$ RH and light intensity of 150 lux in B.O.D incubators till hatching. The photoperiod and scoto period of LD 8:16 h; LD 10:14 h; LD 12:12, LD 14:10, LD 16:8, LD 24:0h and LD 0:24h were maintained during incubation with four replications in each treatment. A control batch of eggs was also incubated under normal laboratory condition for comparison. The hatched larvae were reared under normal laboratory condition. The experiment was conducted in two commercial rearing seasons *viz.,* spring and autumn during 2011-2012. Observations were recorded on different larval growth, cocoon and reproductive parameters.

Results revealed that photoperiodic treatment during egg incubation had significant effect on larval growth, cocoon and reproductive parameters of eri silkworm. Increase of light during egg incubation increased egg hatching, larval weight, cocoon weight and shell weight while incubation period and larval duration decreased. However, exposing the eggs to complete light (LD 24:0h) increased the incubation period with decrease of egg hatching. Significantly the highest egg hatching (88.33%) with shortest incubation period (6.67 days) was registered for the treatment LD 16:8 hr. which was at par with the treatments LD 14:10 h. and LD 12:12 h. Reddy *et al.,* (1998) reported variation in incubation period under low and high light intensities in silkworm, *Bombyx mori* and found that hatching duration increased with increase of light duration and decreased in low light duration. The photoperiod treatment LD 16:8 h was registered better and found to be at par with LD 14:10 h in respect of larval weight, cocoon weight and shell weight with shortest larval duration. Egg incubation in complete darkness (LD 0:24 hrs) exhibited better in respect of moth emergence (71.66%) and fecundity (379 eggs/female). Similar results were also reported by Rao (1988) and Saikia (2001) treating with different photoperiodic duration during larval stage of eri silkworm. However, it was reported that the larvae of *B. mori* grow slowly with the increase of light period during rearing. Thus, it is imperative to conclude that photoperiodic treatment during egg incubation had significant effect on larval growth, cocoon and reproductive parameters of eri silkworm. The photoperiodic treatment LD 16:8 h is found to be optimum for better larval growth and cocoon parameter while total darkness LD 0:24 h is better for moth emergence and fecundity.

Influence of Seasonal Variation on Economic Traits of Eri Silkworm in Different Regions of India

Punjab

Performance of Eri silkworm in different seasons of Punjab at Sericulture Laboratory, Punjab Agricultural University, Ludhiana

Landless farmers and labourers of sub-mountainous area of Punjab are already engaged in rearing of bivoltine races of mulberry silkworm. They have developed their infrastructure for sericulture but rearing is being done only in two seasons i.e. spring (February - April) and autumn (August – October). The interjecting of ericulture during the lean period of mulberry sericulture can improve the financial condition of the cultivators as well as the economy of textile industry in Punjab. Therefore, Virk *et al.,* (2009) have undertaken a study to evaluate the performance of eri silkworm during different seasons in Punjab. The studies on the performance of eri silkworm were conducted in the Sericulture laboratory, Entomological Research Farm, Punjab Agricultural University, Ludhiana during different seasons i.e. from July, 2006 to March, 2007. Meteorological data were obtained from Department of Agro-meteorology located in the same university (Table 10.9). Borduar race of eri silkworm were used for the experiment. Ten replications of 10 larvae/ replicate were kept separately and provided with castor leaves after every 4-6 hours for feeding. Observations were recorded on larval period, pupal period, total developmental period, per cent cocoon formed, larval weight, cocoon weight (green and dry), shell weight and pre-oviposition period. The experiment was repeated till March, 2007, when four generations were completed. During 4th generation (December, 2006 – March, 2007), the different variants of eri silkworm namely yellow plain, yellow zebra, blue plain and blue zebra were reared separately to evaluate their performance.

The eri silkworm completed four continuous generations from July 2006 to March 2007. The smallest life cycle (41 days) was recorded during July-August (first generation) and the longest (88 days) during the winter season i.e. December-March (fourth generation). Ahmed (1957) reported the life cycle as 48-65 days in summer and 91-114 days in winter. Mavi *et al.,* (1997) reared three crops in different seasons and recorded total life cycle to vary from 51 to 93 days in different seasons. The egg period varied from 3.0 to 17.5 days, being the shortest (3 days) during July-August and the longest (17.5 days) during Dec-March (winter season). Similarly, the shortest larval period (19.0 days) was recorded during July-August and it was significantly lesser than winter season (fourth generation) where it was longest (37.0 days). The pupal period was the longest in winter months (28.5 days) during 4th generation (Dec-March) and minimum pupal period (16.0 days) was recorded during July-August (Table 10.10). The pre-oviposition period varied from 3.0 to 6.0 days during different generations. The present results reveal that the temperature and relative humidity variation during July 2006 to December 2007 influence the development of eri silkworm.

The maximum per cent cocoon formation (97.5%) was recorded during second generation (Sep-Oct), followed by first generation (95.6%). The per cent cocoon formation in these two generations was significantly higher than third and fourth generations completed during October to March. The maximum per cent cocoon formation during warm months might be due to suitable temperature and relative humidity for the normal development of eri silkworm and shorter life cycle. Maximum larval weight (46.82 g/10 larvae) was recorded during 4th generation and it was on par with larval weight during second generation (42.36 g/10 larvae) (Table 10.10). The dry shell weight varied from 3.82 to 4.26 g/10 shells and there was no significant difference among different generations. The gain in weight might be due to prolonged development period of eri silkworm during winter months. During July-August, the eri silkworm had the shortest development period and larvae or cocoons had lower weights.

There were no significant differences among the larval period of different variants. However, the larval period varied from 35.11 to 38.80 days, being maximum in blue zebra and minimum in yellow plain (Table 10.11). The maximum larval weight (47.34 g/10 larvae) and green cocoon weight (34.03 g/10 cocoons) was recorded in yellow plain strain and these were on par with blue plain (46.98 g/10 larvae and 33.72 g/ 10 cocoons respectively), strain with no statistical variation among them. The dry cocoon weight was maximum (15.72 g/10 cocoons) in yellow plain and it was significantly higher than all other variants. On the basis of data on larval weight, cocoon weight and shell weight, the yellow plain variant of eri silkworm was the best under Punjab conditions. Similarly, Sehdev (1999) found yellow plain variant was the best as compared to others.

Table 10.9: Meteorological data pertaining to April 2006-March 2007 (monthly mean values) (Virk *et al.,* 2009)

Month	Temperature (^{0}C)		Relative Humidity (%)	
	Maximum	**Minimum**	**Morning**	**Evening**
July, 2006	34.2	27.4	88	67
August, 2006	33.8	26.3	86	68
September, 2006	32.9	23.1	92	61
October, 2006	32.0	18.5	89	46
November, 2006	26.4	12.8	94	45
December, 2006	20.1	7.2	95	54
January, 2007	19.5	4.2	97	48
February, 2007	20.3	10.2	95	64
March, 2007	26.1	12.4	91	44

Table 10.10: Performance of eri silkworm from July 2006 to March 2007 under Punjab conditions (Virk *et al.,* 2009)

Genera-tion	Egg period (Days)	Larval period (Days)	Pupal period (Days)	Pre-oviposition period (Days)	Per cent cocoon formation	Larval weight (g/10 larvae)	Green cocoon weight (g/10 larvae)	Dry cocoon weight (g/10 larvae)	Shell weight (g/10 shell)
I	3.0	19.0	16.0	3.0	95.6	37.82	25.36	10.47	3.82
II	3.5	24.0	20.5	5.0	97.5	42.36	27.45	11.45	4.16
III	9.0	23.5	25.5	6.0	80.82	37.24	24.67	10.14	3.98
IV	17.5	37.0	28.5	5.0	71.25	46.82	30.36	13.18	4.26
CD (p=0.05)	-	4.2	3.6	-	8.72	3.46	2.36	1.64	NS

Table 10.11: Performance of different variants of eri silkworm during December 2006 to March 2007 under Punjab conditions (Virk *et al.,* 2009)

Variant	Larval period (Days)	Larval weight (g/10 larvae)	Green cocoon weight (g/10 cocoons)	Dry cocoon weight (g/10 cocoons)
Yellow plain	35.11	47.34	34.03	15.72
Yellow zebra	38.50	46.24	33.40	14.20
Blue plain	37.50	46.98	33.72	14.36
Blue zebra	38.80	45.42	32.27	13.10
CD (p=0.05)	NS	0.92	0.57	0.34

Madhya Pradesh

Rearing performance of eri silkworm in different seasons of Ujjain district, Madhya Pradesh

To know the rearing performance of eri silkworm, Vaidya and Yadav (2014) conducted an experiment in the department of Zoology, PGMVM college, Ujjain, Madhya Pradesh during three different seasons *viz.,* Autumn (September– October, 2010), winter (November – December, 2010) and spring (January – February, 2011). The abiotic factors *viz.,* maximum and minimum temperature and relative humidity were also recorded for rearing period (Table 10.12). Observations were recorded on larval and cocoon parameters. The results of the experiment revealed that minimum larval duration of 19 days was recorded in spring season followed by autumn (20 days) and winter (26 days). The maximum larval weight of 7.5g was recorded in spring and winter season followed by 7.30 g in autumn. Similarly, maximum larval length of 7.5 cm was recorded in spring followed by 6.20 cm in autumn and 5.5 cm in winter (Table 10.13). The cocoon weight was recorded maximum in spring 3.56g followed by 3.55g in winter and 3.49 g in autumn. Maximum pupal weight of 3.15 g was recorded in spring, while in autumn and winter season, the pupal weight was 3.10g (Table 10.14). Hence the overall results of the experiments indicated that the rearing performance of eri silkworm was better in spring season than winter and autumn season in Ujjain district of MP.

Table 10.12: Temperature and relative humidity in autumn, winter and spring seasons of Ujjain, MP (Adopted from Vaidya and Yadav, 2014)

Seasons	Temperature (^{0}C)	Humidity (%)
Autumn	27 ± 4	75 ± 3%
Winter	20 ± 3	70 ± 3%
Spring	22 ± 3	78 ± 3%

Table 10.13: Larval parameters of eri silkworm in autumn, winter and spring seasons of Ujjain, MP (Adopted from Vaidya and Yadav, 2014)

Larval parameter	Autumn	Winter	Spring
Larval duration	20 days	26 days	19 days
Larval weight	7.30 g	7.50 g	7.50 g
Larval length	6.20 cm	5.50 cm	7.50 cm
ERR	85%	85%	90%
Pupation	88%	89R	91%

Table 10.14: Cocoon parameters of eri silkworm in autumn, winter and spring seasons of Ujjain, MP (Adopted from Vaidya and Yadav, 2014)

Cocoon parameter	Autumn	Winter	Spring
Cocoon weight	3.49 g	3.55 g	3.56 g
Pupal weight	3.10 g	3.10 g	3.15 g
Shell weight	0.50 g	0.50 g	0.52 g
Shell (%)	14.68	14.50	15.19

Karnataka

Seasonal rearing performance of eri silkworm on castor and tapioca under South Karnataka conditions

Castor and tapioca, the two major food plants of eri silkworm are grown abundantly in southern states for castor oil and tapioca tuber. A comparative analysis of rearing performance of eri silkworm was carried out by Rajadurai *et al.,* (2010) on castor and tapioca in different seasons *viz.,* Summer (Feb-May), Rainy (June-Sep) and Winter (Oct-Jan) under South Karnataka conditions during the period 2004-06. Eri silkworm larvae were reared from chawki to spinning stage on healthy castor and tapioca leaves. Seasonal difference of climatological parameters during rearing of eri silkworm was presented in the Table 10.15. The study finds that the castor leaf is superior to tapioca leaf during all the seasons.

In case of castor, the shortest larval duration was found during summer (20.22 days), followed by rainy (21.18 days) and winter seasons (24.16 days). The larval weight (7.27g) was more during winter. The single cocoon weight (3.67g) was found to be more during rainy season. However, silk % (13.35%) and ERR (86.89%) were also more during rainy season (Table 10.20). In case of tapioca, larval duration was 24 days and 06 hours during winter, 24 days and 20 hours during summer and 25 days and 12 hours during rainy season. The larval weight was high during winter (6.8g) followed by summer (6.42g) and rainy season (6.1 g). The single cocoon weight (3.31g) and shell weight (0.45g) were high during summer. Higher ERR (78.23%) was observed during rainy season (Table 10.16). Mani *et al.,* (2002) in a study found that there was slight difference in the qualitative characters of amino acids in castor, kesseru and tapioca in that valine and proline were absent in kesseru and tapioca and the variation in the pattern of these constituents seems to account for relative nutritive inferiority of kesseru and tapioca leaves compared to castor leaves for the growth of eri silkworms. Castor leaf is reported to contain higher content of protein and fat (Joshi, 1981) and this could be one of the reasons for better larval growth and cocoon yield when eri larvae were fed on it. The increase in larval weight, cocoon weight, shell weight and silk (%) on castor leaves may be due to the higher leaf quality, higher rate of food ingestion and food assimilation.

Hence the results clearly indicated that the seasonal difference, climatological factors and host plant selection have profound influence on the rearing efficiency of eri silkworm. Evolving a high leaf yielding castor variety which suits in all the seasons and not compromising much reduction in seed yield should be the first step towards the popularization of ericulture in South India.

Table 10.15: Seasonal difference of climatological parameters during rearing of eri silkworm (2004-2006) at CSR&TI, Mysore (Rajadurai *et al.,* 2010)

Parameters	Summer (Feb-May)	Rainy (June-Sep)	Winter (Oct-Jan)	F-test
Temperature (^{0}C)				
Maximum	35.1	34.5	31.3	5.78[*]
Minimum	19.7	17.6	16.1	4.14[**]
Average	27.40	26.05	23.7	4.26[**]
Relative humidity (%)				
Maximum	60.00	70.00	90.00	3.43[**]
Minimum	30.00	45.00	55.00	3.81[**]
Average	45.00	57.50	72.50	2.37[**]
Rainfall (mm)	16.00	760.00	30.00	3.89[**]

Table 10.16: Seasonal rearing performance of eri silkworm on castor and tapioca (Rajadurai *et al.,* 2010)

Food plant/season	Larval duration (Days)	Larval duration (Hours)	Larval wt (g)	Cocoon wt (g)	Shell wt (g)	Pupa wt (g)	Silk (%)	ERR (%)
Castor								
Summer	20	22	7.13	3.35	0.39	2.96	11.64	80.60
Rainy	21	18	6.88	3.67	0.49	3.18	13.35	86.89
Winter	24	16	7.27	3.49	0.42	3.07	12.03	85.56
Mean	22	11	7.09	3.50	0.43	3.07	12.34	84.35
Tapioca								
Summer	24	20	6.42	3.31	0.45	2.86	13.60	69.95
Rainy	25	12	6.10	2.86	0.33	2.50	11.54	70.85
Winter	24	06	6.80	3.03	0.34	2.69	11.22	78.23
Mean	24	21	6.44	3.07	0.38	2.68	12.12	73.01
Significance between food plants	1.11		0.481	0.339	0.21	0.281	NS	2.893
Significance between seasons	NS		NS	0.415	NS	0.344	NS	3.544
Significance between food plant and season	2.52		1.34	1.26	1.96	1.67	NS	3.21
CV %	5.97		6.93	12.15	21.98	11.68	14.41	13.30

Telangana

Studies on the economic traits of Eri silkworm, *Samia cynthia ricini*, in relation to seasonal variations at Kakatiya University, Warangal, Telangana State

Renuka and Shamitha, (2014) conducted a study on the economic traits of eri silkworm during summer (Feb-March), rainy (July- Aug) and winter

(Nov –Dec) seasons during 2009-2011 at Kakatiya University, Warangal. Optimum temperature and relative humidity of 25-28^{0}C and 80-90% were maintained in the laboratory during the experimental period. Observations were recorded on larval, cocoon parameters and qualitative characteristics of eri silkworm. The morphological characteristics of the eri silkworm observed during the rearing period are the colour of the egg shell was white with a cream-coloured yolk; body colour varied from yellow/cream/blue/green with weak and thorny skin; white flossy cocoons without peduncle, enclosing a brown coloured pupa and chocolate coloured wings in the moth (Table 10.17).

In the investigation of Renuka and Shamitha, (2014) the larval length was greater in winter in all the instars in two years of study, while it was more in summer in third year in all instars except the fifth instar, where it reached maximum in winter. In contrast to cocoon parameters, which were lowest in summer, the larval length was much less during rainy season in most of the instars in the three years of study. However, it is interesting to note that, though length varied in I, II, III and IV instars, there was no notable difference observed when the larvae reached fifth instar in all the seasons studied, which explains that larval length during the first four instars does not play significant role, and that during fifth instar larvae grow to a maximum length (Table 10.18, 10.19 and 10.20).

A systematic study on the effect of seasonal variations has to be envisaged as in the present study, performed at the temperature/ relative humidity were ranging from 18.75 – 30.33°C / 55.50-87.2% during winter, 24 – 34.14°C /54.3-88% during rainy and 20.50-36.66°C / 24-83% during summer seasons respectively. From the tables 21 and 22, it can be deciphered that low temperature and high humidity was seen in winter in contrast to high range of temperature and low RH in summer, while rainy season has shown high temperature as well as high relative humidity. From the studies, it is clear that the larval weights were found to be highest in winter in all the instars, followed by rainy and lowest in summer seasons. In some instances, a notable increase was also seen in winter, which clearly indicates the association of larval weight to RH and temperature. Larval weights also include the water content in the body of the silkworm, which is high due to more RH and low temperature during winter. These values are in confirmation with the cocoon weight, pupal weight, shell weight and shell ratio which were highest in the winter followed by rainy and lowest in the summer season. Thus it can be concluded that the optimum temperature and RH as seen during winter are most suitable for producing good quality eri cocoons, which are economically important.

The results of the Renuka and Shamitha, (2014) study indicated that the cocoon parameters like cocoon weight, pupal weight, shell weight and shell ratio and larval length were highest in winter followed by rainy season. The lowest values were recorded during summer season and castor is the best suited host plant for eri silkworm and show better adaptability during winter season for commercial purposes. It opens an avenue for further studies based on quality and consumption of leaves in various seasons, thereby helping the rural populace in the utilization of castor plantation and help in their economy.

Table 10.17: Qualitative characteristics of eri silkworm (Renuka and Shamitha, 2014)

Egg		Larva			Cocoon		Moth	Voltinism
Colour of egg shell	Colour of yolk	Body colour	Marking of skin	Nature of skin	Colour of cocoon	Shape of cocoon	Wing colour of moth	Multivoltine
White	Cream	Yellow, Cream, Blue, Green	Single spot, double spot, semi zebra	Weak thorn	White	Flossy, no peduncle	Chocolate, greenish	

Table 10.18: Average larval length of Eri silkworm, *Samia cynthia ricini* during 2009 to 2011 (Renuka and Shamitha, 2014)

Year	Season	Instar				
		I	II	III	IV	V
2009	Winter	1.12±0.25	1.9±0.25	3.4±0.35	4.84±0.53	7.34±0.24
	Rainy	1.04±0.25	1.44±0.05	2.42±0.45	4.2±0.5	7.08±0.27
	Summer	0.76±0.11	1.62±0.28	2.6±0.29	4.32±0.19	6.62±0.72
2010	Winter	1.3±0.1	2.5±0.41	4.84±0.58	6.02±0.277	7.62±0.39
	Rainy	1.2±0.18	1.43±0.07	2.42±0.39	4.26±0.29	7.18±0.28
	Summer	0.92±0.19	1.9±0.22	3.02±0.30	4.5±0.38	6.54±0.62
2011	Winter	1.02±0.311	2.02±0.25	3.2±0.33	4.92±0.35	7.8±0.33
	Rainy	1.24±0.20	1.38±0.03	2.3±0.25	4.1±0.44	7.2±0.25
	Summer	1.66±0.75	2.06±0.56	3.74±0.24	5.08±0.53	6.88±0.13

Table 10.19: Average larval weights (in grams) of Eri silkworm, *Samia cynthia ricini* during 2009 to 2011 (Renuka and Shamitha, 2014)

Year	Season	Instar				
		I	**II**	**III**	**IV**	**V**
2009	Winter	0.0062±0.006	0.070±0.010	1.24±0.14	4.84±0.76	7.74±1.64
	Rainy	0.0057±0.0008	0.047±0.002	0.069±0.002	1.09±0.14	4.84±0.73
	Summer	0.0054±0.0031	0.039±0.010	0.061±0.003	0.92±0.233	3.43±0.95
2010	Winter	0.0059±0.006	0.065±0.011	0.098±0.0007	1.19±0.04	5.53±0.22
	Rainy	0.0051±0.0005	0.045±0.002	0.060±0.003	0.97±0.14	4.60±0.72
	Summer	0.0045±0.0034	0.043±0.010	0.050±0.003	0.96±0.06	3.67±0.25
2011	Winter	0.0064±0.0065	0.063±0.008	0.096±0.003	2.57±0.79	5.64±0.36
	Rainy	0.0052±0.0005	0.049±0.002	0.064±0.004	1.12±0.07	4.74±0.85
	Summer	0.0045±0.003	0.043±0.014	0.054±0.004	0.99±0.049	3.69±0.26

Table 10.20: Post cocoon parameters of Eri silkworm, *Samia cynthia ricini* during 2009 to 2011 (Renuka and Shamitha, 2014)

Year	Season	Cocoon weight	Pupal weight	Shell weight	Shell ratio
2009	Winter	3.98±0.008	3.40±0.004	0.58±0.007	14.56%
	Rainy	3.57±0.04	3.08±0.04	0.49±0.01	13.75%
	Summer	3.37±0.03	2.93±0.04	0.43±0.02	12.86%
2010	Winter	3.93±0.01	3.36±0.008	0.56±0.02	14.34%
	Rainy	3.55±0.05	3.04±0.04	0.49±0.01	14.01%
	Summer	3.31±0.02	2.83±0.01	0.45±0.01	13.89%
2011	Winter	3.92±0.052	3.36±0.04	0.53±0.004	13.57%
	Rainy	3.61±0.05	3.10±0.03	0.48±0.01	13.44%
	Summer	3.27±0.08	2.07±0.25	0.43±0.01	13.25%

Table 10.21: The instar –wise temperatures recorded during rearing of Eri silkworm, *Samia cynthia ricini* (2009-2011)
(Renuka and Shamitha, 2014)

Year	Season	Instar – Temperature (^{0}C)									
		I		II		III		IV		V	
		Max	Min	Max	Min	Max	Min	Max	Min	Max	Min
2009	Winter	28.25±2.21	18.75±0.95	30.33±2.08	19.66±1.52	28.5±0.57	24±0.81	30.4±1.51	24.2±1.30	28±0.81	19.28±1.97
	Rainy	29.25±0.5	25.75±1.70	31.66±1.52	26±1	31±2.64	27±1	32.1±1.47	26.5±1	34.14±1.21	26.14±0.37
	Summer	35.37±0.75	23.25±0.5	36.66±0.57	23.66±0.57	36.5±0.70	20.5±0.70	37±1	21±1	36±0.81	21.42±1.81
2010	Winter	28.6±1.67	23.8±1.30	29±1.73	23.33±0.57	30.33±0.57	23.66±0.57	29.75±0.86	23.75±1.70	28.75±1.28	20.62±1.30
	Rainy	27.7±2.24	26±0.81	27.73±0.83	25±1	28.6±1.70	25.3±1.12	29.44±1.87	26.04±1.77	30.35±1.22	25.12±1.15
	Summer	32±1	22.33±0.57	32.5±0.70	24.5±0.70	33.5±0.70	25±1.41	31.5±2.12	22.5±0.70	33.82±1.09	24.2±1.77
2011	Winter	27.66±0.57	22±2	28.66±1.15	20.33±0.57	28.5±0.70	19.5±0.70	28.33±1.52	28.33±1.52	28.75±1.16	18.37±1.99
	Rainy	33.75±2.62	22.5±2.51	33.33±1.15	23.66±2.08	32.33±2.08	24.66±1.152	27.33±2.30	27.33±2.30	30.2±1.93	24±1.63
	Summer	30.33±0.57	19.66±0.57	30.5±0.70	19.5±0.70	33±1.41	23.5±0.70	32.66±1.15	32.66±1.15	31.66±1.93	22.88±1.76

Table 10.22: The instar –wise relative humidity recorded during rearing of Eri silkworm, *Samia cynthia ricini* (2009-2011) (Renuka and Shamitha, 2014)

Year	Season	Instar – Relative Humidity (%)									
		I		II		III		IV		V	
		Max	Min	Max	Min	Max	Min	Max	Min	Max	Min
2009	Winter	80±12.46	51.25±8.61	81.4±5.72	59.2±10.35	84±9.45	57.25±12.47	86.75±5.05	67.25±4.57	86.85±6.22	58.85±11.86
	Rainy	85.25±3.86	60.25±2.21	69.66±6.50	55.33±4.50	79±4.35	59.66±1.15	74.2±6.26	60.4±7.40	81.2±8.89	55.4±3.40
	Summer	77.66±4.61	34±5.2	75±0	24.25±2.06	72.66±11.37	29.33±2.081	74.33±1.15	74.33±1.15	72.14±9.13	28.28±2.49
2010	Winter	87.2±7.15	71.4±7.82	84±0	71.33±6.02	83.5±1	64.5±5.74	79.87±6.91	64.5±5.74	79.87±6.91	55.5±10.07
	Rainy	88±4.61	56.5±10.63	85.66±9.23	56±5.56	86.33±4.93	63.4±15.63	84.63±7.06	63.4±15.63	84.63±7.06	63.45±9.57
	Summer	81.66±11.37	66±5.56	82.5±10.66	73.5±10.75	74 ±2.64	64.33±5.68	72.75±7.36	64.33±5.68	72.75±7.36	57.37±10.75
2011	Winter	82.66±1.52	61.33±7.50	84.66±4.61	65.66±7.76	77±4.69	60±3.46	83.5±8.38	61.25±4.03	81.87±7.86	81.87±7.86
	Rainy	79.5±5.92	79.5±5.92	84.66±4.61	62.33±15.30	88±4.35	54.33±0.78	85.4±3.78	69±5	84.3±7.08	65.4±4.55
	Summer	80.25±13.22	60.75±7.41	80±6.92	60.66±10.01	71.66±6.65	64.33±8.14	83.33±0.57	73.68±2.30	85±6.12	68.44±13.22

Uttar Pradesh

Impact of rearing of *Samia c. ricini* in spring and autumn seasons on different food plants in Uttar Pradesh at Babasaheb Bhimrao Ambedkar University, Lucknow

The eri silk worm feeds on leaves of many food plants like Castor, Kessaree, Tapioca, Barkesseru, Payam, Barpat, Jatropha, Papaya, *Ailanthus altissima*, Gulanch, Gamari etc. All the food plants are not equally good for eri silkworm rearing and eri silkworm shows different types of physiological and behavioural expressions, when reared on different food plants in different seasons. Such studies were not undertaken earlier in Uttar Pradesh. Hence Rajesh Kumar and Gangwar, (2010) conducted a study to screen out best food plants and best season for rearing eri silkworm in Uttar Pradesh climatic conditions. This study was conducted at Babasaheb Bhimrao Ambedkar University, Lucknow by rearing eri silkworm on the leaves of Castor, Tapioca, Barera, Papaya, which are maintained in university campus with uniform culture operations. The temperature and humidity range was in between 25°C - 28°C and 70% - 80%, respectively. Disease free layings (eggs) of eri silk moth having uniform fecundity were reared. For each host plant species three replications were undertaken having one disease free laying in each replication. The host plant species were local varieties of Castor, Tapioca, Barera and Papaya. The data for larval duration, larval weight, ERR%, pupation % and fecundity, cocoon weight, pupal weight and shell weight were recorded. Abiotic factors *viz.,* maximum, minimum temperature, relative humidity and number of rainy days were also recorded during the rearing period (Table 10.24).

The data revealed that larval duration (days) in autumn and spring season was minimum in castor plant (20.00 and 19.00), followed by tapioca (21.12 and 21.00), Barera (22.00 and 22.12), and maximum in the case of papaya (24.00 and 23.00). The observed ERR (%) on Castor plants was maximum (86.0 and 90.0) followed by Tapioca (80.0 and 88.0), Barera (78.0 and 85.0) and Papaya showed minimum ERR (75.0 and 76.0). Likewise, in autumn and spring seasons the cocoon weight (g) was maximum in castor plant (3.56 and 3.62) followed by Tapioca (3.25 and

3.22), Barera (2.95 and 3.20) and minimum (2.80 and 3.00) in case of Papaya. The shell ratio % was calculated and was found maximum in Castor (14.58 and 15.19) followed by Tapioca (14.15 and 13.66), Barera (13.55 and 12.50) and minimum (13.54 and 12.00) in case of papaya plant (Table 10.23). On the basis of results obtained by Rajesh Kumar and Gangwar (2010), we can say that the food plant of different species and seasonal conditions influence the larval growth, larval duration, cocoon weight, pupal weight, shell weight, S.R. %, ERR% and fecundity of eri silk worm. Larval duration, cocoon weight, pupal weight, shell weight, S.R. %, ERR% were higher in eri silkworm fed with castor in spring season, to the extent of 19.00, 3.62, 3.10, 0.55, 15.19, 90.0% respectively in comparison to the other food plants. These favourable results in eri silkworm fed on castor leaf may be due to the higher rate of food ingestion, food assimilation and respiratory activity. The involvement of these factors in increasing the larval body substance has been reported by Hiratsuka, (1920) in *Bombyx mori,* later by Stockner, (1971) in *Philosamia ricini* Hutt. and Dey, (1983) in *Antheraea proylei.* More over the percentage of water content of host plant leaf also might influence the larval body weight. Delvi *et al.,* (1988) and Muthukrishnan and Pandian, (1987) reported that insects could accumulate water from the metabolic food. However, Maribashetty *et al.,* (1999) reported that in silkworm *B. mori* the food consumption has a direct influence on the weight of larvae, cocoon and shell.

From the results of Rajesh Kumar and Gangwar, (2010) it is obvious that the performance of eri silkworm was better on *Ricinus communis* food plant which showed significant increase in the larval weight, silk gland weight, cocoon weight and the shell weight considerably compared to other food plants in spring season of Uttar Pradesh.

Table 10.23: Economic parameters of Eri silkworm reared on different food plants in autumn and spring season (Adopted from Rajesh Kumar and Gangwar, 2010)

Parameters	Food plants				Gain over minimum performer (%)	S.E (±)	C.D. 5 %	C.D. 1 %
	Castor	Tapioca	Barera	Papaya				
Autumn season								
Larval duration (days.hrs)	20.0	21.12	22.0	24.0	16.66	0.351	0.710	1.082
Larval weight (g)	7.45	6.82	6.05	5.35	28.18	0.091	0.201	0.312
ERR (%)	86.0	80.0	78.0	75.0	12.79	0.701	1.412	1.941
Pupation %	88.0	77.0	74.0	68.0	22.72	2.451	5.942	6.524
Cocoon weight (g)	3.56	3.25	2.95	2.80	21.34	0.60	0.178	0.192
Pupal weight (g)	3.12	2.75	2.45	2.43	21.47	0.003	0.007	0.011
Shell weight (g)	0.52	0.46	0.42	0.38	26.92	0.018	0.051	0.074
Shell ratio (%)	14.58	14.15	13.55	13.54	6.92	0.232	0.714	0.885
Spring season								
Larval duration (days.hrs)	19.00	21.00	22.12	23.00	17.39	0.312	0.673	0.912
Larval weight (g)	7.60	6.80	6.25	6.65	25.65	0.172	0.375	0.534
ERR (%)	90.0	88.0	85.0	76.0	15.55	0.751	1.512	1.982
Pupation %	90.0	82.0	75.0	70.0	20.45	1.525	3.564	4.245
Cocoon weight (g)	3.62	3.22	3.20	3.00	17.12	0.392	0.128	0.172
Pupal weight (g)	3.10	2.71	2.50	2.30	25.80	0.004	0.009	0.013
Shell weight (g)	0.55	0.44	0.40	0.36	36.54	0.015	0.041	0.064
Shell ratio (%)	15.19	13.66	12.50	12.00	21.00	0.212	0.623	0.825

Table 10.24: Abiotic factors of rearing during autumn and spring season (Rajesh Kumar and Gangwar, 2010)

Seasons	Temperature		Relative humidity		Rainy days
	Maximum	Minimum	Maximum	Minimum	
Autumn (Sep-Oct, 2006)	30.0	23.00	96.00	70.00	2
Spring (Feb-March, 2007)	31.00	25.00	94.0	75.00	Nil

Rearing performance of different eco-races of eri silkworm *Samia c. ricini* in monsoon season of Uttar Pradesh at Department of Applied Animal Sciences, Babasaheb Bhimrao Ambedkar University, Lucknow

Rajesh Kumar and Elangovan, (2010) conducted a study to screen the suitable eco-races of eri silkworm, *Samia c. ricini* for monsoon rearing in Uttar Pradesh climatic conditions. The study was conducted in a well constructed rearing house of the Department of Applied Animal Sciences, Babasaheb Bhimrao Ambedkar University, Lucknow, Uttar Pradesh during monsoon seasons of year 2006 and 2007. The eco-races of *S. c. ricini* such as Borduar, Titabar, Dhanubhanga, and Mendipathar were selected for the study. The experiment was conducted from second moulting to harvesting of cocoons. The rearing performance of each eco-race was assessed by hatching (%), larval wt (g), larval duration (d), yield / 400 larvae by number and by weight, cocoon weight (g), shell weight (g), shell ratio (%), cocoon shape variability, pupal period (d), pupation rate (%), and leaf silk conversion rate. The Mendipathar eco-race of *S. c. ricini* showed better rearing performance than Titabar, Borduar and Dhanubhanga eco-races (Table 10.25). The rearing parameters such as weight of full grown larvae, yield by number of larvae, yield by weight of larva, cocoon weight, shell weight, shell ratio, cocoon shape variability, pupation rate differed statistically among different eco races of eri silk worm. The temperature and humidity of 2006 monsoon season fluctuated due to the occurrence of rainfall which influenced the rearing performance. Whereas an apparent constant temperature and humidity until spinning during 2007 monsoon season favoured the rearing of all different eco-races of eri silkworm.

The results of present study showed that the rearing performance of different eco-races of eri silkworms during monsoon season was influenced by factors such as temperature and humidity. The results of Rajesh Kumar and Elangovan, (2010) study clearly indicate that there were differences in the rearing performance of different ecoraces of eri silkworm. Several studies have been conducted on phenotypic variability of genotypes under diverse environment and climatic conditions. It is well known that the dynamic environment conditions prevailing in different seasons of the year bring profound changes in growth and development and the expression of economic characters in different silkworm races (Krishanaswami *et al.,* 1970). The highest larval weight, cocoon weight, shell weight and shell ratio were observed during August – September (Debaraj and Sarmah, 2001). Highest larval weight, cocoon weight, shell weight and shell ratio and minimum larval duration during rainy season. The cocoon weight, shell weight and shell ratio and total silk production were high during winter, spring, autumn and rainy seasons while lowest during summer crops (Kar *et al.,* 2004). The larval duration was relatively short during July - August (Monsoon) rearing than winter season, and also recorded minimum pupal period, less cocoon weight and shell weight (Verk *et al.,* 2009). Hata *et al.,* (2005) found highest hatching percentage, larval weight, cocoon weight, shell weight, silk ratio and minimum larval period during August - September rearing season. The highest hatching percentage, average cocoon weight, shell weight and shell ratio were observed during rainy season (Brahma *et al.,* 2006). The maximum cocoon weight, shell weight, shell ratio and highest hatching percentage were observed during September – October rearing season.

The results of the Rajesh Kumar and Elangovan, (2010) study showed that the Mendipathar eco-race showed better adaptability than other eco-races to the condition in monsoon season and suitable for Uttar Pradesh climatic conditions for commercial rearing (Table 10.25).

Table 10.25: Rearing performance of different eco-races of eri silkworm (*Samia c. ricini*) during monsoon season of year 2006 and 2007 (Rajesh Kumar and Elangovan, 2010).

Rearing Parameters	Borduar		Titabar		Dhanubhanga		Mendipathar		C D	
	2006	2007	2006	2007	2006	2007	2006	2007	2006	2007
Hatching %	94.33±0.00	94.35±0.00	93.36±0.00	93.15±0.00	92.65±0.00	93.15±0.00	94.15±0.00	94.00±.00	0.00	0.304
Larval duration (d)	19.45±0.050	19.20±0.050	19.62±0.050	19.50±0.040	20.00±0.492	19.62±0.070	19.75±0.492	19.12±.056	0.586	0.450
Larval weight (g.)	6.88±0.044	7.10±0.026	6.86±0.032	6.80±0.070	6.52±0.045	6.65±0.026	6.95±0.026	6.98±.036	0.087	0.118
Yield /400 larvae by Number	352±3.00	360±2.65	354±2.65	352±2.65	348±2.65	350±3.00	356±4.36	365±3.60	7.460	0.085
Yield / 400 larvae by weight(kg)	1.186±0.096	1.239±0.020	1.153±0.049	1.260±0.026	1.127±0.095	1.156±0.062	1.202±0.044	1.285±.026	0.007	4.848
Cocoon weight (g.)	3.26±0.045	3.47±0.026	3.12±0.017	3.46±0.023	3.05±0.040	3.29±0.017	3.39±0.036	3.520±.0362	0.069	0.057
Shell weight (g.)	0.50±0.036	0.57±0.017	0.48±0.026	0.56±0.020	0.46±0.026	0.52±0.010	0.53±0.017	0.59±.020	0.059	0.059
Shell ratio (%)	15.40±0.026	16.47±0.026	15.43±0.043	16.27±0.034	15.15±0.045	15.88±0.017	15.70±0.045	16.82±.066	0.096	0.032
Cocoon shape Variability	262.51±10.0	258.13±.046	251.39±5.07	256.16±.032	248.63±5.05	247.33±.036	255.80±5.00	259.69±.151	0.106	0.088
Pupal period (d.)	10.50±0.060	10.41±0.046	10.24±0.060	10.33±0.040	10.75±0.030	10.66±0.040	10.62±0.030	10.06±.036	0.092	58.96
Pupation rate (%)	84.15±0.095	85.69±0.815	85.50±0.095	86.73±0.629	82.30±0.131	84.47±0.631	87.28±0.135	88.83±1.207	0.253	0.083
Leaf Silk Conversion rate%	3.05±0.036	3.10±0.026	2.95±0.026	2.98±0.055	2.78±0.020	2.96±0.017	3.09±0.017	3.08±.044	0.047	0.125

Assam and Other North-Eastern Regions

Seasonal performance and reproductive biology of eri silkworm, *Samia c. ricini* feeding on *Ailanthus* species and other promising food plants at Central Muga Eri Research & Training Institute, Lahdoigarh, Jorhat, Assam

Four genotypes of *Ailanthus* are reported in India, *viz.*, *A. excelsa* (Borkesseru), *A. grandis* (Borpat), *A. altissima* and *A. malabarica*. *Ailanthus* plants are not commercially utilized as major host plant of eri silkworm due to non-availability of database on rearing as well as reproductive performance of the silkworm. In view of the above, a study was undertaken by Ahmed *et al.,* (2015) to evaluate the seasonal performance and reproductive biology of eri silkworm feeding on different *Ailanthus* germplasm as well as in combination with other promising food plants. The study was conducted at Central Muga Eri Research & Training Institute, Lahdoigarh during four selected seasons *viz.*, S1 (January-February), S2 (April– May), S3 (July-August) and S4 (September-October) of the year 2012. The leaves of the borpat (*Ailanthus grandis*), borkesseru (*Ailanthus excelsa*) and kesseru (*Heteropanax frgrans*) as well as in combination with castor (*Ricinus communis*) leaves were fed to eri silkworm maintaining three replications with 300 larvae per treatment in six different treatments such as, T1 : Borpat (I-V instar larva); T2 : Castor (I-II instar larva) + Borpat (III-V instar larva); T3 : Borkesseru (I-V instar larva); T4 : Castor (I-II instar larva) + Borkesseru (III-V instar larva), T5 : Kesseru (I-V instar larva) and T6 : Castor (I-II instar larva) + Kesseru (III-V instar larva). In combination treatments, the food plants were changed immediately after second moult out of the larvae. The rearing and reproductive performance of eri silkworm feeding on different food plants and their combinations were assessed by the following parameters *viz.*, fecundity (eggs laid per female moth in numbers), hatching (%), disease free layings (dfl): cocoon ratio, larval wt (g), larval duration (d), Yield / 100 dfls by number and by weight, cocoon weight (g), shell weight (g), shell ratio (%), effective rate of rearing (%), pupal period(d) and pupation rate (%).

Table 10.26: Larval duration and weight of eri silkworm during feeding different combinations of food plants (Adopted from Ahmed *et al.*, 2015)

Treatment	Larval duration (days)					Mature larval weight (g)				
	Jan-Feb	Apr-May	Jul-Aug	Sep-Oct	Mean	Jan-Feb	Apr-May	Jul-Aug	Sep-Oct	Mean
T1	35.33±1.53	23.67±1.15	19.33±0.58	20.67±0.58	24.75±0.96	6.80±0.23	7.13±0.11	7.38±0.23	7.48±0.07	7.20±0.16
T2	31.00±1.00	22.67±0.58	18.33±0.58	20.33±0.58	23.08±0.69	6.98±0.13	7.14±0.10	7.21±0.10	7.61±0.09	7.24±0.11
T3	38.33±0.58	25.67±0.58	22.67±0.58	23.33±0.58	27.50±0.58	6.55±0.11	6.43±0.18	6.88±0.10	7.08±0.15	6.74±0.14
T4	36.67±0.58	24.67±0.58	22.00±1.00	23.00±0.00	26.59±0.54	6.74±0.39	6.55±0.19	6.65±0.32	6.95±0.07	6.72±0.24
T5	42.33±1.53	27.00±1.00	24.33±0.58	26.00±1.00	29.92±1.03	5.83±0.07	5.79±0.11	5.87±0.12	5.36±0.06	5.71±0.09
T6	39.67±0.58	25.33±0.58	23.00±0.00	24.33±1.00	28.08±0.54	5.92±0.10	5.95±0.03	5.91±0.12	5.44±0.08	5.81±0.08
SE	2.31	1.71	1.37	1.55	1.74	0.43	0.29	0.40	0.22	0.34
CD at 5%	5.03	3.73	2.98	3.37	3.78	0.95	0.62	0.87	0.47	0.73

SE: Standard error; CD: Critical difference

T1: *Ailanthus grandis* (I-V instar larva); T2: Castor (I-II instar larva) + *A. grandis* (III-V instar larva); T3: *A. excelsa* (I-V instar larva); T4: Castor (I-II instar larva) + *A. excelsa* (III-V instar larva); T5: Kesseru (I-V instar larva); T6: Castor (I-II instar larva) + Kesseru (III-V instar larva)

Table 10.27: Cocoon characters of eri silkworm feeding different combinations of food plants during different seasons
(Adopted from Ahmed *et al.,* 2015)

Treatment	Single cocoon weight (g)					Single shell weight (g)					Shell ratio (%)				
	Jan-Feb	Apr- May	Jul-Aug	Sep-Oct	Mean	Jan-Feb	Apr-May	Jul-Aug	Sep-Oct	Mean	Jan-Feb	Apr-May	Jul-Aug	Sep-Oct	Mean
T1	2.51	3.18	3.65	3.11	3.11	0.31	0.42	0.41	0.39	0.38	12.49	13.20	11.24	12.70	12.41
	(0.09)	(0.08)	(0.25)	(0.14)	(0.14)	(0.02)	(0.01)	(0.04)	(0.02)	(0.02)	(0.22)	(0.13)	(0.16)	(0.17)	(0.16)
T2	3.08	3.37	4.04	3.00	3.37	0.35	0.44	0.52	0.35	0.42	11.41	13.18	12.92	11.16	12.17
	(0.07)	(0.18)	(0.27)	(0.28)	(0.20)	(0.04)	(0.01)	(0.04)	(0.04)	(0.03)	(0.57)	(0.06)	(0.15)	(0.14)	(0.15)
T3	2.26	2.34	2.48	2.63	2.43	0.29	0.31	0.32	0.30	0.31	12.90	13.39	12.85	11.37	12.63
	(0.32)	(0.10)	(0.16)	(0.26)	(0.21)	(0.07)	(0.02)	(0.03)	(0.01)	(0.03)	(0.22)	(0.20)	(0.19)	(0.04)	(0.14)
T4	2.92	2.78	2.78	2.94	2.86	0.33	0.35	0.35	0.39	0.36	11.42	12.50	12.58	13.17	12.42
	(0.07)	(0.18)	(0.30)	(0.11)	(0.17)	(0.03)	(0.02)	(0.02)	(0.03)	(0.03)	(0.43)	(0.11)	(0.07)	(0.27)	(0.18)
T5	2.04	2.82	2.39	2.40	2.41	0.27	0.33	0.33	0.32	0.31	13.47	11.57	13.92	13.48	13.09
	(0.11)	(0.04)	(0.24)	(0.24)	(0.16)	(0.05)	(0.02)	(0.01)	(0.02)	(0.03)	(0.45)	(0.50)	(0.04)	(0.08)	(0.13)
T6	2.56	2.74	2.63	2.45	2.60	0.30	0.33	0.33	0.32	0.32	11.81	12.02	12.56	13.23	12.41
	(0.25)	(0.04)	(0.13)	(0.10)	(0.11)	(0.01)	(0.03)	(0.02)	(0.02)	(0.02)	(0.04)	(0.75)	(0.15)	(0.20)	(0.18)
SE	0.14	0.09	0.18	0.13	0.14	0.03	0.01	0.02	0.02	0.02	0.74	0.64	1.06	0.59	0.76
CD at 5%	0.30	0.20	0.39	0.29	0.30	0.06	0.02	0.04	0.04	0.04	NS	NS	NS	1.29	NS

Table 10.28: Grainage performance of eri silk feeding different combinations of food plants (Adopted from Ahmed *et al.,* 2015)

Treatment	Seed cocoons kept (no)	Emergence of moth (Number)			Emergence of moth (%)	Un-emerged cocoon (no)	Total coupling obtained (no)	Poor egg (no)	Total dfls obtained (no)	Cocoon:dfl ratio
		Male	Female	Cripple						
T1	3200	1735	1232	157	92.72	76	1121	82	1039	3.08:1
T2	3200	1689	1310	146	93.72	55	1272	65	1207	2.65:1
T3	3200	1605	1083	394	84.00	118	901	95	806	3.97:1
T4	3200	1672	1228	204	90.63	96	1042	87	955	3.35:1
T5	3200	1567	1132	386	84.34	115	956	98	858	3.73:1
T6	3200	1633	1206	256	88.72	105	1097	87	1010	3.7:1

Data represents mean of four seasons.

T1: *Ailanthus grandis* (I-V instar larva); T2: Castor (I-II instar larva) + A. grandis (III-V instar larva);

T3: A. excelsa (I-V instar larva); T4: Castor (I-II instar

larva) + *A. excelsa* (III-V instar larva); T5: Kesseru (I-V instar larva); T6: Castor (I-II instar larva) + Kesseru (III-V instar larva)

Table 10.29: Hatchability (%) of eri silk feeding different combinations of **food** plants (Adopted from Ahmed *et al.*, 2015)

Treatment	Jan-Feb	Apr- May	Jul-Aug	Sep-Oct	Mean±SD
T1	85.0 (1.50)	86.0 (1.32)	88.4 (1.50)	87.0 (1.47)	86.35±1.45
T2	87.3 (5.50)	95.5 (3.00)	95.0 (2.50)	92.0 (4.00)	92.45±3.77
T3	80.5 (1.60)	84.0 (1.45)	82.7 (1.50)	83.0 (1.38)	82.55±1.48
T4	81.1 (2.42)	85.0 (2.56)	84.4 (2.36)	86.6 (2.16)	84.25±2.31
T5	83.0 (2.50)	86.0 (2.56)	84.5 (2.65)	80.3 (2.02)	83.24±2.43
T6	85.3 (1.45)	87.5 (1.56)	85.2 (2.15)	88.4 (1.22)	86.25±1.60
SD	2.63	4.17	4.48	4.11	3.58
CD at 5%	5.23	8.45	8.76	8.26	7.48

Data in the parentheses indicate standard deviation. SE: Standard error; CD: Critical difference

Table 10.30: Month wise performance in grainage parameters of eri silkworm and meteorological data (Sarkar and Sarmah, 2010)

Month	Emergence of moth (%)				Total couplings obtained (%)	Average fecundity (No.)	Hatching (%)	Temperature (^{0}C)		Relative humidity (%)	
	Male	Female	Male	Female				Max.	Min.	Max.	Min.
August	62.33	25.66	3.00	7.66	25.66	311.33	84.66	34.00	23.00	89.00	63.80
September	60.00	32.00	3.00	5.00	32.00	333.33	82.33	32.00	22.00	90.00	65.50
October	52.67	40.00	2.33	5.00	40.00	495.00	90.00	30.00	17.00	88.00	69.50
November	61.33	32.00	1.33	4.33	32.00	367.67	94.33	28.00	11.00	88.50	69.50
December	66.00	28.33	1.33	2.66	28.33	311.33	85.00	25.00	9.00	89.30	67.80
January	67.67	25.33	1.66	3.66	25.33	315.33	83.66	24.00	7.00	83.50	65.80
February	68.33	25.00	1.33	3.00	25.00	340.00	84.66	24.00	10.80	87.30	60.50
March	71.66	26.00	1.00	1.33	26.00	318.33	89.33	32.00	13.00	85.80	50.00
April	68.00	27.00	1.66	3.33	27.00	330.00	84.00	34.00	15.00	90.80	62.50
May	65.33	24.66	2.00	6.33	24.66	274.00	68.00	38.50	25.00	88.80	50.00
June	64.00	22.00	3.33	8.33	22.00	277.67	70.66	38.00	26.00	92.00	70.30
July	58.66	23.33	4.33	11.66	23.33	294.00	81.00	37.00	25.00	90.50	70.80
CD at 5%	5.454	7.072	2.853	1.494	7.072	7.342	5.502	-	-	-	-

Table 10.31: Correlation coefficient of grainage characters with meteorological parameters (Sarkar and Sarmah, 2010)

Grainage characters	Correlation coefficient			
	Temperature (^{0}C)		Relative humidity (%)	
	Max.	Min.	Max.	Min.
Emergence of good male moths	-0.198	-0.416	-0.339	-0.610**
Emergence of good female moths	-0.306	-0.221	-0.092	0.264
Emergence of cripple male moths	0.631**	0.814**	0.570*	0.470
Emergence of cripple female moths	0.666**	0.801**	0.542*	0.412
Coupling obtained	-0.306	-0.221	-0.092	0.264
Fecundity	-0.358	-0.281	-0.190	0.291
Hatchability	-0.621**	-0.650**	-0.392	0.210

The lowest larval period (18.33 ± 0.58 days) was observed during July-August crop in the second treatment i.e. feeding with Castor (I-II instar) and subsequent feeding with Borpat (III-V instar), which was statistically at par with treatment of feeding Borpat from brushing till spinning (T1). The highest larval period of 42.33 days was recorded in the treatment of Kesseru feeding from brushing till spinning (T5) during January- February crop. Significantly the highest seasonal mean larval weight was recorded in T2 (7.24 ± 0.11 g) followed by T1 (7.20 ± 0.16 g) and T3 (6.74 ± 0.14 g), which were statistically at par with one another. The lowest seasonal mean larval weight was also recorded in T5 (Table 10.26). The significant and highest single cocoon weight was recorded in T2 during July-August season (4.04 ± 0.27 g) followed by T1 (3.65 ± 0.25 g). The highest single shell weight was recorded in T2 during July-August season (0.52 ± 0.04 g) followed by T2 during April- May season (0.44 ± 0.01 g). During September-October season, the highest shell ratio (%) was recorded in the treatment T5 (13.48 ± 0.08) followed by T6 (13.23 ± 0.20) and T1 (12.70 ± 0.17) (Table 10.27). The highest moth emergence of eri silkworm was recorded in the treatment T2 (93.72%) followed by T1 (92.72 %) and T4 (90.63%). The highest fecundity of 373 was recorded in T2 during September-October season. The highest hatchability of eri silkworm eggs was recorded in the treatment T2 irrespective of season ranging from 87.3% to 95.5%. The data on seasonal mean also indicated that the significant variations in the parameter which was 92.45± 3.77 %, 86.35 ± 1.45 % and 86.25 ± 1.60% in T2, T1 and T6, respectively (Table 10.28 and 10.29).

While rearing the eri silkworm on different host plants during different seasons, it was observed that, Castor (I-II instar larva) in combination with Borpat (III-V instar larva) was the best combination of food plant in respect of higher larval weight, single cocoon weight, single shell weight, cocoon yield per dfl, ERR, cocoon shell yield per 100 dfl and the lowest larval period. The next best treatment was rearing of eri silkworm on Borpat leaves from brushing till spinning of eri silk. Silkworms fed with Castor in combination with Borpat leaves exhibited heavier larval weight and shorter larval duration during July- August crop which might be due to higher nutritional efficiency as well as high moisture retention capacity of Borpat leaves as compared to other food plants. Besides, the environmental temperature had direct influence on the larval duration which is negatively correlated. The shorter larval duration facilitated the less investment in rearing as well as more crops per annum. The higher ERR or survivability rate of eri silkworm fed with Borpat or Borkesseru leaves might be attributed by the presence of antifungal, antibacterial and anti protozoan properties. The productivity level of eri cocoon shell yield was 8 Kg per 100 dfls in the state sericulture farms. Hence, it indicated the cocoon shell yield improvement of 43.75% in T2 (Castor + Borpat) and 22.86% in T1 (Borpat alone). The better grainage performance of Borpat and Castor fed worms might be due to healthy larval growth and spinning of larva which in turn helped in emergence of healthy moths. The varieties of food plant fed by the larvae have significant effect on grainage performance. The comparative fresh leaf biomass productivity (metric tonnes per ha) was found to be highest in Borpat (33.45) followed by Borkesseru (28.24), Kesseru (21.24) and Castor (11.45). It reveals that Borpat produces higher fresh leaf biomass of 57.48% and 192.14% in comparison with Kesseru and Castor, respectively. In case of Borkesseru, the annual biomass production is more by 33.15% and 146.64% in comparison with Kesseru and Castor. In eri silkworm rearing, more than 80-85% of food is consumed by 4th and 5th instar larvae. Hence, substitution of Castor with Borpat leaves from 3rd instar onwards (late-stage rearing) or rearing with Borpat alone from brushing till spinning would make the sector sustainable and economically vibrant among the primary stakeholders. Further, Borpat is a forest based tree, conservation and utilization of the same in ericulture will also help in mitigation of deforestation and ill effects of climate changes.

Seasonal variation of grainage characters in seed production of eri silkworm at CMERTI, Lahdoigarh, Jorhat, Assam

There is no definite season for eri silkworm rearing. Depending on the availability of eggs and leaf, the silkworm can be reared throughout the year. Seasonal variation is observed in the grainage characters of eri silkworm. The quality of the seed cocoon is greatly influenced by a number of factors such as nutritious leaf, disease free laying, optimum temperature, humidity and hygienic conditions. The embryonic development stage in silkworm is very susceptible to environmental conditions i.e. light, temperature, humidity, vibration etc. (Choudhury, 2005). A temperature of $22\text{-}25^{0}C$ and 80% humidity combined with good ventilation are essential for effective grainage operation. The optimum temperature for eri silkworm egg laying is $22\text{-}24^{0}C$ with high humidity (Sarkar, 1980). Hence Sarkar and Sarmah (2010) made an attempt to study the grainage characters in different months and to evaluate suitable season (s) for eri silkworm seed production. Seed cocoons of Borduar eri silkworm eco-race was taken as experimental material. Third and fourth day old cocoons were collected for the experiment. The experiment was conducted with 3 replications of 100 cocoons (3:2 male female ratio) each and kept them in moth cages. Grainage characters such as emergence of good moths as well as cripple moths and female moths, total coupling, fecundity and hatchability were recorded in each month during August 2006 to July 2007. Month wise meteorological data *viz.,* maximum/minimum temperatures and maximum/minimum relative humidity were recorded. Critical difference among each grainage characters in different months was analyzed using the SPACK software. Correlation coefficient analysis was conducted to determine the possible relation between different grainage parameters and meteorological data.

Eri silkworm moths emerged in the cool hours of the day in shady or dark condition both morning and evening. Emergence was recorded between 5-6 AM and continued up to 9-10 AM. Further emergence started in the evening hours (3-4 PM) and continued up to 7-8 PM. Quantum of emergence was more in evening than morning hours. Generally, male moths emerged earlier than female moths and were also higher in number. Newly emerged female moths were sluggish and crawl on to the edge of the cage for a resting period of about 1 hour till full development of their wings and started coupling. After coupling for 5-6 hours, the female moths started egg laying.

Seasonal variation in moth emergence was recorded in eri silkworm. Significantly, highest good male moth emergence (71.66%) was recorded in the month of March, followed by 68.33% during February. Similarly in case of good female moth, significantly highest emergence was recorded in the month of October (40%) followed by September and November (32%). Correspondingly highest coupling was also observed in the same months, i.e. 40% and 32% respectively. The results indicated that significantly highest cripple male moths emerged during the month of July (4.33%) followed by June (3.33%). Similarly, highest cripple female moth emergence was recorded in the month of July (11.66%) followed by June (8.33%). Significantly highest fecundity was recorded during the month of October (495) followed by November (367.67) and hatching percentage was highest during autumn season i.e. in the month of November (94.33%) (Table 10.30).

Cripple male and female moth emergence were positively correlated with maximum and minimum temperatures in significant level (r = 0.631; r = 0.666 and r = 0.814; r = 0.801 respectively). Hatchability was negatively correlated with temperature (r = -0.621 and r = -0.650) (Table 10.31).

Considering the performance of different grainage characters such as good moth emergence, higher coupling realization, higher fecundity and higher hatchability of eggs, the autumn season (Sep-Nov) was found the best for eri seed production and summer (May to July) was comparatively unsuitable with low coupling recovery, low fecundity, low hatchability of eggs and higher emergence of cripple moths. The emergence of large number of cripple female moths during summer season reduced the quantum of coupling recovery. Remaining seasons are also suitable for eri silkworm seed production. Emergence of both male and female cripple moths depended on fluctuation of temperature and relative humidity. It was observed that, higher the temperature and relative humidity, more the emergence of cripple moths. Higher temperature coupled with higher relative humidity had negative influence on hatchability.

Rearing performance of eri silkworm *Samia cynthia ricini* during different seasons in North Eastern region at Government Eri Seed Grainage, Assam

North Eastern region has four distinct seasons *viz.,* spring (Feb-March), summer (June-July), autumn (Sept-Oct) and winter (Nov-Dec-Jan). As a multivoltine insect eri silkworm which is reared indoor has its exposure to all the changes of the seasons. The optimum condition required for rearing is 24° to 28°C temperature and 85% - 90% relative humidity (Krishnaswamy, 1978). Hence, Deka *et al.,* (2011) conducted a study by rearing the eri silkworm on different types of food plants during 2008 to 2010 at Government Eri Seed Grainage, Assam for selection of food plants for better production and productivity during different seasons with the following treatments.

T1 : Castor (first to fifth instar)

T2 : Kesseru (first to fifth instar)

T3 : Tapioca (first to fifth instar)

T4 : Castor (first to third instar) + Kesseru (fourth to fifth instar)

T5 : Castor (first to third instar) + Tapioca (fourth to fifth instar)

T6 : Kesseru (first to third instar) + Castor (fourth to fifth instar)

T7 : Kesseru (first to third instar) + Tapioca (fourth to fifth instar)

T8 : Tapioca (first to third instar) + Castor (fourth to fifth instar)

T9: Tapioca (first to third instar) + Kesseru (fourth to fifth instar)

The following observations were recorded during the rearing period.

1. **Larval duration**: The larval duration is the period between hatching of eggs and maturity of the larvae and was recorded in each treatment in days.

2. **Weight of full grown larvae**: The weight of full grown larva was taken during fifth instar one day before maturity (ripen). The larvae from each treatment were randomly selected and weighed to find out the average weight of the full grown larvae for each replication separately.

3. **Weight of matured larvae**: The weight of matured larvae was taken when the larvae stops eating, body becomes pale and has excreted last excreta. Ten larvae were picked randomly from each treatment and weighed replication wise and the average was calculated.

4. **Effective rate of rearing**: This is the number of cocoons harvested out of the number of larvae brushed and expressed in percentage

5. **Fecundity**: To record the fecundity, pairs of freshly emerged moths were placed on kharika and kept hanging on a wire. The moths were decoupled after three hours and the females were allowed to lay eggs on the kharika. After three days, the eggs were removed from the kharika and counted treatment wise to record the fecundity or the number of eggs laid by the female.

Cocoon parameters: The following cocoon parameters were recorded:

1. **Colour of Cocoons**: The colour of cocoons is creamy white of barduar ecorace and brick red in Kokrajhar ecoraces.

2. **Single Cocoon weight**: The weight of the cocoons produced by the larvae was recorded treatment wise.

3. **Single Shell weight**: The cocoons were cut open and the weight of the shell was recorded separately.

4. **Silk ratio percentage**: Shell ratio is the amount of silk present in a cocoon shell and expressed in percentage.

$$SR\% = \frac{\text{Weight of the cocoon shell}}{\text{Weight of the cocoon with pupa}} \times 100$$

Table 10.32: Larval duration, larval weight, effective rate of rearing, fecundity, cocoon weight, shell weight and silk ratio of *Samia cynthia ricini* during Spring (Adopted from Deka *et al.,* (2011)

Treatments	Larval duration (days)	Larval weight (g)	ERR (%)	Fecundity (Nos)	Cocoon weight (g)	Shell weight (g)	Silk ratio (%)
T1	17.00	7.38	92.66	478	3.57	0.588	16.47
T2	20.33	6.94	88.20	465	3.49	0.541	15.50
T3	21.9	6.87	88.90	259	3.53	0.550	15.40
T4	22.33	6.05	90.53	374	3.89	0.590	15.16
T5	21.33	7.23	89.46	462	3.51	0.548	15.61
T6	21.00	6.77	83.23	461	3.48	0.500	15.22
T7	21.00	6.68	87.20	451	3.50	0.544	15.54
T8	21.33	6.56	90.43	371	3.49	0.549	15.73
T9	21.33	6.49	84.13	348	3.41	0.511	14.98

Table 10.33: Larval duration, larval weight, effective rate of rearing, fecundity, cocoon weight, shell weight and silk ratio of *Samia cynthia ricini* during Summer (Adopted from Deka *et al.,* (2011)

Treatments	Larval duration (days)	Larval weight (g)	ERR (%)	Fecundity (Nos)	Cocoon weight (g)	Shell weight (g)	Silk ratio (%)
T1	17.33	6.71	88.17	398	2.98	0.380	12.75
T2	19.66	6.07	76.83	387	2.87	0.355	12.36
T3	22.00	5.67	75.30	388	2.59	0.328	12.66
T4	20.33	5.53	79.67	354	2.68	0.331	12.35
T5	20.78	6.57	79.50	389	2.76	0.352	12.75
T6	20.78	6.61	76.57	376	2.49	0.288	11.56
T7	20.33	5.87	77.07	359	2.57	0.289	11.24
T8	20.78	6.34	78.73	361	2.53	0.280	11.06
T9	20.78	5.75	77.30	354	2.61	0.288	11.03

Table 10.34: Larval duration, larval weight, effective rate of rearing, fecundity, cocoon weight, shell weight and silk ratio of *Samia cynthia ricini* during Autumn (Adopted from Deka *et al.,* (2011)

Treatments	Larval duration (days)	Larval weight (g)	ERR (%)	Fecundity (Nos)	Cocoon weight (g)	Shell weight (g)	Silk ratio (%)
T1	20.00	6.48	90.07	468	3.58	0.588	16.42
T2	22.00	6.19	83.60	455	3.56	0.576	16.17
T3	22.60	5.60	86.20	469	3.56	0.566	15.89

Table 10.34: *Contd....*

T4	23.33	5.97	86.87	438	3.90	0.545	16.07
T5	21.66	6.15	85.06	459	3.50	0.556	15.88
T6	22.33	5.93	74.56	461	3.52	0.488	13.86
T7	22.00	5.75	81.70	444	3.49	0.536	15.35
T8	22.00	6.23	83.76	438	3.49	0.539	15.44
T9	22.00	6.12	75.56	446	3.48	0.535	15.37

Table 10.35: Larval duration, larval weight, effective rate of rearing, fecundity, cocoon weight, shell weight and silk ratio of *Samia cynthia ricini* during Winter (Adopted from Deka *et al.,* (2011)

Treatments	Larval duration (days)	Larval weight (g)	ERR (%)	Fecundity (Nos)	Cocoon weight (g)	Shell weight (g)	Silk ratio (%)
T1	26.00	6.47	88.50	458	2.96	0.428	14.29
T2	26.66	5.78	84.40	440	2.87	0.410	14.28
T3	29.00	5.10	85.10	438	2.88	0.405	14.06
T4	27.33	5.47	86.10	439	2.83	0.402	14.20
T5	27.76	5.20	78.80	441	2.73	0.389	14.20
T6	28.33	5.57	81.30	439	2.48	0.348	14.03
T7	27.33	5.32	83.80	440	2.50	0.311	12.44
T8	27.76	5.08	75.40	429	2.59	0.310	11.96
T9	28.23	4.90	71.30	428	2.61	0.313	11.99

Data on the impact of feeding of *Samia cynthia ricini* reared on different food plants and their combinations in different seasons are presented in Tables 10.32 to 10.35. In the investigation of Deka *et al.,* (2011), it was found that shortest larval duration, highest larval weight, highest ERR %, highest fecundity and highest silk ratio % were recorded in T1 group during spring whereas longest larval duration was observed during winter in T3 group. From these results it is obvious that the performance of eri silkworm was better on *Ricinus communis* which is non powdery and highly palatable (Suryanarayana *et al.,* 2003) food plant which showed significant increase in the larval weight, effective rate of rearing, fecundity, cocoon weight, shell weight and silk ratio percentage and minimum larval duration in spring season followed by autumn, winter and summer season respectively. It was also suggested by Mukherji and Guppy, (1970) that interchange of host plants from castor to tapioca and vice versa in rearing of eri silkworm showed better performance when the chawki stages are reared on castor and later stages on tapioca. From this study it can be inferred that, irrespective of season, rearing of Eri silkworm in castor is best in terms of larval duration, healthy larval growth and production of silk. Maximum yield of eri silkworm with significant increase in the larval weight, effective rate of rearing, fecundity, cocoon weight, shell weight, silk ratio percentage and minimum larval duration were recorded during the spring season followed by autumn, winter and summer. The highest silk ratio percentage was recorded in spring season (16.47%) when the silkworm was solely fed on castor leaves and found lowest when the eri silkworm was fed with tapioca leaves from first to third instar and then interchanged with kesseru from fourth to fifth instar larvae in summer season. Further studies on parameters on silk fibre will be helpful for confirmation of silk quality.

Improvement of cocoon parameters of Eri silkworm (*Samia c. ricini*) in their nutritional level during different rearing seasons in Zoology Department of Madhabdev College at Lakhimpur district, Assam

Until recent years few studies have been carried out in respect of consumption and utilization of food during different rearing seasons in silk producing insects like *Bombyx mori*. But in eri silkworm, *S. c. ricini* literature pertaining to few studies are available on the effect of different food plants in respect of quality leaves that directly influence on health, growth and survival. To evaluate the efficiency of food plants on cocoon parameters of eri silkworm in different rearing seasons, Mitalee Baruah, (2012a) conducted an experiment with kesseru by enrichment of leaves with 'Tricovit' foliar spray. To study the nutritional efficiency of eri silkworm the larvae were reared on three food plants *viz.*, Kesseru (normal), Kesseru (treated) and castor (control) separately under laboratory condition at $27.75 \pm 30^{\circ}$ C and $82.8 \pm 7\%$ RH from first to 5th instar in Zoology Department of Madhabdev College during 2009-10. Here the one Kesseru tree used for the investigation was 5 metres in height. The leaves of the trees were treated with 'Tricovit' foliar spray at the rate of 2 teaspoonful powder in 10 liters of water. The solutions were sprayed on both surfaces of the leaves in the morning. After 7 days these leaves became deep green, because 'Tricovit' encourages the process of photosynthesis, nitrogen cycle resulting in healthy leaves. It also supplies Protein, Hormone and mineral contents to the plants, which control different types of diseases. The leaves were fed to the eri silkworm after hatching to the 3rd larval stages. The experiment was laid out in completely randomised design (CRD) for various estimations of nutrition, growth and cocoon characters. Observations were recorded on larval growth and economic cocoon parameters (Fibroin and Sericin content) of *S. c. ricini* reared on 3 different food plants during different rearing seasons like spring, summer, late summer and autumn.

The results showed that among the food plants the highest amount of fibroin content was recorded on Castor (82.46%) which was at par with treated Kesseru (80.99%). While the lowest amount of fibroin content was observed in larvae fed with Kesseru (Untreated) (80.73%). The interaction effects due to season × food plants were not significant. Among the seasons the autumn rearing silk cocoon recorded the highest fibroin content (83.46%), followed by late summer (82.56%), summer (81.19%) and lowest was recorded in spring season (80.84%) (Table 10.36). Among the food plants the lowest sericin content was recorded on Castor (17.64%), followed by Kesseru (treated) (19.04%) and highest on Kesseru (Untreated) (19.55%). Among the seasons, spring rearing cocoon shell was recorded the highest amount of sericin content (19.18%) followed by Summer (19.09%), late summer (18.68%) and lowest in autumn (18.03%) season (Table 10.37).

The mean data of larval duration of eri silkworm on different food plants in different rearing seasons showed shortest larval duration on summer (20 days) followed by late summer (21 days), spring (24.67 days) and the longest on autumn (28.67 day) season. Among the food plants the shortest larval duration was recorded on Castor (20.75 days). Between the two tested food plants, the larval duration was shorter on treated Kesseru (23.50 days) than the untreated Kesseru (26.50 days) (Table 10.38). The results of larval duration and larval weight indicated that the food plants had significant effect on them. The larvae reared on castor recorded the highest value of larval weight and took the shortest duration to complete its larval period. Between kesseru (*H. fragrans*), control and kesseru treated, in later case the shorter duration was recorded to complete its larval period. Similar observations were made by Datta *et al.,* (2000) and reported that the characters of cocoon primarily depends upon the larval weight. The higher values of cocoon weight, shell weight and pupal weight on the food plants of *R. communis* and *H. fragrans* in the present study might be due to their higher larval weight gain at the end of the larval feeding period. From the study it became evident that kesseru treated with 'Tricovit' emerged as the most efficient food plant for rearing of eri silkworm *S. c. ricini*. This might be due to supplying of extra nutrient by spraying foliar spray on the leaves of kesseru, which contain more Carbohydrate, more nitrogen and more minerals. Hence from the study of Mitalee Baruah, (2012a) it can be suggested that 'Tricovit' treated Kesseru leaves can be used to increase the nutrition value of eri silkworm for better rearing performance in environmental condition of Lakhimpur district of Assam.

Table 10.36: Effects of food plants on Fibroin content (%) on the cocoon shell of *S. c. ricini* in different seasons (Mitalee Baruah, 2012a)

Food plants	Spring	Summer	Late summer	Autumn	Mean
H. Fragrans (Untr)	80.09	80.79	80.92	81.15	80.73
H. Fragrans (T)	80.56	80.80	80.99	81.64	80.99
R. communis	81.87	81.99	82.56	83.43	82.46
Mean	80.84	81.19	82.56	83.46	

Table 10.37: Effect of food plants on sericin contents (%) on the cocoon shell of *S.c.ricini* in different seasons (Mitalee Baruah, 2012a)

Food plants	Spring	Summer	Late summer	Autumn	Mean
H. Fragrans (Untr)	19.91	19.82	19.63	18.85	19.55
H. Fragrans (T)	19.44	19.34	19.01	18.36	19.04
R. communis	18.19	18.10	17.40	16.87	17.64
Mean	19.18	19.08	18.68	18.03	

Table 10.38: Effects of food plants on larval duration of *S.c.ricini* in different seasons (Mitalee Baruah, 2012a)

Food plants	Spring	Summer	Late summer	Autumn	Mean
H. Fragrans (Untr)	29	22	23	32	2650
H. Fragrans (T)	24	20	21	29	23.50
R. communis	21	18	19	25	20.75
Mean	24.67	20.00	21.00	28.67	

Seasonal effect on larval weight and shell ratio of eri silkworm fed on castor, kesseru and treated kesseru by foliar spray in Zoology Department of Madhabdev College at Lakhimpur district, Assam

Studies on quantitative aspects of nutrition of any insect are very much essential for better understanding of the insect plant relationship. It is quite obvious that the larva attains its maturity and finally into adulthood by consuming its fullest feeding requirements, and utilizing it at a particular rate to achieve its proper growth so that the energy stored in the body later on helps in performing various metabolic activities which are to be completed in the non-feeding period. Moreover the quality of food which, the insect feeds directly influences the ultimate quantity of product produced by them. The highly nutritious and nutrient balanced food are the prime factors responsible for healthy growth and development of any insect, as it provides the ultimate source of energy to the insects. Thus both these aspects of nutrition (quantitative and qualitative aspects) unanimously contribute a much more prominent knowledge and understanding of the insect plant relationship that exist between them and is of great importance for improvement of diet of the insect through selection of food plants possessing superior nutritive value. To study the nutritional efficiency of eri silkworm, Mitalee Baruah, (2012b) reared the larvae of eri silkworm on three food plants *viz.,* Kesseru (normal), Kesseru (treated) and castor under laboratory conditions at $27.75 \pm 30^{0}C$ and $82.8 \pm 7\%$ RH from first to 5th instar in different seasons. The most important commercial character of cocoon is weight which indicates approximate quantity of silk that can be spun from it. More important than the cocoon weight is the weight of cocoon shell because it is the shell that yields silk. Higher the shell weight, greater the silk yield from it. The shell ratio indicates the quantity of silk that can be spun from a lot of fresh cocoons.

Results of the experiment showed that shell ratio (SR%) recorded was highest on castor (14.96%) and differed significantly from the other two tested food plants (Table 10.40). Between the treated and untreated Kesseru the SR (%) recorded was the highest on the Kesseru treated (14.40%) and lower on the Kesseru (Untreated) (14.16%). Among the seasons the results revealed that highest SR (%) was recorded in autumn (14.80%) followed by summer (14.67%), late summer (14.58%) and lowest in spring (13.98%). The interaction effect of food plant variety with season on SR (%) was recorded to be non-significant. The results of Mitalee Baruah, (2012b) study on silk ratio reared on treated *H. fragrans*, had significant effect on all the cocoon parameters tested. The shell ratio obtained was maximum on *R. communis* while they had obtained minimum values on untreated *H. fragrans*. The highest larval weight was recorded on Castor (10.07 gm). Between the Kesseru untreated and treated, the larval weight is higher in treated (9.06 gm) than untreated one (8.46 gm). Among the seasons, the larval weight recorded was highest in summer (10 gm) followed by spring (9.72 gm), late summer (8.81 gm) and lowest in autumn season with 8.12 gm.

It is quite obvious that the quality of food which the insect feeds directly influence the growth of the larvae. The highly nutritious and nutrient balanced food is the main factors responsible for healthy growth and development of any insect. The results of larval duration and larval weight indicated that the food plants had significant effect on them. The larvae reared on *R. communis* recorded the highest value of larval weight followed by *H. fragrans* (treated) (Table 10.41).

Similar observations were made by Dutta *et al.,* (1996) and reported that the characters of cocoon primarily depends upon the larval weight. Kapil, (1967) also observed similar results on *S. c. ricini.* From their opinions it could be inferred that the occurrence of higher values of cocoon weight, shell weight and pupal weight on the food plants of *R. communis* and *H. fragrans* in the study of Mitalee Baruah, (2012b) might be due to their higher larval weight gain at the end of the larval feeding period. Whereas, the higher values of silk ratio on the food plant of *R. communis* and *H. fragrans* (treated) and *H. fragrans* (natural) could be inferred to the higher rate of utilization of crude protein by the larvae during their feeding period as suggested by Horie, (1978) on *B. mori.* Among the seasons, the SR (%) recorded was highest in the autumn season and lowest in the spring season. This might be due to the presence of comparatively higher amount of crude fibre and crude fat and carbohydrate content of the leaves that were grown in the autumn season than the other seasons. In the available literatures there is no report of any work regarding evaluation of eri silkworm fed with such Kesseru leaves, which are treated with extra foliar spray. Thus the study of Mitalee Baruah, (2012b) suggests that the farmers can use such foliar spray in the Kesseru leaves to increase their nutrition value that can be fed to eri silkworm for better rearing performance.

Table 10.40: Effect of food plants on the shell ratio (%) of eri silkworm in different seasons (Adopted from Mitalee Baruah, 2012b)

Food plants	Spring	Summer	Late summer	Autumn	Mean
H. Fragrans (Untr)	13.74	14.20	14.15	14.56	14.16
H. Fragrans (T)	13.85	14.59	14.39	14.78	14.40
R. communis	14.36	15.23	15.21	15.05	14.96
Mean	13.98	14.67	14.58	14.80	

	S.Ed ±	CD 5%
Host plants	0.24	0.45
Season	0.30	NS
Host plant x season	0.46	NS

Table 10.41: Effects of food plants on the larval weight (in gram) of eri silkworm in different seasons (Adopted from Mitalee Baruah, 2012b)

Food plants	Spring	Summer	Late summer	Autumn	Mean
H. Fragrans (Untr)	9.01	9.41	8.50	6.90	8.46
H. Fragrans (T)	9.55	9.60	8.90	8.20	9.06
R. communis	10.60	11.00	9.04	9.27	10.07
Mean	9.72	10.00	8.81	8.12	

	S.Ed ±	CD 5%
Host plants	0.301	0.694
Season	0.202	0.401
Host plant x season	0.210	0.410

Seasonal variation in the commercial and economic characters of eri silkworm, *Samia c. ricini* under agro climatic conditions of North-Eastern India

In sericulture industry rearing is the most important and critical operation. Silkworm rearing depends upon the prevailing climatic conditions of the place of rearing, availability of essential facilities/ materials like food plants, rearing house, appliances, equipments, etc. The biology of eri silkworm was studied by the earlier workers like Choudhury (1982) and Sarkar, (1983). The Borduar eco-race showed positive characters in all economic parameters thus it was exploited commercially. During the later period attempts were made (in 2003-2008) to explore eri silkworm germplasm resources of the NE region and thus 26 accessions were collected along with one wild race (Chakravorty *et al.,* 2008). In this study also, superiority of the Borduar eco-race was further confirmed. But information on different commercial characters during different seasons of rearing with special reference to agro climatic conditions of North-Eastern India has not been studied so far. Hence, Sarmah *et al.,* (2012) made an attempt with the aim to evaluate the best season for eri silkworm rearing using Borduar ecoraces in terms of larval weight, effective rate of rearing, single cocoon weight, single shell weight and fecundity.

The study was conducted in two consecutive years. The rearing technology developed by the Central Muga Eri Research & Training Institute, Lahdoigarh, Assam, India was followed in the experiment. A promising Borduar ecorace of eri silkworm was utilized by studying its evaluating characters. Characters like fecundity, hatching percentage, larval weight, cocoon weight, shell weight and SR% was considered for evaluation programme. A non-bloomy red variety (NBR-1) of castor (*Ricinus communis*) was used as food plant for the study. The eri moth emerged from the cocoons of the Borduar eco-race was allowed for coupling and lay eggs individually. After hatching of egg, rearing was conducted keeping 100 worms per replication maintaining three replications in seven seasons to collect the information on evaluating characters. Similarly to study suitable season the eri moth emerged from the cocoons of the respective periods were allowed for coupling and lay eggs individually. After hatching of egg, rearing was conducted keeping 150 worms per replication maintaining four replications. The data related to larval weight, ERR (%), single cocoon weight, single shell weight and fecundity were recorded and analysed. Average maximum and minimum temperature and relative humidity data during different seasons were recorded.

The highest larval weight (7.32 g), average fecundity (580 Nos.) and ERR (94.7%) were recorded during October-November. The highest single cocoon weight (4.01 g) was observed during August-September (Table 10.39). Hence, it has been illustrated that autumn is the best season for eri silkworm rearing to improve productivity of eri silk. The overall study revealed that Borduar eco-race showed distinct variation in commercial cocoon characters of eri silkworm in different seasons. Analysis of data showed that the performance of eri silkworm was found to be appreciably good particularly during autumn season (August-September and October-November) compared with the performance in other seasons. Choudhury, (1982) also reported late spring and late autumn are the best seasons for commercial spinning of eri silkworm. Moreover, considering the performance of different grainage characters such as good moth emergence, higher coupling realization, higher fecundity and higher hatchability of eggs the autumn season (September to November) was found to be the best for eri seed production (Sarkar *et al.*, 2008). The commercial rearing of eri silkworm is suggested during August-November season to have better cop harvests. The Borduar eco-race of eri silkworm may be exploited for commercial seed as well as cocoon production to make the eri silk industry more economically vibrant and sustainable.

Table 10.39: Seasonal variations in certain commercial characters of eri silkworm (Adopted from Sarmah *et al.,* 2012)

Sl. No.	Season/ period	Larval weight (g)	ERR (%)	Single cocoon weight (g)	Single Shell Weight (g)	Fecundity (no)	Av.temp. (^{0}C)		Av.RH (%)
							Max	Min	
1	Apr-May	5.09^e	74.0^d	3.23^d	0.49^{ab}	400.25^d	31.41	25.41	75.14
2	May-Jun	7.02^{ab}	91.5^b	3.71^b	0.49^{ab}	467.00^{bc}	29.61	25.75	88.57
3	July-Aug	6.46^c	92.1^{ab}	3.55^{ab}	0.50^{ab}	472.75^b	29.89	26.43	89.0
4	Aug-Sep	6.66^{bc}	92.0^b	4.01^a	0.53^a	483.75^{bcd}	33.20	26.43	88.09
5	Oct-Nov	7.32^a	94.7^a	3.89^{ab}	0.53^a	580.0^a	27.80	22.70	87.98
6	Nov- Jan	5.98^d	92.1^b	3.66^b	0.45^c	332.0^e	21.50	13.56	79.66
7	Feb-Mar	5.92^d	87.9^c	3.62^c	0.47^b	412.5^d	26.50	18.90	70.04

Means in a column followed by same letter are not significantly different (p= 0.05, DMRT)

Orissa

Rearing performance of some ecoraces of eri silkworm (*Samia c. ricini*) in different seasons of Odisha conducted in the Ericulture laboratory, School of Life sciences, Sambalpur University, Burla

The performance of an ecorace of silkworm is largely dependent on combined action of heredity and the environment to which it is exposed during its life time (Harada, 1961). Seasonal change in performance of silkworm races is of vital importance to understand the action of environment and genetic potentiality. It also indicates the extent to which such potential is permitted to be expressed in the ambient environment to which they are exposed.

The ecoraces (types) of *S. c. ricini* are commercially exploited in North Eastern states because of their high silk yield potential. The rearing performance of this variety of Eri has not been attractive (Hota *et al.,* 2005). One of the reasons for the low yield may be attributed to the climatic conditions of Odisha. Ray *et al.,* (2010a) assessed the productivity of the three ecoraces of *S. c. ricini* namely Borduar, Titabar and Mendipathar during different seasons of 2009-2010 and the climatic influence was analyzed. Rearing of these Eri worms was conducted through out the year 2009-2010 covering five seasons i.e. summer (April-June), rainy (July-August), autumn (September- October), winter (November- December) and spring (February-March) in the Ericulture laboratory, School of Life sciences, Sambalpur University, Burla. The meteorological data of Sambalpur district during the rearing period was presented in the Table 10.42.

Rearing and grainage experiments were conducted with five replicates of 30 larvae each in separate bamboo trays to maintain purity of the ecoraces. Commercial traits like cocoon weight (g), shell weight (g), shell ratio (%), pupal weight (g) and ERR (%) were recorded. The data were recorded as mean of five independent replicates (± S.D.) and analyzed by one way ANOVA for each ecorace and seasons; and two way ANOVA for different ecoraces and seasons. It was observed that commercial productivity and reproductive parameters were influenced by the seasonal variations of the rearing environment. Values of productivity parameters like cocoon weight, shell weight, shell ratio percentage and total silk production were found to be highest in winter followed by spring, autumn and rainy seasons while the lowest values were recorded in the summer crop. The two- way analysis of variance (pd" 0.05) showed significant difference in rearing parameters among the different ecoraces of eri silkworm in different seasons. Among the varieties, the rearing efficiency and silk ratio of Borduar ecorace was found to be the highest in comparison to others. The seasonal rearing efficiency showed an increasing trend from summer to winter season with regard to commercial productivity (Table 10.43).

In this project the reproductive and commercial productivity of the three eco races were found to vary with the changes in the climatic and environmental parameters of the rearing environment. Hota *et al.,* (2005) have reported under-performance by the Eri silkworm in Odisha with regards to the larval characters like silk yield, pupation rate, moth emergence and fecundity. In their earlier investigation the pure line ecoraces of *S. c. ricini* showed significant increase over the mixed type in terms of morphological and quantitative characters such as absolute silk content, larval weight, cocoon weight, pupal weight, cocoon shell weight and shell ratio. From the study of Ray *et al.,* (2010a), it is evident that the production during the summer was the least among different seasons. It may be attributed to the lower ERR as well as the lower weight of the cocoon and shell. Studies have been conducted under diverse environment and climatic conditions in Uttar Pradesh to show the phenotypic variations among ecoraces in various seasons (Kumar and Elangovan, 2010). It was reported that the changing conditions of different seasons of the year bring profound variance in growth and development and the expression of characters in *S. cynthia ricini* (Gogoi and Goswami, 1998).

In the study of Ray *et al.,* (2010a), it was found that the winter season is the most favourable season for the rearing of the ecoraces for the production of better silk. Also the performance of the eri silkworm larvae was the best at temperature range of 15°C to 23°C. At very high temperatures such as 43°C during summer, performance of the ecoraces was significantly lower which might be due to the decline in feeding rate of the larvae and their low conversion efficiencies resulting in low silk yield. Further, significant differences were also observed in the productivity parameters of the ecoraces with respect to different seasons of rearing. Thus the productivity parameters of *S. c. ricini* ecoraces are not only controlled by genes, but also are influenced by environmental factors such as temperature and relative humidity. The study shows that the seasonal climatic conditions in the state are quite favourable for the commercial rearing of the three ecoraces and the best season for it is winter.

Table 10.42: Meteorological data of Sambalpur district during 2009-2010 (Adopted from Ray *et al.,* 2010a)

Study Parameters	Summer (Apr – Jun)	Rainy (Jul – Aug)	Autumn (Sep – Oct)	Winter (Nov – Jan)	Spring (Feb – Mar)
Max Temp (°C)	43.6	38.7	33.6	22.4	26.2
Min Temp (°C)	25.3	23.4	20.5	15.2	18.9
Relative Humidity (%)	70.1	81.3	78.4	73.5	71.9
Total Rainfall (mm)	264.18	512.26	491.17	76.33	21.14

Table 10.43: Seasonal finding of commercial traits of three ecoraces of *Samia c. ricini* (Adopted from Ray *et al.,* 2010a)

Ecorace	Seasons	Cocoon wt (g)	Pupal wt (g)	Shell wt (g)	Shell Ratio (%)	ERR (%)
Borduar	Summer	3.2±0.1	2.71±0.1	0.516±0.01	15.75±0.68	94.98±0.31
	Rainy	3.34±0.05	2.79±0.06	0.54±0.013	16.36±0.61	96.1±0.08
	Autumn	3.47±0.08	2.89±0.07	0.58±0.039	16.77±1.1	96.9±0.059
	Winter	3.79±0.01	3.06±0.08	0.73±0.06	19.26±1.0	97.88±0.13
	Spring	3.5±0.07	2.87±0.07	0.65±0.01	18.45±0.34	96.18±0.08
Titabar	Summer	3.1±0.02	2.6±0.02	0.43±0.01	13.72±0.85	94.0±0.032
	Rainy	3.26±0.06	2.8±0.15	0.46±0.02	14.09±0.54	95.12±0.01
	Autumn	3.36±0.17	2.8±0.07	0.51±0.01	15.31±0.81	96.27±0.01
	Winter	3.49±0.08	2.9±0.01	0.58±0.01	16.55±0.52	96.96±0.02
	Spring	3.41±0.06	2.86±0.06	0.54±0.008	16.07±0.55	95.76±0.64
Mendipathar	Summer	2.86±0.07	2.61±0.080	0.25±0.008	8.6±1.1	94.28±0.39
	Rainy	3.03±0.11	2.7±0.126	0.33±0.008	10.95±1.3	94.52±0.02
	Autumn	3.23±0.03	2.8±0.04	0.44±0.008	13.71±1.5	95.26±0.03
	Winter	3.39±0.03	2.9±0.06	0.53±0.013	15.69±1.2	95.48±0.08
	Spring	3.27±0.05	2.82±0.036	0.48±0.011	14.62±1.34	95.45±0.01

Gujarat

Effects of seasonal variation and rearing conditions on eri silkworm *Samia c. ricini* at SK Nagar, Gujarat

Patel and Patel, (2009a) studied the rearing performance of *S. c. ricini* during different periods, *viz.,* August –October, (Crop I), November – January, (Crop II), February – April, (Crop III) and May-July, (Crop IV). Observations on various aspects of life cycle and economic traits like shell weight, cocoon weight and shell ratio were recorded. Out of four crops of eri worms reared during a year, the crop IV during May – July failed which may be due to prevailing unfavourable weather conditions. Among three crops reared successfully, crop II reared between November and January had longest incubation period (15.37 days) and minimum egg viability (86.2%); longest larval period (26.52 days) and maximum larval weight (3.48g); longest pupal period (30.50 days) and highest shell weight (0.41g) as well as cocoon weight (3.03g) and shell ratio (14.43%); shortest adult period (4.75 days), longest total life period (71.32 days); minimum fecundity (77.47 eggs/female) and highest sex ratio (1female : 1.10 male). On the other hand, eri worms reared between February and April (crop III) registered shortest incubation period (5.67 days), highest egg viability (98.17%); shortest larval period (22.07 days) coupled with lowest shell weight (0.16 g) as well as cocoon weight (2.64 g) and lowest shell ratio (6.44%); longest adult period (5.20 days); shortest total life period (48.40 days) and greatest fecundity (271 eggs/ female). Hence the eri silkworms can be reared during November-January season to get highest cocoon yield in Gujarat.

Effect of temperature on biology of eri silkworm *Samia c. ricini* Boisduval under controlled temperature conditions at SK Nagar, Gujarat

With a view to study the effect of different temperatures on life cycle of eri silkworm, Patel and Patel, (2009b) reared one batch of eri worms with four sets in enamel trays on castor leaves. Each set consisted of 25 newly hatched larvae and were reared at different temperatures *viz.*, 15, 20, 25, 30, 35, 40 and 45^0C in BOD incubator during November- January. Observations on date of hatching and death of adult eri worms were recorded. Percent larval survival and growth index were worked out. Laboratory studies conducted in incubator revealed that eri worms did not grow successfully below 20 ± 1^0C and above 30 ± 1^0C. The incubation period was 14.44 ± 1.38 and 11.20 ± 0.58 days and larval duration was 31.04 ± 1.33 and 22.72 ± 0.89 days and its survival was 96 and 92 percent at 20 ± 1^0C and at 25 ± 1^0C, respectively. Similarly, the growth index was 4.23 and 2.96 at corresponding temperatures. The pupal period was 35.12 ± 9.25 and 21.48 ± 6.45 days at 20 ± 1^0C and at 25 ± 1^0C, respectively. The female and male moths lived for 4.88 ± 1.33 and 4.2 ± 0.41 and 3.24 ± 0.83 and 3.04 ± 0.45 days, respectively at 20 ± 1^0C and 25 ± 1^0C. The fecundity was 113.28 ± 27.22 and 85.08 ± 29.9 eggs per female when kept at 20 ± 1^0C and 25 ± 1^0C. Similarly, hatching of eggs was 19.00 per cent and 76 per cent at 20 ± 1^0C and 25 ± 1^0C, respectively. Hence from the study of Patel and Patel, (2009b) it can be inferred that the growth of eri silkworm was good at 20 ± 1^0C.

Tamil Nadu

Influence of weather factors on larval period and silk ratio of eri silkworm in Tamil Nadu

To know the influence of weather factors prevalent in Tamil Nadu on larval period and silk ratio of eri silkworm, Subramanianan *et al.,* (2013) conducted an experiment during September, 2003 to January, 2005 with 20 batches of eri silkworms. Observations were recorded on larval period, ERR (%) and silk ratio (%). The mean maximum temperature, mean minimum temperature and relative humidity during the experimental period was recorded and presented in Table 10.44.

The available information for 20 batches of rearing was pooled and correlated. The mean maximum temperature during the study period of September 2003 to January 2005 was in the range of 26.0 to 37.2°C in the plains and hills of four districts in Tamil Nadu. The mean minimum temperature in the above period ranged from 19.0 to 27.0°C, and the relative humidity was 49 to 78%. Eri silkworm rearing could be done in the above range of temperature and humidity of these selected reference area in Tamil Nadu. The mean larval period during the above mentioned period was ranging from 19 to 24 days and the silk ratio was 11.4 to 16.8 % (Table 10.44). The worms have to be fed for an extra of 5 days on an average. That means the rearing schedule and quantity of leaf required will be having a great influence on ericulture. Again, the additional food ingested should result in increased silk in the cocoon to commensurate the time and material needed additionally. With this in view, correlations were made on the influence of key weather factors on silkworm larval period and the resultant SR percent. There was a significant negative correlation between the mean maximum temperature or mean minimum temperature during the rearing period and larval period of the eri silkworm. The correlation between maximum/ minimum temperature and silk ratio was not significant (Table 10.45). This indicates that lower the temperature during the rearing period, longer is the larval period requiring more leaves for feeding. The correlation between larval period and silk ratio percent was positive and significant.

As the larvae feed for more number of days the silk content in the cocoon also increased proportionately. Hence, during the cooler months the larvae feed for a longer time and produce more silk. Nangia *et al.,* (2000) reported that *Samia cynthia ricini* had 6 generations annually on castor (*Ricinus communis*) in the laboratory. The life cycle took 38-50 days during March-September, 49-61 days during September-November and 114-126 days during November-April. The mean weights of mature larvae, cocoons, pupae and shells were 5.24, 3.75, 3.09 and 0.61g, respectively, during July-August. Similar results were obtained using cassava as the food plant, but the weights of larvae, cocoons, pupae and shells were slightly lower. Sannappa *et al.,* (1999) reported that the larval duration and mature larval weight were correlated with larval survival, effective rate of rearing, cocoon weight, pupal weight, shell weight, shell ratio, adult emergence, fecundity and hatchability. They concluded that increased larval duration and mature larval weight have an influence on rearing performance and cocoon and grainage parameters.

Table 10.44: Influence of weather factors on larval period and silk ratio of eri silkworm (Adopted from Subramanianan *et al.,* 2013)

Batches/ Months	Mean Max. Temp. °C	Mean Mini. Temp. °C	RH %	Larval period	ERR (%)	Silk Ratio%
I. 9-10/03	31.0	27.0	73.0	19	80.2	14.80
II. 11-12/03	30.5	24.0	73.5	25	83.9	15.90
III. 12/03-1/04	29.5	19.5	72.5	21	73.7	13.00
IV. 1-2/04	30.5	22.0	73.8	21	82.6	14.58
V. 2-3/04	32.9	21.5	61.1	21	76.3	15.79
VI. 4-5/04	37.2	26.7	60.8	20	87.1	11.75
VII. 5-6/04	35.1	26.4	58.6	19	58.4	14.68
VIII. 5-6/04	36.9	27.4	60.8	20	57.3	15.23
IX. 6-7/04	34.2	25.4	59.7	20	58.7	11.40
X. 7/2004	26.0	22.0	78.0	23	68.3	15.80
XI. 7-8/04	35.9	26.3	58.6	20	71.7	12.47
XII. 8/2004	33.1	23.1	59.6	22	85.1	14.20
XIII. 8-9/04	36.3	26.9	67.7	19	80.0	13.09
XIV. 9/2004	32.9	22.4	66.2	22	78.8	15.60
XV. 9-10/04	32.9	24.0	65.7	23	76.3	16.80
XVI. 10/2004	30.5	24.6	65.6	23	80.7	16.10
XVII. 11/2004	29.1	24.0	73.0	23	84.3	15.90
XVIII. 11-12/04	31.0	24.0	72.5	23	93.2	15.70
XIX. 12/2004	30.0	19.0	55.1	24	86.0	15.90
XX. 1/2005	31.0	25.5	49.0	22	75.2	15.60

Table 10.45: Correlation between weather and rearing parameters (Adopted from Subramanianan *et al.,* 2013)

Parameters	Larval period	SR%	ERR (%)
Maximum Temp	-0.651**	0.184 NS	-0.269 NS
Minimum Temp	-0.559*	-0.303 NS	-0.337 NS
RH %	0.206 NS	-0.533*	0.233 NS
Larval period	-	0.654**	0.526*

Influence of different seasons on economic traits of eri silkworm in Tamil Nadu

Sakthivel (2014), reared the eri silkworms in different seasons *viz.,* summer (April-May), rainy (September-November) and winter (December-January) on the ruling variety of tapioca MVD1in Tamil Nadu, during 2009 to 2012. The results on the influence of different seasons under Tamil Nadu conditions on the performance of eri silkworm indicated winter season was found the best in respect of effective rate of rearing, larval and cocoon traits of eri silkworm followed by rainy season, whereas least performance was recorded during summer (Table 10.46). During winter the larval duration was relatively longer (29.20 D.H) where as it was 23.16 and 21.13 during rainy and summer seasons respectively. However, Rajadurai *et al.,* (2010) did not find much difference in larval periods in different seasons when reared on tapioca leaves and recorded highest silk % in summer under Southern Karnataka conditions. The seasonal influence was more profound on the ERR of eri silkworm which was reported to be significantly highest (78.50%) during October-January and lowest (53.0%) during March-April (Devaiah *et al.,* 1978). Devaiah and Dayashankar, (1982) recorded less ERR on both castor (65.49%) and tapioca (55.95%) when reared in summer. Deka *et al.,* (2011) recorded better economic traits of eri silkworm including fecundity, effective rate of rearing and cocoon characters in spring season followed by autumn and winter and least in summer.

Kar *et al.,* (2000) reported that lesser rearing period during summer season was associated with lower values of cocoon weight and shell weight and reverse was the case in winter season. Similarly, Dash *et al.,* (1992) recorded highest ERR in tasar silkworm, *Antheraea mylitta* and maximum cocoon production during winter when the temperature and humidity were recorded as 20.27^0C and 65.39%, while there was poor cocoon yield in autumn and rainy seasons when the temperature and RH were ranged between $28.67-31.85^0C$ and 76.51-83.04% respectively. The results of the study by Sakthivel, (2014) revealed, the superiority of the winter for eri silkworm rearing compared to other seasons confirms the earlier reports on tasar silkworm. This could be attributed to the moderate average temperature (20.3-30.6^{0}C) and RH (63.8-71.5%) in Tamil Nadu which are ideal for growth and development of silkworm and cocoon formation whereas high temperature (29.8-38.4^{0}C) and low humidity (48.1-54.7%) prevalent in summer are known to have negative impact on silkworm rearings and hence lower cocoon yield. During rainy season, the economic traits of eri silkworm were better than summer which might be attributed to the favourable temperatures (22.6-33.9^{0}C) and RH (65.8-84.2%).

Kar *et al.,* (2004) recorded that the cocoon weight, shell weight and shell ratio and total silk production by eri silkworm were high during winter, spring, autumn and rainy seasons, while lowest during summer under Orissa conditions. Hazarika *et al.,* (2003) rated the suitable seasons for eri silkworm in the order of autumn > winter > spring > summer under Assam conditions. The study indicates that best performance of eri silkworm crop was recorded in winter season followed by rainy season. Summer exhibited adverse effects both on economic traits and cocoon yield. The eri silkworm rearing in summer therefore could not be advantageous. Moreover, manipulation of temperature and RH suitably inside the rearing house during summer season could help to improve the eri silkworm crop performance.

Table 10.46: Meteorological parameters during eri silkworm rearing in different seasons (Average of three years 2009-10, 2010-11, 2011-12) (Adopted from, Sakthivel, 2014)

Season	Temperature (^{0}C)			Relative humidity (%)			Rainfall (mm)
	Max.	Min.	Avg.	Max.	Min.	Avg.	
Summer (April-May)	38.4	29.8	34.1	54.7	48.1	51.4	117.5
Rainy (Sep-Nov)	33.9	22.6	28.2	84.2	65.8	75.0	516.8
Winter (Dec-Jan)	30.6	20.3	25.4	71.5	63.8	67.6	36.3

Maharashtra

Suitability of climatic conditions in Vidarbha region of Maharashtra for promising strains of eri silkworm

Maharashtra is one of the non traditional states for ericulture due to other cash crops like cotton, sugarcane, grapes, soybean, onion etc. Recently ericulture is gaining importance in the state. In this regard, Wankhade *et al.,* (2014) conducted an experiment to evaluate the suitability of the climatic condition and rearing performance of some promising strains of eri silkworm in Vidarbha region of Maharashtra for commercial exploitation of eri silkworm as composite farming with castor seed production. The rearing of 4 strains of eri silkworm *i.e.* Greenish Blue Zebra (GBZ), Yellow Zebra (YZ), Greenish Blue Plain (GBP), Yellow Plain (YP) was carried out during October-November, 2013 at Karanja (Gh.) tahsil, Wardha district of Vidarbha region of Maharashtra. The Climatic condition of Vidarbha region of Maharashtra is different from the source of origin of eri silkworm. These eri silkworm strains are quite different from each other both in morphology (colour polymorphism) and genetic traits. The evaluation was done considering various parameters i.e. cocoon weight, shell weight, shell ratio, larval duration, ERR %, fecundity and life cycle duration.

During rearing in months of October – November 2013, slight fluctuation in temperature was recorded throughout the period. The average minimum and average maximum temperature ranges from 24°C to 29°C throughout entire rearing period (Table 10.47).

Similarly in case of relative humidity also a narrow range of fluctuation *i.e.* 83% - 91% was recorded (Table 10.47). The results of the rearing show prominent variation in the body weight of larvae among these 4 different strains. The highest larval weight was recorded in Greenish Blue Zebra (7.339 g) and Yellow Zebra (7.253 g) followed by 6.688g in Yellow Plain and 6.665g in Greenish Blue Plain. Thus, higher growth was found in GBZ and YZ whereas slower growth was recorded in YP and GBP. Singh *et al.,* (2012) also reported higher larval weight in Yellow Zebra (YZ) and Greenish Blue Zebra (GBZ) when reared on castor food plants during November-December under North India climatic condition. Priyanki Sarmah and Jogen Chandra Kalita, (2013) also reported highest larval body weight in YZ under climatic condition of Assam. Hence from the study of Wankhade *et al.,* (2014), it can be concluded that GBZ and YZ eri silkworm strains were suitable to the climatic conditions of Vidarbha region of Maharashtra.

Table 10.47: Temperature and Relative Humidity maintained & recorded during rearing of 4 strains of eri silkworm, *Samia c. ricini* in early winter (October-November 2013) (Adopted from Wankhade *et al.,* 2014)

Stage	Temperature °C		Humidity (%)	
	Average Min	Average Max	Average Min	Average Max
I instar	24 °C	27 °C	90%	91%
II instar	26 °C	27 °C	91%	91%
III instar	28 °C	28 °C	91%	91%
IV instar	28 °C	28 °C	88%	88%
V instar	28 °C	29 °C	88%	88%
Spinning-Moth emergence	28 °C	29 °C	83%	83%

Thermal Acclimation in Eri Silkworm

Thermal acclimation in the endogenous respiratory rate of fat body of eri silkworm

Temperature determines life of organism more than many other environmental factors. The cause of this temperature sensitivity lies in the fundamental facts that all organisms are built up by chemical reactions which follow the laws of thermodynamics (Alexandrove, 1977). A perusal of literature reveals that a number of insect species exhibit either 'complete' or 'partial' pattern of respiratory compensation against thermal fluctuation at organismal level. Meyer and Schavb, (1973) have confirmed these results with experiment on oxygen consumption of larval stage of Calliphoridae. However, counter efforts have been made regarding exploring the capacity of thermal acclimation of oxygen consumption at the cellular level of insects. Accordingly, Singh and Singh, (1997) communicated the reports of the thermal acclimatory responses of fat body tissue of both cold and warm adapted eri silkworms.

The eri silkworms were reared in the laboratory at 23 ± 1^0C. The laboratory reared worms were transferred right on the first day of 3rd instar in BOD incubator maintained at 15^0C and at 30^0C with relative humidity of 75%. Thus the period of 3rd and 4th instar larval duration was available to the insects for thermal adaptation. The fat body tissue was dissected and the endogenous respiratory rate of fat body was measured according to methods described by Umbriet *et al.,* (1964). The measurements of the rate of oxygen consumption of the tissue of cold and warm adapted insects were made at 15^0C, 25^0C and 35^0C. The rate of endogenous oxygen consumption of fat body was expressed as μl oxygen consumed/hour/mg protein. The thermal coefficient (Q10) for the endogenous respiratory rate in a particular thermal range was calculated as per VantHoff's equation.

The data of the Table 10.48 demonstrate the effect of temperature on the endogenous respiratory rate of fat body of cold and warm adapted insects measured at three different temperatures (15^0C, 25^0C and 35^0C) during the fifth instar and spinning period. The fat body of cold adapted insects exhibited higher values of respiratory rate than those of warm adapted insects at all temperature of measurements. The degree of difference in the respiratory rate between the two thermal groups of the insects of each stage is observed to decrease gradually with an increase in ambient temperature as evident from Table 10.49. However, percent increase in the endogenous respiratory rate of fat body due to cold adaptation remains almost static irrespective of developmental stage, when measured at 15^0C and 25^0C, but in contrast at 35^0C, it is variable at different developmental stages (Table 10.49). Table 10.50 illustrates the marginal change in the value of Q10 for the endogenous respiratory rate of fat body due to adaptation of insects to low and high temperatures. A tendency of slight decrease in Q10 value is observed due to cold adaptation in the lower and higher thermal range for the insects of all developmental stages.

Almost similar observations can be noticed when the comparison is made between the Q10 values of lower and higher thermal range irrespective of thermal acclimation in each developmental stage. Oxygen consumption is often taken as a measure of overall metabolic rate of an animal. Its measurements have been employed more than any other experimental parameters to monitor changes in insect metabolism associated with temperature (Keister and Buck, 1974). In the study of Singh and Singh, (1997) the endogenous respiratory rate is always higher in cold adapted worms in all developmental stages and at all temperatures. This type of thermal response can be categorised as 'translational' pattern of metabolic compensations, but partial. Similar observations were made by Meyer and Schavb, (1973), in larval stage of Calliphoridae. However, a marginal tendency towards rotation with intersections with rate versus temperature at a high temperature may be noticed. This is further supported by the slight reduction in value of Q10 during cold adaptation (Table 10.50). The extent of increase in the endogenous respiratory rate of the fat body tissue due to cold adaptation presents a gradual decrease with an increase in the ambient temperature (Table 10.49).

It is evident that metabolic strategy employed by the fat body of the insects for thermal acclimation is independent on the stage of its development. Moreover, at each temperature of measurement *viz.,* 15, 25 and 35^0C the degree of difference in the endogenous respiratory rate of fat body between the two thermal categories remains more or less of same magnitude irrespective of developmental stage (Table 10.50). The degree of compensation of the metabolic rate at higher temperature of measurement (35^0C) is lesser than what is seen at lower temperature (15 and 25^0C). Obviously, insect possess a better capacity of cold adaptation than warm adaptation. Perhaps, a high temperature like 35^0C is not suitable for these insects for their normal growth and development (Sachan and Bajpai, 1973). While a 'translational' pattern of thermal acclimation involves a quantitative strategy, the 'rotational' pattern signifies more of a qualitative, one during thermal acclimation of poikilotherms (Prosser, 1973). Since, the pattern of thermal acclimation exhibited by the tissue at each developmental stage is observed to be 'translational' with a slight tendency of rotation. It appears that the metabolic strategy employed by the fat body tissue involves more of quantitative change than qualitative. Probably, this may be the reason why a marginal decrease in Q10 value supports the contention of Rao and Bullock, (1954) regarding an adaptive significance of lowered Q10 of biological rate process during cold acclimation, which results in an increased thermal insensitivity. Besides a critical analysis of data (Table 10.50) demonstrates a diminution in the value of Q10 for the endogenous respiratory rate of fat body with increasing temperature of measurements for any of those thermal categories of a particular developmental stage of the insect. Thus it may be concluded from the aforesaid observations that the fat body tissue of eri-silkworm possesses an excellent capacity of thermal acclimation in its metabolic rate.

Table 10.48: Effect of temperature on endogenous respiratory rate (μl O_2 consumed/hour/mg protein) of fat body of cold and warm adapted eri silkworm (Adopted from Singh and Singh, 1997)

Temperature	Nature of adaptation	Developmental stages					
		Instar V			Spinning period		
		Early	Middle	Late	Early	Middle	Late
15^0C	Cold adapted	0.279±0.013	0.348±0.011	0.740±0.030	0.583±0.020	0.250±0.009	0.129±0.017
	Warm adapted	0.167±0.012	0.213±0.013	0.474±0.014	0.360±0.009	0.158±0.013	0.078±0.008
	Significance	P<0.001	P<0.001	P<0.001	P<0.001	P<0.001	P<0.001
25^0C	Cold adapted	0.523±0.018	0.654±0.210	1.344±0.039	1.030±0.087	0.450±0.030	0.231±0.011
	Warm adapted	0.365±0.016	0.465±0.018	1.010±0.053	0.750±0.024	0.323±0.014	0.160±0.010
	Significance	P<0.001	P<0.001	P<0.001	P<0.001	P<0.001	P<0.001
35^0C	Cold adapted	0.862±0.038	1.050±0.083	2.386±0.067	1.667±0.083	0.743±0.035	0.403±0.013
	Warm adapted	0.662±0.021	0.888±0.031	1.859±0.110	1.408±0.118	0.563±0.034	0.305±0.016
	Significance	P<0.001	P<0.001	P<0.001	P<0.001	P<0.001	P<0.001

Table 10.49: Percent increase in the endogenous respiratory rate of fat body of eri silkworm due to cold adaptation measured at different temperatures (Adopted from Singh and Singh, 1997)

Temperature	Developmental stages					
	Instar V			Spinning period		
	Early	Middle	Late	Early	Middle	Late
15^0C	67.1	63.4	56.1	61.9	58.2	65.4
25^0C	43.3	40.6	33.1	37.3	39.3	44.3
35^0C	30.2	18.2	28.5	18.4	32.0	32.1

Table 10.50: Approximate Q10 values for the endogenous respiratory rate of fat body of cold and warm adapted eri silkworm (Adopted from Singh and Singh, 1997)

Periods	Nature of adaptation	Developmental stages			
		Instar V		Spinning period	
		Thermal range		Thermal range	
		$15\text{-}25^0$C	$25\text{-}35^0$C	$15\text{-}25^0$C	$25\text{-}35^0$C
Early	Cold adapted	1.88	1.65	1.77	1.62
	Warm adapted	2.18	1.81	2.08	1.62
Middle	Cold adapted	1.88	1.60	1.80	1.62
	Warm adapted	2.18	1.91	2.04	1.62
Late	Cold adapted	1.82	1.78	1.79	1.62
	Warm adapted	2.13	1.84	2.05	1.62

Thermal acclimation in eri silkworm (*Samia c. ricini*) to high temperature and low humidity conditions by discontinuous regime

Improvement of eri silkworm tolerance to high temperature was carried out under high temperature (42±1°C) and low relative humidity (50±5%). Eri silkworm rearing at normal temperature (25±2°C; 80±5% R.H.) served as control treatment and was compared to rearing at high temperatures. Directional selection was undertaken with the batches reared at 42±1°C and 50±5% R.H. until the 5th generation. From 6th, 8th, 10th and 12th generations, rearing was conducted as normal rearing (25±2°C; 80±5% R.H.), while for 7th, 9th, and 11th generations directional selection was done as in the case of 1st to 5th generations. Various parameters were used as indices for high temperature tolerance. When the silkworms were reared fom F1 to F12 (12th generation), the survival rate of larva (1st-5th instar) (95.33%) and larva - adult stage (69.33%) including cocooning rate (80.67%) of F12 were the highest compared to F1 and F11. For cocoon yields among F1, F11, and F12, the highest values of fresh cocoon weight (2.7144 g), pupal weight (2.3490 g), shell weight (0.3671 g), shell ratio (13.53%), total cocoon shell weight (17.01 g), and fresh cocoon/10,000 larvae (25. 16 kg) were achieved from F12, which were significantly different to F1 and F11. In the same manner for egg yields, F12 provided the maximum numbers in case of eggs/moth (311.33 eggs), total eggs (6,693.33 eggs) and total hatched eggs (4,842.67 eggs). Of these evaluations between F11 and F1, it was found that F11 was higher than F1 and statistically different in all parameters excluding the percentage of hatched eggs, with no significant difference. These results indicate that the property of high temperature tolerance was improved and heat tolerant property of eri silkworm (SaKKU1) is heritable. This is a first to report on heat tolerance improvement of eri silkworm. Although it is a first trial that was carried out in a laboratory, it can be applied on eri silkworm rearing in the future to cope up with recent global warming trends (Wongsorn *et al.,* 2015).

Influence of Environment on Biological and Economic Characters of Eri Silkworm in Egypt

Effect of Photoperiod in relation to the development and reproduction of the eri silkworm *Samia c. ricini* Boisd in Egypt

A laboratory study was carried out in Egypt to determine the effects of various photoperiods (LD 0:24, 10:14, 11:13, 12:12: 13:11, 14:10 and 24:0) on the development, reproduction and silk production of *Samia cynthia ricini* (Boisd.). No effects on the duration of the larval stage were observed, but the duration of the pupal stage, and pupal and adult weights, were significantly affected. The rearing of larvae under a 10-h photophase resulted in the shortest pupal period and the heaviest cocoons, pupae and adults, while rearing under complete darkness or a long photophase (14 h) resulted in the longest pupal duration and the minimum weight of all stages. Females laid the greatest number of eggs (421.5 eggs/female) when exposed to 10-h photophase, and significantly fewer (224.3 eggs/female) when reared in complete darkness or under 14-h photophase (Ali and Salem, 1978).

Effect of photoperiod on food consumption, digestion and growth of the eri silkworm *Samia c. ricini* Boisd in Egypt

Salem and Ali (1978) conducted a laboratory study to know the effect of photoperiod on food consumption, digestion and growth of the eri silkworm in Egypt. The treatments included were LD 0:24, 10:14, 11:13, 12:12, 13:11, 14:10 and 24:0. The results of the experiment showed that the larvae were found to consume more food when they were reared under 10-h photoperiod. Rate of digestion, amount of food consumed and larval growth rate decreased under photoperiods of 10 h and longer indicated a critical photoperiod (Salem and Ali, 1978).

Effect of temperature on the biology of Eri silk worm in Egypt

A laboratory study was carried out by El-Shaarawy *et al.,* (1975) in Egypt to know the effect of temperature on the biology of eri silkworm, *Samia cynthia ricini* (Boisd.) for ten successive generations on red and green varieties of castor, *Ricinus communis* for two years. The results showed that there was a negative correlation between the duration of the larval stage (which averaged from 21 - 69 days) and the minimum and maximum temperatures (10.7-25.8 deg C and 18.1-33.4 deg C, respectively). The eri silkworm larvae that were reared on red variety of castor resulted in heavier pupae and adults and the number of eggs laid per female was higher than those larvae that were reared on greencastor variety.

Studies on preserving the eri silkworm *Samia c. ricini* Boisd. during the pupal stage under different regimes of low temperature in Egypt

The eri silk moth, *Samia cynthia ricini* (Boisd.) is difficult to rear throughout the year because of its susceptibility to temperature extremes. In the laboratory in Egypt, the ability of the eri silkworm pupae to withstand low temperatures was tested by El-Shaarawy *et al.,* (1983) by exposing them to 15°C for 30 days and to 10 or 5°C for 30, 60 or 90 days. Pupae survived exposure for 90 days to 5 or 10°C without giving rise to adults. If adults emerged after subsequent transfer to 25°C, no mating occurred between individuals that had been exposed to 5°C and no eggs hatched from parents that had been exposed to 10°C for 90 days. Exposure to 10°C for 60 days, or exposure to any of the temperatures for 30 days, usually permitted the completion of the following generation, but the best rates of survival and development occurred after exposure to 15°C for 30 days.

Biochemical changes during the development of eri silkworm pupae exposed to different constant temperatures in Egypt

In an experiment conducted by El-Shaarawy *et al.,* 1982 in the Egypt laboratory during the pupal stage of *Samia cynthia ricini* (Boisd.), accumulation of total lipids increased as the holding temperatures decreased, but no accumulation occurred at temperatures of 20 or 25^0C. The increase in fat content was correlated with a reduction in water and protein contents. At the end of the pupal stage and egg formation stage total lipids decreased and total protein and water content increased. Total protein and amino acids increased during the 1st half of the pupal stage at 20 or 25^0C. Carbohydrates were found in small amounts and were mostly consumed early in pupal development. A steady decline in proteins and amino acids was observed in pupae kept for 90 days at 5^0C. Potassium was found in high and sodium in low quantities, but both these compounds, together with phosphorus, protein and amino acids, increased near the end of the pupal stage.

Effect of temperature on the silk production of the Eri silkworm, *S. c. ricini* Boisd. in Egypt

The effect of different constant temperatures (20, 23, 25 and 30° C) combined with 70% R.H. on the weights of different larval instars, pre pupae, cocoons, pupae and cocoon cortex of the Eri silkworm, *Samia c. ricini* (Boisd.) was studied by Gomaa, 2009 in Egypt. All of these biological aspects were much affected by temperature. The increase in temperature decreased the weight of larvae, pupae and cocoon cortex. Highest quantity of silk was produced, when the rearing temperature was 20^0C. So this temperature is considered as the most suitable temperature for highest silk production in Egypt.

Environmental Influence on Biological and Economic Traits of Other Vanya Silkworms

Bioecological studies of Indian golden silk moth, *Antheraea assamensis* Helfer in North Eastern India

Goswami and Singh, (2012a) collected the parental stocks of wild race and semi domestic race of muga silk moth, *Antheraea assamensis* Helfer from the natural habitat of Meghalaya (India) and studied its potential for breeding in relation to bio-ecological factors in North-East India. The morphological characteristics and behaviour at the different developmental stages of the silk moth are discussed. Colour polymorphism of the larva is observed. The wild silk moth is purely multivoltine in nature and complete four life cycles in a year. After October-November crop it undergoes diapause at the pupal stage and remained in this stage till 2nd or 3rd week of March of the next year. The average fecundity ranged from 132-147 in wild race and from 130 to 150 in semi domestic race during the different rearing seasons. The ERR% of wild race ranged from 12.54 to 48.50 with the highest value during Late Kotia (November-December), while in semi-domestic stock it ranged from 14.48 to 60.12 with the highest value during Baisakhi (April-May). The cocoon shell weight ranged from 1.32 g to 1.51 g in wild race without any significant variations during the different seasons. In semi-domestic race the shell weight ranged from 0.39 g to 0.49 g without any significant variations during the different seasons. From the study of Goswami and Singh, (2012a), it can be inferred that the wild race can be used in breeding for introgression of the genes for higher shell weight to the semi domestic race.

Studies on the long-term preservation method of muga cocoon (*Antheraea assamensis* Helfer) at low temperature

Rajkhowa *et al.,* (2011) conducted a study to develop a suitable long term seed cocoon preservation method for skipping unfavourable Aherua pre-seed crop of Muga silkworm by postponing moth emergence. Muga silkworm seed cocoons of 8 days old were subjected to the low temperatures at $7.5\pm1°C$ and $10\pm1°C$ in BOD incubator for long term preservation of 32 to 52 days following the double step preservation method. The results indicated that muga seed cocoons can effectively be preserved at $10\pm1°C$ for 42 days without affecting the grainage parameters measured for seed production.

Environmental effects on seed cocoon quality and fecundity of Muga silkworm (*Antheraea assamensis* Helfer) at Research Extension Centre, Central Silk Board, Tura, West Garo Hills, Meghalaya, India

Success of sericulture depends on the quality of silkworm eggs (Choudhury, 1982). Therefore, management of quality seed production plays an important role on overall cocoon production. In muga culture the summer and winter pre-seed and seed crops face a heavy fluctuation of temperature and humidity, resulting in crop loss or poor quality cocoons leading to poor grainage performance in the subsequent commercial crops. Therefore, maintenance of ideal environmental conditions becomes crucial for maximization of productivity (Krishnaswamy *et al.,* 1973).

Generally, seed cocoons of muga silk worm are preserved indoor to process for seed production. Normally, the range of temperature 24°C-27°C and relative humidity (RH) 75%-85% are recommended as ideal condition for production of muga seed during grainage operation (Samson and Barah, 1989). The grainage centers at the Government farms with pucca R.C.C buildings are not able to produce sufficient quality seeds especially during summer grainage. About one thousand muga private graineurs mostly with low cost thatched houses and mud plastered walls and a few of the farmers with pucca buildings are engaged in seed production without any facilities to produce the ideal condition for temperature and relative humidity. The temperature and relative humidity in the pucca buildings and thatched houses varies which results in variations in the quality of seeds produced in these three types of grainages.

Therefore, Goswami and Singh (2012b) conducted a comparative study to analyze the effect of ambient temperature and relative humidity prevailing in three different types of grainage halls, on cocooning, grainage performance and fecundity during four different seasons *viz.,* June-July, August-September, December-January and February-March. This study was carried out during 2007 and 2008 under Research Extension Centre, Central Silk Board, Tura, West Garo Hills, Meghalaya, India.

The following grainage houses were designed for the study-

(i) R.C.C. Grainage hall :- 20ftx15ftx12ft,

(ii) Temporary thatched house with bamboo wall in sunny place – 20ftx14ftx10ft (central height) and

(iii) Temporary thatched house in shade with mud plastered wall: 20ftx14ftx10ft (central height)

In each grainage house, 500 healthy and matured larvae were kept in replications for cocooning and pupation. After spinning, the cocoons were consigned till emergence of moths, coupling and egg laying in the respective grainage houses. The ambient temperature and relative humidity in each type of houses were recorded during the different seasons. During winter pre-seed and seed crops, both temperature and relative humidity was increased by using room heater/burning of charcoal and sprinkling of water. A sand bed was arranged on the floor, just below the area of seed cocoon preservation for grainage with regular sprinkling of water to increase humidity in all three types of grainage halls as and when necessary. The cocoon weight, percentage of emerged crippled moths and fecundity were recorded in each grainage house. Data were analyzed to find out the ideal low cost grainage house for the farmers.

The data on the cocoon weight, percentage of crippled moths and fecundity in relation to the different types of grainage houses are shown in

Table 10.51. It was observed that the cocoon weight was significantly higher in thatch house under the shade than that of the RCC hall (control) during summer pre-seed crop (June-July). During this season the crippled moth percentage was significantly lower than that of thatch house in sunlight and RCC grainage hall. Fecundity was also significantly higher in the thatch house under tree shade than those of the other two types of grainage house.

During summer seed crop (Aug-Sept) the cocoon weight was significantly higher in the grainage house under tree shade than that of the other grainage houses. The crippled moth percentage was found to be significantly lower that of other two types of grainage houses. During this season fecundity was higher in the thatch house under tree shade than those of the other two types of grainage houses. It could be observed that during winter pre-seed crop (Dec-Jan) there was no significant difference in the cocoon weight and the crippled moth percentage in all types of grainage houses, while the fecundity was found to be significantly higher in the thatch in sunlight than that of the two types of grainage houses. During winter seed crop (Feb-Mar), the cocoon weight in the thatch house under sunlight showed significantly higher value than that of other two types of grainage houses while, there was no significant difference in the crippled moth percentage in all types of grainage houses. The fecundity in the thatch house in sunlight exhibited higher values than the other two types of grainage houses.

The higher values in cocoon weight and fecundity during summer pre-seed and seed crop may be due to the favourable lower temperature at the thatch house under tree shade than that of the other two types of grainage houses. The lower temperature in the thatch house under tree shade also induces higher percentage of healthy (valid) moths. Higher temperature and the RH reduced the cocoon weight and the fecundity, while higher temperature and the RH caused higher percentage crippled moth. The higher cocoon weight and fecundity observed in the thatch house in Sun light during winter pre-seed and seed crop may be due to the favourable temperature found in such type of grainage house. Among environmental factors, temperature is the most important factor affecting the poikilotherms and has a pervasive effect on physiological aspects of organisms. Growth of caterpillars is strongly temperature dependant and the degree of temperature dependence varies from species to species and optimal temperature for growth is also variable. Temperature is known to play a major role on the growth and productivity in the silkworm. The fecundity is hereditary character and its expression within genotypic limitation of an insect species like *Bombyx mori* is in direct correlation with a number of physiological and ecological factors (Yamaoka *et al.*, 1971). Das *et al.*, (2006) reported that different ranges of temperature and humidity in different seasons significantly affect the oviposition behaviour and egg characters of muga silkworm.

The study by Goswami and Singh (2012b) suggested that for getting quality seed cocoons and to obtain higher valid moths and comparatively higher fecundity, farmers should construct the grainage house under tree shade during summer pre-seed and seed crop and in sunlight during winter pre-seed and seed crop. Adoption of such indigenous and improvised technique for construction of grainage house may help the farmer to minimize their expenditure to obtain quality muga seed.

Effect of rearing seasons on biological and economical traits of *Antheraea mylitta* Drury

The phenotype *A. mylitta* is the result of interaction between genotype and environment in which it develops. *A. mylitta* expresses divergent phenotypic characters in response to varying ecological and climatic conditions. It has been reported that variations in the fluctuations of temperature prevent insects from attaining their physiological potential performance, which involve changes in the consumption and utilization of food, rate and time of feeding behaviour, metabolism, enzyme synthesis, nutrient storages and other physiological and behavioural process.

Sinha and Chaudhury, (1992) asserted that phenological parameters of *A. mylitta* are mostly dependent on the ambient temperature and relative humidity available during the period of larval and pupal development; hence, the developmental pattern of *A. mylitta* is season specific. Dash, (2001) found better cocoon crop performance of *A. mylitta* in winter crops than those of rainy and autumn seasons. Srivastava *et al.*, (1998) reported that environmental conditions are the main cause of variability in *A. mylitta*, because the fecundity, hatchability, cocoon weight, shell weight, absolute silk yield, filament length, denier, and sericin percentage varied significantly during different rearing seasons. Sengupta *et al.*, (2002) found that all the commercial characters of Daba ecorace (BV) of *A. mylitta* viz., cocoon weight, shell weight, SR%, cocoon volume, filament length, denier, average silk recovery percentage and silk yield/1000 cocoons was significantly higher in second crop rearing than the first crop. Similarly, Deka *et al.*, (2011) found better performance of Daba ecorace (TV) of *A. mylitta* in spring season followed by autumn, winter, and summer season.

Table 10.51: Effect of environmental conditions on cocoon quality and fecundity of muga silkworm (Adopted from Goswami and Singh, 2012)

Rearing season	Treatment	Temp.	Av. Tem.	RH (%)	Av. RH	Cocoon wt (g)	% of crippled moths	Fecundity
Pre-seed (June-July)	1.Control (RCC)	29-32^{0}C	31.0	82-90	86	4.2	42	88
	2.Thatch house (in sunlight)	28-30^{0}C	29.0	80-87	84	5.4	38	100
	3.Thatch house (In tree shade)	27-29^{0}C	28.0	72-85	77	5.8	19	120
F value CD1%						65.38**	97.98**	21.62
5%						0.133	2.636	6.343
						0.089	1.769	4.257
Seed crop (Aug-Sept)	1.Control (RCC)	30-34^{0}C	32.0	85-92	89	4.5	33	100
	2.Thatch house (in sunlight)	29-32^{0}C	31.0	80-90	85	5.3	30	115
	3.Thatch house (In tree shade)	28-30^{0}C	29.0	75-85	80	6.0	13	130
F value CD1%						28.69**	84.31**	9.85*
5%						0.222	1.494	5.183
						0.149	1.002	3.478
Pre-seed (Dec-Jan)	1.Control (RCC)	17-22^{0}C	20.0	60-75	68	5.5	10	120
	2.Thatch house (in sunlight)	23-25^{0}C	24.0	72-80	76	6.5	07	145
	3.Thatch house (In tree shade)	15-22^{0}C	19.0	70-80	75	5.7	09	128
F value CD1%						18.83**	1.70(NS)	20.07**
5%						0.184		4.351
						0.123		2.920
Seed crop (Feb-March)	1.Control (RCC)	17-24^{0}C	21.0	68-80	74	5.7	10	120
	2.Thatch house (in sunlight)	24-26^{0}C	25.0	75-85	80	6.3	03	145
	3.Thatch house (In tree shade)	16-22^{0}C	19.0	65-78	72	5.5	05	128
F value CD1%						18.85**	4.71*	10.06*
5%						0.185	2.168	4.856
						0.124	1.455	3.259

* Significant, **Highly significant, NS-Not significant

Dash *et al.,* (1994) assessed cocoon crop performances of the Indian wild tasar silkworm, *Antheraea paphia* on different food plants *viz., T. tomentosa*; *T. arjuna* and *Shorea robusta* during different rearing seasons in agro-climatic condition of Durgapur, Odisha. They evaluated cocoon crop performances in terms of effective rate of rearing (ERR); total cocoons yielded, cocoon weight, pupal weight and shell weight for each category of food plant in each rearing season. Their results showed that cocoon crop performances were better in autumn crop (Sept-Oct) than in rainy (July-Aug) and winter (Nov-Dec) season. The larval form of tropical tasar silkworm can invade diverse habitats and can utilise food sources of many forest tree species and can feed and grow at phenomenal rates. However, under sub optimal environmental conditions, *A. mylitta* D. larvae can survive for delayed period at much slower growth rates. The abiotic and biotic factors of the environment during different seasons greatly influence the life history features of the *A. mylitta* in the form of larval weight, cocoon weight, pupal weight, shell weight, shell percent, percent moth emergence, percent coupling, adult longevity, fecundity, percent hatching and reelability of the silk. Due to outdoor mode of forest based rearing, the silkworm larvae of *A. mylitta* may suffer heavy loss due to predator, parasitoid, and diseases. It is estimated that on an average, 30% of total crop loss due to various diseases is observed per year. Silkworm diseases form major constraint in realizing full crop yield.

Summary and Conclusions

➤ The eri silkworm *Samia cynthia ricini* Boisduval is a cold blooded (poikilothermic), delicate, domesticated insect which cannot tolerate diurnal and seasonal fluctuations in the environmental condition.

➤ The economic characters of eri silkworm are not only controlled by genes, but also known to be influenced by different climatic factors such as temperature, relative humidity, photoperiodic cycle etc.

➤ The eri silkworm and its grainage techniques are quite different from that of other non mulberry silkworms namely, Tasar and Muga.

➤ Temperature, relative humidity, rain fall and photo period are the most important environmental factors that influence the biology of insects. Temperature influences everything that an organism does, Relative humidity affects embryonic development, and rainfall affects both.

➢ Photoperiod also can have profound influence on larval duration and larval weight in eri silkworm apart from its well established effect on induction and termination of diapause in many insect species.

➢ Temperature, humidity, air circulation, gases and photoperiod etc. show a significant interaction in their effect on the physiology of eri silkworm.

➢ Temperature is probably the single most important environmental factor that influences behaviour, development, survival, reproduction, and geographical distribution of insects.

➢ Temperature regimes and different levels of relative humidity are known to play an important role in the life cycle of insect and its adaptability to the local climate.

➢ Most researchers seem to agree that warmer temperatures result in more types and higher populations of insects. It has been reported that the temperature has highly significant effects on all developmental parameters of insect than humidity and rainfall.

➢ With the increase in temperature, the larval growth and development is accelerated resulting reduction in larval duration, cocoons of lower weight and quality, while at low temperature growth and development is slow leading to prolonged larval period, abnormal growth and sensitivity against several diseases.

➢ Silkworms are more sensitive to temperature during 4th and 5th stages. The late age worms particularly the 5th stage larvae cannot tolerate high temperature, and high relative humidity.

➢ The ideal temperature for the physiological activity of eri silkworm ranges from 20°C to 30°C and maximum productivity is observed while the temperature ranged from 23°C to 28°C.

➢ Humidity interacts with free water availability and with the water content of the food plants. Humidity shows mostly indirect effect on growth and development.

➢ Deviations in humidity and temperature levels below and above certain critical limits affect larval growth and development of eri silkworm.

➢ At low humidity larvae lose water from their body in considerable amount which results in inefficient metabolism. Further low humidity also leads to less conservation of energy for pupae and adult stages.

➢ During egg incubation, it is important that humidity should be maintained at 80% on an average for normal growth of embryo. If humidity falls below 70% during incubation, it affects hatching.

➢ Fluctuation in temperature, humidity and other seasonal factors prevents insects from attaining their physiological potential life history performance.

➢ Forest insects display a remarkable range of adaptations to changing environments and maintain their internal temperature (thermoregulation) and water content within tolerable limits, despite wide fluctuations in their surroundings, because impact of temperature is modified by habitat and other physical conditions.

➢ Rainfall alters the functioning of microhabitat, which along with soil and other environmental factors affects foliage and water levels, which consequently affect on performance of insect.

➢ Seasonal and environmental interaction have great role to play on the expression of commercial characters of Eri silkworm.

➢ Season indicates the inter-annual variation in temperature, humidity, sunshine, rainfall, etc. of a particular place, which is governed by different geographical parameters. Identification of best rearing season is a prerequisite for the success of seri-business in a given area.

➢ The variations in climate, nutrient status, feeding duration and larval crowd along with environmental stimuli influence the insect body size and its biological success.

➢ Natural population is often heterogeneous and undergoes continuous selection by the local environmental fluctuations and by geographic transition.

➢ Environment causes effective restrain on biological characters of silkworm like, effective rate of rearing (ERR), single cocoon weight, single shell weight, and shell ratio and filament length.

➢ Host preference of forb chewers does not correspond precisely with host nutritional suitability but can be influenced by ecological factors such as natural enemies, host phenology and microclimate.

➢ In an experiment conducted to know the effect of temperature and relative humidity on rearing performance of the eri silkworm showed that the temperature range of 28±3°C and 70% relative humidity are the best for rearing of eri silkworm for sustainable development of the rearers for Ujjain district of Madhya Pradesh.

➢ In an experiment conducted to know the effect of temperature and relative humidity on moth emergence and fecundity of different eco races of eri silkworm during winter and spring season, recorded highest moth emergence, larval weight and shell ratio in spring season.

- In Uttar Pradesh, during autumn season some eco-races of the eri silkworm had prolonged larval duration because of low temperature, reduced rate of metabolism resulting in the slow growth.

- The major impact of light on performance of an insect, results from the daily rhythm of day and night and the seasonal change in day length or Photoperiod.

- Photoperiod is the environmental rhythm of recurring alternation of light illumination (Photo phase) and darkness (Scoto phase) and is extremely important for biological and physiological process of many insect species.

- Photoperiodism is measured in terms of day length, in the strict sense, is measured as the hours and minutes elapsing between sunrise and sunset.

- In Assam, effect of photoperiod on larval growth of eri silkworm in spring and autumn showed that the larval duration was shortened when eri silkworms were exposed to 16 hrs light followed by 8 hrs dark.

- Under Punjab conditions, yellow plain variant of eri silkworm recorded higher larval weight, cocoon weight and shell weight during April 2006-March 2007.

- The seasons of the different regions influence the larval growth, larval duration, ERR%, pupation%, cocoon weight, pupal weight, shell weight and shell ratio (%).

- Spring is the most ideal season for the rearing of eri silkworm in Ujjain district of Madhya Pradesh, because in this season eri silkworm has recorded higher larval weight, pupal weight, cocoon weight and highest larval length compared to winter and autumn seasons.

- In Karnataka rearing of eri silkworm in rainy season recorded highest single cocoon weight (3.67g), silk ratio (%) (13.35%) and ERR (86.89%). So rainy season is the most suitable season for rearing of eri silkworm.

- Studies on the economic traits of eri silkworm in relation to seasonal variation in Telangana state showed that the highest larval weight of eri silkworm was recorded in winter season followed by rainy and lowest in summer season. In some instances a notable increase was also seen in winter, which clearly indicates the association of larval weight to RH and temperature.

➤ Cocoon parameters like cocoon weight, pupal weight, shell weight and shell ratio and larval length were highest in winter followed by rainy season. The lowest values were recorded during summer season.

➤ In Uttar Pradesh, in an experiment conducted to know the impact of rearing of eri silkworm in spring and autumn seasons, rearing of eri silkworm in spring season recorded the highest larval weight, silk gland weight, cocoon weight and the shell weight.

➤ For commercial rearing, Mendipathar eco-race showed better adaptability than other eco-races in monsoon season of Uttar Pradesh climatic conditions.

➤ Seasonal difference, climatological factors and host plant selection have profound influence on the rearing efficiency of eri silkworm.

➤ In Assam, when eri silkworm fed on leaves of borpat, borkesseru and kesseru as well as in combination with castor, during four selected seasons recorded highest single cocoon weight and highest single shell weight during July-August season.

➤ While rearing the eri silkworm on different host plants during different seasons, it was observed that, Castor (I-II instar larva) in combination with Borpat (III-V instar larva) was the best combination of food plant in respect of higher larval weight, single cocoon weight, single shell weight, cocoon yield per dfl, ERR, cocoon shell yield per 100 dfl and the lowest larval period.

➤ Eri silkworm moths emerged in the cool hours of the day in shady or dark condition both morning and evening. Generally, male moths emerged earlier than female moths and were also higher in number.

➤ Newly emerged female moths were sluggish and crawl on to the edge of the cage for a resting period of about 1 hour till full development of their wings and start coupling. After coupling for 5-6 hours, the female moths start egg laying.

➤ Significantly highest fecundity was recorded during the month of October (495) followed by November (367.67) and hatching percentage was highest during autumn season i.e. in the month of November (94.33%) in Ladoigarh, Jorhat.

➤ In Jorhat region of Assam, eri silkworm seed can be produced throughout the year, but there is variation in grainage performance among the seasons.

➤ Considering the performance of different grainage characters such as good moth emergence, higher coupling realization, higher fecundity and higher hatchability of eggs, the autumn season (Sep-Nov) was found as best for eri seed production and summer (May to July) was comparatively unsuitable with low coupling recovery, low fecundity, low hatchability of eggs and higher emergence of cripple moths in Assam.

➢ The embryonic development stage in silkworm is very susceptible to environmental conditions i.e. light, temperature, humidity, vibration etc. A temperature of 22-25^{0}C and humidity 80% combined with good ventilation are essential for effective grainage operation in Ladoigarh.

➢ In North Eastern region of Assam, rearing performance of eri silkworm showed maximum yield with significant increase in the larval weight, effective rate of rearing, fecundity, cocoon weight, shell weight, silk ratio percentage and minimum larval duration were recorded during the spring season followed by autumn, winter and summer.

➢ Temperature range of 23 to 31°C and relative humidity of 70 to 80% is required for grainage operations in Assam.

➢ Under Assam conditions, suitable seasons for eri silkworm rearing are in the order of autumn > winter > spring > summer.

➢ For eri silkworm rearing, autumn is the best season followed by midsummer and late summer under Nagaland conditions.

➢ To improve the nutritional level of kesseru, the leaves of the trees were treated with 'Tricovit' foliar spray at the rate of 2 teaspoonful powder in 10 liters of water. 'Tricovit' encourages the process of photosynthesis, nitrogen cycle resulting in healthy plants.

➢ In Lakhimpur district of Assam, autumn rearing of eri silkworm recorded the highest fibroin content (83.46%), followed by late summer (82.56%), summer (81.19%) and lowest fibroin content (80.84%) was recorded in spring season.

➢ In Assam, kesseru (treated with tricovit) emerged as the most efficient food plant for rearing of eri silkworm *S. c. ricini*. This might be due to supplying of extra nutrient by spraying foliar spray on the leaves of kesseru, which contain more Carbohydrate, more nitrogen and more minerals.

➢ Autumn is the best season for commercial rearing of eri silkworm to improve productivity of eri silk in Lakhimpur district.

➢ In Orissa, winter season is the most favourable season for the rearing of the eri eco races for the production of better silk.

- At very high temperatures such as 43°C during summer, performance of the eco races was significantly lower which might be due to the decline in feeding rate of the larvae and their low conversion efficiencies resulting in low silk yield.

- The performance of the eri silkworm larvae was the best at temperature range of 15°C to 23°C.

- In an experiment conducted to know the effect of temperature on biology of eri silkworm under controlled temperature conditions at SK Nagar, Gujarat showed that the growth of eri silkworm was good at 20 ± 1^{0}C.

- In Gujarat, eri silkworms can be reared during November-January season to get highest cocoon yield.

- In Tamil Nadu, increased larval duration and mature larval weight have an influence on rearing performance and cocoon and grainage parameters. Winter is the most favourable season for rearing of eri silkworm.

- During the cooler months the larvae feed for a longer time and produce more silk, and hence the rearing of eri silkworm in summer could not be advantageous.

- Higher temperature coupled with higher relative humidity had negative influence on fecundity and hatchability of eri silkworm.

- Low temperature and high humidity was seen in winter in contrast to high range of temperature and low RH in summer, while rainy season has shown high temperature as well as high relative humidity.

- In an experiment conducted to know the promising strain suitable to the climatic conditions of Maharashtra inferred that the Greenish Blue Zebra and Yellow Zebra eri silkworm strains were suitable to the climatic conditions of Vidarbha region.

- A perusal of literature reveals that a number of insect species exhibit either 'complete' or 'partial' pattern of respiratory compensation against thermal fluctuation at organismal level.

- The fat body of cold adapted insects exhibits higher values of respiratory rate than those of warm adapted at all temperature of measurements.

- A tendency of slight decrease in Q10 value is observed due to cold adaptation in the lower and higher thermal range for the insects of all developmental stages.

➢ The fat body tissue of eri-silkworm possesses an excellent capacity of thermal acclimation in its metabolic rate.

➢ In a laboratory study conducted in Egypt to determine the effects of various photoperiods (LD 0:24, 10:14, 11:13, 12:12: 13:11, 14:10 and 24:0) on the development, reproduction and silk production of eri silkworm showed the superiority of the treatment 10:14.

➢ The larvae were found to consume more food when they were reared under 10-h photoperiod.

➢ In an experiment conducted in Egypt, to test the ability of eri silkworm pupae to withstand low temperatures showed that the best rates of survival and development of pupae occurred after exposure to 15°C for 30 days.

➢ When eri silkworm pupae exposed to 5^0C for 90 days a steady decline in proteins and amino acids was observed in Egypt.

➢ Eri silkworm strain (SaKKU1) when reared for 12 generations by exposing them to high temperature (42±1°C) and low relative humidity (50±5%), has developed heat tolerant property and this property is heritable.

➢ Under Egypt conditions, the increase in temperature decreased the weight of eri silkworm larvae, pupae and cocoon cortex. The most suitable temperature for producing the high quantity silk was 20° C.

➢ Indian golden silk moth, *Antheraea assamensis* undergoes diapause at the pupal stage after completion of October-November crop and remain in this stage till 2nd or 3rd week of March of the next year.

➢ In an experiment conducted to know the cocoon quality and fecundity of Muga silkworm (*Antheraea assamensis* Helfer) in three grainage houses in four different seasons, lesser crippled moths and highest fecundity recorded in summer seed crop.

➢ In muga culture the summer and winter pre-seed and seed crops face a heavy fluctuation of temperature and humidity, resulting in crop loss or poor quality cocoons.

➢ Growth of caterpillars is strongly temperature dependant and the degree of temperature dependence varies from species to species and optimal temperature for growth is also variable.

➢ For getting quality Muga silkworm seed cocoons and to obtain higher valid moths and comparatively higher fecundity, farmers should construct the thatch grainage house under tree shade during summer pre-seed and seed crop and in sunlight during winter pre-seed and seed crop.

> Rearing of tasar silkworm, *Antheraea mylitta* during winter when the temperature and humidity were 20.27^0C and 65.39% respectively, recorded highest ERR and maximum cocoon production compared to autumn and rainy seasons.

> Daba ecorace (BV) of *A. mylitta* performed better in spring season followed by autumn, winter, and summer season.

References

Ahmed S A, Sarkar C R, Sarmah M C, Ahmed M and Singh N I. 2015. Rearing Performance and Reproductive Biology of Eri Silkworm, *Samia ricini* (Donovan) Feeding on *Ailanthus* Species and Other Promising Food Plants. *Advances in Biological Research* **9**(1): 07-14.

Alexandrove V Y. 1977. Cells molecules and temperature-Ecological studies. 21 Ed. (Billings W D F, Golley O L, Lamge and Olson J S). *Springer Verlag*, Berlin 171-206.

Ali M A and Salem M S. 1978. Photoperiod in relation to the development and reproduction of the eri silkworm *Philosamia ricini* Boisd. *Agricultural Research Review* **56**(1): 101-108.

Anonymous. 1987. Annual Report. *Central Sericultural Research and Training Institute, Mysore* P.40.

Benchamin K V. 2000. Muga and ericulture: Subsistence to substantial farming. *Indian Silk* Jan-Feb Pp. 51-54.

Bordoloi D, Dutta L C and Bhattacharjee J. 2014. Effect of photoperiod during egg incubation on certain life parameters of eri silkworm, *Samia ricini* Boisd. *Journal of Applied Zoological Researches* **25**(1): 45-47.

Brahma K C, Rao K V S and Saxena N. 2006. Performance of eri silkworm *Samia cynthia ricini* Donovan under Eastern Ghat high land zone of Orissa. *Proceedings, Regional seminar on Prospects and problems of sericulture as an economic enterprises in North West India*, Dehradun Pp. 426-428.

Chakravorty R and Neog K. 2006. Food plants of eri silkworm, *Samia ricini* Donovan, their rearing performance and prospects for exploitation. *Proceedings, National workshop on Eri food plants*, Guwahati Pp 1-7.

Chakravorty R, Singh K C, Sarkar B N, Neog K, Mech D, Sarmah M C, Barah A and Dutta P. 2008. In: *Catalogue on Eri Silkworm (Samia c. ricini) Germplasm*, published by CMER&TI, Jorhat, Assam.

Chakravorty R. 2004. Diversity of eri and muga sericulture and its prospects in Himalayan states. *Proceedings, National workshop on Potential strategies for sustainable development of vanya silk in the Himalayan States*, Dehradun Pp. 9-16.

Choudhury S N. 1982. Eri silk industry. Directorate of sericulture and weaving, Assam.

Choudhury S N. 2005. Biology of silkworms and host plants (Chowdhury S N). Dibrugar, Assam Pp. 133-134.

Das K, Barah A, Das R and Chakravorty R. 2006. Oviposition behaviour and egg characters of muga silkworm *Antheraea assama* WW (Lepidoptera: Saturniidae) during different seasons. Muga silkworm, *Antheraea assama Biochemistry, Molecular Biology and Biotechnology*, Published by RRL-Jorhat, Assam.

Dash A K, Nayak B K and Dash M C. 1992. The effect of different food plants on cocoon crop performance in the Indian tasar silkworm, *Antheraea mylitta* Drury (Lepidoptera: Saturniidae). *Journal of Research on Lepidoptera* **31**(1-2): 127-131.

Dash A K, Patro P C, Nayak B K and Das M C. 1994. Cocoon crop performance of the Indian wild tasar silkworm *Antheraea paphia* (Lepidoptera: Saturniidae) reared on different food plants. *International Journal on Wild Silkmoth & Silk* **1**: 72-74.

Dash A K. 2001. Rainy season energy budget of larva of Indian tasar silk moth *Antheraea mylitta* Drury (Saturniidae) living in wild Sal (*Shorea robusta*) host plant. *International Journal on Wild Silkmoth & Silk* **6**: 1-7.

Datta, Lohit and Chandra. 2000. Effect of castor varieties on growth, nutrition, and cocoon characters of eri silkworm *Samia cynthia ricini* Boisduval. *Ph. D Thesis, Assam Agriculture University*, Jorhat, Assam.

Debaraj Y, Datta R N, Das P K and Benchamin K V. 2002. Eri silkworm crop improvement – A Review. *Indian Journal of Sericulture* **41**(2): 100-105.

Debaraj Y, Sarmah M C and Suryanarayana N. 2003. Seed technology in eri silkmoth-experimenting with other oviposition devices. *Indian Journal of Sericulture* **42**(2): 118-121.

Deka M, Dutta S and Devi D. 2011. Impact of Feeding of *Samia cynthia ricini* Boisduval (red variety) (Lepidoptera: Saturniidae) in respect of larval growth and spinning. *International Journal of Pure and Applied Sciences and Technology* **5**(2): 131-140.

Delvi M R, Radhakrishna P G and Noor Pasha. 1988. Effects of leaf ratio on dietary water budget of the larvae of the silkworm *Bombyx mori* and eri silkworm *Philosamia ricini*. *Proceedings of the Indian Academy of Sciences Animal Sciences* **97**: 197-202.

Devaiah M C and Dayashankar K N. 1982. Effect of different host plants on the economic traits of eri silkworm, *Samia cynthia ricini* Boisduval (Lepidoptera:

Saturniidae). In: *National Seminar on Silk Research and Development*, Central Silk Board, March 10-13, Bangalore, India p. 152.

Devaiah M C, Govindan R and Rangaswamy H R. 1978. Performance of eri silkworm, *Philosamia ricini* Hutt. on castor leaves under Karnataka condition. In: *All India Symposium on Sericultural Sciences*, Bangalore, India p. 48.

Dutta L C, Saikia and Dutta S K. 1996. Nutritional efficiency of two multivoltine breeds of *Bombyx mori* L. native to Assam. *Indian Journal of Sericulture* **35**(1): 32-34.

El-Shaaraway M F, Gomaa A A and Hosny A A. 1982. Biochemical changes during the development of eri silkworm pupae exposed to different constant temperatures. *Mededelingen van de Faculteit Landbouwwetenschappen, Rijksuniversiteit Gent* **47**(2): 563-571.

El-Shaarawy M F, Gomaa A A and El-Garhy A T. 1975. Biological studies on the Eri silk worm, *Attacus ricini* Boisd. (Lepidoptera: Saturniidae). *Zeitschrift fur Angewandte Entomologie* **78**(1): 82-86.

El-Shaarawy M F, Gomaa A A and Hosny A.1983. Studies on preserving the eri silkworm *Philosamia ricini* Boisd. during the pupal stage under different regimes of low temperature. *Mededelingen van de Faculteit Landbouwwetenschappen, Rijksuniversiteit Gent* **48**(2): 393-400.

Engelmann F. 1984. Reproduction in insects. In: *Ecology Entomology* (Eds. Huffaker C B and Rabb R L) John Wiley and Son. New York.

Gangwar S K. 2011. Screening of region and season specific bivoltine silkworm (*Bombyx mori* Linn) hybrid breeds of West Bengal in spring and summer season of Uttar Pradesh climatic condition. *International Journal of Plant, Animal and Environmental Sciences* **1**: 12-15.

Gogoi B and Goswami B C. 1998. Studies on certain aspects of wild Eri silkworm (*Philosamia cynthia* Drury) with special reference to rearing performance. *Sericologia* **38**(3): 465-468.

Gogoi R and Yadav R N S. 1995. Effect of host plants on some biochemical parameters of eri silkworm, *Philosamia ricini* during its development. *Indian Journal of Experimental Biology* **33**: 372-374.

Gohain and Borua R. 1982. Effect o temperature and humidity on development, survival and oviposition in laboratory population of Eri silkworm, *Philosamia ricini* (Boisduval) (Lepidoptera : Saturniidae). *Archives of Physiology and Biochemistry* **91**(2): 87-93.

Gomaa A A. 2009. Effect of temperature on the silk production of the Eri silkworm, *Attacus ricini* Boisd in Egypt. *Journal of Applied Entomology* **74**: 271-274.

Goswami D and Singh N I. 2012b. Environmental effects on seed cocoon quality and fecundity of Muga silkworm (*Antheraea assama* Helfer). *Indian Journal of Entomology* **74**(2): 186-188.

Goswami D and Singh, N. I. 2012a. Bio ecological studies of Indian golden silk moth, *Antheraea assamensis* Helfer (Lepidoptera: Saturniidae). *Munis Entomology & Zoology* **7**(1): 274-283.

Green N P O, Stout G W and Taylor D J. 1996. Biological Science. *Cambridge Low Price Edition* 675-677.

Gupta B K, Sinha A K and Das B C. 1992. Studies on egg yielding capacity of different races of silkworm, *Bombyx mori* L. *Gobios* **18**(4): 173-176.

Harada C. 1961. Heterosis of the quantitative characters in the silkworm. *Bulletin of the Sericultural Experiment Station* **17**(1): 50-52.

Hazarika U, Barah A, Phukon J D and Benchamin K V. 2003. Studies on the effect of different food plants and seasons on the larval development and cocoon characters of silkworm, *Samia cynthia ricini* Boisduval. *Bulletin of Indian Academy of Sericulture* **7**(1): 77-85.

Hiratsuka. 1920. Researches on nutrition of silkworm. *Bulletin of the Sericultural Experiment Station, Japan* **1**: 257-315.

Horie Y. 1978. Quantitative requirements of nutrients for growth of the silkworm *Bombyx mori* L. *Japan Agricultural Research quarterly* **12**(4): 211-217.

Hota M M, Patil P K, Pradhan K C and Sharma Mona. 2005. Ericulture in Orissa-Potential exists. *Indian Silk* **5**:12-13.

Hussain M, Khan S A, Naeem M and Mohsin A. 2011. Effect of Relative Humidity on Factors of Seed Cocoon Production in some Inbred Silk Worm (*Bombyx mori*). *International Journal of Agricultural Biology* **13**: 57-60.

Jaiswal K, Gangwar S K and Rajesh Kumar 2006. Comparative study on rearing performance of different eco-race of eri silkworm (*Philosamia ricini*) in monsoon season of Uttar Pradesh. In: *Proceedings of National seminar on Prospects and problems of sericulture as an economic enterprises in North West India*, Dehradun Pp. 483-485.

Jayswal K P, Dash B D, Sen S K, Subba Rao G and Kumari J P. 1990. Genotype x environment interaction for economic traits in hybrid combinations of silkworm *Bombyx mori* L. *Recent Trends in Sericulture*, Narendra Publishing House, Delhi 103 – 116.

Joshi K L. 1981. Nutritional physiology of Lepidoptera. Evaluation of four dietary regimes for eri silkmoth. *Ph D Thesis, University of Jodhpur*, Jodhpur P.140.

Kapil R P. 1967. Effects of feeding different host plants on the growth of larvae and weight of cocoon of *Philosamia ricini* Hutt. *Indian Journal of Entomology* **25**: 233-242.

Kar J K, Guru B C and Nayak B K. 2000. Influence of season on the life span and commercial traits of eri silkmoth, *Samia ricini* Donovan. *International Journal on Wild Silkmoth & Silk* **5**: 59-60.

Kar J, Guru B C and Nayak B K. 2004. Reproduction and commercial productivity of the cultivated eri silk moth *Samia ricini* Donovan in Orissa. *Bulletin of Indian Academy of Sericulture* **8**(1): 87-92.

Keister M and Buck J. 1974. Respiration of some exogenous and endogenous effects on rate of respiration. In: Rockstein M (eds) *The Physiology of insecta* **6**(2): 470-509.

Kichu N, Imtinaro L and Pankaj Neog 2014. Influence of host plants on rearing performance and cocoon characters of eri silkworm, *Samia cynthia ricini* Boisduval. *Indian Journal of Entomology* **77**(1): 88-91.

Kogure M. 1930. Studies on the voltinism of silkworm. *Bulletin of the Sericultural Experiment Station* **11**: 1-152.

Krishnaswamy S, Narasimhanna M N, Suryanarayan S K and Kumar Raja S. 1973. Sericulture Manual 2. Silkworm rearing. *FAO Agricultural Services Bulletin, Rome* **15**(2): 1-131.

Krishnaswamy S. 1978. New Technology of Silkworm Rearing. *Central Sericultural Research and Training Institute*, Mysore, Bull No. 2.

Kumar R and Elangovan V. 2010. Rearing Performance of Eri Silkworm in Monsoon Season of Uttar Pradesh. *Asian Journal of Biological Sciences* **1**(2): 303-310.

Mahobia C P, Shankar Rao K V, Verma R S, Roy G C and Suryanarayan N. 2005. Potential of eri silkworm rearing in Chhattisgarh. *Indian Silk* **8**: 12-14.

Mani H C, Singh B K and Chakravorty R. 2002. Water soluble and free amino acid (FAA) profiles of the haemolymph of the larva of eri silkworm *Philosamia ricini* Hutt. in relation to the presence of proteins of such nature and FAA in the diets fed by these caterpillars. *Journal of Advanced Zoology* **23**(2): 85-87.

Maribashetty V K, Chandrakala M V, Aftab Ahamed C A and Rao K. 1999. Food and water utilization patterns in new bivoltine races of silkworm *Bombyx mori* L. *Indian Academy of Sericulture* **3**(1): 83-90.

Meyer S G E and Schavb G. 1973. Der respiratorische sttoffwehsel Von Calliphoridenlarven. *Journal of Insect Physiology* **19**: 2183-2189.

Mishra A B. 2011. Study of effect of photoperiod on the movement of food through the gut of *Bombyx mori* Linn. bivoltine larvae. *Bharatiya Vaigyanik evam Audyogik Anusandhan Patrika* **19**: 203-206.

Mitalee Baruah 2012a. Improvement of cocoon parameters of eri silkworm (*Philosamia ricini*) in their nutritional level during different rearing seasons. *Research Analysis and Evaluation* **5**(36): 38-39.

Mitalee Baruah 2012b. Studies on larval weight and shell ratio of Eri silkworm (*Philosamia ricini*) on castor, kesseru and treated kesseru by foliar spray. *IJCAES Special issue on basic, Applied & Social Sciences* **2**: 133-137.

Mubashar Hussain, Sheikh Ahmad Khan, Muhammad Naeem and Ata-Ul-Mohsin. 2011. Effect of relative humidity on factors of seed cocoon production in some inbred silk worm (*Bombyx mori*) lines. *International Journal of Agriculture and Biology* **13**(1): 57-60.

Mukherji M K and Guppy J C. 1970. A quantitative study of food consumption and growth in *Pseudaletia unipuncta* (Lepidoptera: Noctuidae). *Canadian Entomologist* **102**: 1179- 1188.

Muthukrishnan and Pandian T J. 1987. Relationship between feeding and egg population in some insects. *Proceedings of the Indian Academy of Sciences and Animal Sciences* **96**: 171-179.

Nangia N, Jagadish P S and Nageshchandra B K. 2000. Evaluation of the volumetric attributes of the Eri silkworm reared on various host plants. *International Journal on Wild Silkmoth & Silk* **5**: 36-38.

Nuzhat F B A and Delvi M R. 1998. Effect of photoperiodism on growth in the silkworm, *Bombyx mori*. *Journal of Experimental Zoology, India* **1**: 125-129.

Pandey P and Tripathi S P. 2008. Effect of humidity in the survival and weight of *Bombyx mori* L. larvae. *Malaysian Applied Biology* **37**: 37-39.

Pandey Priyanka, Tripathi S P and Shrivastav V M S. 2006. Effect of ecological factors on larval duration of *Bombyx mori* Linn. *Journal of Eco physiology and Occupation Health* **3**(4): 9-14.

Patel B S and Patel G M. 2009a. Effect of rearing conditions on rearing of eri silkworm, *Samia cynthia ricini* Boisduval. *Insect Environment* **14**(4): 171.

Patel B S and Patel G M. 2009b. Effect of temperature on biology of eri silkworm *Samia cynthia ricini* Boisduval. *Insect Environment* **14**(4): 170-171.

Phukan J C D, Singh K C, Neog K and Chakravorty R. 2006. *Ailanthus grandis* Prain (Simaroubaceae – Quassia family) An alternate food plants of eri silkworm, *Samia ricini* (Donovan). *Proceedings, National workshop on Eri food plants,* Guwahati Pp. 102-110.

Priyanki Sarmah and Jogen Chandra Kalita. 2013. A comparative study on six strains of eri silkworm (*Samia ricini* Donovan) based on morphological traits. *Global Journal of Bio-Science and Biotechnology* **2**(4): 506-511.

Prosser C L. 1973. Temperature. In: Prosser C L. ed. *Comparative Animal Physiology*. 3rd edition. Saunders company 362-428.

Rajadurai S, Philip T and Sekar M A. 2010. Seasonal rearing performance of eri silkworm, *Samia cynthia ricini* (Boisduval) on castor and tapioca under South Karnataka conditions. *Indian Journal of Sericulture* **49**(2): 134-137.

Rajesh Kumar and Elangovan V. 2010. Assessment of the volumetric attributes of eri silkworm (*Philosamia ricini*) reared on different host plants. *Indian Journal of Sericulture* **1**(2): 156-160.

Rajesh Kumar and Elangovan V. 2014. Effect of temperature and relative humidity on moth emergence and fecundity of different eco-races of eri silkworm, *Samia ricini* (Donovan). *Indian Journal of Sericulture* **53**(1): 57-62.

Rajesh Kumar and Gangwar S K. 2010. Impact of varietal feeding on *Samia ricini* Donovan in spring and autumn season of Uttar Pradesh. *ARPN Journal of Agricultural and Biological Science* **5**(3): 46-51.

Rajkhowa G, Rajesh Kumar and Rajan R K. 2011. Studies on the long-term preservation method of muga cocoon (*Antheraea assamensis* Helfer) at low temperature. *Munis Entomology & Zoology* **6**(2): 815-818.

Rao K P and Bullock T H. 1954. Q10 as a function of size and habitat temperature in poikilotherms. *American Nature* **88**: 33-34.

Rao K R. 1988. Response of eri silkworm *Samia cynthia ricini* Boisduval (Lepidoptera: Saturniidae) to photoperiods. *Mysore Journal of Agricultural Sciences* **12**: 157.

Ravishankar H M, Reddy D N R, Reddy R N and Baruah A M. 1998. Evaluation of different methods of application of castor leaves of eri silkworm in relation to cocoon and egg production. *Third International Conference on Wild silk moths* p. 118-121.

Ray P P, Rao T V and Dash P. 2010a. Performance of promising ecoraces of Eri silkworm (*Philosamia ricini*) in agro climatic conditions of Western Odisha. *The Bioscan* **5**(2): 201-205.

Ray P P, Rao T V and Dash P. 2010b. Comparative studies on rearing performance of some ecoraces of eri silkworm (*Philosamia ricini* H.) in different seasons. *The Bioscan* **1**: 181-186.

Reddy D N R, Kotikal Y K and Vijayendra M. 1998. Development and silk yield of Eri silkworm *Samia cynthia ricini* (Lepidoptera: Saturniidae) as influenced by the food plants. *Mysore Journal of Agricultural Sciences* **23**: 506-508.

Renuka G and Shamitha G. 2014. Studies on the economic traits of Eri silkworm, *Samia cynthia ricini* in relation to seasonal variations. *International Journal of Advanced Research* **2**(2): 315-322.

Sachan T N and Bajpai S P. 1973. Studies on the biology and effect of seasonal variation on the growth and silk production by eri silkworm *Philosamia ricini*. *Annals of Arid Zone* **12**(2): 39-44.

Sahu M, Bhuyan N and Das P K. 2006. Eri silkworm, *Samia ricini* (Lepidoptera: Saturniidae) Donovan, Seed production during summer in Assam. In: *Proceedings of Regional seminar on "Prospects & problems of sericulture as an economic enterprise in North West India, Dehradun, India* 490-493.

Saikia M, Bhattacharjee J, Dutta L C and Singha T A. 2014. Effect of photoperiod on larval growth of eri silkworm (*Samia ricini* Boisduval) in two different seasons. *Journal of Experimental Zoology, India* **17**(2): 743-745.

Saikia M. 2001. Effect of photoperiod on life parameters of eri silkworm (*Samia cynthia ricini*) Boisduval. *M.Sc (Seri) Thesis, Assam Agricultural University, India* Pp. 107.

Sakthivel N, Qadri S M H and Krishnamoorthy T S. 2004. A new technique for commercial eri silkworm (*Samia c. ricini* Boisd.) seed production. *Proceedings, National workshop on Potential strategies for sustainable development of vanya silk in the Himalayan States*, Dehradun Pp. 171-173.

Sakthivel N. 2014. Effect of feeding methods of tapioca leaves and seasons on economic traits of eri silkworm, *Samia cythia ricini* Boisduval. *Indian Journal of Sericulture* **53**(1): 63-67.

Salem M S and Ali M A. 1978. Effect of photoperiod on food consumption, digestion and growth of the eri silkworm *Philosamia ricini* Boisd. (Lepidoptera: Saturniidae). *Agricultural Research Review* **56**(1): 109-115.

Samson M V and Barah A. 1989. Suggestion for better muga seed production. *Indian Silk* **15**: 13-18.

Sannappa B, Chinnaswamy K P, Raj S S and Chavan S S. 1999. Correlation of larval parameters with economic traits of eri silkworm, *Samia cynthia ricini* Boisduval. *Insect Environment* **5**(1): 7-8.

Sarkar B N and Sarmah M C. 2010. Seasonal variation of grainage characters in seed production of eri silkworm, *Samia ricini* (Donovan). *Indian Journal of Sericulture* **49**(1): 88-91.

Sarkar B N, Sarmah M C and Chakravorty R. 2008. Seasonal variation of grainage characters in seed production of eri silkworm, *Samia ricini* (Donovan). *Indian Journal of Sericulture* **49**(1): 88-91.

Sarkar D C. 1980. Ericulture in India. *Central Silk Board*, Bangalore P.23

Sarkar D C. 1983. Ericulture in India. *Central Silk Board*, Bangalore.

Sarmah M C, Ahmed S A, Sarkar B N, Debaraj Y and Singh L S. 2012. Seasonal variation in the commercial and economic characters of eri silkworm, *Samia ricini* (Donovan). *Munis Entomology and Zoology* **7**(2): 1268-1271.

Sengupta D, Srivastava A K and Ghosh S S. 2002. Studies on commercial and technological characters of different ecoraces of *A mylitta* D commercially available in India. *Annual Report 2001-02 Central Tasar Research & Training Institute*, Ranchi, Jharkhand, India 59 – 60.

Sengupta K. 1988. Bivoltine rearing in the plains and plateaus of South India. *Indian Silk* **27**(7): 25-27.

Sharma A K, Khatri R K, Siddiqui A A and Babulal. 2003. Impact of natural factors for the success of the rearing of silkworm crop. *Proceedings of Recent Researches of Indian Sericulture Industry* p.73-79.

Shiva Kumar C, Sekharappa B M and Arangi S K. 1997. Influence of temperature and leaf quality on rearing performance of silkworm, *Bombyx mori* L. *Indian Journal of Sericulture* 36(2): 116-120.

Shree M P and Nagaveni V. 2002. Development and reproductive performance of eri silkworms (*Samia cynthia ricini* Boisduval) fed with diseased castor leaves (*Ricinus communis* L.). *Sericologia* 42(2): 209-215.

Singh B K and Das P K. 2006. Prospects and problems for development of ericulture in non traditional states. *Proceedings, Regional seminar on Prospects and problems of sericulture as an economic enterprises in North West India,* Dehradun Pp. 312-316.

Singh B K, Debaraj Y, Sarmah M C, Das P K and Suryanarayana N. 2003. Eco-races of eri silkworm. *Indian Silk* May Pp.7-10.

Singh G B and Singh M K. 1997. Thermal acclimation in the endogenous respiratory rate of fat body of eri silkworm *Philosamia ricini. Entomon* 22(1): 83-87.

Sinha A K and Chaudhury A. 1992. Factors influencing phenology of different broods of tropical tasar silk moth, *Antheraea mylitta* Drury (Lepidoptera: Saturniidae) in relation to its emergence and post emergence behaviour. *Environmental Ecology* 10(4): 952-958.

Somaprakash D S and Sathyaprasad K. 2009. Effect of refrigeration on the hatchability of eri silkworm (*Samia cynthia ricini* Boisduval) eggs. *Indian Journal of Sericulture* 48(1): 56-59.

Srivastava A K, Naqvi A H, Roy G C and Sinha B R R P. 1998. Temporal variation in qualitative and quantitative characters of *Antheraea mylitta* Drury. In: *Third International Conference on Wild Silkmoths*, Bhubaneswar, Odisha p. 54-56.

Stockner J H. 1971. Food utilization by the last instar larvae of the silk-moth, *Antheraea proylei* in indoor condition. *The Indian Zoologist* 7(122): 85-87.

Subramanianan K, Sakthivel N and Qadri S M H. 2013. Rearing technology of eri silkworm (*Samia cynthia ricini*) under varied seasonal and host pant conditions in Tamil Nadu. *International Journal of Life Sciences Biotechnology and Pharma Research* 2(2): 130-141.

Sugai E and Takashashi T. 1981. High temperature environment at the spinning stage and sterilization in the males of the silkworm *Bombyx mori* L. *Journal of Sericulture Science of Japan* 50: 65-69.

Suryanarayana N, Das P K, Sahu A K, Sarma M C and Phukan J D. 2003. Recent advances in ericulture. *Indian Silk* 4: 5-12.

Thomas D A and Philogene B J R. 1979. Quality of light effects on immature stages and adults of *Pieris rapae* (L.). *Revue Canadienne de Biologie* **38**: 157-165.

Ullal S R and Narasimhanna M N. 1987. Handbook of practical Sericulture, 3rd Edition, Central Silk Board, Bangalore 1-166.

Umbriet W W, Burns R H and Stauffer J F. 1964. Manometric techniques. A Manual describing methods applicable to the study of tissue metabolisms. *Burgess Publishing Company*, Minnesota.

Vaidya S and Yadav U. 2014. Rearing performance of *Philosamia ricini* (Eri silkworm) in different seasons of Ujjain district. *Environment Conservation Journal* **15**(3): 109-113.

Vaidya S, Yadav U and Bhouraskar J. 2014. Effect of temperature and relative humidity on rearing performance of eri silkworm (*Philosamia ricini*). *Environment Conservation Journal* **15**(3): 189-196.

Wankhade L N, Barman H D, Rai M M and Rathod M K. 2014. Evaluation of some promising strains of Eri Silkworm, *Samia ricini* in climatic condition of Vidarbha region of Maharashtra. *Indian Journal of Pure and Applied Biology* **29**(2): 247-253.

Wongsorn D, Sirimungkararat S and Saksirirat W. 2015. Improvement of eri silkworm (*Samia ricini* D.) tolerance to high temperature and low humidity conditions by discontinuous regime. *Songklanakarin Journal of Science and Technology* **37**(4): 401-408.

Yamamura K and Kiritani K. 1998. A simple method to estimate the potential increase in the number of generations under global warming in temperate zones. *Applied Entomology and Zoology* **33**: 289-298.

Yamaoka, Kageyuki, Hoshino, Masahiro and Hirao Tuneo. 1971. Role of sensory hairs on the anal papillae in oviposition behaviour of *Bombyx mori* L. *Journal of Insect Physiology* **17**(5): 897-911.

Yamashita O and Hasegawa K. 1985. Embryonic diapauses. In: *Comprehensive Insect Physiology, Biochemistry and Pharmacology* (Eds. Kerkut G A and Gilbert L I), Permagon Press, U K.

CHAPTER 11

Participation, Empowerment and Constraints of Rural Women in Ericulture/Sericulture

Women in rural India participate in a variety of economic activities, but their potential is still under-utilised. To examine the participation of rural women in rural industries like sericulture/ericulture, empirical studies become imperative. Withstanding all social and cultural suppressions, a rural woman in India shares abundant responsibilities and performs a wide spectrum of duties in running the family, maintaining the household, attending farm labour, tenting domestic animals and extending a helping hand in rural artisanship and handicraft. This chapter highlights the current scenario of women in the field of sericulture in general and ericulture in particular and focuses on the constraints the rural women are facing.

Ericulture is a labour intensive agro industry and an age-old land-based practice in India with high employment potential and economic benefit to agrarian families. It has a major role in the economic up-liftment of rural poor in the country offering income round the year and plays a vital role in alleviating poverty. It is an important avocation for economic development of rural areas because of the high employment orientation, low capital investment and remunerative production. Ericulture is considered as family enterprise where farm women are actively participating in all the inter related activities. About 53 per cent of ericulture activities are carried out by the women (Sujata *et al.*, 2006). It serves as an important tool for rural reconstruction benefiting the weaker sections of the society. Ericulture not only offers periodical income, but also utilises the untapped potential of family labour for various activities (Lakshmanan *et al.*, 1997).

Women play a pivotal role in sustaining of ericulture in India. Her contribution to the industry is colossal. The gender-wise distribution of human labour employed in ericulture is 41.54 per cent male and 58.46 per cent female in host plant cultivation, and 45.87 per cent male and 54.13 female in silkworm rearing. The employment potential of women in ericulture is 51 per cent. After realizing the magnitude of her contribution to production, the Government of India declared the year 1994 as "Year of Women Sericulture" with a basic intension of providing a due recognition for her role in the industry. Women play a significant role in ericulture as they contribute one third of labour force required for farming operations and allied enterprises. Among different components of ericulture i.e. from soil to silk, approximately 50-60 per cent of different activities like host plant cultivation, silkworm rearing and reeling, twisting, weaving and printing, women involvement is absolutely vital and happening.

Every woman is an entrepreneur as she manages, organizes and assures the responsibility for running her home. It has been increasingly realized that women possess talent which can be harnessed for the productive purposes. It is estimated that women are responsible for 70 per cent of actual farm work and constitute up to 60 per cent of the farming population. It is therefore, not an exaggeration, that women in developing countries in general and India in particular are the back bone of food security and rural economy. Women's contribution to the farm sector has been largely ignored, inadequately understood and grossly underestimated. Despite the fact that women do most of the work related to agriculture and ericulture, the data available on their precise participation are minimal (Roy chowdhury, 2011). Earlier studies have shown that participation of female labour is higher than male in ericulture activities (Meenal and Rajan, 2008).

Every woman on an average works for 7 to 8 hours a day in the field besides their routine duties of cooking, cleaning, fetching water etc. In the recent times, the women are playing important role in agricultural occupation as a manager, decision maker and skilled farm worker. Their role may not be dominant or decisive but they play an important contribution for betterment of the economic conditions of the family. Besides maintaining the home, women also extend their helping hand in managing family farms (Malwale et al., 2010).

Women constitute half of the Indian population. As per 2011 census, the female population of India is 586.5 millions. Of which, around 80% of the total female population live in rural areas. The role played by women in agriculture is enormous. Women accounted for more than 76 per cent of the marginal workers and about 16 per cent of them are main workers. Among main workers as well as marginal workers more women belong to the category of agricultural labourers. Women formed part of highly valuable human resources which with appropriate training and education can bring about phenomena changes in the desirable direction (Sumathi, 2008).

Besides, the Sericulture industry is an income-generating activity which has multiple advantages for women:

- Since this can be done at home, a woman gets a fairer deal from the economy for less effort and she can earn more.

- A woman can combine the work with her other duties which she must execute any way.

- She can control her timing rather than having to be at some other persons back and call.

- She controls her own earnings, the harder she works, more she earns.

- She is ensured of a year-round income.

- She learns to deal with people outside her home/village/community and develops her own personality and self-confidence.

Prospects of Sericulture for Women: The overall prospects, the sericulture offers for women are:

(a) Sericulture increases employment opportunities of women in the silk worm rearing sector in the rural areas. These could take either the form of additional employment (partly mitigating the disguised rural unemployment) or as a substitute over hard-working conditions of farm (or such other) labour. The latter is true especially in the case of women belonging to the poorer classes. The post-cocoon activities of silk reeling and silk weaving also promise a great deal of employment for women in both rural and semi urban areas (Parthasarathy 1994).

(b) Sericulture being labour intensive is eminently suitable to the low resource base social groups (the small and marginal farmer categories, and especially women belonging to these categories, distressed women, and women-headed and women-supported households, etc.). It opens up opportunities for additional income generation and asset formation for these socially deprived groups, including women.

(c) The conditions of work at home in sericulture (though quite hard) are much lighter than farm labour and many other kinds of labour. These are also more convenient for women, allowing optimisation of time and synchronisation with other household activities.

However, the objective of women's development is not only to increase the income and standard of life of women and their families, but also to give recognition of their work, to enable them to develop self-confidence to participate in decision-making process at par with men, and to give them their due status and dignity in the family as well as in the society. Under this perspective, the other possibilities for women that are tied up with sericulture development can be listed as follows:

- Since women have a greater participation in sericulture, an improvement in their decision-making ability in the production process will improve the sericultural productivity and also lead to an improvement in their status.

- Woman being better managers of credit, with a greater access to income from sericulture, the rate of savings leads to a high degree of asset formation among the poor and also to sustainable development.

And most importantly, since under present social set-up, only that labour which produces cash income is valued as productive, a greater access by women to the income from sericulture (arising out of their own contribution) would provide a greater scope for increase in their status in the family and outside

Participation of women in ericultural activities in different states

Castor/mulberry cultivation aspects involve almost similar practices as that of agricultural crop activities. Hence most often rural women are engaged in castor cultivation, the castor/mulberry rearing activities/ operations. These operations are simple, less laborious and involve less drudgery to the rural women and give more income. It is clear that women have contributed more in the ericultural operations for which they are accustomed.

Study conducted in Udalguri district of Assam

In Assam, ericulture is considered as one of the major source of occupation. The families involved in ericulture normally belong to low income group of the society. The men of these families are always remaining busy with their primary sources of occupation and they do not pay more attention in the sericulture activities especially in ericulture. The women normally stay at home for different household activities. In addition to that, they are able to manage the required time for rearing of silkworms to supplement some extent to the family income. Besides, the above activities are very much suited to the physical strength of the women. In major agricultural crops also, women generally contribute more in the light agricultural operations. The reasons behind this may be the lesser physical strength of women than men.

As per 2011 census, total population of Assam is 31.2 million of which male and female are 15.93 million and 15.26 million, respectively. In eri sector more than 1.80 lakh farm families are actively engaged in the state. The rural women in the state contribute substantially to the rural economy through ericulture. Women participate in almost all the activities like leaf harvesting, silkworm rearing, cleaning, collection of dry leaves for spinning cocoons, harvesting of cocoons, spinning of spun yarn and also marketing of pupae, cocoons and fabric production. However, gender based profiling of participation in ericulture have not been studied systematically till date. Further, no efforts were initiated by the researchers to find out the relationships with the socio-economic variables of women with that of their participation in the ericulture. Hence, Mech and Ahmed (2012) had undertaken a study to visualize the participation of women in different activities of ericulture in Assam. It also aimed to focus on relationship with the socio-economic variables of the women and their participation in ericulture.

In Assam, Udalguri district produces huge quantity of eri silk annually. To know the activities that were undertaken by women in this district, survey was conducted in major villages of the district namely Paneri, Gopipur, Khalengduar, Ghagra, Soinajuli, Dimakuchi, Kothalguri, Botiamari and Niz-Japrabari. Primary data on participatory profiles in host plant cultivation and management, rearing of silkworms, spinning of spun yarn, decision making in different activities by men and women, etc. were collected randomly from 104 households actively associated in ericulture through personal contact method using the pre structured interview schedule. The simple percentage was taken to interpret the findings in case of participation of men and women in decision making and in different activities.

Participation of men and women in Host Plant Cultivation and Rearing of Silkworm

In host plant cultivation and management, women do not play significant role. The activities *viz.*, land preparation, pit digging, erection of fencing, transplantation of seedlings, application of fertilizers, plant protection against pest and diseases are hard and need physical strength and skills. That is why, men normally attend to these works. However, in case of some light activities like weeding, watering, etc. for maintenance of nursery and plantation, few of the women were found to participate occasionally. It has been recorded that 21.2 % and 17.3% of women contributes in weeding and watering of nursery and plantations respectively. The study showed that out of the total sampled families involved in ericulture, average participation of women in host plant cultivation and management was only 3.8% against participation of 96.2% men (Table 11.1).

Table 11.1: Participation of men and women in host plant cultivation and their management in ericulture (N=104) (Adopted from Mech and Ahmed, 2012)

S. no.	Activities	Men	Women
1	Nursery development	104 (100.0)	0
2	Watering, weeding, etc. in nurseries	82 (78.8)	22 (21.2)
3	Land development for plantation	104 (100.0)	0
4	Erection of bamboo fencing	104 (100.0)	0
5	Pit digging	104 (100.0)	0
6	Manuring at the pit	104 (100.0)	0
7	Transplantation of seedlings	104 (100.0)	0
8	Watering, weeding, etc. in plantation	86 (82.7)	18 (17.3)
9	Application of fertilizers to the plants	104 (100.0)	0
10	Plant protection against pest and predators	104 (100.0)	0
	Average	100 (96.2)	4 (3.8)

Figure in parentheses indicate percentage

But in silkworm rearing activities, participation of women was significantly higher compared to men in cocoon harvesting, marketing and spinning of spun yarn. While in selection of seed cocoons, women participate to an extent of 74.0%, 66.3% in procuring dfls from seed organization, 81.7% in preparation of laying at own level, 69.2% in leaf harvesting, 83.7% in feeding and cleaning, 76.0% in collection of ripen worms, 78.8% in harvesting of cocoons and 79.8% in removing of pupae from the cocoons. Besides, involvement of women was also high (74.0%) against men (27.0%) in marketing of pupae and cocoons. The highest involvement of women (88.5%) was recorded in spinning of spun yarn where involvement of men was significantly less (11.5%). However, in case of disinfection of rearing house and appliances, involvement of women was comparatively less (41.3%) compared to involvement of men (58.7%). Average participation of women in silkworm rearing, cocoon harvesting, marketing and spinning of spun yarn was 72.1% against participation of 27.9% men (Table 11.2). The similar trend of participation of women was also reported in case of mulberry sericulture (Ahmed, 2003 and Gautam and Sarma, 2003).

Table 11.2: Participation of men and women in silkworm egg production, rearing and post rearing activities (N= 104) (Adopted from Mech and Ahmed, 2012)

S. no.	Activities	Men	Women
1	Selection of seed cocoons	27 (26.0)	77 (74.0)
2	Procuring of dfls	35 (33.7)	69 (66.3)
3	Preparation of laying at own level	19 (18.3)	85 (81.7)
4	Disinfection	61 (58.7)	43 (41.3)
5	Leave harvest and their transportation	32 (30.8)	72 (69.2)
6	Feeding and cleaning	17 (16.3)	87 (83.7)
7	Collection of ripen worms	25 (24.0)	79 (76.0)
8	Harvesting of cocoons	22 (21.2)	82 (78.8)
9	Removing of pupae	21 (20.2)	83 (79.8)
10	Marketing of pupae and cocoon shell	27 (26.0)	77 (74.0)
11	Spinning of spun yarn	12 (11.5)	92 (88.5)
	Average	29 (27.9)	75 (72.1)

Figure in parenthesis indicate percentage

Analysis of Participation of Farm-Women in Sericulture Activities for Ahmednagar District in Maharashtra state

In Ahmednagar district of Maharashtra, women's overall participation was maximum in weeding, fertilizer application, pruning of mulberry leaves, preservation of mulberry leaves and chopping the mulberry leaves (100.00 per cent). In silkworm rearing, storage of cocoons, observation of larval growth and seed treatment operations, women involve to an extent of 96.66, 90.00 and 88.33 per cent respectively. Likewise, women participation in providing qualitative and quantitative leaves for larval growth is 80.00 per cent, disinfection of rearing house is 70.00 per cent and assessing quality of leaves and observing silkworm pest and disease control is 53.33 per cent (Dound *et al.,* 2016). Whereas women participation was less in the operations like irrigation (40.00 per cent), mulberry pest and disease control (30.00 per cent), inter cultivation (20.00 per cent), plant protection measures (15.00 per cent), maintenance of rearing house (13.33 per cent) and selection of improved variety (10.00 per cent). Participation of women is less than 10% in preparatory tillage and transportation of cocoons (6.66 per cent), silkworm pest and disease control (5.00 per cent), preparation of calendar operation for mulberry cultivation, preparation of 2% formalin solution and searching rate of cocoons (3.33 per cent) (Namratha et al., 2009) (Table 11.3).

Table 11.3: Distribution of the women respondents according to their participation in sericulture activities (Adopted from Dound *et al.,* 2016)

Sr. no.	Particulars of Practice	Frequency	Percentage
1.	Mulberry cultivation practices		
a	Preparatory tillage	04	6.66
b	Selection of improved variety	06	10.00
c	Seed treatment	53	88.33
d	Planting	58	96.66
e	Intercultivation	12	20.00
f	Weeding	60	100
g	Fertilizer application	60	100
h	Irrigation	24	40.00
i	Plant protection measures	09	15.00
2.	Assessing quality of mulberry leaves	32	53.33
3.	Participation in preparation of calendar operation for mulberry cultivation	02	3.33
4.	Pruning of mulberry leaves	60	100
5.	Preservation of mulberry leaves	60	100
6.	Chopping the mulberry leaves	60	100
7.	Mulberry pest and disease control	18	30.00
8.	Participation in silkworm rearing	58	96.66
9.	Providing both qualitative and quant. leaves for larval feeding	48	80.00
10.	Observing silkworm pest and diseases	32	53.33
11.	Silkworm pest and disease control	03	5.00
12.	Preparation of 2% formalin	02	3.33
13.	Method of rearing house	08	13.33
14.	Disinfection of rearing house	42	70.00
15.	Storage of cocoons	58	96.66
16.	Searching rate of cocoons in different market	02	3.33
17.	Observing larval growth	54	90.00
18.	Transportation of cocoons	04	6.66
19.	Sale of cocoons	06	10.00

Involvement of Rural women in ericulture in Mandya district of Karnataka

A study was conducted by Sowmya *et al.,* (2012) to know the extent of involvement of rural women in ericulture in Mandya district of Karnataka during 2008-09. A total of 120 respondents from Mandya and Nagamangala taluks were selected for the study, who involved in ericulture enterprise. Rural women to an extent of 82.79 per cent were involved in aspects of ericulture like host plant cultivation, land preparation and fertilizer application. They involve in these operations in all the seasons without fail. The women involvement in incorporation of compost to the field followed by removal of weeds is 74.13% and application of fertilizer to the field is 62.06%. In planting and intercultural operations and leaf plucking, involvement of women was cent per cent. Most often rural women are involved in leaf plucking from castor plants or harvesting of whole shoots and sorting of leaf for feeding. Among the silkworm rearing practices, in preparing the house for silkworm rearing aspect, women participation in cleaning the house premises is cent per cent, followed by cleaning and sun drying, the rearing trays/mountages (86.25%) and disinfection of rearing room (84.48%). In management of worms during instars 100% of women participation was observed in feeding of worms, regulation of temperature and humidity, management of worms during moulting, identifying and disposing of the diseased worms followed by dusting of bed disinfectants (87.98%). In harvesting and grading of cocoons cent per cent of rural women were most often involved in picking the riped worms from tray for reeling, placing the worms on the mountages for reeling, harvesting of cocoons from mountages and cleaning and storage of mountages (Table 11.4).

Involvement of rural women in sericulture activities in West Bengal, Meghalaya and Tripura

To know the extent of women participation in sericultural activities and the best suited technologies to women, a study was conducted in traditional sericultural districts namely Murshidabad and Malda of West Bengal and in the non-traditional sericultural states namely Meghalaya and Tripura. Data were collected on age group, religion, level of literacy, land holding, facilities available for sericulture, cosmopolitanism and involvement in other business etc. at farmer's level by personal interview method for Murshidabad and Malda districts in West Bengal and by postal interview for Meghalaya and Tripura through field functionaries during the year 2010 (Roy Chowdhuri *et al.,* 2011).

Table 11.4: Involvement of rural women in ericulture activities in Karnataka (Adopted from Sowmya *et al.,* 2012) (n=120)

Activities	Most often		Often		Least often	
	No.	Per cent	No.	Per cent	No.	Per cent
Castor cultivation						
I. Land preparation and fertilizer application						
a. Clod crushing	19	32.79	17	29.33	22	37.93
b. Removal of weeds	43	74.13	15	25.87	0	0.00
c. Uniform distribution and incorporation of compost to the field	48	82.79	10	17.24	0	0.00
d. Application of fertilizers to the field	36	62.06	19	32.77	3	5.17
II. Planting, intercultural operations and leaf plucking						
a. Planting the cuttings in the field	41	70.68	12	20.68	5	8.65
b. Hand weeding	22	37.93	30	51.72	6	10.39
c. Pruning of garden	42	72.41	16	27.59	0	0.00
d. Leaf plucking from castor plants or harvesting of whole shoots	58	100.0	0	0.00	0	0.00
e. Carrying of leaf from mulberry garden to home	23	39.65	21	36.25	14	24.13
f. Storing the leaf for feeding	58	100.0	0	0.00	0	0.00
Silkworm rearing						
I. Preparing house for silkworm rearing						
a. Cleaning the house premises	58	100.0	0	0.00	0	0.00
b. Disinfection of rearing room	49	84.48	9	15.56	0	0.00
c. Cleaning and sun drying the rearing trays/ mountages	50	86.25	8	13.79	0	0.00
II. Management of worms during instars						
a. Feeding the worms	58	100.0	0	0.00	0	0.00
b. Dusting of bed disinfectants	51	87.98	7	12.06	0	0.00
c. Regulation of temperature and humidity	58	100.0	0	0.00	0	0.00
d. Management of worms during moulting	58	100.0	0	0.00	0	0.00
e. Identifying and disposing of the diseased worms	58	100.0	0	0.00	0	0.00
III. Harvesting and grading	58	100.0	0	0.00	0	0.00
a. Picking the riped worms from tray for reeling						
b. Arranging the mountages for spinning	42	72.41	13	22.41	3	5.18
c. Placing the worms on the mountages for reeling	58	100.0	0	0.00	0	0.00
d. harvesting of cocoons from mountages	58	100.0	0	0.00	0	0.00
e. Cleaning and storage of mountages	58	100.0	0	0.00	0	0.00

In host plant cultivation 11 activities were taken into consideration such as, digging/tilling, cutting preparation, plantation, weeding, pruning etc. 40% of the respondents were engaged in digging/ tilling of plots, 35% in mulberry plantation, 43.7% in weeding, 36.2% and 42.5% in application of manures and chemical fertilizers respectively, 51.2% in leaf harvest, 48.7% in leaf transport from field to rearing house, 21.3% in pruning and 33.7% in cleaning of mulberry plots. However, none of the respondents performed propagation of cuttings and irrigation of mulberry plots. In Murshidabad and Malda districts women did not participate in mulberry cultivation practices except digging, weeding and leaf harvest. But in Meghalaya and Tripura states, almost all the mulberry cultivation practices were performed by women except preparation of cuttings and irrigation in mulberry gardens (Table 11.5).

Table 11.5: Participation of women in sericulture activities (% of share) (Adopted from Roy Chowdhuri *et al.,* 2011)

Activities	Murshidabad	Malda	Meghalaya	Tripura	Over all
Digging/ tilling of plot	15	0	100	45	40.00
Preparation of mulberry cuttings	0	0	0	0	0.00
Plantation of mulberry	0	0	100	40	35.00
Weeding	25	0	100	50	43.75
Irrigation	0	0	0	0	0.00
Application of manure	0	0	100	45	36.25
Application of chemical fertilizers	0	0	80	90	42.5
Leaf harvesting	15	0	100	90	51.25
Leaf carrying from mulberry plots	0	0	100	95	48.75
Pruning of mulberry plants	0	0	45	40	21.25
Cleaning of mulberry plots	0	0	100	35	33.75
Disinfection of rearing house	100	100	100	100	100.00
Disinfection of rearing appliances	100	100	100	100	100.00
Procurement & transportation of dfls	0	0	40	20	15.00
Incubation of dfls	100	100	100	100	100
Black boxing of dfls	100	100	100	100	100
Brushing of dfls	100	100	100	100	100
Feeding of worms (Chawki)	100	100	100	100	100
Feeding of worms (Late age)	100	100	100	100	100
Moulting care	100	100	100	100	100
Application of bed disinfectants	100	100	100	100	100
Bed cleaning	100	100	100	100	100
Mounting care	100	100	100	100	100
Marketing/cocoon transaction	0	0	0	0	0
Average Score	**48.12**	**45.83**	**81.88**	**68.75**	**51.14**

In silkworm rearing and related activities, woman folks were engaged more in all the activities except procurement of dfls and marketing of cocoons/silk. Women were involved fully in all the activities related to silkworm rearing such as, disinfection of rearing house and appliances, incubation of silkworm layings, brushing, feeding of chawki and late age worms, bed cleaning, mounting care etc in all the locations studied. While, involvement of women in procurement of dfls was recorded at 40% and 20% in case of Meghalaya and Tripura respectively. Data recorded on knowledge of women on improved technologies of sericulture revealed that, they do not have proper knowledge about improved technologies and have low level of literacy. But their knowledge was very good on disinfection of rearing house and appliances, use of silkworm bed disinfectants, diseases of silkworm, maintenance of hygiene during rearing and maintenance of temperature and humidity in rearing room (Table 11.6). The spearman's rank correlation for consistency among 20 respondents from each area in scoring of individual items was calculated between all pairs of them. It was found that all rank correlations for each of the 4 locations studied was highly significant. Hence it is inferred that judgment of each of the respondents to be consistent and thoughtful.

Table 11.6: Distribution of sericulture respondents by their knowledge on improved sericulture technologies (Adopted from Roy Chowdhuri *et al.*, 2011)

Activities	Murshida-bad	Malda	Meghalaya	Tripura	Over all
Name of improved mulberry variety	3.9	3.8	2.8	2.9	3.3
Recommended dose of FYM/ha/yr	4.4	4.0	4.3	3.9	4.1
Recommended dose of chemical fertilizers/ha/yr	4.4	4.6	4.8	4.7	4.6
Diseases of mulberry plants	3.2	3.3	2.9	3.1	3.1
Mulberry diseases and pest control measures	3.8	3.6	3.0	3.4	3.5
Application of plant growth regulators	3.8	3.7	3.1	3.2	3.4
Disinfection of rearing house & appliances	2.1	2.5	2.2	2.4	2.3
Name of silkworm breeds/hybrid	3.3	3.4	2.3	2.9	3.0
No. of crops/year	3.0	3.0	3.4	2.9	3.1
Diseases of silkworm	2.2	2.3	3.4	3.4	2.8
Silkworm disease & pest control measures	3.0	2.9	3.3	2.9	3.0
Use of silkworm bed disinfectant	3.0	3.0	2.4	2.6	2.8
Maintenance of temp. % RH in rearing house	3.0	2.8	3.0	2.9	2.9
Maintenance of hygiene during rearing house	3.0	2.6	3.2	3.1	3.0
Uzifly control	3.0	2.8	4.8	3.8	3.6
Average Score	**3.3**	3.2	3.2	3.2	3.2

Women empowerment and contribution in sericulture: It was estimated that during silkworm rearing, about 32% of the total day works were spent by the women respondents in different sericulture activities. Information recorded from the respondents shows that on an average total annual income from sericulture were around Rs. 18,500/- of which, they have saved the expenditure on man days for mulberry cultivation and silkworm rearing by using the family labour which on an average was Rs. 30,200/- per annum.

Role of women in ericulture activities in Tirupattur taluk of Vellore district in Tamil Nadu: In Tirupattur taluk of Vellore district in Tamil Nadu, data were collected from 360 respondents by personal interview method (Sumathi, 2008). Out of 360 respondents, 210 were engaged in farming and 90 women were engaged in ericulture activities in the taluk.

The results showed that the tasks performed by the farm women under land preparation were very negligible. It must be due to the fact that these activities would have necessitated arduous effort with more physical strain on the part of the doers. Thus the farm women had performed this task in a lesser magnitude. Majority of the respondents have performed the job of sowing the seeds. About three-fourth of them have performed the tasks like selection of seeds and the treatment of those seeds in an effective way. Most of the respondents have played the role of transplanting the seedlings followed by thinning and gap filling. In earthing up operations, 91.42% women were involved, while in pulling out the seedlings from the nursery the involvement is up to 85.17% and 43.33% respondents were involved in transplanting the seedlings. The role performed by the women with respect to weedicide application and plant protection measures was of lesser magnitude. The probable reason might be due to more risk involved in these roles and should have made the farm women physically weak to take up the roles.

Hundred per cent of the women were involved in harvesting operations. More than seventy per cent of the respondents have performed the tasks like collection and heaping the produce, bundle out and carry off to the yard. None of them were engaged in draining the water. Most of them have performed the tasks like drying and storage. The other practices *viz.,* threshing (96.19%), winnowing (77.14%), bagging (76.19%) and transportation (24.76%) were carried out by majority of the respondents. Very meagre per cent of the women have been involved in marketing the produce and transporting the produce to market centres (Table 11.7). It might be due to the fact that they were men oriented practices/ activities.

Table 11.7: Role of farm women in Host plant cultivation practices (n=210) (Sumathi, 2008)

S.No	Role performance	Farm women	
		Number	**Per cent**
1.	Land preparation		
i)	Ploughing, puddling and levelling	-	-
ii)	Stubble collection	55	2619
iii)	Application of manures	25	11.90
iv)	Cleaning of field boundaries	30	14.28
v)	Farming ridges and furrows	-	-
2.	Nursery preparation		
i)	Selection of seeds	167	79.52
ii)	Seed treatment	152	72.38
iii)	Sowing the seeds	210	100.00
iv)	Irrigating the nursery	-	-
v)	Plant protection in nursery	-	-
3.	Transplanting		
i)	Pulling out the seedlings from the nursery	180	85.71
ii)	Bio-fertilizer application	62	29.52
iii)	Transporting the seedlings	91	43.33
iv)	Transplanting the seedlings	210	100.00
4.	Inter cultivation		
i)	Irrigation	-	-
ii)	Cleaning the irrigation channels	60	28.57
iii)	Thinning and gap filling	210	100.00
iv)	Earthing up	192	91.42
v)	Detrashing	102	48.57
vi)	Weedicide application	-	-
vii)	Top dressing	51	24.28
viii)	Plant protection measures	-	-
5.	Harvesting		
i)	Draining the water	-	-
ii)	Harvesting	210	100.00
iii)	Collection and heaping	152	72.38
iv)	Bundling	150	71.42
v)	Carrying to the yard	162	77.14

In Vellore district of Tamil Nadu, in silkworm rearing practices, the results were not significant. The reason being 27.77 per cent of the women were engaged in picking the leaves, feeding the larvae, disease management and cleaning the chandrika and rearing room (Table 11.8).

Table 11.8: Role of farm women in ericulture (Sumathi, 2008)

S.No	Role performance	Farm women	
		Number	Per cent
1	Picking the leaves	25	27.77
2	Feeding the leaves	25	27.77
3	Disease management	25	27.77
4	Cleaning the chandrika	25	27.77
5	Cleaning the rearing room	25	27.77

Participation of women in ericulture in Maharashtra and Andhra Pradesh

In Ramtek Panchayat Samiti of Nagpur district of Maharashtra, moderate number of women were participating in ericultural activities, which include sowing and intercultural operations (Malwe *et al.,* 2010). For running the family smoothly, here majority of rural women participated in non-agricultural activities like growing shops, teacher ship, jobs, tailoring, embroidery etc. In Andhra Pradesh detailed study was conducted by Savithri and Sujathamma, (2003) to determine the extent of women participation, involvement in decision making, and the health problems they encounter while practising ericulture in different districts using a pre-tested schedule. Results showed that participation of men and women in ericulture is not similar, and their role in decision making do not commensurate their involvement. As per the recent study conducted by Purusottam Dash *et al.,* (2015) in rural India, women are involved in ericulture sector from host plant cultivation to dyeing, printing of fabrics, thereby deriving higher returns than men.

Role of Women in Decision Making in Different Activities of Ericulture

Decision-making and proper planning is considered as an important part of ericulture. Success of silkworm crops are mostly dependant on appropriate decisions involved in selection of variety of host plant and silkworm breed, selection of rearing seasons, assessment for requirement of dfls/layings for capacity utilization, time of disinfection, feeding of silkworms, etc. In most of these activities, women are playing a major role in decision making. It has been found that among the 104 farmers, 48.9% women alone take decisions compared to 33.6% men in various activities of ericulture. Participation of women in decision making varies from 26.0 % in adoption of host plant cultivation technologies to 71.2% in training needs on ericulture. The involvement of women in decision making was recorded as high as 58.6% in taking up rearing, 52.0% in selection of rearing seasons, 54.8% in assessment of quantity of dfls/layings to be brushed, 55.8% in adoption of improved rearing technologies, 52.0% in marketing of cocoons, 53.8% in marketing of pupae and 55.8% in procurement of improved spinning machine (Table 11.9). On the other hand, in some activities *viz.,* selection of variety of host plants, adoption of improved technologies for host plant cultivation and plant protection, role played by women in decision-making is comparatively less in comparison with men. The Joint decision taken by both men and women was recorded as 17.4%.

Table 11.9: Participation of men and women in decision making in different activities of ericulture (N=104) (Adopted from Mech and Ahmed, 2012)

S. no.	Activities	Men (%)	Women (%)	Both (%)
1	Raising of eri food plants	35.6	48.1	16.3
2	Selection of variety of eri host plants	44.2	35.6	20.2
3	Adoption of host plant cultivation technologies	55.8	26.0	18.2
4	Plant protection	50.0	27.0	23.0
5	Rearing of eri silkworms	26.0	58.6	15.4
6	Rearing seasons	21.0	52.0	27.0
7	Assessment of quantity of dfls/layings to be brushed	23.1	54.8	22.1
	Adoption of rearing technologies	26.0	55.8	18.2
8	Marketing of cocoons	36.5	52.0	11.5
9	Marketing of pupae	30.8	53.8	15.4
10	Procuring of improved eri Spinning machine	30.8	55.8	13.4
11	Contact with the departments	40.4	46.2	13.4
12	Training need	16.3	71.2	12.5
13	Average	33.6	48.9	17.4

In ericulture enterprises, decision by the women themselves was very meagre, restricted to the activities such as, quality of leaves for silkworms, maintenance of hygiene in rearing houses, disease incidence in silkworms, moulting care and quality of cocoons (Raju *et al.,* 1997). In other sericulture activities such as, mulberry cultivation as a whole, assessing the availability of mulberry leaves and the quantity of silkworm layings, cocoon/silk yarn marketing etc. were taken by their male family members. It was estimated that only 27% of the woman respondents recorded decisions on sericulture enterprises jointly along with their male members of the family, while remaining 73% were taken by their male family members. Distribution of family income on various aspects and decision taken thereof by the women in sericulture enterprise was mainly on their livelihood, health care, children education and other exigencies. Regarding decision taken on distribution of income to various family expenditures, it was observed that 30% of the total respondents took their decision jointly on sericulture activities, 100% in monetary distribution on family expenditure and around 44.8% both for sericulture activities and monetary distribution (Table 11.10).

Table 11.10: Empowerment of women in decision making (Adopted from Roy Chowdhuri *et al.,* 2011)

Empowerment	No. of respondents	%
Decision on the sericulture activities	24	30
Decision on monetary distribution on family expenditure	80	100
Both sericulture activities and money distribution	56	70

On the basis of above findings, it may be concluded that the rural women play a significant role in ericulture. Although, their participation is minimum in host plant cultivation, but involvement of the women was noticed in farm activities like nursery preparation, transplanting, inter cultivation, harvesting and post-harvest operations. Women participation was maximum in silkworm rearing and post rearing activities. Further, the women play major role in taking decisions on different activities. The study could also be concluded that participation of women in ericulture have significant association with her marital status and annual income. Hence, women extension agents should take effort to create awareness and to impart knowledge and skill on production, processing and marketing technologies through periodical training, field trips, group discussion, demonstration and campaign. It would definitely help to increase the role performance of farm women in farm and allied activities. The results imply that in many parts of the world there is an increasing trend in the involvement of women in agriculture and other allied sectors and they have more responsibilities especially in the Indian agricultural sector.

Relationship of Socioeconomic Variables of Women with Their Participation in Ericulture: Mech and Ahmed, (2012) made an attempt to find out the relationship between the socioeconomic variables of women with their extent of participation in ericulture. While in Mandya taluk of Karnataka, the women were interviewed through semi structured schedule, informal discussions and focussed group discussions. The women ericulture farmers were looked at by their age, marital status, type of family, size of family, caste, educational level, main occupation, experience in ericulture, total land holding, total mulberry land holding, work category of ericulture farmers, membership to self-help group, total income and income from ericulture. It can be visualized that the age and family size of the women is not correlated with their extent of participation. Although, the physical strength enables the young age women to perform more works, but their involvement in ericulture is not encouraging. Similarly, the physical strength of the old age women does not permit frequent movement to operate the different activities of ericulture. With regard to the family size, most of the women belong to nuclear family constituted of 5-8 members. Joint families with more members are very less in the present society. Similarly, education, caste and size of land holding are found to have negative correlation with their extent of participation. The reason may be due to poor economic conditions and illiteracy, compel the women to adopt the ericulture as a subsidiary occupation for supplementing the family income. On the other hand, the socioeconomic conditions of educated women are much better than the illiterate or low educated women. The educated women normally want to be busy with government or semi government job or to be associated with other enterprises. With regard to the caste, the fact in ericulture is that, involvement of tribal people is more compared to the involvement of the people belonging to higher caste or general caste. Further, the women having larger size land holdings are always busy with different agricultural operations as that gives higher income (Table 11.11).

On the other hand, marital status and annual income were found to be correlated significantly. The reason may be the married women of every household are always more responsible for running a family smoothly. They want to contribute to the family income through different means of indoor activities within their capacity. Similarly, the association of women having high annual income is also high because they can easily afford the cost for required infrastructure *viz.,* rearing house, rearing appliances and engagement of hired labour.

Table 11.11: Relationship with socioeconomic variables of women and their participation in ericulture (N =77) (Adopted from Mech and Ahmed, 2012)

S. no.	Socio economic variables	Coefficient of correlation 'r'
1	Age	0.571NS
2	Educational status	-0.448NS
3	Caste	-0.808 NS
4	Marital Status	1.000*
5	Size of land holding	-0.617 NS
6	Annual income	0.989*
7	Family size	0.581 NS

*Significant at 5 percent level & NS = Non significant

Majority of the women were found to be illiterate. Bulk of them had taken up ericulture as a main occupation with an experience of more than 20 years. Ericulture was practiced as livelihood even with very low land holding or without owning a land. Since, this occupation is not interlinked with levels of education, ericulture is more ideal for rural women with no schooling. This could also be an indication of the reality that ericulture is an apt source of revenue for women in rural vicinity as a source of livelihood and an excellent living for rural women. Moreover, it is one of the finest means for rural development in general and women development and women empowerment in particular.

Knowledge of Women Sericulturists on Improved Silkworm Rearing Practices in Karnataka: Knowledge is a pre-requisite to do anything, lack of sufficient knowledge about any idea/practice prevents an individual to avail its benefits. Perfect knowledge about an idea or practice helps an individual to relate it to his / her needs in terms of profitability and productivity. To assess the knowledge level of women sericulturists on improved silkworm rearing practices, Lakshminarayan *et al.,* (2013) have undertaken a study in 15 randomly selected villages of Chickballapur district in Karnataka during 2009-2010. Twelve improved silkworm rearing practices given in Table 11.12 were presented to women sericulturists to assess their knowledge level. A total of 100 farm women practicing sericulture were interviewed.

Forty per cent of women sericulturists had low level of overall knowledge on silkworm rearing practices, whereas, 36 per cent had medium and 24 per cent had high levels of overall knowledge on silkworm rearing practices (Table 11.13). It is disheartening to note that as high as 76 per cent of the women sericulturists had low to medium levels of overall knowledge on silkworm rearing practices (Ramakrishna Naika *et al.,* 2011). Majority of women sericulturists had correct knowledge on bed spacing of silkworms (52 %) and mountages (51%). Whereas, less than half of the women sericulturists had correct knowledge on rearing house construction (49 %), silkworm breeds (48 %), uzifly management (46 %), disinfection (45 %), hygiene (43 %), incubation (40 %), black boxing (39 %), shoot rearing (30 %), bed cleaning (30%) and mounting method (29 %). It can be inferred that more than half of the women sericulturists had incorrect knowledge on almost all the selected improved silkworm rearing practices (Table 11.13).

Table 11.12: Specific knowledge level of women sericulturists on improved silkworm rearing Practices (n=100) (Adopted from Lakshminarayan *et al.,* 2013)

Silkworm technologies*	Correct knowledge	
	Number	Per cent
Bed spacing of silkworms	52	52.0
Mountages	51	51.0
Rearing house construction	49	49.0
Silkworm breeds	48	48.0
Uzi fly management	46	46.0
Disinfection	45	45.0
Hygiene	43	43.0
Incubation	40	40.0
Black boxing	39	39.0
Shoot rearing	30	30.0
Bed cleaning	30	30.0
Mounting method	29	29.0

*Multiple responses possible

Table 11.13: Overall knowledge level of women sericulturists on silkworm rearing practices (n=100) (Adopted from Lakshminarayan *et al.,* 2013)

Category	Correct knowledge	
	Number	**Per cent**
Low (< 5.85 score)	40	40.0
Medium (6.69 to 7.91 score)	36	36.0
High (> 7.91 score)	24	24.0
Total	100	100.0

Mean =6.88; Standard deviation = 2.06

Technology adoption in sericulture and employment generation by women in Tamil Nadu

Sericulture with its short gestation period and low investment is gaining popularity and plays a lead role in providing employment opportunities to the rural mass. It not only fetches periodical income, but also ensures employment for the family labour especially for women. With the introduction of improved technologies and its diffusion, the area under mulberry and the number of farmers adopting the recommended technologies are rapidly increasing in Tamil Nadu.

Tamil Nadu is one among the traditional and major silk producing states in the country with an annual production of 1898 tonnes of raw silk during 2015-16. One of the major reasons for this rapid increase is adoption of improved technologies by farmers. This has influenced the income and employment generation from sericulture. Though there are reports on employment generation, studies on impact of technology adoption on employment generation are scanty. Hence, the study was undertaken by Meenal and Rajan (2008) with the objectives *viz.,* to investigate the extent of human labour used for various activities such as mulberry cultivation, silkworm rearing etc., by adopters and non-adopters and to analyse the share of family labour involved and the participation of women labour in various activities. Gobichettipalayam taluk in Erode district of Tamil Nadu was purposively selected for the study as sericulture is practiced by a large number of farmers. Thirty farmers who have adopted the improved technologies and 30 farmers who had not adopted the technologies were randomly selected. Personal interview method was employed for collection of data. The survey was conducted during the year 2005-2006. Details regarding the male, female labour, family and hired labour involved in various activities were collected and male to female labour participation ratio was worked out.

Total employment generated per acre per year among the adopters was 565.25 man days and 467.72 man days among the non-adopters. Out of this, 16.86% was utilized for mulberry garden establishment, 18.81% for garden maintenance and 64.33% for silkworm rearing by adopters, whereas, it was 17.76%, 20.20% and 62.04% respectively by non adopters. Employment generated by adopters was high when compared to non-adopters. This is because of the adoption of improved mulberry cultivation and silkworm rearing technologies by adopters. As a result of adoption, the leaf yield/acre increased which encouraged the adopters to brush more number of dfls. This again required more labour utilization. The data on composition of family labour and hired labour in various activities of sericulture are presented in Table 11.14. It is very clear that 23.30%, 16.38% and 49.91% of family labour was utilized for garden establishment, garden maintenance and silkworm rearing, respectively by farmers who had adopted the improved technologies. Similarly, among non-adopters, 22.86%, 21.13% and 59.04% of labour engaged in garden establishment, garden maintenance and rearing respectively was from family source. The male: female labour participation ratio was worked out to find out the involvement of women in sericulture. Data revealed that the male: female ratio was 1:1.89 and 1:1.88 in mulberry garden establishment and maintenance respectively in the case of adopters whereas, it was 1:1.90 and 1:1.44 respectively in case of non-adopters (Table 11.15). Involvement of women was high in application of manure and fertilizer, planting, weeding, leaf harvest and transport etc. In silkworm rearing, generally men are involved in shoot harvesting and cocoon marketing. Women remain indoor involving themselves in feeding, picking up of matured larvae, harvesting of cocoons etc.

The study clearly indicated that, the human labour employment increased with the adoption of improved technologies. There is a direct relationship between the hired labour use and technology adoption. The study also revealed that, sericulture is women friendly and involvement of women in various sericulture activities was higher compared to men. Thus, mulberry sericulture can solve many of the pressing problems such as unemployment, rural migration and poverty.

Table 11.14: Human labour utilization in mulberry sericulture (man days/acre/year) (Adopted from Meenal and Rajan, 2008)

Labour utilisation	Adopters				Non-adopters			
	Garden establishment	Garden maintenance	Silkworm rearing	Total	Garden establishment	Garden maintenance	Silkworm rearing	Total
Family labour								
Male	17.13	12.04	80.86	110.03	15.20	14.28	72.90	102.38
	(17.97)	(11.33)	(22.24)	(19.47)	(18.30)	(15.11)	(25.12)	(21.89)
Female	5.08	5.37	100.63	111.08	3.79	5.68	98.41	107.88
	(5.33)	(5.05)	(27.67)	(19.65)	(4.56)	(6.01)	(33.91)	(23.07)
Hired labour								
Male	15.90	24.81	58.51	99.22	13.48	24.29	47.85	85.82
	(16.68)	(23.33)	(16.09)	(17.55)	(16.23)	(25.92)	(16.49)	(18.35)
Female	57.20	64.09	123.63	244.92	50.59	50.03	71.02	171.64
	(60.01)	(62.16)	(34.00)	(43.33)	(60.91)	(52.95)	(24.47)	(36.70)
Total	95.31	106.31	363.63	565.25	83.06	94.48	290.18	467.72
	(100.0)	(100.0)	(100)	(100)	(100)	(100)	(100)	(100)

Table 11.15: Male and female labour participation in various sericulture activities (Adopted from Meenal and Rajan, 2008)

Activity	Male labour	Female labour	Total labour	Male: female ratio
Adopters				
Mulberry garden establishment	33.03	62.28	95.31	1:1.89
Mulberry garden maintenance	36.85	69.46	106.31	1:1.88
Silkworm rearing	139.37	224.26	363.63	1:1.61
Non-adopters				
Mulberry garden establishment	28.68	54.38	83.06	1:1.90
Mulberry garden maintenance	38.77	55.71	94.48	1:1.44
Silkworm rearing	120.75	169.43	290.18	1:1.40

Constraints faced by the women in Sericulture / ericulture activities in India

For many social reasons, the role of women in sericulture remains unrecognized and unrewarded even though the percentage of their work share in sericulture activities are notable. Therefore, the concept that contribution of women in the society is restricted mostly to house hold activities is not fair. It is fact that the women in the society particularly the rural women are actively involved in almost all the activities in their family works and in assisting the male member of family to uplift the economy. Sericulture being a rural enterprise participation of women is higher. Low level of literacy, lack of awareness on the improved technologies of mulberry cultivation and silkworm rearing, poor level of perception on the technologies and also lack of empowerment are the major constraints faced by the women in sericulture. The women and silk are the two sides of the same coin and hence quoted as "silk is by the women, for the women and of the women". More than 60 per cent of the various ericultural activities are being carried out by women, across different ericulture practicing regions. In spite of deep involvement and participation in ericulture activities, women's socio-economic status remains low and she is disempowered. Women's participation and their empowerment often run into troubles and they are kept at bay. The main barriers could be caste system, poverty, low educational status, limited work opportunities, low income, lack of assets and access to credit.

1. **Social constraints:** Socio-cultural constraints are those related to the status of women in the family- the class and caste background, the

occupational status of the household, family size, and the educational level. But socio-cultural constraints do not seem to affect women's participation in the case of the small and the marginal households which depend on sericulture for their livelihood. Many households in different parts of India and in both traditional and new areas have been observed where family women from nuclear household belonging to these socioeconomic groups actively participate in silkworm rearing.

2. **Technological Constraints:** Women, by and large, do not have sufficient knowledge and information about techniques of silkworm rearing. Technical know-how, which should instil a sense of confidence among the women, is somehow inaccessible. The flow of information from the male extension staff usually stops at the men, and does not trickle down to the women members of the houses and wherever it does what filters down in the form of do's and don'ts rather than any explanation about the production process, strategy and use of inputs. Women carry out all the sericultural activities starting from brushing till harvesting, without male support. Sericulture becomes a routine risk-free crop, and nothing is regarded as too 'skilled' enough to demand men's attention. In other cases, men, on the other hand, keep themselves in some of the more 'skilled' activities like brushing, chawki rearing and moult setting, disease control, etc. Women are more involved in feeding and disinfection of the equipment.

3. **High Level of Participation:** It is well accepted that the participation of women in sericulture and their contribution to cocoon production is substantial. And this is in addition to the usual household chores requiring four to eight hours a day, depending on the region, season, socio-economic status of the household and the family type.

4. **Gender-based Division of Labour:** The division of labour, however, is rather discriminatory. Women's lack of access to modern technology is reflected clearly in the gender based, rain fed areas of Karnataka, where sericulture is predominantly in the women's domain; in most other areas, women are relegated to the relatively less skilled activities.

5. **Low Access to Technology and Extension:** Extension workers are mostly male. Except for certain parts of India, women elsewhere have less access to the support provided by the male extension staff. Interactions of women with outside male extension workers are

usually not acceptable in the rural society. The male extension workers in turn have their own inhibitions and bias. There is an implicit belief that the information provided by the extension workers to male members of the family would automatically trickle down to the female members.

6. **Lower Status in Production Decisions:** In the technical process of production, women merely do what they are asked to do, not because they believe it is the right thing to do- be it in relation to spacing of worms in the rearing tray or how to sort out the diseased worms. In decisions, pertaining to the scale of production, timing of production, input mixing and technology choice, women's participation is even less.

7. **Low Access to Support Services:** While the women usually carried out the bulk of the work (especially the physically tiring activities), the men assumed the overall responsibility. The 'land bias' put the man (usually the owner of the land, and the one who dominated the cultivation of crop) as the main sericulturist. The women as always remained as a mere 'support'. Moreover, the male was regarded as the 'head', not only of the household, but also of the sericultural production unit. The non-recognition of the woman as the main sericulturist deprived her of direct access and sometimes even her right to the various institutional support services.

8. **Low Access to Market and Income:** The existing social and cultural impediments, coupled with the inconvenience of travelling long distances and her hostile environment at the cocoon market, reduce women's access to market and income. Thus the control over the fruits of the women's labour is exercised mostly by the male members of the family.

9. **Low Access to Credit:** Credit, as means of strengthening the resource base of women, coupled with other facilitating conditions, has the potential of enabling vast section of women to realise the opportunities provided by sericulture. Populist political interference in the institutional credit system, the loan melas, the loan waivers etc., however, has all contributed to the emergence of a vitiated environment. Bankers are hesitant and look at even genuine credit proposals with multiplied suspicion, which together with the overload of paperwork and formalities makes credit flow to sericulture a difficult proposition. Given the patriarchal system of society, women, by and large, are deprived of direct ownership of land. On the other

hand, efforts at securing co-titleship on land for women face insurmountable problems. Firstly, this is considered a social stigma on the husband and has serious implications for the familial harmony. Secondly, the land records are hardly up-to-date. The cost of transferring ownership from the dead forefathers to the present generation is prohibitive. Without land as security, the bankers are hesitant to advance credit to women. The Reserve Bank of India has recently waived the requirement of land as security for loan up to Rs.20,000/- this is yet to become operative in the field. Bankers still insist upon 'collaterals' (Geetha, 2010).

10. **Rich Farmer Bias:** There is an implicit assumption that the rich farmers are always better adopters of new technology. This usually leads to the extension staff concentrating more on the rich farmers, often to neglect the poor and the deprived (including women). The expectation is that the new technology would trickle down to the poor through the so-called demonstration effect.

11. **Technology bias:** The technology bias of the research and extension staff usually led to greater priority being placed on the high productive high-input oriented technology, irrespective of the fact that such technology could not be afforded by the poor. Often there was a lack of concern for appropriate technologies that matched the low-resource base, the lower risk taking ability and the overall production strategies of the poorer sections. These biases also worked against women, especially those belonging to the poorer section.

12. **Marketing constraints:** Apart from all the general problems, women sericulturists face particular problems with regards to marketing of cocoons, transportation, availability of market information, interference of middle men, etc. Being uneducated, women sericulturists without any male person's support, are unable to go to market due to a number of reasons. They are easily prone to be exploited by the middle men. Hence proper marketing and transport facilities can be arranged to make more women take active role in sericulture (Shobha *et al.*, 1998).

13. **Lack of Separate Rearing House:** It was observed that majority of the women farmers do not have a separate rearing house. As a result, they use to brush very less quantity of dfls (only 5-15 dfls per crop) in a corner of their living house.

14. **Lack of Adequate Plantation:** Although, the plantation is a primary requirement for rearing of eri silkworm, majority of the farmers do

not have systematic plantation for rearing of silkworms. They use to collect leaves from naturally grown castor or kesseru plants from different places nearer to their home. As a result, the farmers cannot take up the rearing in every season regularly. These farmers can rear only 2-3 crops instead of 5-6 crops in a year.

15. **Lack of Improved Technological Knowledge:** Most of the women are not aware of the improved rearing technologies *viz.,* high yielding silkworm variety, disinfection, improved rearing technology, etc. This results in low production of cocoons and generates low income.

16. **Inadequate Time:** In case of the women belonging to very low-income group, it is obligatory to engage themselves in agricultural field for certain works like transplanting of rice, weeding, harvesting, etc. Sometimes, they work in others' agricultural fields by taking wages. As a result, this group of women cannot participate continuously in the ericulture (Mech and Ahmed, 2012).

In Mandya district of Karnataka, the women are facing institutional or culturally induced constraints. They are the products of prevailing social and cultural norms (Geetha, 2011) (Table 11.16).

Table 11.16: Constraints faced by women in ericulture (Geetha, 2011)

Constraints	Frequency		Percentage %	
	Positive	**Negative**	**Positive**	**Negative**
Landlessness	63	37	63	37
Inadequate access to credit	65	35	65	35
No separate rearing house	68	32	68	32
Lack of technological know how	41	59	41	59
Lack of family labour	85	15	85	15
Unaffordable lease amount on land	67	33	67	33
Land located at distance	67	33	67	33
Non-profitable activity	100	-	100	-
Health problems	99	1	99	1
Lack of sericulture inputs	84	16	84	16

Despite the constraints, the women participate in the activities of sericulture, thus provide ample scopes for their development through awareness, capacity building through imparting training/ demonstration of technologies, processes, techniques etc. and guiding for empowerment so that the society will be socio-economically uplifted and the country as well (Roy choudhury, 2011).

Women in Sericulture: Need for a Shift in Focus: A large percentage of the rural women, must earn outside the house in addition to looking after their own home and family. In rural areas most women, except a handful from extremely well-off families, work outside the home. Even if a family has some land, the women out of necessity must work on others' fields to supplement their family income. Yet, the women have very little say regarding the earning or any decision regarding where she has to work or how much she has to be paid. With the introduction of more farm machinery, women's labour is getting further and further pushed into the unskilled category. This entire process lead to continued marginalisation of women (Sandhyarani, 2006). If sericulture is taken up as a true development activity, these problems/mistakes/processes/trends should not be repeated. If possible, mechanism to prevent or even reverse these processes should be incorporated into the development process of this industry. Most importantly, the potential that this industry has for women's development should be realised. By women's development one means that process by which women have the freedom and opportunity to develop their own social and economic lives. Other related issues, women's and children's health, are affected by the time the mother is available for child care and attention to her own needs. Control over her earnings ensures a higher-living standard to the family, and therefore, a sustained economic and social development of the family is possible (Jesia, 1994).

Action Plan for the development of women: In view of the significant and critical position occupied by women, the development of women through sericulture has been identified as one of the paramount objectives of the National Sericulture Project (Bhat and Sharmila, 1994). The most important features of the action plans are

(a) Improving women's access to resources, especially land:

- By promoting joint ownership (co-title ship) of family land
- By allotting ownership titles, while distributing government land among the poor, either on women's names or jointly on the man's and woman's names.

(b) Improving women's access to extension and technology:

- By organising special training programmes for women
- By appointing more and more women extension and Para-extension staff

- By organising gender sensitisation orientation programmers among the department staff

- By and large, the training programmes are proving to be useful in improving women's skills. However, they are too inadequate. There is a need to expand the women's training component in a big way. Again there are a number of areas where major improvements are required (proper trainee identification, more demonstration-based programmes as opposed to classroom based curriculum etc.).

(c) **Organising women's groups in sericulture:** So as to facilitate input supply, credit flow and absorption of technology, and above all, to promote the development of self-confidence among women through mutual support.

Formation of women's groups constitutes probably the most challenging but effective means of empowering women. However, action under this component has been meagre, and experience with the formation of women's groups shows mixed results. Nevertheless, there is a great scope for improving women's access to technology, input and credit through these groups.

(d) **Encouraging women to come to the cocoon markets:**

In order to improve women's access to income, their access to market is being promoted through a number of schemes:

- By improving the facilities and amenities at the markets

- By reserving separate space for cocoon lots brought to the market by women

- By giving priority to women in auctioning etc.

However, in order to reduce the hurdles, minimise the gender disparity and thereby to bring her into the main stream, the government machineries such as government departments, financial institutions, NGOs and other related agencies took many measures i.e. reservation for women, provision of credit, incentives and subsidies, market facilities, invention of women friendly technologies, appointment of women para extension workers, conducting training programmes, facilitate organize in groups etc, but many a times it goes a futile exercise. The extent of these constraints needs to be studied in order to plan a programme for horizontal and vertical expansion of women's participation in ericulture. Therefore, participation, involvement and empowerment efforts of ericulture women need to address these social constraints. That is to say, changes must be brought about in social norms, traditions and ideologies to increase the capacity of ericulture women. According to Giddens (1979), these changes could be brought about in three ways. These include communication for enabling understanding, unifying society through legitimising the norms and domination of the actors for coercing or inducing others.

In ericulture, the present social structure witnesses unequal power relations among men and women. This is the root cause of current social constraints. These constraints can be reduced only by increasing women's access to resources. Women's increased access to resources enables challenging power relations, social norms and thereby helps in enhancing the social transformation of women's agency and gendered roles and responsibilities in public and private spheres. In a country like India, constraints for engaging in gainful economic activity hit women harder than it would affect men. Both structural and operational factors block women's participation in economic activities on a large scale. Lack of knowledge of latest technological innovations in the ericulture sector is coupled with restrictions which are created and perpetuated by patriarchal value structures. Hence, any participation, involvement and empowerment intervention in ericulture/sericulture industry should be provided using the bottom-up approach, giving a space for women's participation, backed with institutional change.

Both state and central governments may initiate gender specific schemes by establishing ericulture women self help groups with 'Eri fund' that could provide easily manageable loans for ericulture activities besides transferring new knowledge related to ericulture and reducing some major constraints experienced by women.

Trainings: Rural women were most often involved in majority of activities of agriculture and allied enterprises either as manual work, as supervisors or as decision makers. Hence, massive empowerment programme of the rural women needs to be launched to develop them as progressive entrepreneurs in agriculture and allied enterprises. Table 11.17 reveals that, the participation of respondents in training programme had highly significant association with their knowledge level on improved silkworm rearing practices, whereas, age, family size, material possession and social participation of women sericulturists had no association with their knowledge level on improved silkworm rearing practices. Training provide defreezing of old behaviour and refreezing of new behaviour. Hence, there is a significant association with the participation in training programmes and knowledge level of women sericulturists. The results of

Lakshminarayan *et al.,* (2013) revealed that, as high as 76 per cent of the women sericulturists had low to medium levels of knowledge on improved silkworm rearing practices. It was encouraging to note that women's participation in training programmes had association with the knowledge level of women sericulturists on improved silkworm rearing practices. Hence, ample opportunities need to be provided by the Farm Universities, Karnataka Sericulture Department, Central Silk Board and other agencies to the women sericulturists to undergo training for increasing their knowledge on the improved silkworm rearing practices.

Table 11.17: Association between personal and socio psychological characteristics of women sericulturists and their knowledge level (Adopted from Lakshminarayan *et al.,* 2013)

Characteristics	Chi-square value
Age	0.181 NS
Family size	3.333 NS
Social participation	2.905 NS
Material possession	2.902 NS
Participation in training programmes	11.124**

NS = Non – significant ;

** = Significant at 1 per cent level

As the women contributes a lot in socio-economic development of rural sector through ericulture, the rural women should be assisted financially for construction of rearing houses, development of adequate plantation, procurement of rearing appliances, etc through different sericultural development schemes. Besides, they should be properly trained on improved technologies through imparting training programme for further enhancing their participation (Meenal and Rajan, 2008).

Savithri and Sujathamma, (2003) gave some suggestions to empower the women in sericulture

1. Evaluation of the important technologies which play a crucial role in improving the productivity and quality;

2. Conducting training programmes at the village level to ensure greater and effective participation of women to transfer the technologies to upgrade the skill;

3. Improvement of access to credit for women sericulturist in on farm sector by organizing credit camps, and special sessions exclusively for women with local manages;

4. Development of alternative, simple and safe technologies to overcome the health problems; and

5. Identification of work postures, work methods, tools and appliances based on economic conditions to reduce manual drudgery and improve work efficiency.

It is suggested that sericulture offers wide scope for economic empowerment of women, especially in the rural sector. Hence, well defined, time framed and practical oriented programmes can be evolved at all the levels and their implementation at the grass-root level can be ensured to improve the socio-economic status of women in sericulture. Formation and activation of women self help groups (SHGs) can be utilized as a launch pad for promoting women's participation in sericulture sector (Smita, 1994). Adoption of sericulture by women will make them empowered and will help transforming India into a developed country (Purusottam Dash *et al.,* 2015).

Summary and Conclusions

➢ Ericulture is a labour-intensive agro industry and an age-old land based practice in India with high employment potential and economic benefit to agrarian families.

➢ Ericulture is considered as family enterprise where farm women are actively participating in all the inter related activities. About 53 per cent of ericulture activities are carried out by the women.

➢ The gender-wise distribution of human labour employed in ericulture is 41.54 per cent male and 58.46 per cent female in host plant cultivation, and 45.87 per cent male and 54.13 female in silkworm rearing.

➢ By realizing the magnitude of women's contribution towards sericultural production, the Government of India declared the year 1994 as "Year of Women Sericulture" with a basic intension of providing a due recognition for her role in the industry.

➢ Every woman is an entrepreneur as she manages, organizes and assures the responsibility for running her home. It has been increasingly realized that women possess talent which can be harnessed for the productive purposes.

➢ Every woman on an average works for 7 to 8 hours a day in the field besides their routine duties of cooking, cleaning, fetching water etc.

- In the recent times, the women are playing important role in agricultural occupation as a manager, decision maker and skilled farm worker.

- Sericulture industry is an income-generating activity with multiple advantages for women *viz.,* she gets fairer deal for her products, she can combine the work with her other duties, she controls her own earnings, she can earn throughout the year and more importantly She learns to deal with people outside her home/village/community that develops her own personality and self-confidence.

- Woman being better managers of credit, with a greater access to income from sericulture, the rate of savings leads to a high degree of asset formation among the poor and also to sustainable development.

- In Assam, the families involved in ericulture normally belong to low income. In eri sector more than 1.80 lakh farm families are actively engaged in the state. The men of these families are always remaining busy with their primary sources of occupation and they do not pay more attention in the sericulture activities especially in ericulture.

- The rural women in Assam contribute substantially to the rural economy through ericulture. The women stay at home and manage the required time for rearing of silkworms to supplement some extent to the family income.

- Women generally involved in the light agricultural operations, because of their lesser physical strength than men. Average participation of women in host plant cultivation and management was only 3.8% against participation of 96.2% men.

- In silkworm rearing activities, participation of women was significantly higher compared to men in cocoon harvesting, marketing and spinning of spun yarn.

- Average participation of women in silkworm rearing, cocoon harvesting, marketing and spinning of spun yarn was 72.1% against participation of 27.9% men.

- In Ahmednagar district of Maharashtra, women's overall participation was maximum in weeding, fertilizer application, pruning, preservation and chopping the mulberry leaves.

- Participation of women is less than 10% in preparatory tillage and transportation of cocoons, silkworm pest and disease control, preparation of calendar operation for mulberry cultivation, preparation of 2% formalin solution and searching rate of cocoons.

➤ In Mandya district of Karnataka, 82.79 per cent of women were involved in aspects of ericulture like host plant cultivation, land preparation and fertilizer application.

➤ To know the involvement of rural women in sericultural activities, data were collected in the traditional sericultural districts namely Murshidabad and Malda of West Bengal by personal interview method, and in the non-traditional sericultural states namely Meghalaya and Tripura by postal interview method.

➤ In Murshidabad and Malda districts women did not participate in mulberry cultivation practices except digging, weeding and leaf harvest.

➤ But in Meghalaya and Tripura states, almost all the mulberry cultivation practices were performed by women except preparation of cuttings and irrigation in mulberry gardens.

➤ Silkworm rearing, about 32% of the total day works were spent by the women respondents in different sericulture activities.

➤ In Tirupattur taluk of Vellore district in Tamil Nadu, the tasks performed by the farm women under land preparation, transporting and marketing the produce to market centres were very negligible.

➤ In Ramtek Panchayat Samiti of Nagpur district of Maharashtra, moderate number of women participated in ericultural activities like sowing and intercultural operations.

➤ In Andhra Pradesh, participation of men and women in ericulture is not similar, and their role in decision making do not commensurate their involvement.

➤ Decision-making and proper planning is considered as an important part of ericulture. Success of silkworm crops are mostly dependant on appropriate decisions involved in selection of variety of host plant and silkworm breed, selection of rearing seasons, assessment for requirement of dfls/layings for capacity utilization, time of disinfection, feeding of silkworms, etc.

➤ In various activities of ericulture, 48.9% women alone take decisions compared to 33.6% men.

➤ In some activities *viz.,* selection of variety of host plants, adoption of improved technologies for host plant cultivation and plant protection, role played by women in decision-making is comparatively less in comparison with men.

➤ To increase the participation of women in ericulture, women extension agents should take effort to create awareness and to impart knowledge and skill on production, processing and marketing technologies through periodical training, field trips, group discussion, demonstration and campaign.

➤ The physical strength of the old age women does not permit frequent movement to operate the different activities of ericulture.

➤ The educated women normally want to busy with government or semi government job or to be associated with other enterprises.

➤ In a study conducted to know the relationship of socioeconomic variables of women with their participation in ericulture, involvement of tribal people is more compared to the involvement of the people belonging to higher caste or general caste.

➤ Knowledge is a pre-requisite to do anything, lack of sufficient knowledge about any idea/practice prevents an individual to avail its benefits. It is disheartening to note that as high as 76 per cent of the women sericulturists had low to medium levels of overall knowledge on silkworm rearing practices.

➤ Sericulture with its short gestation period and low investment is gaining popularity and plays a lead role in providing employment opportunities to the rural mass. It not only fetches periodical income, but also ensures employment for the family labour especially for women.

➤ In a study conducted to know the impact of silkworm rearing technology adoption on employment generation in Tamil Nadu by adopters and non-adopters shows that total employment generated per acre per year among the adopters was 565.25 man days and 467.72 man days among the non-adopters.

➤ The women and silk are the two sides of the same coin and hence quoted as "silk is by the women, for the women and of the women". More than 60 per cent of the various ericultural activities are being carried out by women, across different ericulture practicing regions.

➤ Low level of literacy, lack of awareness on the improved technologies of mulberry cultivation and silkworm rearing, poor level of perception on the technologies and also lack of empowerment are the major constraints faced by the women in sericulture.

➤ In Mandya district of Karnataka, the women are facing institutional or culturally induced constraints. They are the products of prevailing social and cultural norms.

- In view of the significant and critical position occupied by women, the development of women through sericulture has been identified as one of the paramount objectives of the National Sericulture Project.

- Formation of women's groups constitutes probably the most challenging but effective means of empowering women.

- By reducing the constraints faced by the women and thereby to bring her into the main stream, the government machineries such as government departments, financial institutions, NGOs and other related agencies took many measures i.e. reservation for women, provision of credit, incentives and subsidies, market facilities, invention of women friendly technologies, appointment of women para extension workers, conducting training programmes, facilitate organize in groups etc.

- Massive empowerment programme of the rural women needs to be launched to develop them as progressive entrepreneurs in agriculture and allied enterprises.

- Training provides defreezing of old behaviour and refreezing of new behaviour. Hence, there is a significant association with the participation in training programmes and knowledge level of women sericulturists.

- Formation and activation of women self help groups (SHGs) can be utilized as a launch pad for promoting women's participation in sericulture sector.

References

Ahmed S. 2003. Participation of women groups in development of Tasar Culture. *Indian Silk* **48**(8): 19-20.

Bhat D V and Sharmila K K. 1994. Peripatetic Training: A useful, tool for training women sericulturists. *Indian Silk* **32**(11): 61-63.

Dound R V, Bhosale B V and Kadam P M. 2016. Analysis of Participation of Farm-Women in Sericulture Activities for Ahmednagar District. *International Journal of Tropical Agriculture* **34**(6): 1679-1681.

Gautam S and Sarma A. 2003. Women empowerment through sericulture in Himachal Pradesh. *Indian Silk* **48**(9): 18-20.

Geetha G S. 2010. Socio-economic profile of farm women in sericulture activities – A Case study. *Mysore Journal of Agricultural Sciences* **44**(4): 872-876.

Geetha G S. 2011. Institutional and Social constraints of women in sericulture - A case study in Karnataka. *Indian Journal of Sericulture* **50**(2):160-165.

Giddens A. 1979. Central Problems in Social Theory: Action, structure and contradiction in social analysis. *Macmillan Publications*, London.

Jesia R. 1994. Women in Sericulture: Need for a Shift in Focus. *Indian Silk* **32**(11): 7-8.

Lakshmanan S, Ganapathy Rao R, Jayaram H and Geethadevi R G. 1997. Labour composition in sericulture. *Indian Silk* **35**(12): 19-21.

Lakshminarayan M T, Narayana Reddy R, Banuprakash K G and Jahir Basha C R. 2013. Knowledge of women sericulturists on improved silkworm rearing practices *Mysore Journal of Agricultural sciences* **47**(2): 435-436.

Malwe P, Parshuramkar S and Soor M. 2010. Participation of rural women in income generation activities of the family. *Journal of Soils and Crops* **20**(1): 128-132.

Mech D and Ahmed S A. 2012. Participatory profiles of women in Ericulture in Assam State of India. *European Journal of Applied Sciences* **4**(4): 177-181.

Meenal R and Rajan R K. 2008. Technology adoption and employment generation – Analysis in Tamil Nadu. *Indian Journal of Sericulture* **47**(1): 108-110.

Narmatha N, Uma V, Arun L and Geetha R. 2009. Level of participation of women in livestock farming activities. *Tamil Nadu Journal of Veterinary and Animal Sciences* **5**(1): 4-8.

Parthasarathy B. 1994. Karnataka: Hopes Galore. *Indian Silk* **32**(11): 15-18.

Purusottam Dash, Subhashree Dash and Sasmita Behera 2015. Women in developing sustainable livelihood system through sericulture in rural India. *Odisha Review* No. December: 31-35.

Raju L, Nataraju M S and Niranjanamurthy 1997. Women in sericulture: An analysis. *Indian Silk* December pp. 31-34.

Ramakrishna Naika, Manjunatha Gowda, Narayanaswamy K C, Vijayendra M, Rao L N, Shivananda K S and Nagaraja N. 2011. Knowledge level of SC / STs on pests and diseases of mulberry cultivation and silkworm rearing technologies. *Compendium of Abstracts of the Golden Jubilee National Conference* on Sericulture Innovations: Before and Beyond: 220.

Roy Chowdhuri, Umasankar, Das N K, Sahu P K and Majumdar M K. 2011. Studies on involvement of women and their contribution share in sericulture activities. *Journal of Crop and Weed* **7**(2): 37-40.

Sandhya Rani G. 2006. *Women in Sericulture*. Discovery Publishing House, New Delhi Pp. 127.

Savithri G and Sujathamma P. 2003. Participation of women in sericulture. *Bulletin of Indian Academy of Sericulture* **7**(2): 102-106.

Shobha V, Vimala Devi V and Jyothi. 1998. Socio-Economic Status of Women in Sericulture, Kurukshetra. *India's Journals of Rural Development* **46**(9): 46-48.

Smita P. 1994. Management Training Needs for Women Co-operatives. *Indian Silk* **32**(11): 55-59.

Sowmya T M, Narasimha N, Swetha B S and Pushpa P. 2012. Extent of involvement of rural women in agricultural and subsidiary enterprises. *Mysore Journal of Agricultural Sciences* **46**(1): 120-124.

Sujata B, Lakshminarayan Reddy S, Sankar Naik S and Sujathamma P. 2006. A study on adoption of recommended mulberry cultivation practices by sericulturists in Chittoor District of Andhra Pradesh. *Indian Journal of Sericulture* **45**(2): 142-148.

Sumathi P. 2008. Role of women in faming, allied and off-farm activities. *Agriculture Update* **3**(3&4): 247-250.

CHAPTER 12

Value Addition and by Products of Eri Silkworm

India, the only country which produces all four varieties of silk, i.e. Mulberry, Tasar, Muga and Eri. Among the non mulberry silks, only eri silk production is in increasing trend. The rural agro-based ericulture industry occupies cultivation of food plants, rearing of silkworms, conducting silk reeling, twisting, dyeing, weaving etc., and provides continuous employment to rural people in India. During the eri silkworm rearing processes other than main product a number of secondary wastes are developed. Effective utilization of secondary waste helps in achieving higher income and development of many new products and additional income. The benefits of ericulture, apart from being a source of silk yarn for weaving the traditional products like chaddars and pupae for consumption have been realized to a great extent after the recent advances in cocoon production techniques and introduction of simple but highly efficient spinning devices and looms. In addition, the wide variety of processes, designs and products developed in eri and its blends with other natural fibres have contributed to its steady growth. The eri silk sector is poised for a revolution that could probably have no parallels in the non-mulberry silk industry. The chapter discusses the spinning of eri silk, its value addition and by products in detail.

Silk obtained from the sources other than mulberry are termed as Non-mulberry or vanya silk. Among the non-mulberry silks, only eri silk production is in increasing trend. Ericulture is a science and technology of silk production in which different processes are involved. Silk is the main product of ericulture. During its production process a number of secondary wastes will develop in each stage which is of great value and gives additional income if properly utilized. In recent years, effective utilization of waste at different levels to churn out commercially acceptable by-products is gaining popularity as it brings in considerable value addition and supplements the overall income from ericulture. Effective utilization of waste would not only fetch them additional revenue but also pave way for development of new by products and generate additional employment (Sarmah, 2010). Silk wastes are used in different fabric making, and also as biomaterial in surgical and pharma industry. In ericulture starting from host plant cultivation to fabric production the waste generated is being utilized efficiently in many of the countries. Both on-farm and off-farm sectors of the culture have potential to convert their wastes into useful by-products of commercial value. There are several industrial units in countries like China for converting ericultural wastes into useful products that are useful not only in research and medicinal lines, but also for the common people.

There has been remarkable growth of these by-product extracting activity since 1970s, because of realization on cost benefit ratio. Many by-products presently discarded as wastes, can be put to better use for human benefit. The total utilization of silkworm host plants, wastes of rearing, grainages, cocoons, pupal matter through different marketable products and such an integrated exploitation of the total resources will certainly make the ericulture more sustainable. Indian sericulture though it is in the second position in production of raw silk eighty to ninety percent of waste obtained during various processes is not effectively utilized because of lack of awareness in waste utilization technology and market facilities. But in countries like Korea, Thailand these secondary wastes are being exploited in huge way which is fetching additional income to the farmers.

In general, insect foods are well-known as a protein source in the country, especially in the North and Northeast. Most of the tribal communities of NE India, *viz.,* Garo, Naga, Bodo, Missing, Rabha, Kachari etc. prefer pre-pupal stage for consumption. People of the Ahom community consume eri pupa (chrysalid) in the mature stage (Chaoba Singh and Suryanarayana, 2005). Pre-pupa is removed after the cocoon has been completely formed. The edible pupae are the by-product of commercial silk production, and obviously any insect that can produce two or more useful products simultaneously increases its economic and environmental efficiency. The pupae of *Bombyx mori* are used as food and animal feed in various Asian countries. In addition to the mulberry silk moth, there are more than a dozen species of 'wild' silk producers (Lepidoptera) of commercial interest, the pupae of which are also used as food or animal feed. As per the Sirimungkararat *et al.,* (2010), a total of 194 species of edible insects are reported in Thailand. There are 81 species of edible forest insects. Of the edible insects, Coleoptera represents the major group with 61 spp, followed by Lepidoptera (47 spp.), Orthoptera (22 spp.), Hymenoptera (16 spp.), Hemiptera (11 spp.), Homoptera (11 spp.), Odonata (4 spp.), Isoptera (2 spp.) and others.

Nutritive value of eri silkworm pupae in comparison to others

The pupa of eri silkworm has high moisture content of 74.66%, compared to *Bombyx mori* (65.13%) and *Attacus ricini* (70.14%). The protein and ash contents were 18.44% and 1.58% respectively for *Samia c. ricini*, 11.99% and 0.79% respectively for *B. mori* and 15.97% and 1.36% respectively for *A. ricini*. However, the lipid content (20.10%) in *B. mori* (domestic strain), was much higher than that in *S. c. ricini* (4.24%) and *A. ricini* (11.09%) (wild strains). Proximate analysis of pupae has shown that they contain other substances like hormones, trace elements and vitamins, thus indicating that they could be a good protein source for various purposes. The dried pupae of eri silkworm contain about 25% oils and 50% protein. De-oiled pupa contains about 11% nitrogen.

Uses of eri silkworm

Eri silkworm itself is nutritious. It does not make annoying taste nor does have any odour. It does not need water to drink. Its mortality rate is less than other silkworms (Vaidya *et al.,* 2014). The eri silkworm is an ideal example of sustainable agriculture, which produces silk with unique thermal properties and a pupa that is a high-protein food or animal feed. It also gives other rearing residues that can be used for fishpond culture. Because of high protein content (66 percent), eri food products have been developed using more than eight recipes, which have been registered as intellectual property. Eri silkworms are safe "green" edible insects because no chemicals are used in the rearing process. Moreover, eri products could generate supplementary income for farmers. Publicity campaigns should provide more information for consumers. The eri silkworm has the potential to support government food security policies in the context of supplying edible insects as protein sources for communities in Thailand (Sirimungkararat *et al.,* (2010). Silkworm pupae have been used in Chinese traditional medicines since ancient times. Pharmacological studies have shown that silkworm pupae increase immunity, protect the liver and prevent cancer. The pupae can be sold as fertilizer, biogas, feed and for other agricultural purposes. Judicial utilization of silkworm as a source of food is practiced in many countries of the world. Different recipe preparation like boiled pupa, fried pupa, chilli pupa, pupa masala etc are also prepared from eri silkworm. It can be used as feed for livestock, poultry, piggery etc. Pupal oil can be used in soap making industry. The litter of eri silkworm may be used as manure and it is said that the ash of eri litter is an effective insecticide. It also contains carotene (Sarmah *et al.,* 2012).

Mishra *et al.,* 2003 analysed the human consumption pattern of pupae of non-mulberry silkworms eri and muga along with mulberry silkworm. Overall consumption was the highest for eri pupae (87.7%) followed by muga (57.4%) and mulberry pupae (24.6%) irrespective of age group and gender. Amongst the three major communities predominant in the villages of Assam, the highest consumption was in the ahom community (eri 91% and muga 63%). The energy contents of the silkworm pupae were in the range of 706 to 988 kJ. Proximate composition suggested that these unconventional food items with high cultural acceptability and nutritive value may be utilized in formulating potential alternate recipe for malnourished population as well as nutritious delicacy for others. The composition (%) for non-mulberry and mulberry silkworm pupae was in the range of total protein (12 to 16%), total fat (11 to 20%), carbohydrate (1.2 to 1.8%), moisture (65 to 70%) and ash (0.8 to 1.4%). The amino acid scores of eri pre pupae and pupae protein were 99 and 100, respectively, with leucine as the limiting amino acid in both cases. Human net protein utilisation (NPU) of pre pupae and pupae was 41 as compared to 62 in casein. Protein digestibility corrected amino acid score (PDCAAS) was 86. The high protein content in the defatted eri silkworm meal with 44% total essential amino acids makes it an ideal candidate for preparing protein concentrate isolates with enhanced protein quality that can be used in animal nutrition (Longvah *et al.,* 2011).

Silk powder from eri silkworm has hygroscopicity, moisture desorption and deodorization indicating its suitability for high performance silk products and suggest that it has an attractive potential for skin care and as anti-odor agent (Ito *et al.,* 2017)

Eri cocoon characteristics

The cocoon of eri silkworm is very unique in its characteristics and different from other silkworms. Eri cocoons are open mouthed with a discontinuous filament, which makes them suitable only for spinning. Eri silk cocoons are made up of uneven fibres and it cannot be reeled. Approximately, 90 per cent of eri cocoons are hand spun in Assam, India. The soft cocoons are better for mechanical spinning and slightly hard and bigger cocoons are hard for mechanical spinning. Eri cocoon is white or brick red in colour and less lustrous than other silks. Eri fresh cocoon weighs 3-5 g and shell weighs 0.4-0.6g giving 11 to 14% shell. Size of eri cocoon is 4.8 x 2.5 cm. Danier of eri fibre is 2 to 3 with a tenacity and elongation of 3 to 3.5 g/d and 20 to 21 per cent respectively. Eri cocoons are marked on the basis of weight. There is no quality test for eri cocoons before transaction. Cocoon quality plays an important role in spinning of eri silk cocoons into spun yarn. Quality and colour of the eri cocoons depends on the race and rearing technology. The shape, build, colour and weight of the cocoons are some of the important parameters that a breeder has to focus in evolving races to achieve the desired objectives.

The colour is a racial character and it is due to the presence of pigments in the sericin layer of the bave. Choudhary (1982) reported that the variation in colour is due to the impermeability of cell wall and silk gland as a result of which pigments pass out along with the excrements. Colour of cocoons depends on pigments absorbed from leaves of the host plants. Eri silkworm larvae fed on castor leaves generally gives plain white or brick red coloured cocoons depending upon the strain. Similarly, tapioca and Kesseru fed cocoons give creamy white cocoons. Saikia (2008) observed highest white coloured cocoons on Barkesseru fed larvae. The extent of tightness or firmness indicates the shell texture and hardness of cocoons. Good quality cocoons are firm, compact and slightly elastic. It was revealed that host plants had significant effect on the compactness of cocoons. The cocoon spun by the eri silkworm fed on castor leaves were rated as soft cocoons or more moderate, while the cocoons obtained from the eri silkworm fed on host plants other than castor were rated as more hard cocoons (Saikia, 2008).

Eri silk characteristics

Eri silk is thick with a dull sheen and it has a soft, cotton-like feel. It is fine, dense, strong and absorbs moisture. It is durable and it has good elasticity. It wrinkles less and drapes well. It is soft, dull and has wool like finish. Eri silk is heavier and darker than other silks. Eri yarn is used for making excellent winter garments. Eri silk fabric is coarse. A microscopic observation of the eri silk shows continuous striations along the fibre. The cross section shows two elongated triangles facing each other on the flat side with rounded corners surrounded by sericin. Eri silk is widely used for preparing warm clothing like 'Eri chaddar', quilts and scarves, but other products like kurtas, maxis, dokhans etc. are also made from eri silk. Eri fabrics are warm and more durable than mulberry silk. It is also resistant to perspiration, dust etc. Further, the texture improves by use and wash and the colour also become brighter. Eri spun yarn has a slightly higher whiteness index compared to mulberry spun silk and also possesses better dye absorption capacity.

The raw silk is highly valued material not only as textile material but also it has many applications in the field of medicine, pharmaceuticals, cosmetic industry. Since long silk fibre is being used as suture material as it does not cause inflammatory reactions and is absorbed after healing of wounds. Recently regenerated silk solutions have been used to form a variety of biomaterials, such as gels, sponges and films, for medical applications. Silk scaffolds have been successfully used in wound healing and in tissue engineering of bone, cartilage, tendon and ligament tissues. The silk fibroin films facilitate epithelisation, remodelling of connective tissues and collagenisation. The fibroin powder is known for wound dressing. Silk protein shampoos, body lotions, moisturizers are commercially popular in Thailand. Silk protein sericin is known for antibacterial, uv resistant and moisture management properties and used in making bioactive textiles. Silk base laminates, renewable silk fibre reinforced polymeric composites are becoming popular in house building and interior decoration sectors because of its high flexural strength, silky and glossy appearance. Handicrafts made up of silkworm cocoons are more popular in almost all silk producing countries. The multicoloured silk paper is the new innovation and it is used in preparation of beautiful flowers. Effective utilization of secondary wastes of sericulture industry fetches good income and has a high potential in international market. Developing technologies and setting up of centres for effective utilization of secondary wastes help not only creating employment and also additional income to rural small and marginal farmers and there by rural development.

Properties of the silk: Silk contains 70-75% fibroin and 25-30% sericin protein. The biochemical composition of fibroin can be represented by the formula $C_{15}H_{23}N_5O_6$. It has the characteristic appearance of pure silk with pearly lustre. It is insoluble in water, ether or alcohol, but dissolves in concentrated alkaline solutions, mineral acids, and glacial acetic acid and in ammonical solution of oxides of copper. Sericin, a gummy covering of the fiber is a gelatinous body which dissolves readily in warm soapy solutions and in hot water, which on cooling forms a jelly with even as little as 1% of the substance. It is precipitated as a white powder from hot solutions by alcohol. It can be dyed before or after it has been woven into a cloth. The weight in gram of 900m long silk filaments is called a denier which represents size of silk filament.

Properties of silk

- Natural colour of Eri silk is brick red or creamy white or light brown. Mulberry silk is white, yellow or yellowish green in colour, while the muga silk is light brown or golden in colour and tasar silk is brown (Table 12.1).

- Silk has all desirable qualities of textile fibres, viz. strength, elasticity, softness, coolness, and affinity to dyes. The silk fibre is exceptionally strong having a breaking strength of 65,000-lbs/sq. inch.

- Silk fibre can be elongated to 20% of its original length before breaking.

- Density is 1.3-1.37g/cm^3.

- Natural silk is hygroscopic and gains moisture up to 11%.

- Silk is poor conductor of heat and electricity. However, under friction, it produces static electricity. Silk is sensitive to light and UV- rays.

- Silk fibre can be heated to higher temperature without damage. It becomes pale yellow at 110^{0}C in 15 minutes and disintegrates at 165^{0}C.

- On burning it produces a deadly hydrocyanic gas.

Use of silk: Silk is used in the manufacture of following articles:

- Garments in various weaves like plain, crepe, georgette and velvet.

- Knitted goods such as vests, gloves, socks, stockings.

- Silk is dyed and printed to prepare ornamented fabrics for saris, ghagras, lehengas and dupattas.

- Jackets, shawls and wrappers.

- Caps, handkerchiefs, scarves, dhotis, turbans.

- Quilts, bedcovers, cushions, table-cloths and curtains generally from Eri-silk or spun silk.

- Parachutes and parachute cords.

- Fishing lines.

- Sieve for flour mills.

- Insulation coil for electric and telephone wire.

- Tyres of racing cars.

- Artillery gunpowder.

- Surgical sutures.

Eri silk fabric characteristics such as smoothness, firmness, fullness, crispiness and hardness have strong impact on consumer preference for particular textile products. Fabric handle is a function of fibre properties, yarn structure (single, ply, cable filament etc), fabric geometry and finish given to the fabrics (Kavitha *et al.,* 2006).

Table 12.1: Fibre properties (Adopted from Kariappa *et al.*, 2011)

S.No	Particulars	Mulberry	Brick red eri	White eri
1	Fibre fineness (micron)	14.33	16.51	15.05
2	Std. deviation of fineness	3.95	4.31	4.26
3	Whiteness index	63.96	43.07	54.90
4	Yellowness index	5.59	10.84	7.8
5	Specific gravity of fibre	1.296	1.307	1.305
6	Fibre bundle strength (g/t)	42	29.4	31.1
7	Fibre elongation %	9.4	11	13.2
8	Moisture content (Wet)	10.69	969	10.55
9	Moisture regain (Dry)	9.66	8.83	9.54
10	Denier	1.8	2.37	2.08

Tenacity: The maximum specific stress recorded in extending a textile specimen to breaking pint is known as tenacity of the specimen. This tenacity equals to specific stress at break. The silk filament is strong compared to other natural fibres. This strength is due to its linear beta-configuration and highly crystalline polymer system. These two factors permit many more hydrogen bonds to be formed in a much more regular manner. When wet, silk loses strength. This is due to water molecules hydrolyzing a significant number of hydrogen bonds and in the process weakens the silk polymer. The tenacity of white eri fibre is 2.82, red eri fibre is 2.97 and that of mulberry is 3.91.

Elongation: Silk is considered to be more plastic than elastic because it is a crystalline structure and does not permit the amount of polymer movement which could occur in a more amorphous system. Hence, if the silk material is stretched excessively, the silk polymers, which are already in a stretched state, slide past each other. The process of stretching ruptures a significant number of hydrogen bonds. When stretching ceases, the polymers do not return to their original position, but remains in their new positions. This disorganizes the polymer system of silk, which is seen as a distortion and wrinkling or creasing of silk textile material. The handle of the silk is described as a medium, and its crystalline polymer system imparts a certain amount of stiffness to the filaments. This is often misinterpreted, in that the handle is regarded as soft, because of the smooth, even and regular surface of silk filaments.

Spinning of eri silk

The eri silkworm spins open mouthed, unreelable cocoon with discontinuous filament and used only for spinning. The cocoons are graded as soft small and bigger hard for hand spinning. The cocoon shells should be dried under the Sun or in ushna kotis or hot air ovens for longer preservations/storage.

Methods of stifling: There are different methods in stifling of the eri silk cocoon.

Sun drying: It is economical and widely practiced. Prolonged exposure to hot sun denatures the sericin leading to difficulty in cooking and affects spinning.

Hot air drying: It is preferred to sun drying due to uniform stifling. Spinning is done with takli, the traditional method and by improved spinning appliances.

Takli spinning: It consists of a spindle with a disc like base. The spinner holds the cocoon cake in the left hand and feeds the raw silk with right hand to the spindle. The spindle is rotated to impart twist.

Improved spinning: Spinning done by machines is called improved spinning. The improved spinning machines have been developed on continuous spinning principle with simultaneous drafting, twisting and winding. The different machines available for spinning are the das spinning wheel, Trivedi and Chaudary spinning wheel, Central silk board, Bangalore spinning machine and CSTRI (Central Silk Technological Research Institute, Bangalore) motorized Spinning-cum-Twisting machine. The das spinning wheel is pedal driven with flyer system. The improved spinning machine developed by central silk board, Bangalore can be operated both by pedalling and electricity. The production on these machines will be 25-30 g per hour for 20 metric counts yarn. Eri silk has been successfully spun employing Japanese and Italian technologies into yarn of counts $2/20^s$ to $2/120^s$. Eri silk fibre performs extremely well on both the systems of spun silk manufacturing.

Silk yarn spun on CSTRI motorized spinning-cum-twisting machine shows unequal distribution of slubs and snarls, but the advantage of this machine is it gives a fancy appearance and texture for the yarn. The courser and uneven Ahimsa silk yarn is most suitable for handloom sector and appropriately used as shot weft. Eri silk with greater tensile strength with better elongation, resistance to abrasion, excellent hand-feel properties and moderate pilling can be compatible to interweave with cotton, art silk, terrycot, polyester and filature silk to produce designer's fabrics (Sannapapamma and Naik, 2015). The spinning of eri cocoon involves the following process

Degumming: In this process about 11% sericin is removed to loosen the fibres. The methods employed for degumming are, the cocoons are boiled in banana leaf or paddy straw ash and pieces of green papaya, which dissolves the sericin. The second method involved is boiling the cocoons in 10 per cent Na_2Co_3 solution enclosed in a cloth bag for 45 minutes.

In another method, the cocoon shells were tied in a piece of perforated cloth along with a stone and dipped into a vessel containing the degumming solution made by dissolving one g of soap and 0.3 g of soda per litre of water, and boiled for one hour. The cocoons are then washed thoroughly, squeezed by hand and again boiled in plain water for one hour. The cocoons are then removed from boiling water. Excess water may further be removed using a hydro-extractor. The individual cocoon shell should be stretched open and 3-5 such stretched cocoons put together into a cake.

Before spinning, loose the cakes by hand, draw the filaments twist them and wind onto the bobbin until the cake is exhausted. Then while rotating the spindle, put the end of the yarn on to the loosened filaments of the next cake to continue spinning. Degumming loss in eri is a mere 8% as against 37% in case of mulberry silk resulting in a lower production costs in spinning.

Spinning: The rowings are spun on the ring spinning frame into fine yarn in the range of 60 to 210 metric counts. The single spun yarn from the ring frame is then doubled and twisted. Finally the spun silk is cross reeled on a reeling frame to make standard hanks. The hanks are dressed, folded and packed in bundles (Sarkar, 1988).

By-products

In spite of worldwide augmentation of eri product diversification to enhance the cost benefit ratio through value and employment addition, India has not kept that pace in the ericultural industry. The wastes presently felt as by-products, can be put to better use in generating the value-based products to make the ericulture industry commercially more feasible. The full utilization of silkworm food plants and cocoons as different marketable products and such an integrated operation can certainly make the Indian ericulture industry more practical. Further, the cost of primary end product of ericulture i.e. the silk can be proportionately brought down by converting the wastes as useful by-products and also can help in elevating the socio economic status of the rural ericulturists. The profitable conversion of wastes to high value utilities through phyto and post harvest technologies, the collaboration of eri-scientists with related industries, to locate functional activities for potential applications can reduce the production cost, pollution, recycles the resources to cater the demands of ever growing population. Further, the suitability of ericultural end products for human diet, animal feed, soap, glycerine, pharmaceuticals, bio-gas, organic manure, chlorophyll, carotene, phytol, n- tricantanol, pectin, fibre, paper and art crafts need exploration priority. Additionally, the chitin, shinki-fibroin, serra-peptidase and glucose-amine available in silkworm pupae and moths can be functional in neurological, post-surgical, ophthalmic, hepatitic, pancreatic, anti-histamic and anti-carcinogenic drugs. The potential of silk protein as tissue regenerating bio-material, wound healing bio-adhesive, ultra violet screening cosmetic and bio-active fabric deodorant show its application possibilities (Table 12.2).

Ericulture offers a wide range of by-products, having diverse utility either directly or indirectly in agriculture and village industry as well. Eri host shoots are used as fuel, for paper preparation and in hut roofs. Castor stems are good source of cellulose, used in making wallboards, newsprint etc. The castor oil is used in paints, varnishes, as fuel of jet planes and also in medicines. The castor cake after oil extraction has high carbon and nitrogen and can be used as manure. The detoxified cake can be used as animal feed and as human food in West Africa (Weiss, 1971). The leaves of castor plant are fed to cattle, goat, sheep etc. The green leaves are used as leafy vegetable during winter in countries like Korea (Raymond, 1961 and Weiss, 1971). The bark from jatropa, tapioca, champa and *Ailanthus* sps are used as purgatives. The bark of *Ailanthus* yields lactic acid and leaves are used as tonic (Narayanaswamy *et al.,* 1990). Components like chlorophyll, carotene, phytol, pectin are extracted from the eri silkworm faeces for use in the manufacture of edible colours, candy, wine, ice cream, medicine for gastric, ulcer, liver and blood diseases. The tricantanol is used as a plant growth regulator. In ericulture we get by-products from reeling industry and silkworm rearing sector.

Silkworm pupae are consumed as food, utilized in extraction of oil, which finds use in soaps, paints etc. the pupal powder has amino acids, vitamins etc, which is used as feed for poultry, fish, swine, cattle and as manure (Reddy *et al.,* 1998). The by product utilization may also emerge as a small to medium scale industry, which could be managed by unemployed and under employed women. These projects in the years to come would make ericulture an economically viable enterprise to withstand competition from other cash crops (Reddy *et al.*, 1998).

Table 12.2: Estimated by-products from subsectors of ericulture from rain fed castor crop on dry weight

Particulars	Quantity (Kg/ha)
Castor stem biomass (%)	36.93
Eri silkworm excreta	90.09
Eri silkworm leaf litter	24.11
Chlorophyll from eri worm excreta	
i) Chlorophyll - a	0.12
ii) Chlorophyll - b	0.26
iii) Total Chlorophyll	0.47
Eri pupae	
i) Male	51.46
ii) Female	86.93
iii) Mixture	66.73
Eri pupal oil	
i) Male	13.32
ii) Female	16.54
Spent eri moths	
i) Male	14.04
ii) Female	23.18
Larval exuviae	1.69

Reeling Industry

Eri pupae – A protein rich delicacy

Eri silkworm pupa is an immediate by-product of reeling industry, obtained after reeling. Consumption of eri silkworm larvae and pupae is an age-old tradition in many parts of the world. The tradition also exists in Asia including the North-Eastern parts of India. Eri pupa is a delicacy and dietary staples for many tribes in this region. The eri pupae have a high calorific value of 460 kcals/100g on dry weight basis and 133 kcals/100g on fresh weight basis, which is higher to cow's milk (69 kcals/100 ml), eggs (163 kcals/10g), chicken (120 kcals/100 g), white sugar (385 kcals/100 g) and raw carrot (42 kcals/100 g). Further, the eri pupae are rich in protein (53.3%), fats (25.6%) and carbohydrates (4.4%).

The estimated quantity of eri pupae obtained from ericulture using leaves from one hectare of castor garden was found to be 266.39 and 68.80 kg on fresh and dry weight. Eri silkworm pupal protein has a great value in baking industry, for the preparation of protein rich biscuits (Majhi *et al.,* 1991). Food items prepared out of pupae are pupal fry, pupal masala, pupal pickle, pupal tarkara, boiled pupae etc. The traditional pupal recipes basically involve frying, deep frying, baking and boiling with spices. A famous dish named "Onla-Onkhri'was prepared in Assam by removing the skin of the pupae (Baruah, 1999). The pupae in these cases are used fresh and the food prepared is highly perishable. Value addition can be enhanced by suitable preservation methods and by conversion of silkworm pupae into convenient processed products for wider market acceptability in different regions. However, the only difficulty is collection of pupal wastes from various large and small reeling units scattered in different parts of the country in odourless condition. So that it can be utilized for various purposes.

Pupal protein contains all essential amino acids including high concentration of leucine, which is important in human nutrition. The pupa also has sufficient quantities of calcium, iron and other minerals required for growth and development, all together making it comparable to meat and fish. Consumption of pupae can substantially supplement the protein starved rural populace in addition to income generation through silk production.

Priyadarshini and Revanasiddaiah (2013) analysed the pupal development of eri silkworm, *Samia c. ricini* and estimated crude and purified protein percentage from de-oiled pupal powder at the different hours (0 hours, 72 hours, 144 hours, 216 hours). It was found that crude protein and protein concentrate was gradually increased from 0 hours to 216 hours both in male and female pupae. However, female pupae exhibited 45.1 g of protein concentrate at 216 hours of development when compared to male pupae (39.3 g). The presence of high protein concentrate in the pupae may be due to the fact that at the beginning of development the protein was utilized for histolysis and the increase of protein at 216 hours may be due to histogenesis of adult organs hence, the gradual increase of protein was observed from 0 hr-216 hrs.

Silkworm pupae as human food: Silkworm pupae are highly nutritious and possess genuine food of high nutritive value. The detailed chemical analysis of *S. c. ricini* pupae reveals that they contain water, protein, fat, glycogen, chitin, ashes and other constituents. Deoiled pupa is a valuable source of amino acids. In some parts of North eastern India and China, the silkworm pupae are regarded as delicious food and are eaten after the silk has been reeled off (Roychoudhury and Joshi, 1995).

Silkworm pupae as fish and poultry feed: Refined protein of silkworm pupae is superior to that of fish meal and about equal to that of beef. The de-oiled pupae can be used as a fish and poultry feed. The highest growth rate of catla carps was recorded when the eri pellets of silkworm pupae were used as feed, indicating that silkworm pupae is an effective feed for culturing catla fish (Jayaram and Shetty, 1980). Whereas the de-oiled silkworm pupae are suitable as a better feed for culturing Masheer carps (Nandeesha *et al.,* 1989).

Silkworm pupae being rich in protein, fat and vitamins can be used as poultry feed. It also contains vitamins B_2, B_1, nicotinic acid and folic acid and thus serves as an excellent source of these nutrients for livestock particularly to fowls (Bose and Majumdar, 1990). However, inclusion of higher levels of silkworm pupal meal in the diets of poultry results in gizzard erosion and ulceration and hence, its inclusion in poultry feed should be at lesser proportion (Jain, 1988).

Uses of pupal powder: Application of deoiled pupae as manure significantly increases the leaf yield, shoot and root weight of mulberry. The pupal residue serves as a good source of nitrogen and phosphoric acid. Further, de-oiled pupal powder application to mulberry results in increase of crude protein and decrease of mineral content in the leaves. The pupal powder can also be used as a moulting material in Bakelite industry. The protein rich oil free pupal powder is utilized in preparing dog biscuits.

Pupae oil

Another important by-product from the reeling industry is the eri pupal oil, which has a refractive index (at 30°C) of 1.47, comparable to other common vegetable oils. Rancidification in pupal oil is slower because of its low saponification value (150.88). These properties of eri pupal oil have vast application in food and oleo chemical industries.

Female pupae yield more oil (16.58 kg/ha) than male pupae (13.41 kg/ha). Pupal oil can be mixed with linseed oil (25:75) and with chrysalis oil (25%) and can be used for manufacture of paints, varnishes and for preparation of emulsion solutions in jute industry to soften the jute fibres (Mishra and Dash, 1992 and Mathur *et al.,* 1988). Pupal oil is a good source of sterols. The white fat (stearic acid) obtained after hydrogenation, forms an excellent raw material for manufacture of high-grade soap, candles and textile industries, apart from being used in manufacture of low grade soaps (Mathur *et al.,* 1988 and Bose and Majumdar, 1990). The pupal oil can be used for burning purpose. Several countries have switched over to use of pupal oil for their textile industries. The refined pupal oil may be utilized as an alternative to edible oil and dalda. If, the odour and colour are removed, white oil can be utilized for skin hide tanning process.

A study was undertaken by Thingnganiag *et al.,* (2012) to provide value addition to silkworm pupal oil as an alternative source of edible oil for the food and feed industry by carrying out a short-term nutritional and toxicological evaluation of eri silkworm pupae oil using Wistar NIN rats. Growth performance of rats fed with sunflower oil (Control) and eri silkworm pupae oil (Experimental) was compared. Histopathological examination of the various tissues showed no signs of toxicity even after feeding the eri silkworm pupal oil for 18 weeks. Serum cholesterol and triglyceride was significantly reduced while high-density lipoprotein cholesterol was significantly increased which is attributed to the high α-linolenic acid content of eri silkworm oil. The study showed that eri silkworm pupae oil is safe and nutritionally equivalent to commonly used vegetable oils. Eri silkworm pupae can be harvested to provide cost effective alternative edible oil that can be used to nutritional advantage in the food and feed industry.

Eri pupal oil extraction and properties: Extraction of pupal oil from eri silkworm pupae can be done by using different solvents either under hot and cold conditions. Hot extraction method yields more pupal oil and more so with chloroform as solvent (29.00%). The oil so obtained can be purified by steam distillation or by filtering through activated charcoal or fuller's earth. However, the purification method depends upon the cost-effectiveness of the process. The freshly obtained pupal oil is brownish, fishy in odour and slightly acidic in nature.

Silk wastes: Another by product of reeling industry is silk waste. Though the recovery of pupae is almost constant per unit quantity of cocoons, the silk waste recovery depends upon, the type of reeling machine, quality of cocoon, skill of reeler, reeling water etc. Silk waste produced in reeling industry ranges from 25 to 40 % depending upon the reeling establishments. Realization from the sale of silk waste plays an important role in determining the cost structure of raw silk. It is estimated that 6-10 per cent of the total cost can be recovered from the sale of silk waste. Based on the process in which waste is generated in silk industry, silk waste is categorized into two major categories, *viz.,* cocoon waste (floss, double cocoons, pierced cocoons, stained cocoons etc.) and thread waste (re-reeling waste and twisted waste) (Naik *et al.,* 1992 and Sonwalkar, 1988).

Use of silk waste: Spun silk: Silk wastes are utilized for blending with other natural and manmade fibres for the production of blended yarns. Mulberry cocoon waste alone yields 93.4% yarn, while eri cocoon waste yields 81.2% of spun yarn (Sonwalkar, 1990). The short fibre, by-products of spun silk industry obtained during the process of dressing and preparing silvers, is utilized for spinning yarn of coarse count, called nail yarn.

Preparation of handicrafts: Silk waste can be very well utilized in rural and urban handicraft industries. Handicrafts preparation depends upon the skill and artistic talents of the persons, cost and availability of indigenous material etc. Handicrafts like greeting cards, garlands, bouquets etc. can be prepared from cocoon wastes. Wall hangings, carpets, purses, hankies, scarves etc. can be prepared from silk waste (Magdi *et al.,* 1989).

Silkworm rearing

The major by products of silkworm rearing are the unfed host plant leaves and faeces, which together constitute eri silkworm litter. This litter has found several uses apart from silkworm itself being used for few utilities.

Quantification of silkworm litter: It is estimated that 45 per cent of the total leaves fed to the eri silkworm goes as waste in the form of unfed leaves and shoots. The silkworms ingest only 40 per cent of the leaf spread in the trays. Of the material ingested by the silkworm, only about 55 per cent is digested and the rest is converted as silkworm faeces. Eri silkworm reared on the leaves from one hectare of castor crop yields 189.38 and 90.82 kg excreta on fresh and dry weight basis. The analysis of dried silkworm litter reveal water content of 8.08 per cent, crude protein 14.78 per cent, crude fat 2.14 per cent, crude fibre 19.70 per cent and soluble non-nitrogenous substances 44.57 per cent. Compared to soybean cake (48.22%) the crude protein content in silkworm litter is lower but higher than wild grass (7.86%). The nitrogen content of silkworm litter is around 3.06 per cent (Mathur *et al.,* 1988). The mean excreta and litter production in ericulture varied from 130.70 ± 0.87 to 151.70 ± 0.62 and 255.00 ± 0.62 to 265.00 ± 1.07 kg/ha, respectively during different rearing seasons (Reddy *et al.,* 2000). The eri silkworm leaf litter contains higher NPK than cow dung, thus it can be used as organic manure in crop husbandry. Thus leaf litter with greater potentiality can be further explored in manifold on farm activities.

Silkworm litter as manure: Eri silkworm excreta can be effectively used as manure (Madan and Vasudevan, 1989). The litter produced by IV and V instars of eri silkworm from 100 layings is equivalent to dung produced by two to three cows per day (Mishra and Dash, 1992). The silkworm litter contains some phosphoric acid and potash besides nitrogen and can be used as an organic fertilizer. Direct application of fresh litter to the fields is less effective and often causes spread of silkworm diseases while rearing. Hence it is recommended to use it in the form of compost. Litter as manure can be used directly for agriculture or vegetable crop production. Application of silkworm litter as manure, resulted in better yields of radish (4.00 kg/plot) compared to biogas slurry (3.4 kg/plot) but lower than chemical fertilizers (5.225 kg/plot) indicating its potential utility as a manure for other crops (Madan *et al.*, 1997).

Silkworm litter for mushroom cultivation: Silkworm litter is rich in nitrogen serves as a good source for the growth of edible mushrooms (Reddy *et al.*, 1998; Baruah, 1999). Silkworm litter can be mixed with paddy straw to grow the edible mushrooms. However, higher proportions of paddy straw (1:3 or 1:4) are essential to get better yields. The protein, carbohydrate, ash, fat, crude fibre and mineral content of mushrooms that were obtained from silkworm litter and paddy straw is similar to that the mushrooms that were reared using paddy straw alone (Madan *et al.*, 1997 and Reddy *et al.*, 2003).

Extraction of chlorophyll from the eri silkworm excreta: In sericulturally advanced countries like Japan and China, the chlorophyll is being extracted on commercial scale from eri silkworm excreta, which finds use in medicine and cosmetics (Narayanaswamy *et al.*, 1993). In China, four chlorophyll derivatives are isolated from silkworm excreta, among them two substrates *viz.*, hydroxyl pheophytin-a and pheophytin-b have therapeutic value in drug industry. In a study conducted by Narayanaswamy *et al.*, (1993), 0.12, 0.26 and 0.47 kg of chlorophyll a, b and total chlorophyll were recorded in the excreta when the eri silkworms were reared on castor leaves from one hectare of castor garden. Chlorophyll content gradually increased from larval stages I to V (Bai and Revanasiddiah (2005). Further, the chlorophyll extracted from the excreta of eri silkworm was used as food additive and medicine (Ziran, 1998). Paste chlorophyll extracted from the silkworm excreta can be used for large scale production of copper chlorophyllin, sodium, pectin, carotene, phytol, tricantanol that can be used in the manufacture of medicines, cosmetics and as food additive (Revanasiddaiah and Yashodabai, 2000).

Eri silkworm excreta as substrate for biogas production

In ericulture industry silkworm excreta is an important by product, which finds its use in various fields including biogas production. Eri silkworm excreta and litter can be used for biogas production and as manure (Naik, 1994). This has necessitated quantifying the eri silkworm excreta produced per unit number of worms, its chemical composition, biogas production from a unit excreta and its combination with other organic wastes for efficient utilization.

When silkworm litter is used for biogas production, the fermentation proceeds rather rapidly and also maximum quantity of gas is produced when silkworm litter is mixed with other organic wastes (Naik, 1994). Silkworm litter alone yielded maximum biogas of 1593 ml/day over a period of 6 weeks. In contrary, cow dung alone gave 812 ml/day only. However, the quantity of gas produced per gram of total solids destroyed was maximum in silkworm litter with cow dung (2450 ml/g) whereas cow dung alone resulted in 1910 ml/g of solids destroyed. The amount of carbon and nitrogen in the material to be digested affects the anaerobic process. The addition of silkworm litter, which is rich source of total nitrogen containing up to 2.55 per cent may simulate organisms causing increased gas production. The nitrogen and potassium content in silkworms excreta was highest (3.57 and 1.95%), compared to cow dung (1.75 and 0.78%). The bio-digested slurry of silkworm litter incorporated with cow dung contains more nitrogen (2.7%), phosphorus (0.97%) and potassium (1.36%) than cow dung alone (1.70%, 0.70 and 0.67% N, P, K respectively). The use of litter in bio gas plant helps to prevent further spread of silkworm diseases.

Ramakrishna Naika and Reddy, (2001) conducted an experiment in the laboratory to determine the amount of gas produced by eri excreta and coffee husk along with cow dung from four races of eri silkworm *viz.,* plain white, blue, zebra and semi zebra. The excreta was collected daily and weighed separately for each race. Various treatments were established by mixing fresh excreta of fifth instar eri worms, coffee husks and cow dung i.e. T1= cow dung (400 g), T2= coffee husk (400 g), T3= eri excreta (400 g), T4= eri excreta + coffee husk in 1:1 ratio (200 g eri excreta + 200 g coffee husk), T5= eri excreta + coffee husk in 1:2 ratio (130 g eri excreta + 270 g coffee husk), T6= eri excreta + coffee husk in 1:3 ratio (130 g eri excreta + 400 g coffee husk), T7= eri excreta + cow dung in 1:1 ratio (200 g eri excreta + 200 g cow dung). Biogas generated was recorded up to 45 days and was analyzed for methane, carbon dioxide and other gases

(Jones, 1970). Nitrogen, phosphorous and potassium of digested slurry were determined as per the procedure of Jackson (1973).

The results of the experiment show that the fresh excretal weight was comparatively high (876.69 g) in semi zebra race and minimum (757.94 g) in white race (Table 12.3). This difference could be due to varied larval durations. The total larval durations among the races were 27 days (semi zebra), 26.5 days (zebra), 26 days (blue) and 25 days (white). Increase in feeding duration might have added to the increased excretal weight in respective races. The faecal matter production was directly related to the food intake (Ramdev and Rao, 1979). Differences of N, P and K content in the excreta of four eri silkworm races were not observed. Chinnaswamy *et al.,* (1996) quantified the average excreta of the 5 larval instars of 4 races of the eri silkworm. The maximum amount (79.16 g) was produced by white zebra race and the minimum (75.88 g) by white plain race. Almost 98% of the excreta was produced during the 4th and 5th instars in all races. The study throws light on scope for further exploitation of eri pupae as food of variety and an industrial raw material for extracting the components.

Average biogas production over a period of six weeks was more (1588.72 cc) with eri silkworm excreta. This might be due to easy bio digestibility, more methanogenic bacteria and optimal C/N ratio (Nagarajan *et al.,* 1990, Kalimuthu and Rajasekaran, 1992). Average gas production was minimum (341.44 cc) in coffee husk treatment, may be due to the presence of slow degradable high lignin content (Elias, 1972). The gas production per gram of substrate ranged from 5.12 cc (coffee husk) to 23.83 cc (eri excreta). The efficiency of gas production mainly depends on the lignin content of the raw material (Hobson *et al.,* 1981). The methane content of biogas was significantly more (60.44%) in the treatment three consisting of eri silkworm excreta alone (Table 12.4).

The nutrient composition of eri excreta, cow dung and coffee husk increased after the biodegradation (Table 12.5). The increase in nitrogen content after digestion may be due to the utilization of carbon by microorganisms which lowers the C/N ratio and thus increases the per cent nitrogen. The slurry from eri silkworm excreta treatment had significantly higher nitrogen content (3.45%) which was on par with eri excreta + cow dung 1:1 ratio (3.39%). It is evident that eri silkworm excreta alone or with cow dung can be utilized in biogas production which also yielded rich manure after bio digestion.

Table 12.3: Amount of excreta (g) produced by eri silkworm races on fresh weight basis and its chemical composition (%) (Adopted from Ramakrishna Naika and Reddy, 2001)

Race	Total faecal weight/100 larvae	Nitrogen	Phosphorus	Potassium
White	757.94	3.12 ± 0.06	0.89 ± 0.04	1.70 ± 0.08
Blue	797.12	3.26 ± 0.07	0.91 ± 0.05	1.70 ± 0.06
Zebra	848.56	3.10 ± 0.05	0.94 ± 0.04	1.76 ± 0.04
Semi zebra	876.69	3.22 ± 0.06	0.93 ± 0.03	1.81 ± 0.06

The values are average of five replications

Table 12.4: Biogas production over 6 weeks and its composition (Adopted from Ramakrishna Naika and Reddy, 2001)

Treatments	Average gas production (cc)	Gas/g of the substrate (cc)	Methane (%)	Carbon dioxide (%)	Other gases (%)
Cow dung	824.77	12.35	54.52	38.62	6.86
Coffee husk	341.44	5.12	45.00	40.50	14.50
Eri excreta	1588.72	23.83	60.42	37.96	1.62
Eri excreta+coffee husk in 1:1 ratio	432.44	06.49	56.75	40.19	03.06
Eri excreta+coffee husk in 1:2 ratio	411.50	06.17	53.94	42.01	04.05
Eri excreta+coffee husk in 1:3 ratio	379.50	05.69	50.58	40.73	08.69
Eri excreta+cow dung in 1:1 ratio	1507.27	22.61	58.83	39.95	01.22
F-test	Sig	NS	Sig	Sig	Sig
CD at 5%	5.458	-	2.197	2.411	1.305

Significant at 5% probability level

Table 12.5: Nutrient status of bio digested slurry of eri silkworm excreta and other substrates (Adopted from Ramakrishna Naika and Reddy, 2001)

Treatment	Nitrogen (%)	Phosphorus (%)	Potassium (%)
Cow dung alone	1.34	0.82	0.89
Coffee husk alone	1.86	0.47	1.13
Eri excreta alone	3.45	1.06	1.96
Eri excreta+coffee husk in 1:1 ratio	2.18	0.52	1.20
Eri excreta+coffee husk in 1:2 ratio	1.94	0.49	1.17
Eri excreta+coffee husk in 1:3 ratio	1.87	0.48	1.16
Eri excreta+cow dung in 1:1 ratio	3.39	1.04	1.94
F-test	Sig	Sig	Sig
CD at 5%	0.253	0.092	0.185

Significant at 5% probability level

The benefits of ericulture, apart from being a source of silk yarn for weaving the traditional products like chaddars and pupae for consumption have been realized to a great extent after the recent advances in cocoon production techniques and introduction of simple but highly efficient spinning devices and looms. In addition, the wide variety of processes, designs and products developed in eri and its blends with other natural fibres have contributed to its steady growth. The eri silk sector is poised for a revolution that could probably have no parallels in the non-mulberry silk industry.

Summary and Conclusions

➢ Silk obtained from the sources other than mulberry are termed as Non-mulberry or vanya silk. Among the non-mulberry silks, only eri silk production is in increasing trend.

➢ The rural agro-based ericulture industry involves cultivation of food plants, rearing of silkworms, conducting silk reeling, twisting, dyeing, weaving etc., and provides continuous employment to rural people in India.

➢ Ericulture is a science and technology of silk production in which different processes are involved. Silk is the main product of ericulture. During its production process a number of secondary wastes will develop in each stage which is of great value and gives additional income if properly utilized.

➢ In recent years, effective utilization of waste at different levels to churn out commercially acceptable by-products is gaining popularity as it brings in considerable value addition and supplements the overall income from ericulture.

➢ In ericulture starting from host plant cultivation to fabric production the waste generated is being utilized efficiently in many of the countries.

➢ Indian sericulture though it is in the second position in production of raw silk eighty to ninety percent of waste obtained during various processes is not effectively utilized because of lack of awareness in waste utilization technology and market facilities.

➢ In general, insect foods are well-known as a protein source in the country, especially in the North and North eastern states of India.

➢ People of the Ahom community consume eri pupa (chrysalid) in the mature stage.

➤ The eri silkworm is an ideal example of sustainable agriculture, which produces silk with unique thermal properties and a pupa that is a high-protein food or animal feed. It also gives other rearing residues that can be used for fishpond culture.

➤ Eri silkworms are safe "green" edible insects because no chemicals are used in the rearing process. Moreover, eri products could generate supplementary income for farmers.

➤ Silk contains 70-75% fibroin and 25-30% sericin protein. The biochemical composition of fibroin can be represented by the formula $C_{15}H_{23}N_5O_6$.

➤ In addition to the mulberry silk moth *Bombyx mori*, there are more than a dozen species of 'wild' silk producers (Lepidoptera) of commercial interest, the pupae of which are also used as food or animal feed.

➤ The pupa of eri silkworm has high moisture content of 74.66%, compared to *Bombyx mori* (65.13%) and *Attacus ricini* (70.14%). The protein and ash contents were 18.44% and 1.58% respectively for *Samia c. ricini*, 11.99% and 0.79% respectively for *B. mori* and 15.97% and 1.36% respectively for *A. ricini*.

➤ Pupal oil can be used in soap making industry. The litter of eri silkworm may be used as manure and it is said that the ash of eri litter is an effective insecticide.

➤ The amino acid scores of eri pre pupae and pupae protein were 99 and 100, respectively, with leucine as the limiting amino acid in both cases.

➤ Silk powder from eri silkworm has hygroscopicity, moisture desorption and deodorization indicating its suitability for high performance silk products and suggest that it has an attractive potential for skin care and as anti-odor agent.

➤ Eri cocoons are open mouthed with a discontinuous filament, which makes them suitable only for spinning. Eri silk cocoons are made up of uneven fibres and it cannot be reeled.

➤ Eri cocoon is white or brick red in colour and less lustrous than other silks. Eri fresh cocoon weighs 3-5 g and shell weighs 0.4-0.6g giving 11 to 14% shell. Size of eri cocoon is 4.8 x 2.5 cm. Danier of eri fibre is 2 to 3 with a tenacity and elongation of 3 to 3.5 g/d and 20 to 21 per cent respectively.

➤ The colour is a racial character and it is due to the presence of pigments in the sericin layer of the bave.

- Eri silkworm larvae fed on castor leaves generally gives plain white or brick red coloured cocoons depending upon the strain. Similarly tapioca and Kesseru fed cocoons give creamy white cocoons.

- The cocoon spun by the eri silkworm fed on castor leaves were rated as soft cocoons or more moderate, while the cocoons obtained from the eri silkworm fed on host plants other than castor were rated as more hard cocoons. Generally, soft cocoons fetch higher rates as compared to hard cocoons.

- Eri silk is thick with a dull sheen and it has a soft, cotton-like feel. It is fine, dense, strong and absorbs moisture. It is durable and it has good elasticity. It wrinkles less and drapes well. It is soft, dull and has wool like finish. Eri silk is heavier and darker than other silks.

- Eri fabrics are warm and more durable than mulberry silk. It is also resistant to perspiration, dust etc. Further, the texture improves by use and wash and the colour also become brighter. Eri spun yarn has a slightly higher whiteness index compared to mulberry spun silk and also possesses better dye absorption capacity.

- The raw silk is highly valued material not only as textile material but also it has many applications in the fields of medicine, pharmaceuticals and cosmetic industry.

- The maximum specific stress recorded in extending a textile specimen to breaking pint is known as tenacity of the specimen. The tenacity of white eri fibre is 2.82, red eri fibre is 2.97 and that of mulberry is 3.91.

- Silk is considered to be more plastic than elastic because it is a crystalline in structure and does not permit the amount of polymer movement which could occur in a more amorphous system.

- Eri silk has been successfully spun employing Japanese and Italian technologies into yarn of counts $2/20^s$ to $2/120^s$.

- Eri silk with greater tensile strength with better elongation, resistance to abrasion, excellent hand-feel properties and moderate pilling can be compatible to interweave with cotton, art silk, terrycot, polyester and filature silk to produce designer's fabrics.

- Spinning of eri cocoons involves degumming process, in which 11% sericin is removed to loosen the fibres.

- Degumming loss in eri is a mere 8% as against 37% in case of mulberry silk resulting in a lower production costs in spinning.

➤ Ericulture offers a wide range of by-products, having diverse utility either directly or indirectly in agriculture and village industry as well. Eri host shoots are used as fuel, for paper preparation and in hut roofs.

➤ The castor oil is used in paints, varnishes, as fuel of jet planes and also in medicines. The castor cake after oil extraction has high carbon and nitrogen and can be used as manure.

➤ Silkworm pupae are consumed as food, utilized in extraction of oil, which find use in soaps, paints etc. The pupal powder has amino acids, vitamins etc, which is used as feed for poultry, fish, swine, cattle and as manure.

➤ Eri silkworm pupa is an immediate by-product of reeling industry, obtained after reeling.

➤ The eri pupae have a high calorific value of 460 kcals/100g on dry weight basis and 133 kcals/100g on fresh weight basis, which is higher to cow's milk (69 kcals/100 ml), eggs (163 kcals/10g), chicken (120 kcals/100 g), white sugar (385 kcals/100 g) and raw carrot (42 kcals/100 g). Further, the eri pupae are rich in protein (53.3%), fats (25.6%) and carbohydrates (4.4%).

➤ Eri silkworm pupal protein has a great value in baking industry, for the preparation of protein rich biscuits. A famous dish named "Onla-Onkhri' was prepared in Assam by removing the skin of the pupae.

➤ In an analysis conducted to estimate the crude and purified protein percentage from de-oiled pupal powder from 0 hours to 216 hours showed that the crude protein and protein concentrate gradually increased from 0 hours to 216 hours both in male and female pupae.

➤ Silkworm pupae being rich in protein, fat and vitamins can be used as poultry feed. It also contains vitamins B_2, B_1, nicotinic acid and folic acid and thus serves as an excellent source of these nutrients for livestock particularly to fowls.

➤ Female pupae yields more oil (16.58 kg/ha) than male pupae (13.41 kg/ha). Pupal oil can be mixed with linseed oil (25:75) and with chrysalis oil (25%) can be used for manufacture of paints, varnishes and for preparation of emulsion solutions in jute industry to soften the jute fibres.

➤ Silk wastes are utilized for blending with other natural and manmade fibres for the production of blended yarns. Mulberry cocoon waste alone yields 93.4% yarn, while eri cocoon waste yields 81.2% of spun yarn.

➢ The major by-products of silkworm rearing are the unfed host plant leaves and faeces, which together constitute eri silkworm litter. This litter has found several uses apart from silkworm itself being used for few utilities.

➢ The mean excreta and litter production in ericulture varied from 130.70 ± 0.87 to 151.70 ± 0.62 and 255.00 ± 0.62 to 265.00 ± 1.07 kg/ha, respectively during different rearing seasons.

➢ The eri silkworm leaf litter contains higher NPK than cow dung, thus it can be used as organic manure in crop husbandry.

➢ In China, four chlorophyll derivatives are isolated from silkworm excreta, among them two substrates *viz.,* hydroxyl pheophytin-a and pheophytin-b have therapeutic value in drug industry.

➢ In an experiment conducted to quantify the eri silkworm excreta by using different races of eri silkworm, higher excreta (876.69 g) was recorded in the semi zebra race and minimum was recorded in white race (757.94 g). Almost 98% of the excreta was produced during the 4th and 5th instars in all races.

References

Bai S Y and Revanasiddiah H M. 2005. Extraction of chlorophyll from the faeces of eri silkworm *Philosamia ricini. Environment and Ecology* **23**(3): 628-630.

Baruah A M. 1999. Utilization of eri silkworm litter in mushroom culture. *M.Sc (Seri.) Thesis, UAS, Bangalore* p. 59.

Bose P C and Majumdar J K. 1990. Biochemical composition of pupal waste and its utilization. *Indian Silk* **29**(2): 45-46.

Chaoba Singh K and Suryanarayana N. 2005. Principles of Ericulture. *Central Tasar Research and Training Institute, Central Silk Board, Ranchi* pp. 84–85.

Chinnaswamy K P, Leina M J and Hariprasad K B. 1996. Quantification of eri silkworm, *Samia cynthia ricini* Boisduval excreta. *Insect Environment* **2**(3): 103-104.

Choudhary S N. 1982. Eri silk industry. Directorate of Sericulture and Weaving, Govt of Assam, Guwahati pp. 1-177.

Elias L G. 1972. Chemical composition of coffee berry by products. In: Coffee pulp (Eds. Brham J E and Bressani R). *International Development Research Centre,* Canada pp. 11-16.

Hobson P N, Bousfield S and Summers R. 1981. Methane production from agricultural and domestic wastes. *Applied Science Publishers,* London.

Ito M, Takaki M, Azuma K, Okamatsu S and Nakasone T. 2017. An evaluation of functionality of silk powder from eri silkworm. *Journal of Silk Science and Technology of Japan* **25**: 27-34.

Jackson M L. 1973. Soil Chemical Analysis. *Prentice Hall,* New Delhi.

Jain A K. 1988. Reassessment of nutritive value of silkworm pupae meal and its toxic effect on broilers. *M.Sc (Poultry) Thesis*, UAS, Bangalore p. 128.

Jayaram M G and Shetty H P C. 1980. Studies on the growth rates of catla, rohu and common crap fed on different formulated feeds. *Mysore Journal of Agricultural Sciences* **14**: 598-606.

Jones A R. 1970. An introduction to gas liquid chromatography. *Academic Press, London.*

Kalimuthu K and Rajasekaran P. 1992. Bio digestion of silkworm litter for biogas production. *Indian Journal of Sericulture* **31**: 17-22.

Kariyappa P M, Damodara Rao and Somashekar T H. 2011. Evaluation of low stress, surface properties and total hand value of mulberry, white eri and red eri spun silk fabric. *Indian Journal of Sericulture* **50**(2): 180-187.

Kavitha K, Padamashwini V R, Giride R, Neelakandan and Senthil Kumar M. 2006. Studies on low stress mechanical properties of Chitosan treated wool fabrics. *Journal of the textile Association*, July-August, 2006.

Longvah T, Mangthya K and Ramulu P. 2011. Nutrient composition and protein quality evaluation of eri silkworm (*Samia ricini*) pre pupae and pupae. *Food Chemistry* **128**(2): 400-403.

Madan M and Vasudevan P V. 1989. Silkworm litter: use as nitrogen replacement for crop cultivation and substrate for mushroom cultivation. Biological Wastes **24**: 201-216.

Madan M P, Vasudevan and Sharma S. 1997. Cultivation of *Pleurotus sajorcaju* on different wastes. *Biological Wastes* **22**: 241-250.

Magdi S P, Vijayendra M K, Hadimani V V, Mahadeveppa D and Patil C S. 1989. Versofalice fashionable handicrafts from silk waste. *Indian Silk* **27**(12): 47-49.

Majhi S K, Sinha U S P and Thangavelu K. 1991. Use of by-products of non-mulberry sericulture and silk industry. *Indian Silk* **30**(7): 13-14.

Mathur S K, Mukhopadhyaya B K and Ganguli 1988. Utilization of by products of mulberry silkworm. *Indian Silk* **27**(11): 23.

Mishra C S and Dash M C. 1992. Utility of sericulture wastes and by-products in agriculture. *Indian Silk* **31**(7): 38-40.

Mishra N, Hazarika N C, Narain K and Mahanta J. 2003. Nutritive value of non-mulberry and mulberry silkworm pupae and consumption pattern in Assam, India. *Nutrition Research* **23**(10): 1303-1311.

Nagarajan P, Rajasekaran P, Radha N V and Jayaraj S. 1990. Bio digestion of silkworm larval litter for biogas production. *Madras Agricultural Journal* **77**: 32-35.

Naik R. 1994. Quantification, nutritional composition of eri silkworm excreta and its use in biogas production. *M.Sc. (Seri.) Thesis*, UAS, Bangalore, p.79.

Naik S V, Hariraj G, Lakshmipathaiah B N and Nadiger G S. 1992. The silk waste raw material for spun silk industry. *Indian Silk* **31**(8): 15-17.

Nandeesha M L, Basavaraja N, Keshavanath P, Varghese T J, Sudhakar N S, Srikantha G K and Ray A K. 1989. Formulation of pellets with sericulture waste and their evaluation in crap culture. *Indian Journal of Agricultural Sciences* **59**(9): 1198-1205.

Narayanaswamy K C, Bhat G G, Reddy D N R and Prabhakar M K. 1990. Medicinal value of food plants of non-mulberry silkworms. *Indian Silk* **29**(5): 23-25.

Narayanaswamy K C, Prabhakara M K, Reddy D N R and Vijayendra M. 1993. Sericulture – a multifaced agro based industry. *Farmer and Parliament* **28**(6): 11-13 & 27-30.

Priyadarshini P A and Revanasiddaiah H M. 2013. Estimation of crude and purified protein from de-oiled pupae of ERI silkworm, *Philosamia ricini*. *International Journal of Current Microbiology and Applied Sciences* **2**(8): 215-220.

Ramakrishna Naika and Reddy D N R. 2001. Possible utilization of eri silkworm excreta and coffee husk in biogas production. *Indian Journal of Environment and Ecoplan* **5**(2): 77-80.

Ramdev V P and Rao P J. 1979. Consumption and utilization of castor by *Achaea janata. Indian Journal of Entomology* **41**: 260-266.

Raymond W D. 1961. Castor beans as food and fodder. *Tropical Science* **3**: 19-24.

Reddy D N R, Baruah A M and Reddy R N. 1998. Effective utilization of eri silkworm wastes. *The Third International Conference on Wild Silkmoths*. Bhubaneshwar p.109.

Reddy R N, Reddy D N R and Mallesha B C. 2003. Bio potentiality of silkworm rearing wastes in mushroom culture. *Advances in Agricultural Biotechnology* 47-52.

Revanasiddaiah H M and Yashodabai S. 2000. Extraction of paste chlorophyll from the faeces of different larvae of bivoltine silkworm, *Bombyx mori* L. *Proceedings of National seminar on Tropical Sericulture – Non Mulberry Sericulture, Silk Technology, Sericulture Economics and Extension* Eds: Chinnaswamy K P, Govindan R, Krishnaprasad N K and Reddy D N R, UAS, Bangalore **3**: 79-80.

Roychoudhury N and Joshi J C. 1995. Silkworm pupae as human food. *Indian Silk* **34**(7): 10.

Saikia P. 2008. Cocoon and yarn parameters of eri silkworm *Samia ricini* Boisduval reared on certain host plants. M.Sc. (Agri) Thesis, Assam Agricultural University, Jorhat, Assam.

Sannapapamma K J and Naik S D. 2015. Ahimsa silk union fabrics - a novel enterprise for handloom sector. *Indian Journal of Traditional Knowledge* **14**(3): 488-492.

Sarkar D C. 1988. Ericulture in India. *Central Silk Board*, Bangalore P.51.

Sarmah M C, Ahmed S A and Sarkar B N. 2012. Research and technology development, by product management and prospects in Eri culture - A review. *Munis Entomology and Zoology* **7**(2): 1006-1016.

Sarmah M C. 2010. The Silkworm Saga. *The Assam Tribune, Sunday Reading* 26th September, 2010.

Sirimungkararat S, Saksirirat W, Nopparat T and Natongkham A. 2010. Edible products from eri and mulberry silkworms in Thailand. Forest insects as food: humans bite back. *Proceedings of a workshop on Asia-Pacific resources and their potential for development, Chiang Mai, Thailand* 19-21 February, 2008 p. 189-200.

Sonwalkar T N. 1988. Spun silk yarn manufacture. In: *"Production and properties of Indian Silk"*, published by CSTRI, Bangalore p. 16-20.

Sonwalkar T N. 1990. Hand spinning of pierced/cut mulberry and eri cocoon. *Published by CSTRI*, Bangalore.

Thingnganing Longvah, Korra Manghtya and Qadri. 2012. Eri silkworm: a source of edible oil with a high content of α-linolenic acid and of significant nutritional value. *Journal of the Science of Food and Agriculture* **92**(9): 1988-1993.

Vaidya S, Yadav U and Bhouraskar J. 2014. Effect of temperature and relative humidity on rearing performance of eri silkworm (*Philosamia ricini*). *Environment Conservation Journal* **15**(3): 189-196.

Weiss E A. 1971. Castor, sesame and safflower. *Leonard Hill Books*, London p.227-308.

Ziran H. 1998. Utilization of sericultural resources in China. *The Third International Conference on Wild Silkmoths*. Bhubaneshwar p. 1-6.

CHAPTER 13

Role of Extension Strategies in Eri Culture

The introduction of new ericultural technology initiated the transformation of Indian ericulture and there by created a large potential for increasing sericulture production in India. A number of new technologies have been made by the scientists of research institutes which are boon for the development of sericulture industry. Unless all these innovations reach the field the development would not takes place. By realizing the need of extension activities to create awareness on the new innovations to the farmers, extension activities are being conducted regularly by the seri extension personnel. Since knowledge on the improved seri techniques and their adaption are influenced by various factors like education, economic conditions of the farmers. A systematic analysis after every training course is a must to achieve the intended overall goal and objectives of the training programme. This chapter discusses in detail about the trainings conducted by CSB, capacity building programmes, Studies on the impact of selected sericultural technologies, Training needs and source consultancy pattern of commercial chawki rearing centres and Faculty performance and course coverage in ericulture training etc. Since due to lack of information in ericulture, some of the extension strategies in sericulture were mentioned.

Ericulture is a farm enterprise, gaining increasing popularity every year. It is attractive to all the farmer's *viz.,* large, medium and small. It provides employment throughout the year to both men and women. Castor being a long standing crop with less gestation period, the returns are quick, frequent and better when compared to many cash crops. It is this enterprise, which provides better employment opportunities in different sectors *viz.,* seed production, chawki rearing, silkworm rearing, silk reeling, silk twisting, printing and dyeing, fashion designing etc. The success of eri silkworm rearing largely depends upon the quality of young age silkworms or chawki which not only ensures higher productivity of cocoons but also the quality. Further, ericulture is an environment friendly farm occupation. In modern era, however, systems of intervention have been evolved to make diffusion more rapid and systematic. Extension education is one such system. All the innovations do not get diffused in a community in the same manner. Those that are perceived as having greater relative advantage, compatibility and simplicity are more likely to get a wider and more rapid acceptance. Extension teaching methods are communication techniques employed for educational purposes. There are many methods one has to carefully choose and combine them in his work. Selection and use of extension methods are governed by a number of factors such as subject matter, audience characteristics, educational goals, time and facilities, as well as the competence and experience of the extension worker in using these methods (Sethu Rao, 2000, Dwarakinath *et al.,* 1994 and Ram Mohan Rao and Kamble, 2009). One of the powerful tools in ericulture extension teaching methods is demonstration. There are two kinds of demonstrations, result and method demonstration. Result demonstration is based on the principle 'seeing is-believing'. Ericulture farming community is mostly with low literacy. It is a powerful tool, particularly for working with low literacy groups, since it establishes visual evidence of the superiority of the recommended technology under farmers own conditions. Success of any new technology depends on its acceptance or adoption by the sericulturists. And the user acceptance is much dependent on carefully drawn and implemented extension programme (Gopala and Krishna, 1993).

Training is the process of assisting a person for enhancing efficiency and effectiveness in work by imparting and updating professional knowledge through improving skills and inoculating positive attitude towards work and people. Farmers have limited resources and their objective is to maximize farm returns from the resources available with them. Hence, in order to operate the farm business efficiently, training plays an important role. It is important that the extension officers are to be trained in modern ericulture technologies so as to maintain the quality and stability in production. Training can bridge enormous gap between the yield achieved in lab and land. Training effectiveness is analysed obviously in terms of enhanced skill and knowledge to achieve the desired goals, through increased learning behaviour. Central sericultural research and training institute (CSR&TI), Mysore has been involved in imparting skills to extension personnel, farmers and students through different training programmes. Even though several impact studies in the field of ericulture training and extensive studies in other fields were undertaken by different workers for evaluating training programmes, a more systematic analysis of every training programme is required for taking up remedial and corrective measures to improve the training programmes further (Rahmathulla *et al.,* 2007).

Trainings conducted by Central silk Board

Central Silk Board conducts mainly three types of training courses namely, Long-term Structured courses (3-15 months), Short-Term Capsule courses (2–45 days) and Adhoc courses of different duration, conducted on specific request of the sponsoring agencies on cost basis. The pre-cocoon training programs in non mulberry sector are mostly conducted at Central Tasar Research & Training Institute, Ranchi and Central Muga, Eri Research & Training Institute, Lahdoigarh. The post-cocoon training courses are conducted mainly at Central Silk Technological Research Institute, Bangalore.

The Central Silk Board (CSB), Ministry of Textiles, GOI, realizing the importance of Training and Skill development in the overall development of the silk industry has gradually evolved into a strong training organization with modern infrastructure and facilities in each of its nine R&D institutions located in different Sericulture zones. CSB offers capacity building, training and skill development courses /modules with the following objectives:

- To develop and strengthen Human Resource in Sericulture by infusing/ imparting knowledge and skill in the field of sericulture through Distant Education.
- To create awareness about opportunities of employment and livelihood in sericulture.
- To involve women in developmental aspects of sericulture.
- To stimulate unemployed youth and school drop-outs to take up sericulture as an occupation.
- To impart knowledge to the people interested and involved in sericulture.
- To stimulate the entrepreneur for their participation in sericulture.

Capacity Building & Training

The R&D institutions of CSB, spread across the country, covering all activities on the silk value-chain pertaining to all the four silk sub-sectors, are intensively involved in training, skill seeding and skill enhancement on a sustainable basis. From the year 2015-16 onwards, CSB's capacity building and training initiatives have been restructured under the following five heads to be implemented and monitored by the Capacity Building & Training Division:

(i) **Skill Training & Enterprise Development Programmes (STEP):** Under this category a variety of short-term training modules focusing on Entrepreneurship development, In-house and industry Resource Development, Specialized Overseas Training, popularization of sericulture technologies, lab to land technology demonstration programmes, training impact assessment surveys etc have been planned to be taken up.

(ii) **Establishment of Sericulture Resource Centre (SRC):** These training cum facilitation centres would be established to select Mulberry bivoltine and Vanya clusters with a unit cost of Rs.3.50 lakhs to act as an important link between Extension Centres of R&D labs and the beneficiaries. The purpose of these SRCs is - technology demonstration, skill enhancement, one-stop shop for Seri-inputs, doubt clarification and problem resolution at cluster level itself.

(iii) **Capacity Building & Training by R&D Institutes of CSB:** In addition to conducting structured long-term training programme (Post Graduate Diploma in Sericulture) the R&D institutes of CSB will also conduct technology-based trainings both for farmers and other stakeholders besides organizing Krishi Melas, Farmer's day, farmer's interaction workshops etc. for empowering the farmers and other industry stakeholders.

(iv) **Capacity Building in Seed Sector:** Silkworm seed is the most critical sector that drives the entire silk value chain. The quality of seed determines the quality of industry output. Therefore, addressing the capacity building and training needs of this sector is of paramount importance. It is proposed to conduct a variety of training programmes to cover industry stakeholders like, Pvt. Silkworm Seed Producers, Adopted Seed Rearers, Managers and work force attached to Govt. owned grainages.

(v) **Information, Education and Communication (IEC):** IEC is meant for supporting Capacity Building and training initiatives by popularizing recommended technologies through brochures, pamphlets, handouts, booklets etc. This component also proposes to produce technology based instructional videos, study materials and documentary films to show case the industry.

It may not be out of place here to mention the author's own experience in training. As a part of the project, "Eri Culture – An Additional income generation source for the resource poor farmers of the Mahabubnagar district of Andhra Pradesh" author has conducted four training programmes to different farmers groups and two training programmes especially to self-help group women have been conducted. The impact of project was evident immediately. Some castor growing farmers have initiated rearing of eri silkworms (Plates 41-48).

Plate 41 As a part of the training programme explaining the results of the project to Chief General Manager, NABARD, Dr. Mohanaiah garu

Plate 42 Explaining the eri silkworm rearing technology to farmers

Plate 43 Training programme conducted to the self-help group women

Plate 44 Releasing the pamphlet on "*Amudam akutho labhadayakanga eri pattupurugula pempakam*"

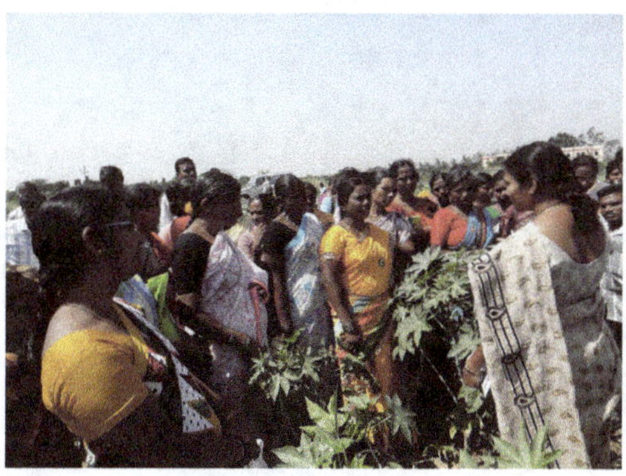

Plate 45 Demonstration of different defoliation methods to self-help group women

Plate 46 Joint Director (Sericulture) explaining the rearing methodology to self-help group women

Plate 47 Collector and District Magistrate, Mahabubnagar visit to Ericulture laboratory

Plate 48 QRT team members visit to Ericulture laboratory

Knowledge gap of silkworm rearers of Jammu Division of Jammu and Kashmir State

Sericulture is a subsidiary occupation for about 25,000 rural families in the Jammu and Kashmir state. Eri silkworm rearing was taken up in 10 districts of Jammu division. On the basis of number of silkworm rearers in each districts, the districts were categorized into three categories with (i) less than 500 rearers ii) 500-1000 rearers iii) above 1000 rearers. To know the knowledge gap of silkworm rearers of Jammu division Lyaqet Ali *et al.,* (2016) selected three districts namely Poonch, Reasi and Rajouri for the purpose of study. Data were collected from the selected respondents with the help of semi- structured interview schedule method.

Table 13.1: Knowledge gap of silkworm rearers in different practices of sericulture management

(Adopted from Lyaqet Ali *et al.,* 2016)

Practices	District wise mean knowledge Score			Overall mean know-ledge score	Maxi-mum obtain-nable	Know-ledge gap (%)
	Rajouri (n=170)	Poonch (n=36)	Reasi (n=34)			
Mulberry and its management	4.14	4.60	4.79	4.30	7.00	38.57
Management of Young age worms	2.82	2.57	2.82	2.78	4.00	30.50
Disinfection and disinfectant	2.42	2.30	2.29	2.39	4.00	40.25
Management of late age rearing	2.66	2.44	2.58	2.50	4.00	37.50
Disease and insect pest management	3.24	3.72	2.99	3.39	4.00	15.25

The data presented in the Table 13.1 depicts the knowledge gap related to mulberry and its management, management of young age worms, disinfection and disinfectant, management of late age rearing and disease and insect pest management of silkworm. There was significant difference between the desired knowledge scores of respondents. The knowledge gap was quantified by subtracting the actual knowledge score from the desired knowledge score. In Rajouri, Poonch and Reasi districts the maximum mean knowledge score related to mulberry and its management was 4.14, 4.60 and 4.79 and minimum mean knowledge score related to disinfection and disinfectant were 2.42, 2.30 and 2.30, respectively. Overall maximum mean knowledge score was 4.30 related to mulberry and its management and overall minimum mean knowledge score was 2.39 related to disinfection and disinfectant. The overall maximum difference was in case of disinfection and disinfectant which was 40% followed by 39% about mulberry and its management and 38% respondents about management of late age rearing. About 31% respondents had knowledge about management of young age rearing and the lowest difference was in case of disease and insect pest management.

Studies on the impact of selected sericultural technologies in Kolar district, Karnataka

Shobha Rani *et al.,* (2011) have undertaken a study to assess the prevailing cultivation and rearing practices in relation to sericulture and corresponding yield gaps as well as to identify major factors that contribute to the differential yield levels among the farmers in Kolar district of Karnataka.

The study was conducted for two years during 2006-2008 in seven villages representing major traditional sericultural area of Karnataka. Benchmark survey was conducted to identify the crucial technologies responsible for the cocoon production before commencement of the work. Soil fertility management, maintenance of separate chawki garden, effective disinfection of rearing house and equipments, incubation and black boxing of eggs, shelf method of rearing with shoot feeding and use of bed disinfectants for late age silkworms were accordingly identified as crucial technological gaps. Soil and leaf samples from 108 farmers were collected and analyzed to assess the soil fertility and leaf nutritional levels among target farmers of the study area. Further, for intensive study, 21 among these 108 farmers were selected based on their willingness to adopt the suggested technologies. The selected farmers were grouped into High (H), Medium (M) and Low (L) cocoon yield farmers based on the cocoon yields of >70 kg, 50-60kg and <50 kg/100 dfls, respectively. Regular interactions on one to one basis and periodical group discussions among these farmers were held for two years along with motivational efforts for technological interventions as were enlisted for respective farmers and groups.

Extent of adoption of suggested technologies by different categories of farmers: In case of the high group of farmers, 100% adoption was observed for certain crucial technologies like application of farmyard manure at recommended dose, separate chawki rearing room, egg incubation, black boxing and use of bed disinfectants (Table 13.2). Nearly 86 per cent of the farmers adopted technologies like application of bio-fertilizers, separate chawki garden, effective disinfection of rearing house and equipments. Shoot feeding technology was adopted by 71 per cent of the farmers.

In case of medium group of farmers, high adoption rate was observed for technologies such as soil test-based fertilizer application (86%), separate chawki garden (71%), separate chawki rearing room (85%), egg incubation (100%), black boxing (100%) and using of bed disinfectants (100%). Similarly proper disinfection of rearing house and equipments was found with 100 per cent. Among other technologies, application of bio-fertilizers and shoot feeding technology were implemented by about 29 per cent of farmers each.

Among the low group of farmers, 100 per cent adoption rate was observed for technologies such as egg incubation and black boxing. Using of bed disinfectants was recorded to an extent of 71 per cent, while the application of farm yard manure in required quantity, application of bio fertilizers were practiced by 43 per cent of the farmers. Technologies like separate chawki garden, effective disinfection of rearing house and equipments, chawki rearing room and shoot feeding technology were adopted by 29, 57 and 29 per cent of farmers, respectively.

It is evident from the earlier reports that higher levels of productivity could be achieved by adoption of new technologies. An analysis of data from the present study showed that increase in cocoon yields was more in medium group of farmers (19 kg) which could be attributed to full adoption of suggested technologies. 100 per cent adoption with respect of farmyard manure was recorded in high group of farmers followed by medium and low groups of farmers in that area. Success in silkworm rearing mainly depends on the quality of castor leaves fed and application of requisite dosage of farmyard manure is one of the quickest means of increasing foliage yield and quality (Chaluvachari *et al.,* 2002). Application of FYM could be promoted effectively among the high and medium group of farmers while, among low group the limitation was its limited availability and high cost of the input.

Table 13.2: Extent of adoption of suggested technologies by the different categories of target farmers (Adopted from Shobha Rani *et al.,* 2011).

Technologies	High group of farmers	Medium group of farmers	Low group of farmers
Application of farm yard manure (20t/acre)	100	86	43
Application of bio-fertilizers	86	29	43
Separate chawki garden	86	71	29
Effective disinfection of rearing house and equipments	86	100	57
Separate chawki rearing room	100	86	29
Egg incubation	100	100	100
Black boxing	100	100	100
Shoot feeding	71	29	29
Bed disinfectants	100	100	71

The average yield/100 dfls which was 74.71 kg in high group of farmers before adoption of technologies went up to 82.5 kg/100 dfls after adoption. In middle and low group of farmers average yield/100 dfls which were 54.57 kg and 45.5 kg/100 dfls before adoption of technological interventions increased to 73.85 kg and 58.11 kg/100 dfls respectively (Table 13.3).

Table 13.3: Cocoon yields after technological interventions among the target farmers (Adopted from Shobha Rani *et al.,* 2011).

Farmers code	Yield during bench mark survey (kg/100 dfls)	I year	II year
H*	74.71	76.17	82.5
M*	54.57	72.05	73.85
L*	45.5	56.0	58.11

H = High group of farmers; M = Medium group of farmers; L = Low group of farmers

Overall improvement of cocoon yield of 8-19 kg/100 dfls was recorded in the study and proper adoption of improved sericulture technologies helped the farmers to harvest good cocoon yields resulting in an increase in their income. Adoption of improved technologies among the medium performing farmers was more encouraging with increase of 19 kg of cocoons/100dfls owing to the adoption of technologies. Hence from the study of Shobha Rani *et al.,* (2011) it can be inferred that, there is great scope for further strengthening the technologies support through need based extension approaches.

Training needs and source consultancy pattern of commercial chawki rearing centres in Kolar district, Karnataka

The training enhances the knowledge and skills and thereby helps in producing quality silk. The identification of training needs is to be carried out on regular basis because of changing nature of technologies. Further, it is often said that knowledge gap and training needs go hand in hand. This hypothesis needs to be tested in real field situation to take advantage of the relationship in designing training strategies. Keeping this in view, Madhu Prasad *et al.,* (2010) conducted a study to assess the training needs of 22 commercial chawki rearing centre owners in Kolar district of Karnataka.

Training needs of commercial chawki rearing centre owners

It was observed from the data that out of 22 areas where the commercial chawki rearing centre owners expressed training needs were disinfection and hygiene maintenance (71.66%), chawki rearing appliances (61.29%) and chawki rearing house including environment management (48.38%). These practices were considered as highly technical, scientific and complex in nature, which require latest knowledge and skills (Table 13.4). The results of correlation analysis revealed that education, occupation, area under host plant and economic motivation of the respondents were found positive and had significant relationship with training needs. The possibility could be that majority of the chawki rearing centre owners had college level education, occupation was ericulture, had more area under castor, hence they were looking towards earning more profits (Table 13.5).

Table 13.4: Distribution of respondents according to their training needs
(Adopted from Madhu Prasad *et al.,* 2010)

S.No	Area of training need	Most needed		Needed		Not needed	
		No.	%	No.	%	No.	%
I	**Cultivation technology of host plant**						
1	Training schedule	4	6.45	9	14.52	49	79.03
2	Manures and fertilizer schedule	20	32.25	18	29.03	24	38.71
3	Irrigation	-	-	3	4.184	59	95.16
4	Leaf harvest	9	14.52	23	36.10	30	48.39
5	Pruning	33	53.23	22	35.48	7	11.29
II	**Chawki rearing technology**						
1	Disinfection and hygiene maintenance	43	71.66	17	27.42	2	3.23
2	Chawki rearing appliances	38	61.29	20	32.26	4	6.45
3	Handling of silkworm eggs	-	-	13	20.97	49	79.03
4	Surface sterilization of silkworm	-	-	8	12.90	54	87.10
5	Incubation of silkworm eggs	6	9.68	11	17.74	45	72.58
6	Black boxing	-	-	9	14.52	53	85.48
7	Brushing	3	4.84	8	12.90	51	82.26
8	Box type of chawki rearing	-	-	4	6.45	58	93.55
9	Chawki rearing house including environment management	30	48.38	24	38.70	8	12.90
10	Leaf quality and feeding	13	20.97	20	32.26	29	46.77
11	Spacing	-	-	7	11.29	55	88.71
12	Bed cleaning	2	3.23	6	9.68	54	87.10
13	Moulting care	-	-	6	9.68	56	90.32
14	Chawki transportation	3	4.84	7	11.29	52	83.87

Table 13.5: Relationship between characteristics of respondents with their training needs (Adopted from Madhu Prasad *et al.,* 2010)

S. No	Characteristics	Correlation co-efficient ('r' values)
1	Age	-0.089
2	Education	0.311**
3	Occupation	0.29**
4	Land holding	0.081
5	Family size	0.095
6	Annual income	0.073
7	Area under castor	0.316**
8	Experience in chawki rearing enterprise	-0.051
9	Extension contact	0.163
10	Extension participation	0.170
11	Risk orientation	0.034
12	Economic motivation	0.495**

The findings of the Madhu Prasad *et al.,* (2010) study revealed that majority of the commercial CRC owners felt that they need training in disinfection and hygienic maintenance on chawki rearing appliances, pruning in chawki garden including environment management and manures and fertilizers schedule in chawki garden for increasing their knowledge and skills. Out of 12 characteristics education, occupation, area under host plant and economic motivation had positive and significant relationship with training needs. Hence the extension personnel should plan and organize the training programmes based on their needs, by considering significantly related characteristics with training needs and further, strengthening the most consulted source of information which will help in higher quality cocoon production.

Faculty performance and course coverage in ericulture training

Rahmathulla *et al.,* (2007) conducted a study during 2003-2006 at CSR&TI, Mysore and evaluated the trainees that have undergone a training programme. The data were collected from 160 trainees with the help of structured questionnaire after the completion of the training programme. The faculty performance was measured from response of trainees for specific training programme on five-point scale *viz.,* poor, average, good, very good and excellent. Similarly, the training course or subject coverage as per course curriculum for theory and practical was calculated by collecting information from trainees trained in different training programmes. The pre and post training test scores were also prepared for knowing the knowledge improvement of the trainees.

The data were subjected to regression analysis in order to understand the factors influencing the faculty performance. The regression equation is

$$Y = a + b_i x_i + e$$

Where,

Y = Dependent variable (Faculty performance index)

a = Intercept

b_i = Coefficient of ith independent variable

x_i = Independent variable

e = Error term

Faculty performance index was calculated based on the perception of the trainees on the performance of faculty including usage of audio visual aids. Correlation studies were carried out to understand the influence of different variables on faculty performance. The independent variables used in the study were experience of the trainees, qualification of trainees, post test score, course coverage and training environment.

Performance of the faculty as assessed by the trainees: It could be seen that the faculties were very good in maintaining cordial relation with trainees (77.5%), capable of clarifying the doubts raised by the trainees and possessed ability to make subject matter simpler and clear (81.9% each). The trainees had also expressed that the faculties were very good in imparting skills (86.9%), subject preparation for the class (86.2%), using modern audio-visual aids (88.8%) and in presentation (67.5%). This data also indicate that there is a scope for further improvement in all the selected areas of faculty skills (Table 13.6).

In correlation studies it could be observed that the course coverage, faculty performance, experience of trainees, qualification of trainees and training environment are not significantly correlated with post test score. Obviously, Faculty performance and Course coverage are independent and hence not significantly correlated with experience of trainees, qualification of trainees and training environment (Table 13.7). Regression analysis shows that post test score of the trainees and course coverage index are highly associated with performance of the trainer (Table 13.8).

Table 13.6: Faculty performance of the training programme as assessed by trainees (n=160)

(Adopted from Rahmathulla *et al.*, 2007)

S. No	Particulars	Poor	Average	Good	Very good	Excellent
1	Cordial nature	15 (9.37)	21 (13.12)	45 (28.12)	50 (31.25)	29 (18.12)
2	Clarification of doubts	05 (3.13)	24 (15.00)	50 (31.25)	65 (40.62)	16 (10.00)
3	Possess ability to draw attention of trainees	09 (5.63)	16 (10.00)	60 (37.50)	40 (25.00)	35 (21.87)
4	Ability to make the subject simple and clear	04 (2.50)	12 (7.50)	55 (34.37)	65 (40.62)	24 (15.00)
5	Capable of imparting skills	08 (5.00)	13 (8.12)	45 (28.12)	60 (37.50)	34 (21.25)
6	Subject preparation and knowledge	07 (4.38)	15 (9.38)	40 (25.00)	65 (40.62)	33 (20.62)
7	Use of modern audio-visual aids	06 (3.78)	12 (7.50)	35 (21.88)	70 (43.75)	37 (23.13)
8	Presentation skill	22 (13.75)	30 (18.78)	35 (21.87)	45 (28.13)	28 (17.50)

Data in parentheses are percentages.

From the study of Rahmathulla *et al.,* (2007), it can be concluded that the trainees were satisfied with the performance of the trainer and high course coverage index indicating that subjects were well covered during the training programme. The study results also indicate the influence of faculty performance on course coverage and improvement in knowledge of trainees.

Table 13.7: Influence of different variables on faculty performance, course coverage and post test score (Adopted from Rahmathulla *et al.,* 2007)

S. No.	Variable	Correlation coefficient		
		Faculty performance	Course coverage	Post test score
1	Faculty performance	-	0.416*	0.14
2	Course coverage	0.416*	-	0.542*
3	Post-test score	0.342*	0.542*	-
4	Experience of trainees	0.018	0.100	0.431*
5	Qualification of trainees	0.028	0.098	0.367*
6	Training environment	0.125	0.023	0.334*

Table 13.8: Regression analysis of different variables with faculty performance (Adopted from Rahmathulla *et al.,* 2007)

S. No.	Variable	Regression coefficient
1	Post test score of trainees	0.784* (0.184)
2	Course coverage during the course	0.604* (0.202)
3	Age of the trainees	-0.155 (0.246)
4	Qualification of trainees	0.038 (0.980)
5	Training environment	0.040 (0.112)

A study on the knowledge and adoption level of improved sericulture practices by the farmers of Chittoor district

A number of new technologies have been made by the scientists of research institutes which are boon for the development of sericulture industry. Unless all these innovations reach the field the development would not takes place. By realizing the need of extension activities to create awareness on the new innovations to the farmers, extension activities are being conducted regularly by the seri extension personnel. Since knowledge on the improved seri techniques and their adaption are influenced by various factors like education, economic conditions of the farmers, a study was undertaken by Priyadarshini and Vijaya Kumari

(2013) to know the impact of education and knowledge of improved practices on the adaption of innovative seri technologies in 16 villages of Chittoor district of Andhra Pradesh state.

Data were collected from 90 farmers by direct personal interview method using a questionnaire on educational levels, knowledge on modern ericultural technologies and their adoption. Knowledge on different sericulture practices were classified as Full knowledge (F), Partial Knowledge (P), No Knowledge (N) and the percentage was calculated. The farmers following all the recommended practices fully were considered as full adapters (FA), those who adapt them partially as partial adapters (PA) and those who were not adapting all practices as non-adopters (NA).

The study revealed that 6.6% of the respondents were educated above secondary level, 35.5% respondents were up to secondary level, 32.2% respondents were educated up to primary level and the remaining 25.5% farmers were illiterates. The knowledge on mulberry cultivation practices was high especially on variety (100%), spacing (100%) and adoption level was also high (60 & 40% respectively). 75% farmers had full knowledge on application of Farm Yard Manure, 25% had partial knowledge and the adoption was 60% (FA), 40% (PA) respectively. For the practice of plant protection 50% of the farmers had full knowledge, 30% had partial knowledge and 20% had no knowledge and their adoption levels were 45% (FA), 35% (PA), 20% (NA) respectively (Table 13.9).

Table 13.9: Knowledge and Adoption Level of Farmers on Mulberry Cultivation (Adopted from Priyadarshini and Vijaya Kumari, 2013)

S.No.	Technology	Knowledge Index (%)			Adoption Index (%)		
		F	P	N	FA	PA	NA
1	Mulberry variety	100	0	0	100	0	0
2	Spacing	100	0	0	60	40	0
3	FYM	75	25	0	60	40	0
4	Fertilizer	4.5	40	13	37.5	32.5	30
5	Plant protection sprayer	50	30	20	45	35	20
6	Drip irrigation	50	38	13	50	30	20

Data on knowledge status of the respondents on new silkworm rearing technologies reveal that 92.5% respondents had full knowledge, 7.5% had partial knowledge on model rearing house but 80% of the farmers had adapted fully, 20% of farmers had partial adaption. 90% of respondents had full knowledge on shoot rearing, 10% respondents had partial knowledge but the farmer's adaption level was 75% (FA), 12.5% (PA) and 12.5% (NA). For the control of Uzifly IPM up to 90% of farmers had full knowledge and the remaining 10% of them had partial knowledge but the adaption of the practice was 90% (FA) and 10% (PA) respectively (Table 13.10).

The study revealed that the farmers who are well educated and have good knowledge on the improved sericultural technologies have adopted all the sericulture practices. The farmers who had partial knowledge on new technologies in sericulture have adopted few technologies and are getting lesser yields, compared to those who have full knowledge and adoption of the practices. Hence, education level of the people and proper extension activities for the transfer of new technologies from lab to land play a major role in the adoption of new practices. The extension activities should be in such a way that the farmers should get convinced about the benefits of improved techniques.

Table 13.10: Knowledge and Adoption Level of Farmers on Silkworm Rearing Technology (Adopted from Priyadarshini and Vijaya Kumari, 2013)

S.No.	Technology	Knowledge Index (%)			Adoption Index (%)		
		F	P	N	FA	PA	NA
1	Rearing	92.5	.5	0	80	20	0
2	Disinfection	70	30	0	62.5	5	12.5
3	Shoot rearing	90	10	0	75	12.5	12.5
4	Bed spacing	80	15	5	75	15	10
5	Bed cleaning	85	15	0	80	10	10
6	Bed disinfectants	95	5	0	95	5	0
7	Mountages	80	20	0	70	15	15
8	Uzifly IPM	90	10	0	90	10	0

Factors determining the training needs of extension officials in different South Indian states

The training needs of the farmers and the extension personnel vary from region to region depending upon the climatic conditions and the existing resource base at the farmer's level. In this context, Srinivasa *et al.,* (2007) carried out a study to understand the training needs of extension officials and sericulturists in different South Indian states *viz.,* Karnataka, Andhra Pradesh and Tamil Nadu. The primary data were collected from the selected extension officials and farmers on their personal characters, knowledge on sericulture technology and the training needs felt and unfelt by them with the help of a structured schedule by personal interview method. In all 110 extension officials from Karnataka, 36 from Tamil Nadu and 35 from Andhra Pradesh and total 90 farmers (30 from each state) were interviewed in the study at random. The data collected on education (5 point scale), designation (5 point scale), knowledge level (2 point scale) and the training undergone (2 point scale) were scored by different scales. In order to know the factors influencing the training needs of extension officials and farmers, regression equation was fitted.

Training need index (TNI) and Knowledge index (KI) were calculated as

$$TNI = \frac{\text{Total training need score obtained}}{\text{Total training score obtainable}} \times 100$$

$$KI = \frac{\text{Total knowledge score obtained by each respondent}}{\text{Total training score obtainable}} \times 100$$

Based on the knowledge and skill level of the technology at individual level, the technology which is not known or partially known, possessing partial or no skill for adopting the same by an individual was considered as a need which is not expressed (unfelt) by the respondents. The survey was conducted in Karnataka, Andhra Pradesh and Tamil Nadu covering a total sample size of 181 officials and 90 farmers. Of the total sample collected, 142 represented grass root level workers and 39 represented supervisory staff.

Training needs of extension staff in the study area: The training needs were classified as the felt and unfelt needs. The felt needs are those which are expressed by the respondents and the unfelt needs are those which are calculated by the investigators based on the knowledge and skill level of the respondents in the respective technology. The analysis of the data revealed that the felt training needs are less as compared to the unfelt training needs (Table 13.11). Extension workers felt that they need to be educated in the machine tools that could be utilized in sericulture (51%) followed by calculation of economics of sericulture (50%) and vermin compost preparation (42%). Identification of worms under moult (38%), shoot rearing, visit to sericulture areas, soil sampling and reclamation and improved mulberry varieties are some of the other important technical inputs that the extension workers wanted to include in the curriculum for training them (Rahmathulla *et al.*, 2003). The unfelt needs of the officials were more than those expressed by them in almost all the technologies considered.

Effect of personal and technical factors of officials on their training needs: In order to understand the impact made by different personal and technical factors on technical needs of the extension officers of South India, the factors were regressed on the training needs expressed by them (Table 13.12). The factors such as education level of extension officers, their designation, experience and knowledge level have significantly influenced them in expressing their training needs in different states of South India. Further, it could be seen that knowledge index has negative influence in Karnataka, which indicates that the trained personnel have minimum training needs. All other factors showed non-significant effect in Karnataka. In the case of Tamil Nadu, the variables namely education level, designation and experience have significantly influenced their training needs. As in Tamil Nadu, in Andhra Pradesh too, education was found to be influencing training needs as they were having low level of education as compared to their counterparts in Karnataka. Moreover, the extension workers of Karnataka were found attending almost all training programmes available in sericulture.

Effect of personal and technical factors of farmers on their training needs: The impact of personal and technical factors on training needs of the farmers of South India indicate that the factors such as, age of the farmers and poor knowledge level have significantly influenced the training needs. In the case of Tamil Nadu, none of the factors has significantly influenced their training needs whereas, in Andhra Pradesh, barring the characters *viz.*, age of the farmers and land holdings, all other selected variables have influenced training needs. The selected variables have explained 64.90%, 72.20% and 89.50% of variation in Karnataka, Tamil Nadu and Andhra Pradesh respectively. This indicates that 35.10% variation in Karnataka, 27.80% in Tamil Nadu and 10.50% in Andhra Pradesh was not captured by the variables selected for the study and this variation is controlled by the variables other than those selected in the study (Table 13.13).

Table 13.11: Training needs of extension staff in the study area (Adopted from Srinivasa *et al.,* 2007)

S.No.	Technology	Training needs of extension officials (n=181)		Training needs of farmers (%) (n=90)
		Felt need (%)	Unfelt need (%)	
	Moriculture aspects			
1	Soil sampling methods	37	44	39
2	Reclamation based on soil test results	32	54	39
3	Mulberry varieties and their package of practices	38	70	58
4	Nursery raising technology	24	52	27
5	Vermi compost preparation	42	59	52
6	Mulberry pests and diseases control	26	61	24
7	Intercropping in mulberry	2	72	24
8	Drip irrigation technique	25	61	30
	Sericulture aspects			
1	Incubation and black boxing	23	67	45
2	Disinfection & hygiene	21	42	60
3	Maintenance of temperature and humidity	1	31	35
4	Brushing of loose eggs	21	49	40
5	Chawki rearing	27	58	54
6	Spacing of worms	26	50	30
7	Shoot rearing	39	47	43
8	Identification of worms under moult	15	73	13
9	Silkworm pest and disease control	30	62	45
	Other related aspects			
1	Visit to successful sericulturists/ research organizations of sericulture	36	63	60
2	Discussion with successful sericulturists	33	93	55
3	Using improved mountages	38	68	57
4	Mechanization	51	63	46
5	Sericulture economics	50	77	45

Table 13.12: Effect of personal and technical factors of officials on their training needs (Adopted from Srinivasa *et al.*, 2007)

S. No	Particulars	Regression co-efficient		
		Karnataka (n=110)	Tamil Nadu (n=36)	Andhra Pradesh (n=35)
1	Age (years)	-0.230 (0.389)	-1.741 (0.743)	-0.085 (0.449)
2	Education (Score)	-0.100 (0.478)	1.128* (1.140)	1.350* (0.471)
3	Designation (Score)	0.104 (1.680)	8.512* (3.869)	-2.364 (2.312)
4	Experience (Years)	-0.028 (0.390)	1.737* (0.453)	0.229 (0.299)
5	Knowledge Index (Score)	-0.211* (0.104)	0.122 (0.115)	-0.163 (0.185)
6	Training undergone (Score)	-3.529 (3.302)	2.508 (5.207)	6.899 (3.473)
	Constant	74.426	40.367	23.978
	R^2	55.00	41.60	36.00

* Significant at 10% level of significance

Figures in parentheses indicate the standard error values

Table 13.13: Effect of personal and technical factors of sericulturists on their training needs (Adopted from Srinivasa *et al.*, 2007)

S. No	Particulars	Regression co-efficient		
		Karnataka (n=110)	Tamil Nadu (n=36)	Andhra Pradesh (n=35)
1	Age (years)	0.689* (0.157)	0.675 (1.150)	-1.59 (0.422)
2	Education (Number of years of schooling)	0.291 (0.573)	1.980 (4.879)	4.058** (0.561)
3	Family size (No.)	1.277 (0.805)	5.991 (8.904)	5.257** (1.149)
4	Land holding	-0.570 (0.370)	-3.142 (8.413)	-0.490 (4.323)
5	Experience (Years)	-0.063 (0.204)	1.364 (3.222)	2.021** (0.399)
6	Training undergone (Score)	0.889 (2.642)	3.142 (8.413)	5.004** (1.901)
7	Knowledge index for moriculture technologies	0.215* (0.105)	-1.364 (3.222)	5.997** (2.486)
8	Knowledge index for silkworm rearing technologies	-0.10 (0.98)	3.114 (13.983)	5.463** (2.231)
9	Overall knowledge index	-0.96 (0.123)	-0.081 (0.176)	-11.316** (4.611)
	Constant	6.565	-59.411	-5352
	R^2	64.90	72.20	89.50

* Significant at 10% level of significance

** Significant at 5% level of significance

Figures in parentheses indicate the standard error values

Hence from the study of Srinivasa *et al.,* (2007), it can be inferred that the sericulture training programmes could be conducted throughout the year with theory and practical in equal proportion. The training needs are almost same for extension personnel and the farmers in the study area. Training needs will depend on a person's education level, training undergone, experience in sericulture and knowledge level.

Impact of demonstration of technology package in sericulture extension and future extension strategies in Kolar district, Karnataka

Success of any new technology depends on its acceptance/ adoption by the sericulturists. The user acceptance is much dependent on carefully drawn and implemented extension programme (Dolli *et al.,* 1998; Gopala and Krishna, 1993). In this regard, Rama Mohana Rao and Kamble (2009) made an attempt to study the impact of sericulture technology package demonstration and also to suggest new strategies in sericulture extension. Central Silk Board introduced a programme during 1998 to conduct demonstration of improved sericultural technologies through research extension centres for improving the yield levels of sericultural farmers. The newly developed mulberry cultivation and silkworm rearing technologies were demonstrated with the lead farmers identified for each group. The other participant farmers who observe the results of technologies were persuaded to repeat the same on their farm for their yield improvement.

The farmers whose average cocoon yield is below 40 kg/100 disease free layings (dfls) were selected for the group demonstrations with the target of reaching their yield above 60 kg at the end of the programme. The new technologies under demonstrations were maintenance of chawki garden, vermicomposting, application of azotobacter bio-fertilizer, foliar spray of triacontanol and IPM against Tukra in mulberry cultivation and bleaching powder and chlorine dioxide disinfection, wrap up method of chawki rearing, shoot rearing, vijetha bed-disinfectant, raksha rekha and uzi trap in silkworm rearing. The study was conducted in four villages of kolar district of Karnataka with a total of 80 farmers during April 2001 to March 2003. Data were recorded for mulberry and silkworm cocoon yield after every crop.

Results of the demonstration show that in phase I of the demonstration programme, lead farmers have recorded 19.83 to 30.27% and 20.49 to 21.88% improvement in mulberry and silkworm cocoon yield respectively. In case of cocoon yield, improvement was 97.97 to 102.8% and 81.83 to 87.34% in respect of lead farmers and followers respectively

(Table 13.14). In the II phase of demonstration, improvement was 11.04 to 14.3% and 12.14 to 16.27% in respect of lead farmers and followers respectively. The study clearly indicates the impact of demonstration of technology package in increasing the mulberry and cocoon yield significantly (Table 13.15).

New strategies in sericulture extension

Focus on women in sericulture: Women play a major role in sericulture both in terms of operations performed and time invested (Dwarakinath *et al.,* 1994). Because of this involvement in sericulture operations, management and decision making, these women required to be knowledgeable in the relevant operations they perform. This need has been overlooked for too long. Recently, some programmes have been started in this direction, but this is not sufficient and it is absolutely necessary that she is brought into the ambit of training and extension activities in sericulture development for the sustainability of sericulture in the rural areas.

Integrated farming system approach: The conventional farming system which mainly focuses on the growing of traditional crops for sustenance will no longer benefit the rural mass because of market price. Possibility of increasing the size of holding being extremely less, diversification in the farming sector to high value crops and enterprises can hasten farmer's income and livelihood (Dandin and Jayaram, 2003).

Diagnostic approach: There are two situations in which sericulture extension work is conducted, they are traditional areas and new areas. The sericulture extension strategy in these two areas will have to be distinctly different. In the traditional areas, farmers have not only accepted sericulture but have also gained considerable personal experiences. Mainly for this reason, they have some hardened attitudes about how to manage sericulture. They are complacent on the ground that they know all that is to be known in the matter. This becomes a major obstacle for the farmers in learning. The extension strategy here should start with a diagnostic effort to identify the specific production constraints and educate the farmers in identifying and remedying them. This is essentially a selected technology approach to sericulture development.

Establish the presence of technology: Any new technology becomes the engine of growth in the development. As such, it is necessary that the presence of this technology is felt, seen and talked about in spheres where it matters. Display of information at extension centres in the form of pictures, descriptions and messages is a must.

Table 13.14: Impact of demonstration of new sericulture technologies in Kolar district of Karnataka (2001-2003) (Mean of 10 crops of Phase –I) (Adopted from Rama Mohana Rao and Kamble, 2009)

Village/ Farmer	Mulberry yield/acre/crop (kg)			Cocoon yield/100 dfls (kg)		
	Before adoption	After adoption	Improvement (%)	Before adoption	After adoption	Improvement (%)
Kaiwara						
Lead farmer	2420	2900	19.83	35.00	70.98	102.80
Follower	2274	2740	20.49	33.47	60.86	81.83
M Mangala						
Lead farmer	2240	2918	30.27	35.00	69.29	97.97
Follower	2272	2794	22.98	32.63	61.13	87.34
Grand mean	2302	2838	23.12	34.03	65.57	92.48
't' value		9.851*			8.989**	

Table 13.15: Impact of demonstration of new sericulture technologies in Kolar district of Karnataka (2003-2005) (Mean of 10 crops of Phase –II) (Adopted from Rama Mohana Rao and Kamble, 2009)

Village/Farmer	Mulberry yield/acre/crop (kg)			Cocoon yield/100 dfls (kg)		
	Before adoption	After adoption	Improvement (%)	Before adoption	After adoption	Improvement (%)
M Hosahalli						
Lead farmer	3000	3429	14.30	37.56	74.32	97.87
Follower	2784	3237	16.27	34.13	66.09	93.64
Kuruburahalli						
Lead farmer	3088	3429	11.04	41.11	68.39	66.36
Follower	2981	3343	12.14	37.96	64.59	70.15
Grand mean	2963	3360	13.43	37.69	68.35	82.01
't' value		7.899*			6.955**	

Summary and Conclusions

➢ Ericulture is a farm enterprise, gaining increasing popularity every year. It is attractive to all the farmers groups.

➢ Ericulture provides better employment opportunities in different sectors *viz.,* seed production, chawki rearing, silkworm rearing, silk reeling, silk twisting, printing and dyeing, fashion designing etc.

➢ In modern era, however, systems of intervention have been evolved to make diffusion more rapid and systematic.

- Those that are perceived as having greater relative advantage, compatibility and simplicity are more likely to get a wider and more rapid acceptance.
- Extension teaching methods are communication techniques employed for educational purposes.
- Selection and use of extension methods are governed by a number of factors such as subject matter, audience characteristics, educational goals, time and facilities, as well as the competence and experience of the extension worker in using these methods.
- A number of new technologies have been made by the scientists of research institutes which are boon for the development of sericulture industry.
- Unless all these innovations reach the field, the development would not take place. By realizing the need of extension activities to create awareness on the new innovations to the farmers, extension activities are being conducted regularly by the seri extension personnel.
- One of the powerful tools in ericulture extension teaching methods is demonstration.
- The training enhances the knowledge and skills and thereby helps in producing quality silk. The identification of training needs is to be carried out on regular basis because of changing nature of technologies.
- Training is the process of assisting a person for enhancing efficiency and effectiveness in work by imparting and updating professional knowledge through improving skills and inoculating positive attitude towards work and people.
- Training can bridge enormous gap between the yield achieved in lab and land.
- Training effectiveness is analysed obviously in terms of enhanced skill and knowledge to achieve the desired goals, through increased learning behaviour.
- The training needs of the farmers and the extension personnel vary from region to region depending upon the climatic conditions and the existing resource base at the farmer's level.
- Central Silk Board conducts mainly three types of training courses namely, Long-term Structured courses (3-15 months), Short-Term Capsule courses (2–45 days) and Adhoc courses of different duration.
- In an experiment conducted to test the knowledge of silkworm rearers in Jammu division of Jammu and Kashmir state that 31% respondents had knowledge about management of young age rearing.

➢ Overall maximum mean knowledge score was 4.30 related to mulberry and its management and overall minimum mean knowledge score was 2.39 related to disinfection and disinfectant.

➢ Studies conducted on the impact of selected sericultural technologies in the selected farmers groups of Kolar district of Karnataka inferred that higher levels of productivity could be achieved by adoption of new technologies.

➢ The average yield/100 dfls which was 74.71 kg in high group of farmers before adoption of technologies went up to 82.5 kg/100 dfls after adoption. In middle and low group of farmers average yield/100 dfls which were 54.57 kg and 45.5 kg/100 dfls before adoption of technological interventions increased to 73.85 kg and 58.11 kg/100 dfls respectively.

➢ Adoption of improved technologies among the medium performing farmers was more encouraging with increase of 19 kg of cocoons/100dfls owing to the adoption of technologies.

➢ In Kolar district of Karnataka, majority of the commercial CRC owners felt that they need training in disinfection and hygienic maintenance on chawki rearing appliances, pruning in chawki garden including environment management and manures and fertilizers schedule in chawki garden for increasing their knowledge and skills.

➢ Faculty performance and course coverage in ericulture training at CSR&TI, Mysore stated that the trainees were satisfied with the performance of the trainer and high course coverage index indicating that subjects were well covered during the training programme.

➢ A study was undertaken to know the impact of education and knowledge of improved practices on the adoption of innovative seri technologies in 16 villages of Chittoor district of Andhra Pradesh state. Knowledge on different sericulture practices were classified as Full knowledge (F), Partial Knowledge (P), No Knowledge (N) groups.

➢ The farmers who are well educated and have good knowledge on the improved sericultural technologies have adopted all the sericulture practices. The farmers who had partial knowledge on new technologies in sericulture have adopted few technologies and are getting lesser yields.

➢ Education level of the people and proper extension activities for the transfer of new technologies from lab to land play a major role in the adaption of new practices.

- The extension activities should be in such a way that the farmers should get convinced about the benefits of improved techniques.

- The training needs of extension officials and sericulturists in different South Indian states *viz.,* Karnataka, Andhra Pradesh and Tamil Nadu were classified as the felt and unfelt needs.

- The felt needs are those which are expressed by the respondents and the unfelt are those which are calculated by the investigators based on the knowledge and skill level of the respondents in the respective technology.

- The factors such as education level of extension officers, their designation, experience and knowledge level have significantly influenced them in expressing their training needs in different states of South India.

- The training needs are almost same for extension personnel and the farmers in the study area. Training needs will depend on person's education level, training undergone, experience in sericulture and knowledge level.

- Training programmes could be conducted throughout the year with theory and practical in equal proportion.

- Success of any new technology depends on its acceptance/ adoption by the sericulturists. The user acceptance is much dependent on carefully drawn and implemented extension programme.

- The farmers whose average cocoon yield is below 40 kg/100 disease free layings (dfls) were selected in two phases for the group demonstrations with the target of reaching their yield above 60 kg at the end of the programme in Kolar district, Karnataka.

- In phase I of the demonstration programme, lead farmers have recorded 19.83 to 30.27% and 20.49 to 21.88% improvement in mulberry and silkworm cocoon yield respectively.

- In the II phase of demonstration, improvement was 11.04 to 14.3% and 12.14 to 16.27% in respect of lead farmers and followers respectively.

- Women play a major role in sericulture both in terms of operations performed and time invested. Hence it is absolutely necessary that she is brought into the ambit of training and extension activities in sericulture development for the sustainability of sericulture in the rural areas.

- There were two situations in which sericulture extension work is conducted, they are traditional areas and new areas.

➢ In the traditional areas, farmers have not only accepted sericulture but have also gained considerable personal experiences. Mainly for this reason, they have some hardened attitudes about how to manage sericulture.

➢ The extension strategy here should start with a diagnostic effort to identify the specific production constraints and educate the farmers in identifying and remedying them.

References

Chaluvachari U D, Bongale, Manjunath M S and Anantharaman M N. 2002. Soil fertility, leaf quality and cocoon crop performances in bivoltine seed area Anekal in Karnataka. *Bulletin of Indian Academy of Sericulture* **6**: 57-62.

Dandin S B and Jayaram H. 2003. Mulberry based integrated farming system. Lead paper, *National Conference on Tropical Sericulture for Global Competitiveness*, CSRTI, Mysore, 5-7, Nov, 2003 p 29-31.

Dolli S S, Singhvi N R, Iyengar M N S and Geethadevi R G. 1998. Technology: Generation, Validation and Feasibility. *Indian Silk* **37**(2): 25-27.

Dwarakinath R, Sethu Rao M K and Hanumappa P. 1994. Guide to sericulture extension. Ramesh M N (Ed.), *Central Silk Board*, Bangalore p 45-46.

Gopala M and Krishna K S. 1993. Problems in adoption of recommended sericulture practices. *Indian Silk* **32**(4): 53.

Lyaqet Ali, Kher S K, Slathia P S and Rakesh Kumar. 2016. Knowledge Gap of Silkworm Rearers of Jammu Division of Jammu and Kashmir State. Paper presented in the International Conference on *"Natural Resource Management: Ecological Perspectives"*, 18-20, February, 2018, SKUAST, Jammu.

Madhu Prasad V L, Ramakrishna Naika, Usha Ravindra and Gokul Raj M P. 2010. Training needs and source consultancy pattern of commercial chawki rearing centre owners in Kolar district. *Agriculture Update* **5**(3&4): 325-327.

Priyadarshini M B and Vijaya Kumari N. 2013. A study on the knowledge and adoption level of improved sericulture practices by the farmers of Chittoor district. *International Journal of Agricultural Science and Research* **3**(2): 43-46.

Rahmathulla V K, Geethadevi R G and Srinivasa G. 2003. Evaluation of bivoltine training programme for extension workers of South India. *Indian Journal of Sericulture* **42**(2): 137-141.

Rahmathulla V K, Srinivasa G, Vindhya G S, Rajan R K and Kamble C K. 2007. Faculty performance and course coverage in sericulture training programme – An analysis. *Indian Journal of Sericulture* **46**(2): 136-139.

Rama Mohana Rao P and Kamble C K. 2009. Impact of demonstration of technology package in sericulture extension and future extension strategies. *Indian Journal of Sericulture* **48**(2): 178-181.

Sethu Rao M K. 2000. Extension and Sericulture Development – An overview. National Conference on Strategies for Sericulture Research and Development, Lead paper, *Central Sericultural Research and Training Institute*, Mysore, 16-18 November 2000: 62-64.

Shobha Rani M, Dinesh Kumar G D, Krishna Murthy R, Raveendra H R and Bongale U D. 2011. Studies on the impact of selected sericultural technologies in Kolar district. *Agriculture Update* **6**(3&4): 170-174.

Srinivasa G, Himantharaj M T, Vindhya G S, Rajan R K and Kamble C K. 2007. Factors determining the training needs of extension officials and sericulturists in South India. *Indian Journal of Sericulture* **46**(2): 130-135.